Desert and River in Nubia

Geomorphology and
Prehistoric Environments at
the Aswan Reservoir

Desert and River in Nubia

Geomorphology and
Prehistoric Environments at
the Aswan Reservoir

Karl W. Butzer
and Carl L. Hansen

With contributions by Egbert G. Leigh, Jr.,
Madeleine Van Campo, and
Bruce G. Gladfelter

The University of Wisconsin Press   1968
Madison, Milwaukee, and London

Published by The University of Wisconsin Press
Box 1379, Madison, Wisconsin 53701
The University of Wisconsin Press, Ltd.
27–29 Whitfield Street, London, W.1
Copyright © 1968 by the Regents
of the University of Wisconsin
All rights reserved
Printed in the United States of America by
Kingsport Press, Inc., Kingsport, Tennessee
Library of Congress Catalog Card Number 67–20761

To the Artists, Cartographers, and Engineers
of the French Expedition of 1798–99

Who, amid incredible difficulties,
laid the foundations for scientific
exploration of Upper Egypt

Foreword

The international campaign to save the monuments of Nubia was inaugurated by UNESCO in 1960, in order to document or remove the archeological wealth of the High Dam Reservoir upstream of Aswan. Salvage excavations financed by a host of nations were well under way late in 1961. Although the Predynastic cultures of Nubia were among the archeological objectives of the UNESCO appeal, the Paleolithic prehistory and the record of Pleistocene environments had not been included in the original program. In May of 1961, Charles A. Reed and I drew up a proposal to the National Science Foundation for an expedition to study the Pleistocene paleoecology of Nubia. Reed assumed the burdens of organization and made a preliminary reconnaissance in Egypt during January of 1962. The resettlement project for the displaced Nubian population on the Kom Ombo Plain, north of Aswan, called for immediate attention since this area was then being prepared for cultivation by land grading and canal digging. In view of the urgency of prehistoric investigation here, the original plans of the expedition were broadened to include Kom Ombo with its rich archeological record.

Consequently, when the Yale Prehistoric Nubia Expedition began its work in September, 1962, the Kom Ombo Plain was its first objective and the archeologists did not excavate in Nubia during the first season. The earth science work, for which Carl L. Hansen and I were responsible, was extended, however, to Egyptian Nubia, the Kurkur Oasis, and a part of the Red Sea Hill country. Hansen, then a graduate student in geography at the University of Wisconsin, was my field assistant. The other members of the 1962–63 expedition included Martin A. Baumhoff (University of California, Davis) and Heinz Walter (Ethnologisches Museum, Berlin) as senior archeologists, David S. Boloyan as assistant archeologist, and Egbert G. Leigh, Jr., and Thomas E. Lovejoy as assistant biologists. After completion of the first season in April, 1963, the Yale Expedition continued to work in Nubia during the winters of 1963–64 and 1964–65, with a new staff except for Reed and Walter. Although the total results of the 1962–63 expedition were only conceivable as a team effort, the fractionation of work by different individuals over wide areas and focused on a variety of distinct problems made it seem preferable that publication should proceed independently. The following study is devoted to the geomorphological investigations carried out by Hansen and me during the 1962–63 season.

This contribution to an understanding of the late Tertiary and Pleistocene history of southern Egypt was supported, in part, by NSF grant G–23777 and U.S. Department of State grant SCC–29629, both to the Peabody Museum of Natural History (Charles A. Reed), Yale University, and by NSF grant GS–678 to the Department of Geography (Karl W. Butzer), University of Wisconsin. Despite the vital support of these foundations, the present study could not have been realized without the constant support of the University of Wisconsin which, through salaries, laboratory equipment and supplies, and the services of its Cartographic Laboratory, has made its own contribution to the Nubian Monuments Campaign. I am particularly grateful to the Wisconsin Alumni Research Fund, which made much of the laboratory work possible, and to the Soils Department, which provided various facilities and kindly analyzed water-soluble salts from a variety of samples, as well as to the Center for Climatic Research, which made certain laboratory equipment available. And the University of Wisconsin Press has generously made it possible to present a fully-documented study of the Pleistocene geomorphology, stratigraphy, and environments—a record that will be available to scientists long after Nubia has been permanently flooded by the Aswan Reservoir.

It is a pleasure for me to be able to express my gratitude to the many individuals who have made this book possible: Charles A. Reed, who through several eventful seasons carried many responsibilities as expedition director; Martin A. Baumhoff, colleague and friend in the field; Thompson Webb, Jr., director of the University of Wisconsin Press, whose interest made this book a reality; Randall D. Sale of the University of Wisconsin Cartographic Laboratory and his able assistants, particularly Jeanne Tsou-Liu, who drafted the maps and diagrams; and Bruce G. Gladfelter (University of Chicago) who participated in the laboratory studies with dedication. Opportunities for exchange of information among the different expeditions were provided by the Bellagio Conference (August, 1964), sponsored by the Society for American Archaeology; the Burg Wartenstein Symposium on systematic investigation of the African later Tertiary and Quaternary (July, 1965), sponsored by the Wenner-Gren Foundation; the symposium on Nile Valley prehistory at the seventh International Quaternary Association Congress, Boulder (September 1, 1966), organized by Joe Ben Wheat (University of Colorado); and by colloquia held at the geography departments of the Universities of Göttingen, Stuttgart, and Tübingen in July, 1965.

Valuable discussion or information was provided by Jean de Heinzelin (University of Ghent); Henry Irwin (Harvard University); Roland Paepe (Geographical Service of Belgium); Karl Kromer, the late Ludwig Ehgartner, and Manfred Bietak (Naturhistorisches Museum, Vienna); B. W. Sparks (Cambridge); B. F. Kukachka (Wood Products Laboratory, University of Wisconsin); and Leslie G. Freeman (University of Chicago). P. E. L. Smith (Université de Montréal) and Robert J. Fulton (Geological Survey of Canada) read those parts of the report dealing with sites studied by the expedition of the National Museum of Canada. David L. Thurber (Lamont Geological Observatory) generously made Th^{230}/U^{234} assays on shell samples from the Red Sea coast; Gerd Lüttig kindly submitted two shell samples to the Bundesamt für Bodenforschung, Hannover, for radiocarbon determinations; and C. S. Tucek (Isotopes, Inc.) supervised the commercial radiocarbon determinations. Particular thanks are also due to Charles A. Reed (now University of Illinois, Chicago Circle), Mintze Stuiver (Yale University), and P. E. L. Smith for making their unpublished radiocarbon dates available. Richard Klein (Northwestern University) kindly translated the paper by I. S. Chumakov (Soviet Academy of Sciences, Moscow), reproduced here with Dr. Chumakov's permission.

Several of my coworkers have contributed directly to this manuscript. Carl L. Hansen collaborated in all phases of the work and contributed directly to the writing of several chapters and the appendices. The data on the fossil mollusca was provided by Egbert G. Leigh, Jr. (Princeton University), with the valuable collaboration of J. P. E. Morrison (Smithsonian Institution). I am particularly grateful that the difficult and tedious study of the fossil pollen was undertaken by so competent a specialist as Madeleine Van Campo (École Pratique des Hautes Études, Paris), with Philippe Guinet (École Pratique des Hautes Études) and Jacqueline Cohen (École Pratique des Hautes Études). Some of the pollen data has already been made available for inclusion as an appendix. The final results will be published separately at a later date.

While we were in the field, Chicago House, Luxor, with its former director, Robert G. Hughes, and its resident staff, provided an unqualified hospitality on several occasions. And last, but not least, Christoph Northoff, who first introduced me to the mysteries of Abu Simbel, was a true friend in need throughout the season.

Karl W. Butzer

Chicago, Illinois
March, 1967

Table of Contents

1 INTRODUCTION 3
The Nile Basin, 3. The Pleistocene geological and prehistoric record of the Nile Valley, 6. The objectives of the present study, 8. Stratigraphy and chronology, 11. Sediment and soil nomenclature, 13.

2 GEOMORPHIC EVOLUTION OF THE KOM OMBO PLAIN 17
Introduction, 17. Early geological work, 21. Prehistoric salvage work during 1962–63, 23. Bedrock lithology, 23. Structure, 27. Geomorphic units, 29. The Pliocene deposits, 33. The early Pleistocene Proto-Nile system, 43. Age of the Aswan Pediplain, 49. The "high" gravel terraces, 52. The "middle" gravel terrace and the Fatira beds, 57. The "low" gravel terraces, 60. Comparative terrace sequences at Aswan, Edfu, Luxor, and Qena, 63. Stratigraphic conclusions, 65. Contemporary geomorphic activity and soil development in the Kom Ombo drainage basin, 65. Paleoclimatic interpretation of Pleistocene features, 78.

3 LATE PLEISTOCENE AND HOLOCENE SEDIMENTS OF THE KOM OMBO PLAIN 86
Introduction, 86. The Korosko Formation (Basal Sands and Marls), 87. The Masmas Formation (Older Floodplain Silts), 97. The Gebel Silsila Formation (Younger Channel Silts), 107. The Ineiba Formation (Lower Wadi Alluvium), 116. The Shaturma Formation (Upper Wadi Alluvium), 121. Description of the type sections, 123. Deposits and stratigraphy south of Darau, 128. Deposits and stratigraphy of the Gebel Silsila area, 132. Channel stratigraphy at the Khor el-Sil sites, 141. Channel stratigraphy of the Sebil area, 143. Late Pleistocene nilotic sediments north of the Kom Ombo Plain, 146. Stratigraphic and paleoclimatic conclusions, 148.

4 GEOGRAPHICAL FACTORS CONDITIONING PREHISTORIC SETTLEMENT OF THE KOM OMBO PLAIN 153
Introduction, 153. Early Paleolithic industries of southern Egypt, 153. Early Paleolithic settlement in the Kom Ombo area, 157. Middle Paleolithic industries of southern Egypt, 158. Middle Paleolithic settlement in the Kom Ombo area, 160. Late Paleolithic industries of southern Egypt, 163. Physical setting of occupation sites at Gebel Silsila 2A, 168. Physical set-

ting of Gebel Silsila 2B, Area I, 170. Physical setting of Gebel Silsila 2B, Area II, 177. Physical setting of occupation sites at Khor el-Sil, 177. Physical setting of occupation sites along the Sebil Channel, 180. Physical setting of the Dar es-Salam site, 181. Late Paleolithic geography of the Kom Ombo Plain, 181. Neolithic industries of southern Egypt, 187. Late prehistoric settlement of the Kom Ombo Plain, 189.

5 GEOMORPHIC EVOLUTION OF EGYPTIAN NUBIA 196

Introduction, 196. Early geological work, 200. The Yale Prehistoric Nubia Expedition 1962–63, 202. Bedrock lithology, 203. Structure, 205. Geomorphic units, 210. The Tertiary pediplains, 221. The Pleistocene pediments, 224. The nature of the Nile and wadi gravels, 226. Stratigraphy of the Nile and wadi gravels, 231. Pleistocene terraces between Adindan and Ballana, 232. Pleistocene terraces between Wadi Or and Arminna, 236. Pleistocene terraces between Tushka and Qatta, 240. Pleistocene terraces between Tumas and Korosko, 241. Pleistocene terraces between Korosko and Dakka, 244. Pleistocene terraces between Kushtamna and Aswan, 247. Pleistocene dissection of the Nubian Nile, 250. Red paleosols, 252. Early and middle Pleistocene deposits of the Sudanese Nile Valley, 254. Contemporary geomorphic processes in Nubia, 258. Stratigraphic and paleoclimatic conclusions, 262.

6 LATE PLEISTOCENE AND HOLOCENE SEDIMENTS OF EGYPTIAN NUBIA 266

Introduction, 266. The Korosko Formation, 266. The Masmas Formation, 272. The Gebel Silsila Formation and its subdivisions, 274. The Ineiba and Shaturma Formations, 280. Other deposits of local origin, 282. Stratigraphy of late Pleistocene and Holocene deposits in Khor Adindan, 284. Late Pleistocene deposits at Ballana and Qustul, 290. Stratigraphy of late Pleistocene and Holocene deposits in Wadi Or, 294. Late Pleistocene deposits between Abu Simbel and Amada, 307. Stratigraphy of late Pleistocene and Holocene deposits between Korosko and el-Madiq, 314. Late Pleistocene and Holocene deposits between Seiyala and Aswan, 319. Late Pleistocene and Holocene deposits of the Sudanese Nile Valley, 323. Stratigraphic and paleoclimatic conclusions, 327.

7 THE KURKUR OASIS 334

Introduction, 334. Early work on the Kurkur Pleistocene, 335. The work of the 1963 Yale Expedition, 337. Bedrock lithology, 338. Structure, 341. General geomorphology, 342. The groundwater resources, 343. Evolution of the Kurkur pediments, 350. The Plateau Tufa, 355. Age of the Plateau Tufa and the local cuestas, 361. Landslide and breccia formations, 362. The older Wadi Tufas (Tufas I and II), 366. The intermediate Wadi Tufas (Tufa III complex), 370. The younger Wadi Tufas (Tufa IV), 378. The Red Silts, 381. Paleoclimatic interpretation, 383. Stratigraphic conclusions, 388. Prehistoric occupation of the Kurkur area, 389.

8 THE COASTAL PLAIN OF MERSA ALAM 395

Introduction, 395. Early work on the coastal Pleistocene, 396. Bedrock lithology, 399. Structure, 400. General geomorphology, 400. Alluvial terraces of Wadi Alam–Khariga, 402. Alluvial terraces of lower Wadi Sifein, 406. Pleistocene littoral deposits near Mersa Sifein, 408. Pleistocene littoral deposits Mersa Alam–Mersa Samadai, 413. Observations south of Ras Samadai, 416. Pleistocene sea-level stratigraphy, 418. Stratigraphy of the littoral and alluvial beds, 421. Paleoclimatic interpretation, 423. Conclusions, 429.

9 TOWARDS A HISTORY OF THE SAHARAN NILE 431

Cenozoic evolution of the Saharan Nile, 431. Cretaceous and Cenozoic changes of climate in Egypt and the northern Sudan, 436. Origin and nature of the "Ethiopian" flood silts, 443. Pleistocene environmental changes in Ethiopia, 449. Problems and conclusions, 453.

Appendices

A MECHANICAL AND CHEMICAL SAMPLE ANALYSES 461
Karl W. Butzer and Carl L. Hansen

B HEAVY MINERALS OF THE LATE PLEISTOCENE NILOTIC DEPOSITS: A PRELIMINARY REPORT 467
Karl W. Butzer and Carl L. Hansen

C QUARTZ-GRAIN MICROMORPHOLOGY 473
Karl W. Butzer and Bruce G. Gladfelter

D CLAY MINERALS 483
Karl W. Butzer and Carl L. Hansen

E ISOTOPIC DATES 495
Karl W. Butzer and Carl L. Hansen

F MODERN AND FOSSIL MARINE MOLLUSCA OF THE RED SEA LITTORAL AT MERSA ALAM 499
Egbert G. Leigh, Jr.

G FOSSIL MOLLUSCA FROM THE KOM OMBO PLAIN 509
Egbert G. Leigh, Jr., and Karl W. Butzer

H FOSSIL MOLLUSCA FROM THE KURKUR OASIS 513
Egbert G. Leigh, Jr.

I FOSSIL POLLEN FROM LATE TERTIARY AND MIDDLE PLEISTOCENE DEPOSITS OF THE KURKUR OASIS 515
Madeleine Van Campo, Philippe Guinet, and Jacqueline Cohen

J	PLIOCENE INGRESSION INTO THE NILE VALLEY ACCORDING TO NEW DATA I. S. Chumakov, translated by Richard G. Klein	521
K	GEOLOGICAL MAP OF EGYPTIAN NUBIA (1:166,000)	523
	BIBLIOGRAPHY	537
	List of recent intermediate- to large-scale topographic maps of southern Egypt	550
	INDEX	553

Tables

1–1	Geological time scale	5
1–2	Late Cenozoic time scale	11
1–3	Textural classes of sediments	14
2–1	Sediment characteristics of some Pliocene and Pleistocene deposits of the Kom Ombo Plain	37
2–2	Geomorphic evolution of the Kom Ombo area during the late tertiary and early to middle Pleistocene	66
2–3	Morphometric gravel analyses	80
3–1	Morphometric gravel analyses	91
3–2	Sediment characteristics of different facies of the Masmas Formation	102
3–3	Sediment characteristics at the type sites of the Korosko, Masmas, Ineiba, and Shaturma Formations	124
3–4	Sediment characteristics of the Masmas Formation at Gebel Silsila 2B	133
3–5	Sediment characteristics at occupation site Gebel Silsila 2B, Area I	139
3–6	Sediment characteristics at occupation site Gebel Silsila 2B, Area II	139
3–7	Geomorphic evolution of the Kom Ombo area during the Late Pleistocene and Holocene	149
3–8	Sediment characteristics of miscellaneous samples from the Kom Ombo Plain	151
4–1	Coarse aggregates from Square C–2, Level 2, Gebel Silsila 2B, Area I	174
5–1	Accordance of summits in Egyptian Nubia	212
5–2	Paleosols and soil sediments in Egyptian Nubia	253
5–3	Geomorphic evolution of Egyptian Nubia during the late Tertiary and early to middle Pleistocene	263
6–1	Morphometric gravel analyses	271
6–2	Sediment characteristics of deposits in Khor Adindan	285
6–3	Sediment characteristics of deposits in Wadi Or	300
6–4	Sediment characteristics of deposits between Arminna Temple and Wadi Umm el-Hamid	309
6–5	Geomorphic evolution of Egyptian Nubia during the late Pleistocene and Holocene	328
7–1	Morphometric gravel analyses	358
7–2	Tentative geomorphic evolution of the Kurkur area during the late Tertiary and early Pleistocene	363
7–3	Sedimentological characteristics of deposits from the Kurkur Oasis	373
7–4	Geomorphic evolution of the Kurkur area during the late Pleistocene and Holocene	389

8–1	Morphometric gravel analyses	404
8–2	Pleistocene levels of the Red Sea in the Mersa Alam area	419
8–3	Correlation of middle and late Pleistocene wadi and littoral phenomena	422
A–1	Textural data (in per cent) for noncarbonate residues of samples cited in text tables	463
B–1	Relative percentages of selected heavy-mineral groups from modern sediments of the major Nile drainage basins	468
B–2	Preliminary heavy-mineral analysis of sediments from the Korosko Formation (in per cent)	469
B–3	Preliminary heavy-mineral analysis of sediments from the Masmas Formation (in per cent)	470
B–4	Preliminary heavy-mineral analysis of sediments from the Gebel Silsila Formation (in per cent)	471
C–1	Quartz-grain micromorphology: regional or stratigraphic means of frosting and rounding indices	475
C–2	Quartz-grain micromorphology (Kom Ombo Plain)	479
C–3	Quartz-grain micromorphology (Nubia)	480
C–4	Quartz-grain micromorphology (Kurkur Oasis)	481
C–5	Quartz-grain micromorphology (Red Sea Coast)	481
F–1	Modern marine mollusca of the Mersa Alam Littoral	502
G–1	Late Pleistocene mollusca from the Kom Ombo Plain	510
I–1	Fossil pollen spectra from the Kurkur Oasis	517
I–2	List of fossil pollen from the Kurkur Oasis	518

Illustrations

1-1	The Nile Basin today and in Mid-Tertiary times	4
1-2	Egypt and the Northern Sudan	10
2-1	Settlement expansion on the Kom Ombo Plain, 1902–65	18
2-2	The Kom Ombo Plain in Hellenistic times	19
2-3	Structure of the Kom Ombo region	25
2-4	Inclined Nubian Sandstone strata near New Tushka	29
2-5	Geomorphology of the Aswan–Kom Ombo area	30
2-6	The Etbai Upland north of the Kom Ombo Plain	31
2-7	Edge of the Gallaba Gravel Plain north of Faris	33
2-8	Paleogeographic reconstructions of the Kom Ombo region	34
2-9	West-East section of the Fatira Terrace	38
2-10	Detail of Pliocene section, Fatira Terrace	38
2-11	Pliocene and Pleistocene beds of the Fatira Terrace	39
2-12	Late Tertiary to Middle Pleistocene geology of the Kom Ombo Region .	41
2-13	North-South section of Gebel Silsila and the Fatira Terrace	46
2-14	West-East section of Gebel Silsila and the Silsila Gap	48
2-15	Gravels of the southern Burg el-Makhazin	49
2-16	Profile of sandstone uplands at New Farqanda	50
2-17	The Pleistocene sequence at New Masmas	50
2-18	The drainage basins of Wadis Shait, Natash, and Kharit	51
2-19	Pleistocene deposits east of the Wadi Ellawi embouchure	54
2-20	Pleistocene deposits at New Shaturma	54
2-21	Soil profile of High Terrace II between New Korosko and New Shaturma	55
2-22	High Terrace gravels between New Korosko and New Shaturma . .	56
2-23	Pleistocene deposits at El-Nasser	57
2-24	Polygonally cracked clays deposited by recent spates in Wadi Kharit, near New Masmas	61
2-25	Pleistocene deposits east of New Tushka	62
2-26	Pleistocene deposits west of New Abu Simbel	62
2-27	Modern floodplain silts near Fatira exhibiting dehydration cracks and vertisol phenomena	72
2-28	Patinated lag gravels and the effects of salt-weathering *in situ* in the Fatira beds near Gebel Silsila	75
3-1	Late Pleistocene and Holocene deposits of the Kom Ombo Plain . .	88
3-2	Microstratigraphy at New Korosko	89
3-3	Topography and profile locations at New Korosko	90
3-4	Microstratigraphy at New Shaturma and New Sebua	92

3-5   Topography and profile locations at New Shaturma and New Sebua . . 94
3-6   The Masmas Formation as exposed in tributary wadi at New Ballana I . 98
3-7   Cross section of the Masmas Formation northwest of Gebel Silsila Station 100
3-8   Different facies of the Masmas Formation northeast of Gebel Silsila Station 101
3-9   The Masmas Formation in Wadi Shurafa 103
3-10 Nile silts of Masmas Formation underlying wadi alluvium of Shaturma Formation in Wadi Shurafa 104
3-11 Polygonal crack network developed in Masmas Formation near New Abu Simbel and fossilized by carbonates and gypsum 106
3-12 Deflated, irregular gypsum crust formed over Masmas silts near New Abu Simbel 107
3-13 The Masmas and Gebel Silsila Formations at Gebel Silsila 2 . . . 109
3-14 Inclined bed-load deposits of the Gebel Silsila Formation south of Darau 110
3-15 Surface lithology of a part of the Nile Basin 111
3-16 Cobble bed load and fluted sandstone on west face of Gebel Silsila at 101.5 meters 113
3-17 The Ineiba and Shaturma Formations near the mouth of Wadi Ellawi . 117
3-18 The rubefied sandy facies of the Ineiba Formation exposed in the bed of Wadi Kharit 12 kilometers upstream of New Arminna 120
3-19 Part of type section of Masmas and Ineiba Formations at New Masmas . 126
3-20 Intercalated gravelly sand and clayey silt of the Masmas Formation near New Ballana I 129
3-21 Channel gravel bar of the Gebel Silsila Formation south of Darau at Nag el-Darira 132
3-22 Silt and sand strata of the Gebel Silsila Formation exposed in the Silsila Gap 133
3-23 Geology and topography of the Gebel Silsila 2 area, showing orientation of the Fatira Channels A and B 134
3-24 Yardangs of Gebel Silsila deposits preserved along western bank of former Channel A at Gebel Silsila 2A 136
3-25 General view of deflated Gebel Silsila 2A area, looking south along western bank of Channel A 137
3-26 North-South profile through Gebel Silsila 2B, Area I 138
3-27 West-East profile through Gebel Silsila 2B, Area I 138
3-28 The Gebel Silsila Formation at the Khor el-Sil archeological sites . . 142
3-29 Section of the Khor el-Sil alluvia 143
3-30 Alluvial fill cut by the irrigation drain at Khor el-Sil 144
3-31 The Sebil Channels and related flood benches A and B between Sebil Qibli and Matana Bahari 146
3-32 Abandoned late Pleistocene channel near Sebil Bahari 147
4-1   Major Early and Middle Paleolithic sites in southern Egypt 155
4-2   The occupation floor of Gebel Silsila 2B, Area I 173
4-3   Microstratigraphy of the Late Paleolithic site Khor el-Sil 2 179
4-4   Reconstruction of the Kom Ombo Plain at the time of Late Paleolithic occupation 184

4–5	Predynastic sites in southern Egypt	190
4–6	Cross section of the Nile floodplain and late Pleistocene deposits north of Darau	195
5–1	Tentative structural map of Egyptian Nubia	206
5–2	West-East profile through basin and dome structures at approximately 22° 49′ N	208
5–3	134- to 137-meter Nile platforms at Seiyala, veneered with eolian sand or late Pleistocene wash	208
5–4	Geomorphology of Egyptian Nubia	211
5–5	Erosional surfaces east of Amada Temple	218
5–6	The Korosko Hills seen from Ineiba	219
5–7	The Aswan Pediplain west of Abu Simbel at about 200 meters elevation	220
5–8	Lower Nubia	233
5–9	Autochthonous Nile and wadi gravels in southern Egypt	234
5–10	Surficial deposits and geomorphology at Adindan West	235
5–11	Surficial deposits and geomorphology at Adindan East	236
5–12	Gravels of the Wadi Allaqi stage north of Khor Adindan	237
5–13	Surficial deposits and geomorphology at Qustul (Kimam Goha)	237
5–14	Lower and Middle Pleistocene Nile terraces between Ballana and Arminna	238
5–15	Surficial deposits and geomorphology 1.7 kilometers downstream of Arminna Temple	239
5–16	Surficial deposits and geomorphology north of Wadi Tushka (west bank)	241
5–17	Surficial deposits and geomorphology at Amada Temple	242
5–18	The confluence of Wadi Korosko and Wadi Guhr el-Daba	243
5–19	The + 15 meter wadi terrace of the Wadi Korosko stage at the confluence of Wadis Korosko and Guhr el-Daba	244
5–20	Gravels of the Dihmit stage south of Khor Dihmit at 140 meters elevation	248
6–1	Late Pleistocene nilotic deposits in southern Egypt	267
6–2	Marls and marly silts of the Korosko Formation at Km 0.21 in Wadi Or	268
6–3	Marly sands of the Korosko Formation overlain by sands of the Ineiba Formation and detrital colluvium of the Shaturma Formation	269
6–4	Homogeneous sands of the Korosko Formation, intercalated with cobble gravels in lower Wadi Umm el-Hamid	270
6–5	Vertisol phenomena developed in dark silts of the Masmas Formation at Km 1.2 in Khor Adindan	273
6–6	Sandy gravels of the Darau Member resting on bedrock in Wadi Tushka West	275
6–7	Interdigited wadi beds (Shaturma Formation) and nilotic silts (Kibdi Member) near Km 0.91 in Khor Adindan	277
6–8	Cross section of Khor Adindan at Km 1.12	278
6–9	Interbedded nilotic silts of the Arminna Member and wadi alluvium of the Ineiba Formation, resting on the Masmas Formation at Km 1.12 in Khor Adindan	279

xx • ILLUSTRATIONS

6–10	Stratigraphic section of late Pleistocene deposits at Ballana (Nag el-Nuqta)	279
6–11	General view of Wadi Or (North Branch), looking towards Nile	294
6–12	Cross sections of Wadi Or at Km 1.28 and Km 1.97	295
6–13	Cross section of Wadi Or at Km 3.92	295
6–14	Stratigraphic sequence in Wadi Or (North Branch)	296
6–15	Section at Km 1.53, Wadi Or (North Branch)	296
6–16	Section of alluvial deposits at Km 2.55 in Wadi Or	297
6–17	Near Km 3.15 of Wadi Or	298
6–18	Fossil reddish paleosol near Km 2.1	299
6–19	Late Pleistocene deposits 500 meters south of Arminna Temple	309
6–20	Khor Hamra section, mixed nilotic and wadi facies intersected at the mouth of a wadi 4.5 kilometers southwest of Ineiba Station, opposite Nag Shurbagi	312
6–21	Late Pleistocene deposits in Wadi Korosko	316
6–22	Intergrading subaqueous and fluvial facies in lower Wadi Umm el-Hamid	318
7–1	Geological section from Kurkur to the Nile Valley	339
7–2	Bedrock geology of the Kurkur area	340
7–3	The South Well area of Kurkur Oasis	345
7–4	Longitudinal profile of Northwest Wadi	346
7–5	Longitudinal profile of upper Wadi Kurkur	347
7–6	Longitudinal profile of the watershed between Tufa Wadi (Wadi Kurkur) and False Wadi (Wadi Abu Gorma)	348
7–7	The modern oasis and location of archeological sites	349
7–8	Generalized geomorphology of the Kurkur area	351
7–9	Pediment I and Pediment II, seen from 376 meters elevation in the False Wadi headwaters	352
7–10	Pediment pass between North Wadi and Wadi Umm Seiyala, looking north from Km 4.5	354
7–11	Surficial deposits of the Kurkur Oasis	356
7–12	Plateau Tufa resting on Chalk east of North Well	358
7–13	Looking north across South Wadi, with Plateau Tufa resting on Chalk at horizon	359
7–14	Section of False Wadi	364
7–15	Chalk slump blocks, in part surmounted by Tufa I, on north side of False Wadi	365
7–16	Block of bulbous travertine and organic tufa resting at base of Tufa I terrace, south flank of False Wadi	367
7–17	Transverse section through Tufa I channel, now intersected by Tufa Wadi	368
7–18	Sections in Northwest Wadi at Km 1.38 and Km 0.65	369
7–19	Typical facies of Tufa IIIa in Central Wadi (Km 0.5 South), with well-stratified, marly lime sands underlying current-bedded gravel	371
7–20	Former spring vent from Tufa IIIb terrace southwest of North Well	372
7–21	Conglomerates of + 6 meter (Tufa III) terrace on south bank of Wadi Kurkur	373
7–22	Section of Central Wadi at former North Well	374

7–23	Exposure of Tufa IIIc west of North Well	375
7–24	Section of North Wadi at Km 1.92	376
7–25	Tufa IV (+ 3 m) terrace at Km 1.61 in North Wadi	379
7–26	Detail of Tufa IV terrace showing surface crust	380
7–27	Red Silts accumulated in crevice of Plateau Tufa southeast of former North Well	382
8–1	Geomorphology of the Red Sea coastal plain at Mersa Alam	401
8–2	South bank of Wadi Alam at Km 0.8	403
8–3	North bank of Wadi Umm Khariga at Km 6.7	403
8–4	Wadi Alam and Wadi Umm Khariga confluence	404
8–5	Wadi Umm Khariga at Km 11.3	405
8–6	Longitudinal section of lower Wadi Sifein	407
8–7	Wadi Sifein terraces at Km 1.68	407
8–8	The Low and Middle Terraces of Wadi Sifein at Km 1.37	409
8–9	Littoral and alluvial deposits at Mersa Sifein (South Shore)	410
8–10	The 3.2-meter reef at Mersa Sifein, capped by veneer of Low Terrace gravel	410
8–11	Estuarine conglomerate at Mersa Sifein at low water	411
8–12	Littoral and alluvial deposits 500 meters south of Mersa Sifein	412
8–13	Littoral and alluvial deposits at Km 5.5, south of Mersa Alam	414
8–14	Wadi gravels overlying lagoonal evaporites at Km 5.5, south of Mersa Alam	415
8–15	Fossil coral reef dissected by Wadi Ambaut and overlain by a + 10 meter wadi terrace at Km 14.5, south of Mersa Alam	417
8–16	Fossil coral reefs at Sharm Skeikh (South Shore)	418
D–1	X-ray diffractograms from the Korosko Formation	486
D–2	X-ray diffractograms from the Masmas Formation	486
D–3	X-ray diffractograms from the Gebel Silsila Formation	486
D–4	X-ray diffractograms from the Ineiba Formation: the Sinqari Member; the Malki Member; and the Omda Soil	489
D–5	X-ray diffractograms from various red paleosols and colluvial soils	491
D–6	X-ray diffractograms from Wadi Tufa III, Kurkur Oasis	493
I–1	Some fossil pollen grains from the Plateau Tufa, Kurkur Oasis	520
K–1	Surficial geology of the Nile Valley from Adindan to Abu Simbel	525
K–2	Surficial geology of the Nile Valley from Arminna to Tushka	526
K–3	Surficial geology of the Nile Valley from Masmas to Qatta	527
K–4	Surficial geology of the Nile Valley from Tumas to Sinqari	528
K–5	Surficial geology of the Nile Valley from Malki to Sebua	529
K–6	Surficial geology of the Nile Valley from Madiq to Qurta	530
K–7	Surficial geology of the Nile Valley from Allaqi to Girf Husein	531
K–8	Surficial geology of lower Wadi Allaqi	532
K–9	Surficial geology of the Nile Valley from Mariya to Abu Hor	533
K–10	Surficial geology of the Nile Valley from Kalabsha to Dabud	534

Desert and River in Nubia

Geomorphology and
Prehistoric Environments at
the Aswan Reservoir

1

Introduction

The Nile Basin

The Nile is the longest river of the world, measuring 6,670 kilometers from the headwaters of the Kagera River in Rwanda to the shores of the Mediterranean in Egypt. Yet, even today, the Nile Basin is a very complex hydrographic system, consisting of at least three distinct river systems (Fig. 1–1): the White Nile, between Lake Victoria and Khartum; the Blue Nile, between Lake Tana and Khartum; and the Saharan Nile, between Khartum and the Mediterranean Sea. To this, the Atbara–Khor Gash system might also be added as a separate unit.

Within the headwater regions of both the White Nile and the Blue Nile basins, large but shallow lakes form great reservoirs. Lake Victoria, with a surface elevation of 1,100 meters, has an area of 68,800 square kilometers and a maximum depth of 79 meters. Lake Tana has an elevation of about 1,750 meters, an area of 3,060 square kilometers, and a depth of almost 100 meters. From these headwater lakes perched high within the block-fault mountains and plateaus of the great Rift system, the two Niles descend rapidly to the plains of the Sudan, crossing the Sudan-Uganda border near the 600-meter contour and the Sudan-Ethiopia border near the 500-meter contour. Subsequent gradients to Khartum and ultimately to the Mediterranean Sea are weak, as the waters traverse the comparatively stable African Shield on their northward course. The floodplain at Cairo (143 km from the sea) is at 21 meters elevation; that at Aswan (990 km from the sea) is at 93 meters; and that at Khartum (2,910 km from the sea), at 380 meters. Consequently, the Saharan Nile descends only 380 meters over a longitudinal stretch of 2,900 kilometers, compared with drops of 720 meters in 3,000 kilometers for the White Nile from the exit of Lake Victoria and 1,370 meters in 1,500 kilometers for the Blue Nile from Lake Tana.

Spanning some 35 degrees of latitude (approximately 3° S to 32° N), the Nile basin includes the tropical rainforests of East Africa, the savan-

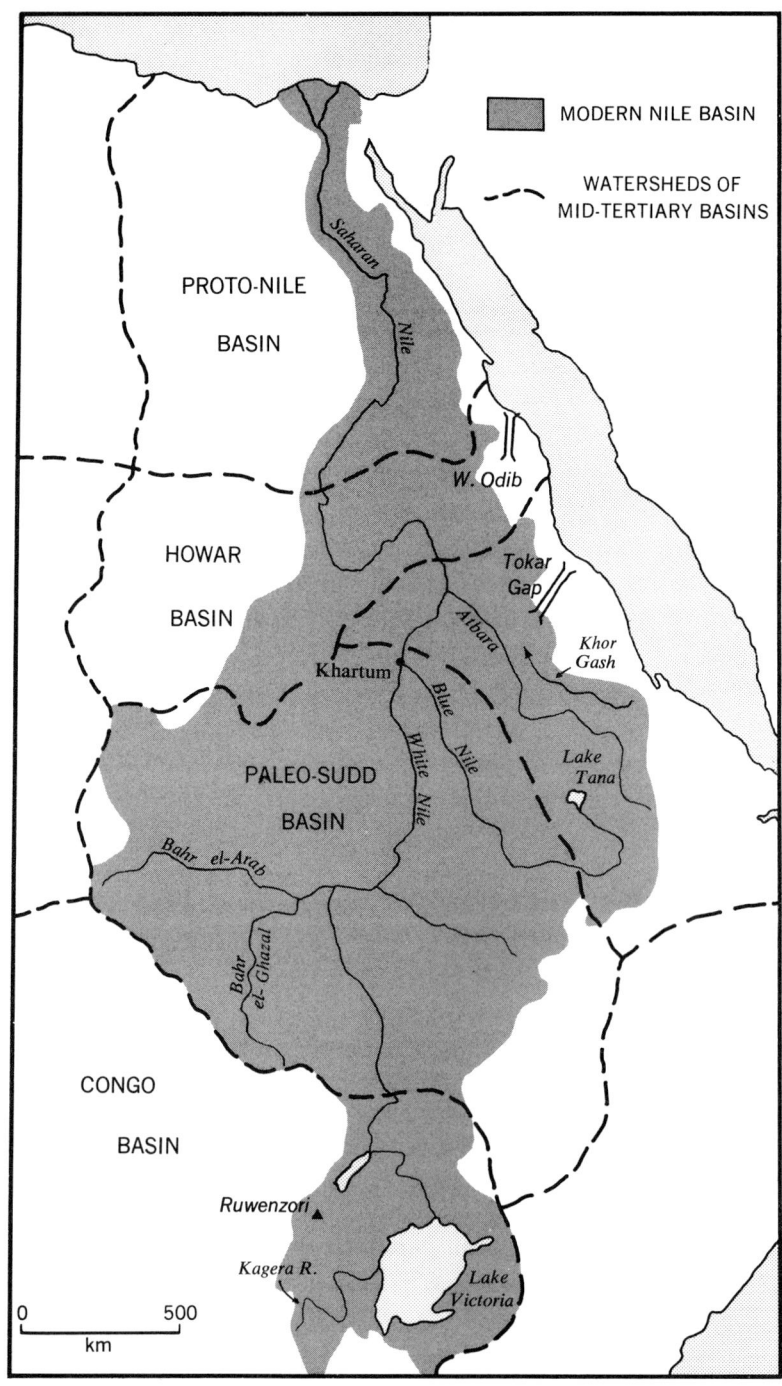

Fig. 1–1. The Nile Basin today and in Mid-Tertiary times (modified after de Heinzelin, 1963). Map: UW Cartographic Lab.

nas of the Sudan, and the deserts of Egypt and Nubia. At the same time, this basin ranges from sea level to the glaciated summits of Mt. Ruwenzori (4,750 m), traversing a succession of environments ranging from lowland deserts to montane forests and alpine meadows. The physical complexity of the Nile Basin is not confined to climate and vegetation. The topography and geological history of this zone are far from uniform, and few hydrographic systems have emerged in so complicated a fashion, through successive linking of diverse drainage basins.

In mid-Tertiary times (see Table 1–1), prior to the major faulting of the great Rift valleys, the present Nile Basin was divided into five or more separate hydrographic units (Fig. 1–1) (see de Heinzelin, 1963). Those parts of the modern basin found south of the Sudan-Uganda border appear to have drained westward into the Congo Basin, and modern hydrographic conditions were not fully established here until middle or late Pleistocene times. Much of the Sudan and western Ethiopia formed a great interior basin, the Paleo-Sudd, which, judging by massive alluvial and lacustrine sediments cored in the center of this depression, may have maintained an internal drainage pattern until the early Middle Pleistocene. Northern Ethiopia and much of the eastern

Table 1–1. Geological time scale

Era	Period	Epoch	Beginning dates (B.P.)[a]
Cenozoic	Quaternary	Holocene	0.01
		Pleistocene	2.75 ± 0.75
	Tertiary	Pliocene	13 ± 2
		Miocene	25 ± 2
		Oligocene	36 ± 3
		Eocene	58 ± 3
		Paleocene	63 ± 4
Mesozoic	Cretaceous		135
	Jurassic		180
	Triassic		230
Paleozoic	Permian		280
	Carboniferous		345
	Devonian		405
	Silurian		425
	Ordovician		500
	Cambrian		600
(Precambrian)			

[a] Approximate dates for beginning of units are in millions of years.

Sudan were drained towards the east by one or more river systems, until the primary uplift of the Ethiopian Plateau and the Red Sea block-fault ranges in Miocene times. The northernmost Sudan probably formed a separate drainage basin, centering around Wadi Howar and the Abu Hamed bend of the Nile and draining northeastward via Wadi Odib. Both the Ethiopian and Howar basins may possibly have continued to drain towards the Red Sea as late as mid-Pleistocene times, however, and rock types derived from the Howar Basin make their first appearance in Nubia as late as 17,000 years ago, although Ethiopian silts are already recorded some 50,000 years ago. Finally, there is the Proto-Nile Basin, which includes the Nile in Egypt and Nubia as far south as the Third Cataract. There is evidence that a great river already emptied into the Mediterranean Sea west of Cairo during late Eocene times. This "Proto-Nile" drained Egypt and Lower Nubia, establishing by early or middle Miocene times a course similar to that of the modern Nile. During the early part of the Pleistocene, this Proto-Nile still received the bulk of its waters from the eastern hill country, north of Aswan, rather than from a southerly source.

Consequently, although the Egyptian and Nubian Nile is an ancient river, the Nile hydrographic system as we know it today is of comparatively recent age. The White Nile and the Blue Nile basins certainly did not merge with the Saharan Nile before the early Pleistocene, and a *typical* summer-flood regime was established in Egypt no earlier than 25,000 years ago. Crustal warping and fracturing, aided by volcanic activity, continued to modify the gross hydrography of the Nile Basin well into Pleistocene times.

The Pleistocene Geological and Prehistoric Record of the Nile Valley

The significant hydrographic changes undergone by the Nile Basin in recent geological times proceeded simultaneously with geomorphic events leading to erosion or sedimentation. Superimposed upon these changes of tectonic origin were a sequence of major climatic fluctuations that induced aggradation-degradation cycles on a broad, regional scale. And, near the coast, the base level of erosion has been frequently modified through worldwide eustatic changes of sea level. For these reasons, the geomorphic record of the Nile Basin is both varied and complex.

Our present information concerning the geomorphology of the Nile Basin indicates that, wherever the Pleistocene record is well preserved, it is indeed complex and rewarding. But preservation of all but late Pleistocene deposits is poor outside of the tectonic basins, e.g., the Egyptian Nile, the Kom Ombo Plain, the Sudd Basin, and some of the grabens of the East African lake district. Even here, early Pleistocene deposits are seldom readily accessible at the surface. Denudation by water and wind has been dominant elsewhere, particularly along the gentle swells that mark the watersheds of the Tertiary drainage basins (Fig. 1-1). As a result, the Nile Basin does not preserve the continuous suite of exposed Pleistocene deposits which would allow easy regional—let alone extraregional—correlation to be made. Instead, intensive studies are required in each local situation. Once such detailed investigations have been completed in strategic areas of each of the Tertiary basins, large-scale correlation within the present Nile Basin should be possible.

Potentially, the Nile Basin has much to contribute to our understanding of African stratigraphy and prehistory. As Caton-Thompson (1946) has already pointed out, the modern Nile provides one of the few continuous links between the tropics and the middle latitudes. It may ultimately be possible to correlate paleoclimatic events as well as prehistoric cultures along the length of the Nile corridor. So far, however, the available information does not allow any speculation as to whether or not this goal can be realized. Before 1961, a number of detailed local studies were already available from the lake district of East Africa and a reconnaissance survey had been carried out along the Egyptian Nile Valley. But investigation in the Sudan and in Ethiopia was quite rudimentary, being limited to some casual observations and a few small-scale surveys. Since the early 1940's, little interest had been shown for Pleistocene field work in the Nile Basin except in East Africa.

The High Dam Project at Aswan threatened to inundate permanently the Pleistocene deposits and erosional surfaces of a vital 600-kilometer segment of the Nile Valley in Nubia and, as a corollary, to destroy the late Pleistocene record of the Nubian resettlement area on the Kom Ombo Plain. Because neither of these regions had been thoroughly studied before, it has been gratifying that several expeditions have volunteered to document or salvage the Pleistocene geomorphology and prehistoric archeology of the Aswan Reservoir. Since serious work began late in 1961, interest in the Pleistocene problems of the Nile Valley has grown, and teams of geomorphologists and prehistorians are still in the field or planning new surveys in Upper Egypt as well as in Upper Nubia. Publication of the results of the different prehistoric missions will presumably provide renewed impetus for further field study by interdisciplinary groups.

The Objectives of the Present Study

The Nile Valley in Egypt and the northern Sudan is eminently suited for paleoclimatic correlation between the poleward and equatorward margins of the subtropical high-pressure belts. The modern river derives almost 80% of its waters from Ethiopia, where the Atbara, the Blue Nile, and the Sobat[1] are fed by the summer monsoonal rains (June through October). These rains give rise to the Nile flood, which reaches its crest in Egypt between mid-August and mid-October, leaving an annual increment of silt on the alluvial flats of its extenuated floodplain. Pleistocene fluctuations of climate in Ethiopia could therefore be expected to leave a sedimentary record far downstream along the desert tracts of the Nile Valley.

At present, the tributary wadis of Egypt and Nubia contribute practically no sediments to the Nile floodplain. The rare rainstorms and local wadi floods, related to westerly frontal-cyclonic disturbances (November through March) north of the Sudanese border or to various monsoonal disturbances (July through August) south of the border, perform little more than local redeposition of wadi-bed materials. Because the long-term mean annual rainfall in Upper Egypt and Lower Nubia is less than 5 millimeters, the modern hydrography of the desert wadis is, indeed, almost defunct. But the prevailing hyperarid climate of Egypt was interrupted by several moist Pleistocene "pluvials." These were accompanied by deposition of local sands and gravels in the form of wadi terraces, wadi fans at the edge of the valley, and wadi materials redistributed far downstream along the bed of the Nile itself.

From the point of view of Pleistocene paleoclimates, it is important to establish whether or not the subtropical pluvials of Egypt (related to the circumpolar westerlies) were synchronous with the tropical pluvials of Ethiopia (related to the equatorial easterlies). This question should be susceptible to geological study in Egypt, where the interrelationships of nilotic and wadi sediments must somehow be evident. The late Pleisto-

---

1. The 30-year means for several key gauging stations of the Nile Basin (Hurst, 1944: 46) give the following daily mean discharge (in million cubic meters): mouth of the Atbara, 32; Blue Nile at Khartum, 141; mouth of the Sobat, 36; White Nile above Sobat confluence, 39. Since about two thirds of the Sobat drainage is derived from the Ethiopian uplands via the Baro, Gila, and Akobo, the total Ethiopian drainage amounts to about 197 of 248 units if we assume that evaporative losses balance precipitation runoff in the central Sudan.

cene record is fairly complete and most suitable for study south of about latitude 25° N, where the Nile Valley is moderately narrow, where both the nilotic and wadi deposits lie above present floodplain level, and where eustatic fluctuations of base level have had little or no impact in late Pleistocene times (see Butzer, 1959a).

This problem of nilotic and wadi interrelationships, presupposing as it does the establishment of a late Pleistocene and Holocene stratigraphy (Chaps. 3 and 6), is one of the first and foremost objectives of the present study. The second problem is almost a corollary to the first. It concerns the evolution of the Proto-Nile Basin into an integral part of the modern hydrographic system. Only a balanced evaluation of all classes of geomorphic evidence from the mid-Tertiary to the present day can hope to unravel the history of the Nile in southern Egypt (Chaps. 2, 5, and 9). To this end, the study area was extended to include the Kurkur Oasis, in the Libyan Desert (Chap. 7), as well as a sector of the Red Sea littoral at Mersa Alam (Chap. 8) (Fig. 1–2). In addition, Pleistocene studies were carried out in the Red Sea hill country by Hansen (1966). Although these two underlying questions could not be completely resolved, the subsequent chapters will show that much has been documented and that new insights have been gained.

A third objective of the present study involves the geomorphic processes and landforms in an arid environment such as southern Egypt. A variety of lithologies—including limestones, sandstones, and complex igneous and metamorphic rocks—were available in the study areas, and our primary concern with geomorphic history provided frequent time markers with which to gauge landscape evolution (Chaps. 2, 5, and 7). A comparative landform study for the Nubian sandstone and for the complex Precambrian rocks is the theme of Hansen's doctoral dissertation (1966), and a similar study of other lithological types at the Kurkur Oasis has already been published (Butzer, 1965).

A final goal of the study has been the prehistory of southern Egypt, a topic in which the senior author has a long-standing interest. The prehistoric geography and chronology of Paleolithic occupation in the Nile Valley is a complex subject and one not about to be exhausted. One problem in particular that demanded attention long before 1961 was the age of the Sebilian cultures of the Kom Ombo Plain, first defined by Vignard in 1923. The Sebilian occupation of the Nile Valley was, on rather tenuous geological grounds, thought to date from the terminal Pleistocene. In this context, the Sebilian industries, with their evidence of intensive utilization of a riverine environment, seemed rather precocious, and several recent authors have insisted on a post-Pleistocene age. The work of the Yale mission has served to unravel much of the story of these Late Paleolithic inhabitants of the Nile Valley (Chap. 4). And in doing so we have gained new perspectives on man-land relationships in late Pleistocene times.

Fig. 1–2. Egypt and the northern Sudan. Study areas shaded. Map: UW Cartographic Lab.

Stratigraphy and Chronology

Since the subsequent discussion will emphasize stratigraphy, it seems appropriate to provide tables of the conventional geological time scale (Tables 1-1, 1-2) and to outline the bases for stratigraphic correlation in southern Egypt.

It should be emphasized that isotopic dating in the late Cenozoic of Egypt is almost entirely limited to radiocarbon, which has a *de facto* time span of 40,000 years. A fairly good chronometric framework now exists for the period between 7,000 and 20,000 years ago. Radiocarbon determinations within this time range have proved to be reasonably consistent in the case of shell and charcoal, although determinations from inorganic carbonates have met with mixed success only. Radiocarbon dates from

Table 1-2. Late Cenozoic time scale

Epoch	Stage	European substage	Beginning dates
Holocene	Upper	Subatlantic	800 B.C. ± 100
	Middle	Subboreal	3000 B.C. ± 150
		Atlantic	5600 B.C. ± 150
	Lower	Boreal	8300 B.C. ± 200
Pleistocene	Upper	Würm	70,000 B.P. ± 5000
		Eem	90,000 B.P. ± 5000
	Middle	Warthe	120,000 B.P. ± 5000
		Treene	
		Saale	
		Holstein	300,000 B.P. ± 25,000
	Lower	Elster II	
		Cortonian	
		Elster I	
		Cromerian	1,000,000 B.P. ± 150,000
	Basal (Villafranchian)		2,750,000 B.P. ± 750,000
Pliocene	Upper	Astian	
	Middle	Plaisancian	
	Lower	Pontian	13,000,000 B.P. ± 2,000,000
Miocene	Upper	Sarmatian	
	Middle	Tortonian	
		Helvetian	
	Lower	Burdigalian	
		Aquitanian	25,000,000 B.P. ± 2,000,000

the late-prehistoric and historical time range are often badly at variance with well-established chronologies. Consequently, specific dates based on radiocarbon determinations must be considered as "radiocarbon years," which in Egypt are proving to be on the young side wherever contaminations of carbonate samples by "dead" carbonates can be excluded.[2]

Prior to the time range of radiocarbon dating, stratigraphic and chronologic data are based almost entirely on geomorphic or geological estimates. Paleontological evidence is lacking except for pollen or macrobotanical remains found in a few deposits, and such paleobotanical evidence as there is does not provide conclusive arguments for adopting one stratigraphic scheme or another. Archeological dating has been attempted in Egypt by some—by Reed (1966), for example, but the bases for assigning archeological designations or industries to a particular geological unit are often more questionable than the geomorphic assumptions used to assign a stratigraphic age to that formation in the first place. In short, we must continue to rely strongly on relative stratigraphic dating based on internal geological evidence.

Fundamental prerequisites for establishing a relative stratigraphy are availability of interrelated deposits and consistent interpretation of all such interrelationships. The stratigraphic schemes presented in each of the subsequent chapters are valid or probable for the local areas in question. Where not supported by radiocarbon determinations or by correlation of distinctive sediment facies, however, external correlations are meant as suggestions only.

The Pleistocene subdivisions used are specifically of local application only. A fourfold subdivision of Basal, Lower, Middle, and Upper is employed, analogous to that of Table 1–2. No absolute dates are implied, however, and no specific correlations with the European substages are envisaged. The general, noncapitalized terms early, middle, and late Pleistocene are most commonly used in the text, with "early" referring to the local Basal and Lower Pleistocene. "Late" Pleistocene is used for features dating from the last 60 millennia or so of the Pleistocene, as based on or extrapolated from radiocarbon determinations.

The late Tertiary stratigraphy adopted in Table 1–2 follows the synthesis of Tertiary stratigraphy in the Mediterranean area by Papp and Thenius (1959), revised on the basis of more recent conferences (see Thenius, 1960; Aguirre, 1964). Because there are few good stratigraphical links with the Mediterranean Sea, few late Tertiary events in southern

---

2. Radiocarbon dates from samples collected by the writers are discussed and evaluated in Appendix E.

Egypt can be stratigraphically dated with any confidence. The tentative correlations suggested in the subsequent chapters will be discussed in the light of the available evidence.

Cultural or typological terminologies are seldom ideal, since they often tend to confuse culture, technology, and time. The terminologies applied to Egyptian prehistory are no exception, and the problem here is intensified by the question of whether European–Near Eastern or African classifications are more appropriate. The general cultural designations used in the subsequent chapters are employed with reluctance, as an expedient until a more suitable nomenclature has been decided upon. Thus, the term Lower Paleolithic is applied to the Acheulian-type bifacial industries,[3] while Middle Paleolithic is used in reference to the flake industries of Levalloisian facies.[4] Late Paleolithic seems an expedient designation for the array of diminutive flake, blade, and microlithic industries that lack pottery,[5] although "Epi-Paleolithic" is used in lieu of Mesolithic to describe such industries whenever they appear to be younger than 8000 B.C.

Sediment and Soil Nomenclature

A variety of field and laboratory methods were applied to the study of sediments and soils, and several of these are outlined in Appendices A–D. The nomenclature used for sediment or soil description generally follows that of Butzer (1964b: Chap. 10) and can be briefly summarized here.

Stratification refers to disposition of beds as horizontal (topset) or inclined (foreset if inclined in the direction of sediment transfer, backset if inclined in the opposite direction). Current bedding implies alternating horizontal and inclined beds.

Several degrees of *consolidation* may be present. Loose sediments are noncohesive in the wet state or the dry state; unconsolidated sediments are cohesive but soft or pliable in the wet state. Consolidated sediments are neither soft nor pliable in the wet state, but edges are fairly brittle. Cemented sediments can only be broken up by hammer.

---

3. The "Upper Acheulian" and "Acheulio-Levalloisian" of Caton-Thompson (1946, 1952) and the "Lower Paleolithic" of Guichard and Guichard (1965).

4. Including the "Mousterian" of Sandford and Arkell (1929, 1933, 1939), the "Lower" and "Upper Levalloisian" of Caton-Thompson (1946), and the different "Middle Paleolithic" industries of Guichard and Guichard (1965).

5. Including the Sebilian of Vignard (1923), the Epi-Levalloisian, Khargan, and Aterian of Caton-Thompson (1946, 1952), and a large number of industries recently defined by archeological investigation in southern Egypt (see Chap. 4).

Texture refers to the grade or size of particles constituting a particular sediment. Individual particle sizes are defined as follows by the modified Atterberg scale:

cobbles	over 60 mm
coarse pebbles	20–60 mm
medium pebbles	6–20 mm
fine pebbles (granules)	2–6 mm
coarse sand	0.2–2.0 mm
medium sand	0.06–0.2 mm
silt and fine sand	0.002–0.06 mm
clay	under 0.002 mm

In a more general way, materials greater than 2 millimeters in diameter are referred to as gravel, those between 0.06 and 2 millimeters, as sand. The textural class of a sediment is defined after the relative proportions of gravel, sand, silt, and clay have been determined, using the following table (Table 1–3). Sediment textures described in the subsequent chapters refer exclusively to textural classes as defined in this table. In addition, classes of coarse and medium sand are distinguished, depending on the preponderance of the one or the other, in a "pure" sand or in a mixed, sandy sediment. A term such as silty coarse sand is sufficiently clear, but coarse-sandy silt (in hyphenated form only) would refer to silt with a significant admixture of coarse sand. Similarly, medium-sandy silt implies

Table 1–3. Textural classes of sediments
(after Wentworth, 1922)

Grade component	Percentage	Textural class
Gravel	> 80	Gravel
Gravel	> sand > 10, others > 10	Sandy gravel
Sand	> gravel > 10, others > 10	Gravelly sand
Sand	> 80	Sand
Sand	> silt > 10, others > 10	Silty sand
Silt	> sand > 10, others > 10	Sandy silt
Silt	> 80	Silt
Silt	> clay > 10, others > 10	Clayey silt
Clay	> silt > 10, others > 10	Silty clay
Clay	> 80	Clay

silt with medium sand. Cobble, coarse, medium, or granule gravel is distinguished on the basis of mean pebble diameter or of dominant pebble grade.

Sand, silt, or clay grades refer to physical size and do not necessarily connote a particular type of material. Sands in the Nile Valley consist of quartz with a certain admixture of heavy minerals, while, at the Kurkur Oasis, they include calcite with an admixture of quartz. Clay minerals are generally distinguished from "clay" as a textural term. The word silt (singular) has a specific grade connotation, although silts (plural) refers only in a general way to fine-grained sediments of mixed calibration.

Sorting is here used to refer to the concentration into particular size grades of particles with diameters of under 2 millimeters. Semilogarithmic particle sizes are used (2–6, 6–20, 20–60, 60–200, 200–600, and 600–2000 microns, where a micron is equal to 0.001 millimeter). When 90% of a sample falls in one or two size grades, sorting is "good"; in three or four size grades, it is "moderate"; and in five or six size grades, it is "poor" (Payne, 1942).

Sediment and soil color are given according to the *Munsell Soil Color Charts* (1954). Since soil or sediments in Egypt are dry in the natural state, dry colors are given exclusively.

Structure refers to the macroscopic aggregation of soil or sediment particles. Blocky structure involves regular, cube-shaped aggregates that may be refitted and have angular, subangular, or subrounded edges. The individual aggregates range in diameter from less than 5 millimeters (very fine) to over 50 millimeters (very coarse). Granular structure involves crudely spheroidal aggregates that do not refit well, ranging in size from less than a millimeter (very fine) to over 10 millimeters (very coarse). Sediments with prismatic structure break up into columnar units with well-defined vertical faces, ranging in length from less than 10 millimeters (very fine) to over 100 millimeters (very coarse). In the case of platy or platelike structure, sediments break up into laminar or sheetlike aggregates, from less than 1 millimeter thick (very fine) to over 10 millimeters thick (very coarse). Loose, single-grain structure refers to quite noncohesive, usually sandy material. Slickensides, in the case of sediments or soils, refers to polished or finely fluted contact surfaces cutting diagonally through clayey materials. The various structural characteristics are a result of the kind and amount of clay minerals present as well as of the organic matter.

Profile *thickness* for soils or sediments are given in absolute figures rather than in depth below surface. Numbers given refer to typical maximum values, or, in the case of variable thickness (e.g., "40 to 60 centimeters"), they express the typical range represented.

Soil *horizons* encountered among the paleosols or the modern desert soils in southern Egypt include:

(B) —chemically weathered and discolored zone of finer texture than the parent material, without illuvial soil products—Numerical subscripts may be used to indicate subdivisions;

(B)C —transitional horizon with visible but incomplete weathering;

C_1 —partly disintegrated parent material with some evidence of oxidation;

C_2 —intact parent material;

Ca —horizon of diffuse lime precipitate or zone of calcareous concretions, nodules, or crusts, often forming mixed horizons such as (B)/Ca or C/Ca;

Sa —zone of salt accumulation, usually halite (NaCl) or gypsum;

P —seasonally waterlogged, pseudo-gley horizon of mottled reddish, yellowish, and grayish color resulting from alternating oxidation and reduction;

(B)E —soil sediment derived from an eroded (B)-horizon.

Gravel-shape analyses are based on the modified Lüttig method described by Butzer (1964b: 161 f.). The index of rounding (ρ) is expressed as the percent of smoothed, convex circumference of a pebble, obtained by visual estimation. Based on the mean value of ρ, the following classes are distinguished: 0–10% angular; 11–20%, subangular; 21–40%, subrounded; 41–60%, rounded; over 60%, well rounded. The coefficient of variation (CV) is defined as $1000\sigma/\text{mean}$, where σ is the standard deviation. This coefficient is used to express the homogeneity of the index of rounding, with the following classes: 0–25%, very homogeneous; 25–50%, homogeneous; 50–75%, heterogeneous; over 75%, very heterogeneous. The detrital component refers to the percentage of pebbles with an index of rounding equal to, or less than, 8%, giving an estimate of the proportion of local rubble in a sediment. The other indices employed include L (major axis or length of pebble), l (minor axis or width of pebble), and E (the height of pebble). In the case of flattened pebbles, the ratio E/L is under 50%, the ratio E/l is under 65%; in the case of nonflattened pebbles, E/L is over 60% and E/l, over 75%. The possible implications of these indices are discussed by Butzer (1964b: 162 ff.).

2

Geomorphic Evolution of the Kom Ombo Plain

Introduction

The Nile Valley immediately north of Aswan is physically and culturally an extension of Nubia. The alluvial land is less than one kilometer wide, with sandstone cliffs rising abruptly to 100 meters and more above the river. Lateral wadi gorges are deeply cut into the former tableland. Some 30 kilometers downstream of Aswan, however, the desert bluffs recede on either side of the floodplain. Massive gravels now tower to the horizon on the west, and rolling plains of drab silts undulate far to the east. For over 25 kilometers, the railroad passes through endless fields of lush sugarcane. Then again the sandstone closes in precipitously, forcing the Nile through a 400-meter-wide gorge on the west, while the railroad follows the broader gap to the east of Gebel es-Silsila. This landmark—called *Silsilis*, "the Mountain of the Chain," by the Greeks—delimits the northern edge of the great plain of Kom Ombo.

The present sugar fields were part of a forbidding desert as recently as 1900—an airless lowland of brown alluvium, soaking in the sun as dust devils spiraled endlessly on the floating eastern skyline. Coming around the Nile bend by boat, the nineteenth-century traveller saw the temple of ancient *Ombos* perched precariously above the river bank, with arid wastes stretching beyond into the distant valleys of Kharit and Shait. The only sizeable tracts of recent alluvium were found between the arms of the Nile meander, on the west bank opposite Ombos (Fig. 2-1). The total cultivated area was about 43 square kilometers, and the sedentary population in the late nineteenth century was little more than 20,000 in the several administrative districts: Darau, 9,200; Mansuriya, 3,400; Aklit, 4,600; Fatira, 1,100; and Faris, 2,000 (Boinet, 1899). Such was the limited agricultural base of the Kom Ombo Plain before 1902, when the

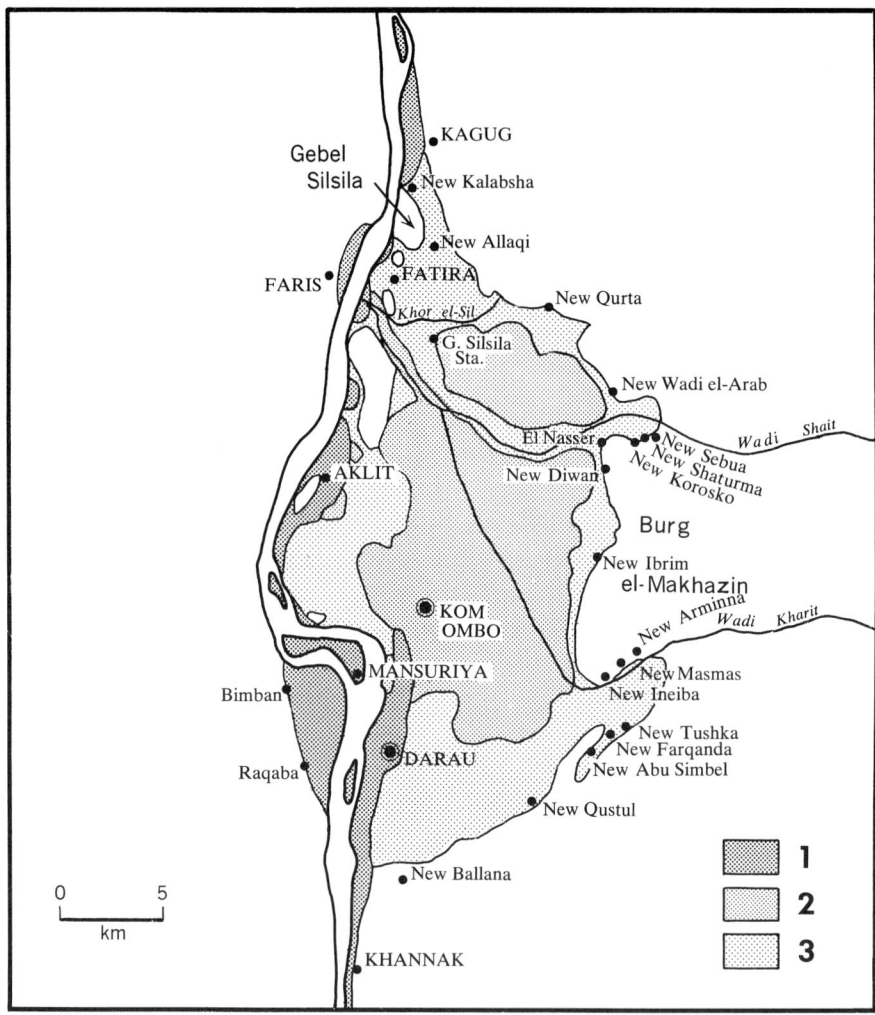

Fig. 2–1. Settlement expansion on the Kom Ombo Plain, 1902–65. *1:* cultivated lands, 1902; *2:* expansion of cultivation by Wadi Kom Ombo Company, 1903–29; *3:* Nubian resettlement areas graded and irrigated, 1961–65. Based on various maps of Survey of Egypt and other, partly unpublished maps. Nineteenth century administrative villages (*nahiehs*) capitalized, after Boinet (1899). Map: UW Cartographic Lab.

completion of the first Aswan Dam opened vast possibilities for large-scale desert irrigation.

The Kom Ombo region played an insignificant role as a frontier zone through most of historical times (Fig. 2–2). There are few records before

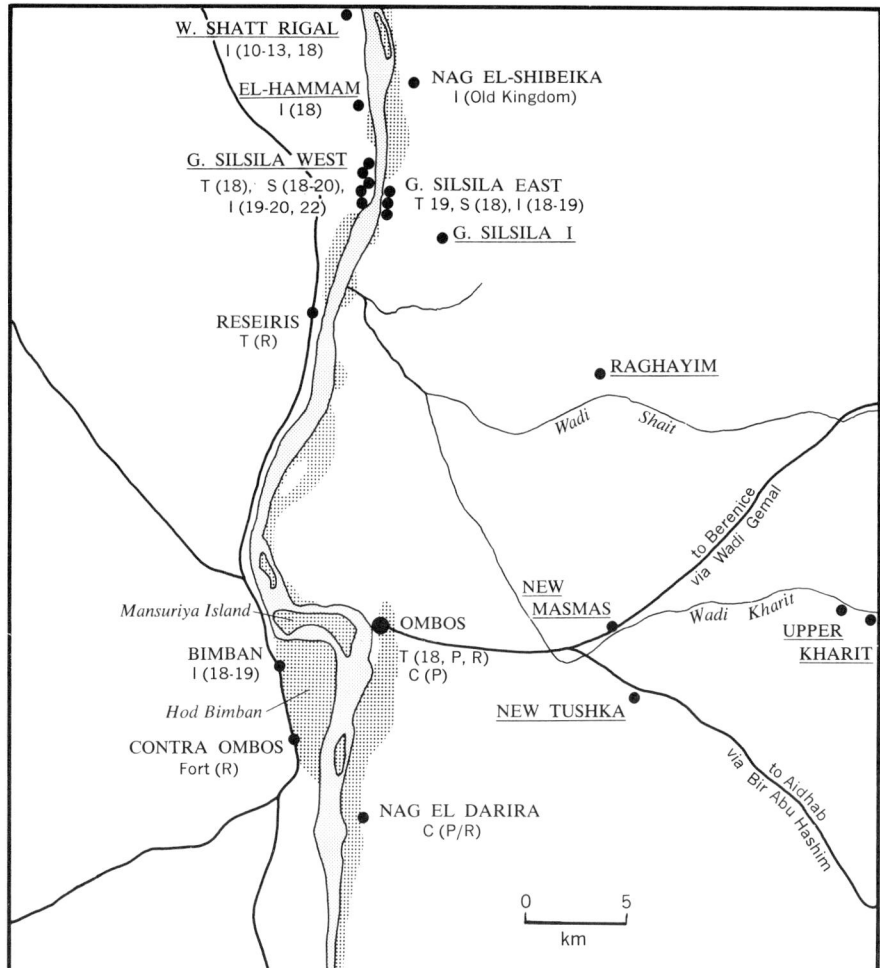

Fig. 2–2. The Kom Ombo Plain in Hellenistic times. (*T*) temples; (*S*) shrines and rock tombs; (*I*) inscriptions and stelae; and (*C*) cemeteries. Age identified according to dynasties (e.g., 18th–19th Dynasties), (*P*) Ptolemaic times, and (*R*) Roman times (after Porter and Moss (1937) and others). Predynastic and possible A-Group sites underlined, including rock drawings. Major caravan routes in Hellenistic and Byzantine times after Murray (1925), Meredith (1958), and others. Floodplain in gray. Map. UW Cartographic Lab.

the New Kingdom, when Hatshepsut (1503–1482 B.C.) quarried sandstone opposite Gebel Silsila and Thutmosis III (1482–50 B.C.) built a now-lost temple on the site of Ombos (Porter and Moss, 1939: 200). Large-scale quarrying activities were continued on both banks of the river at Gebel Silsila well into the Twentieth Dynasty (1200–1085 B.C.), providing building stone for Thebes and other cities of Upper Egypt (Porter and Moss, 1937: 208 ff., 220–21). Some 3,000 laborers were employed here by Ramses II (1304–1237 B.C.) during the construction of the Ramesseum.

In Ptolemaic times, Ombos replaced Aswan as the nome capital of the southernmost province of Egypt and occupied a key position at the head of a caravan route to the Red Sea port of Berenice, via Apollonos in Wadi Gemal. Between 180 B.C. and 30 B.C., an attractive temple was built to adorn the new metropolis and was dedicated jointly to the falcon-headed deity, Haroeris, and the ancient crocodile-god, Sobek. A town flourished just to the north of the temple enclosure, and the last Roman emperor to add a small sanctuary here was Caracalla (211–17 A.D.). The incursions of the Blemyes and other border-folk in the later third century invoked several Roman military campaigns and destroyed local prosperity. Diocletian (285–305 A.D.)—who built a fort, Contra Ombos, near Raqaba—managed to stabilize the frontier at Philae by settling the Nobatae in northern Nubia in the years following 297 A.D. A bishopric since the early fourth century, Ombos nevertheless played a very minor role in Coptic ecclesiastical history and was abandoned in early Islamic times, before the twelfth century.

Between the eighth and the fifteenth centuries, Upper Egyptians, Nubians, and Beduin clashed repeatedly in this shatter zone, and trade was the only economic pursuit of any stability. Darau emerged as a significant market town, at the head of the Eastern Desert caravan route from Berber in the Sudan. Camel herds, destined for Middle and Lower Egypt, were watered and pastured here before being driven further north to Farshut. Pilgrims en route to Mecca followed another caravan trail from Darau over Bir Quleib and Bir Abu Hashim to the former Red Sea port of Aidhab. Bimban, on the opposite bank of the Nile, served as a relay point on the caravan routes between Esna and Farshut on the one hand, Aswan or Kurkur and the Sudan on the other. The impoverished sedentary population was ultimately either displaced or absorbed by infiltrating Beduin, so that by the time of the French Survey of 1799, the agricultural lands along the river were largely occupied by former Ababda from the Eastern Desert.

The contemporary role of Kom Ombo began in 1903 with the foundation of the Wadi Kom Ombo Company by a group of international financiers. Making use of the new Aswan Reservoir waters and of steam-driven, high-pressure pumps, the company developed the undulating Pleistocene silt plain which lies 15 meters above modern flood level north and east of Ombos into a great commercial sugar and cotton enterprise. By 1914, over 100 square kilometers had been put under cultivation, and some 20,000 colonists, mostly impoverished Lower and Middle Egyptians, settled on the virgin lands. Further expansion of cropland to a total area of over 140 square kilometers was made in the 1920's (Fig. 2–1). The second raising of the Aswan Dam led to relocation after 1933 of some Nubian groups from el-Diwan to Dar es-Salam, near Darau. By 1947, the new market and administrative center of Kom Ombo, 3 kilometers northeast of ancient Ombos, had a population of over 40,000. The remaining high Pleistocene silts on the northern, eastern, and southern peripheries of the plain (totaling an area of 130 square kilometers) were claimed between 1961 and 1964 as part of the final irrigation project (Fig. 2–1). 48,000 Nubians evacuated from the Aswan Reservoir were resettled on these newly won lands during 1963–64 (Horton, 1964; Fernea and Kennedy, 1966).

Early Geological Work

Systematic geological observations on the Kom Ombo Plain were initiated a century ago by Leith Adams (1864), who in January, 1863, observed cross-bedded (Pliocene) sands at Muneiha and high (Pleistocene) silts with *Corbicula* near Fatira. Early in May of 1882, Georg Schweinfurth (1901) discovered prehistoric kitchen middens while on a geological and botanical reconnaissance of the northern plain. Rich concentrations of bone and molluscs—including Nile oyster and fossil *Unio*—were found in what Schweinfurth described as a former Nile channel southeast of Gebel Silsila. Massive alluvial clays over marls were recognized along Wadi Shait, where that dry watercourse cuts deeply into the Pleistocene silts of the plain. Schweinfurth believed in a great prehistoric lake or floodplain ponded back by former cataracts on both sides of Gebel Silsila. His map shows the presence of extensive high gravels along the west bank of the Nile and further identifies the Burg el-Makhazin hills (Fig. 2–12), between the embouchures of Kharit and Shait, as a mass of ancient gravel.

In the 1890's, during the planning stages of the first Aswan Dam, sporadic observations were made near the temple of Ombos. Sir William Willcocks (1894: 208) and Edward Hull (1896) found nilotic muds with mollusca up to 12 or 15 meters above floodplain, and Willcocks discovered further middens south of Gebel Silsila. J. de Morgan (1897: 42) was the first to specifically report stone artifacts from a midden near Fatira, however. A geological survey followed in the autumn of 1904, at which time H. J. L. Beadnell (1905) recognized the graben structure of the depression and postulated extensive Pleistocene lakes, fed by Wadis Kharit and Shait, in order to explain the great spread of alluvial sands and clays. Both Willcocks and Beadnell noted that the nilotic beds north and south of Gebel Silsila are at similar levels, precluding the existence of a barrier or cataract at that point.

Shortly after World War I, Edmond Vignard (1923), a chemist employed by the sugar refinery at Kom Ombo, discovered a sequence of new Paleolithic industries—described as Sebilian I, II, and III—from a series of platform-like ridges some 3 kilometers west of Kom Ombo, between Bayara and Sebil. Working intensively in his spare time, this ardent prehistorian salvaged numerous other Paleolithic sites southeast and northeast of Kom Ombo during the levelling and irrigation projects of the 1920's. Vignard (1923, 1935, 1955a) interpreted these cultures as occupation sites on the banks of a wadi delta emptying into a progressively shrinking lake that had been dammed up behind Gebel Silsila for most of Paleolithic times. Desiccation of the lake and the alleged bogs (*tourbières*) was attributed both to failure of the eastern watercourses and to incision of the Nile through the supposed Silsila barrier. Momentous as the archeological implications of the Sebilian cultures were, questions were soon raised as to their exact age and their relationships with the Paleolithic and Neolithic cultures of Europe and of North Africa.

The well-known Pleistocene survey of the Nile Valley by K. S. Sandford and W. J. Arkell investigated the western third of the Kom Ombo Plain during part of the winter of 1929–30 and the late spring of 1931. Valuable observations were made on the Pliocene and Pleistocene exposures along the Nile, which were mapped at approximately 1:180,000 (Sandford and Arkell, 1933: 4). The massive west-bank gravels were identified as part of a proto-Nile system, fed by Wadis Kharit and Shait, flowing westwards into the Libyan Desert and thence northwards to Esna (*ibid.*: 22). Widespread gravels 30 meters above present floodplain were thought to be downwarped in the center of the plain (*ibid.*: 10, 27, 33), while a 15-meter stage with Acheulian artifacts was believed to form a channel fill adjacent to the sterile +30 meter gravel unit (*ibid.*: 32). Sandford and Arkell do not seem to have traversed the eastern part of the plain because the Burg el-Makhazin is recorded as Nubian Sandstone on their 1:1,000,000 geological map. Said (1962: 89) repeats this error with a variation, describing the Makhazin hills as Chalk.

The late Pleistocene alluvial deposits of Kom Ombo were designated as "Sebilian Silts" by Sandford and Arkell (1933: 43–44, 49–52), but no unequivocal association with the Sebilian Silts was shown for geologically stratified sites in Upper Egypt. These authors attributed the nilotic alluvium to riverine marshes. They emphasized that the Sebilian III sites were found *on* the silts, in longitudinal arrangements suggesting riverbank settlement during temporary stages of equilibrium as the Nile incised its bed into the Sebilian Silts at the close of the Pleistocene. The Sebilian I was believed (but not conclusively shown) to be contemporary with the deposition of the silts, the Sebilian II belonging to the incipient phase of downcutting.

Prehistoric Salvage Work During 1962–63

In 1961, the UNESCO campaign to salvage archeology in the areas to be flooded by the projected High Dam at Aswan focused new attention on the prehistoric heritage of southern Egypt. Although downstream of the High Dam, the Kom Ombo Plain was also to be affected. Most of the Nubians were scheduled for resettlement in those parts of the plain that had not already been brought under irrigation by the Kom Ombo Estate prior to World War II (Fig. 2–1). Bulldozing and grading activities began in 1961, and a year later two missions received concessions for prehistoric salvage in the areas threatened. A group from the National Museum of Canada and the University of Toronto, under the direction of Dr. P. E. L. Smith (1966), excavated several sites and collected materials from some 40 others. For seven weeks in 1963, this group also included Dr. Robert J. Fulton of the Geological Survey of Canada. The Yale expedition excavated two major stratified sites near Gebel Silsila and collected a large number of surface sites. Both expeditions had previously carried out a joint archeological survey.

Attached to the Yale group to study the Pleistocene setting of the plain, we spent a total of four months working in the Kom Ombo area between October 3, 1962 and March 29, 1963. Bedrock, surficial geology, and geomorphology of a 1,500-square kilometer sector focused on Kom Ombo were mapped at 1:100,000 (the largest scale maps available) (Figs. 2–8, 3–1). Over 25 major Abney-level transects were run at 1:1,250, employing spot elevation marks of the Survey of Egypt. Detailed 1:2,500 maps were prepared of significant areas, and 246 sediment samples were collected and subsequently analyzed in the laboratory.

Bedrock Lithology

The oldest rocks of the Kom Ombo area, constituting part of the Basement Complex, are exposed at Aswan. Rather widespread is the coarse-grained, often porphyritic granite used for building and ornamental stone. These pink granites are considered to be mainly of migmatitic origin (Gindy, 1954). Gneisses and schists are also common. Both the granite and the metamorphics are intruded by sheets and dykes of finger-grained magmatic granites, as well as by several generations of dykes: lamprophyric and other basic rocks (striking NNE), felsites (striking NNW and WNW), and pegmatites (striking NE) (Kolbe, 1957). The migmatite and the metamorphics are probably Precambrian and are certainly no younger than Carboniferous in age, although some of the youngest minor intrusions may date from the Tertiary (Gindy, 1954; Said, 1962: 50–54, 59–63).

Resting disconformably on the planed surface of the Basement Complex is the Nubian Sandstone, the dominant bedrock facies of the Kom Ombo region. An excellent generalized profile, established east of Aswan in Wadis Abu Haggag and Abu Subeira, has been given by Kolbe (1957). Essentially applicable to Kom Ombo as well, the following stratigraphic sequence rests on the kaolinized surface of the Basement Complex:

a) *Lower Sandstone* (20 to 40 m). Massive arcoses, kaolinitic sandstones, or shales, intercalated with coarse-grained sandstones or conglomerates.

b) *Intermediate Sandstone* (20 to 30 m). Alternating fine-grained beds of (*i*) very-pale-brown, light-gray, light-brown, or pink sandstones; (*ii*) white, yellow, or red shales; (*iii*) red clayey sandstones; (*iv*) weak-red, ferruginous sandstones with ripple marks; and (*v*) as many as three lenses of red, oolitic iron ore, primarily hematite. The upper strata of this intermediate member contain marine and freshwater mollusca (*Inoceramus, Isocardia,* and *Cyprina* as well as three species of *Unio*), the marine worm *Geolaria filiformis,* and plant remains believed to belong to the Upper Cretaceous (Attia, 1955).

c) *Upper Sandstone* (60 to 100 m). Very-pale-brown, cross-bedded, massive, coarse-grained sandstones, with some shaley beds near the base. This member forms the typical bedrock at Kom Ombo.

McKee (1963), who interprets these strata near Aswan as lagoonal or estuarine formations, has published textural analyses. The sandstones generally show a single, strong grade-size maximum between 0.25 and 0.5 millimeters.

Younger rocks, preserved fragmentarily on top of the sandstone uplands to the northeast and southeast, are exposed as disassociated downfault blocks within the graben of the Kom Ombo Plain (Fig. 2–3). Lying

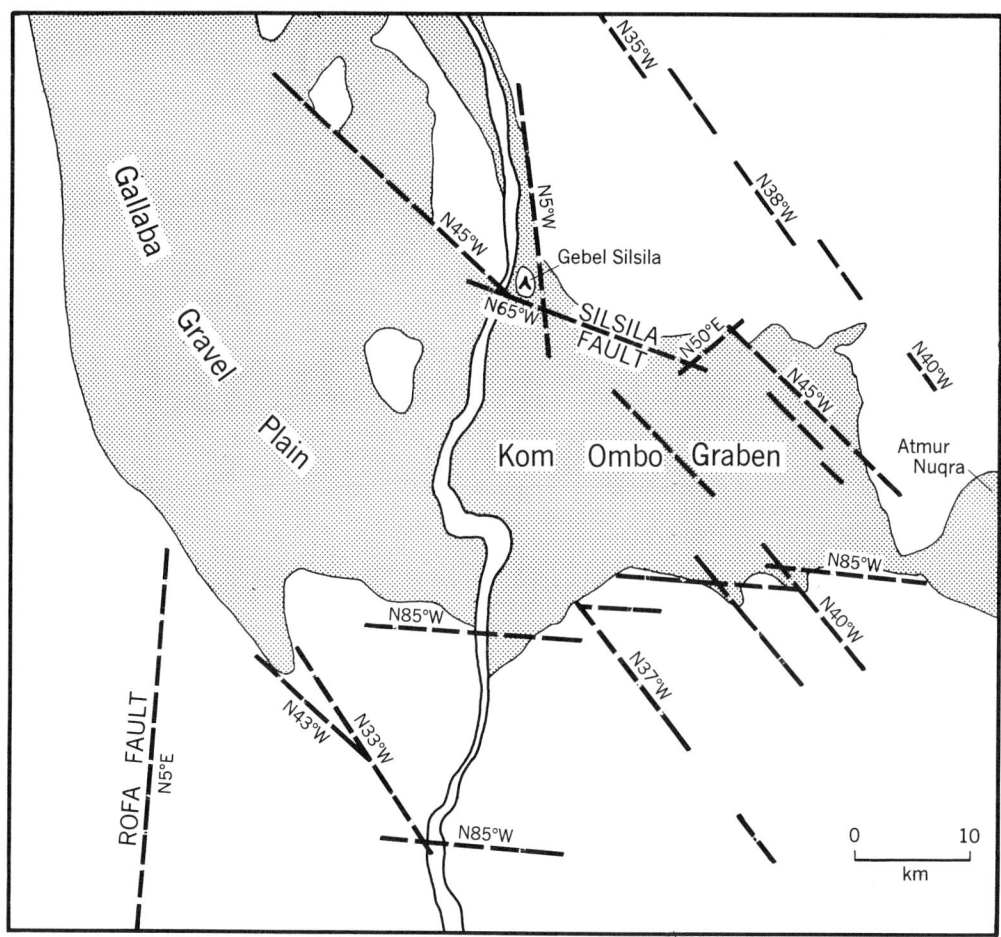

Fig. 2–3. Structure of the Kom Ombo region. Major faults shown with strike orientation; major areas of Pleistocene alluviation shaded; Chalk hills indicated in black. Map: UW Cartographic Lab.

close to the northern edge of the plain or along the Nile, these terminal Cretaceous to early Tertiary strata record considerable vertical displacements. Based on Beadnell (1905), Sandford and Arkell (1933: 4, 10–11), and personal observations, the later sequence can be reconstructed as follows:

d) *Dakhla Shale* (thickness unknown). Light-gray, laminated shales, at times calcareous (No. 90, Table 2–1). Danian.[1]
e) *Chalk* (25 m). White, thin- to medium-bedded, chalky limestone. Paleocene.
f) *Esna Shale* (over 5 m). Light-gray shales. At Gebel Miyahi these shales embed blocks of
g) *Nummulitic Limestone* as a breccia attaining over 15 meters thickness.

Both *f* and *g* are Lower Eocene in age (Berggren, 1964). The Pliocene strata will be considered in further detail below.

The exact relationship of the Dakhla Shale to the local Nubian Sandstone is not clear. In 1962 the Yugoslav firm *Geoistrazivanja*, under contract from the Wadi Kom Ombo Company, drilled a hole just east of the Cassel Canal, 100 meters north of the narrow-gauge railroad running eastwards out of Kom Ombo, at a surface elevation of 101 meters. Cores were taken at 50-meter intervals, and we were allowed to examine the samples then extracted on October 20, 1962:

− 50, − 100, − 150, and − 200 m: sandy silt or clayey silt;
− 250 m: reddish-yellow shale;
− 300, − 350 m: light-gray, laminated shale, somewhat calcareous;
− 400 m: light-gray to gray, organic, laminated shale;
− 450 m: highly bituminous shale with sandy laminae.

Despite subsequent identification of the − 250 meter coring as Nubian Sandstone by *Geoistrazivanja*, we do not believe that the sandstone had in fact been reached. The laminated shales between about − 300 and − 450 meters are markedly similar to the Dakhla Shale exposed near Faris and Fatira. Granite was apparently struck at − 850 meters (*Geoistrazivanja, in litt.*, June 29, 1964). The structural implications would appear to be momentous.

---

1. For a further discussion of the stratigraphic problems associated with the Dakhla Shale and the Chalk units see Chapter 7.

Structure

The Nile Valley has been appropriately described as a tectono-erosional valley, in which fluvial erosion has in varying degree been modified or controlled by older structural features (Yallouze and Knetsch, 1953). The Basement Complex has been fractured several times, probably soon after intrusion of the sheets or dykes of magmatic granodiorite and again in Tertiary times. Structural features may either be common to both the Basement Complex and the sedimentary mantle, or they may be inherited in the Nubian cover as gentle undulations, joint systems, or lines of weakness, possibly activated during the Alpine orogeny.

Although the overall deformation of the sedimentary cover in southern Egypt is slight but marked, the Basement is crossed by major tear faults, commonly intruded by aplitic dykes. When a river such as the Nile cuts down to the Basement, it will, if not already influenced by the projected structures of the sedimentary strata, readjust its course by preferring the exhumed shatter zones of the Precambrian base. Yallouze and Knetsch (1953) have further shown that the wadi tributaries of the Nile show systematic, linear geometric patterns clearly related to the structural orientation of the sedimentaries, and, ultimately, to the structural grain of the Basement. In the Kom Ombo area these directions are N–S and NW–SE, corresponding to the East African and Erythrean fault systems of Said (1962: 31 ff.).

Beadnell (1905) first recognized the Kom Ombo Plain as a tectonic depression, bounded by a major WNW–ESE fault to the north and a hinge line to the south with beds dipping northward at 5% to 10%. Sandford and Arkell (1933: 3, 10) estimated the throw of the northern bounding fault at over 200 meters and postulated another fault along the southern margin of the plain. In a general way, these earlier observations are correct, but the detailed picture is rather more complicated.

The tributary wadi drainage to the Nile is primarily related to a great number of complex fractures striking from N 35° to 45° W (Figs. 2–3 and 2–12). This pattern represents the dominant grain of the local bedrock, crossed locally by minor transverse faults striking N 45° to 50° E. Although Yallouze and Knetsch (1953) were able to demonstrate a secondary north-south grain (N 5° E to N 15° W) in the Kom Ombo–Aswan area, fault systems with this strike are difficult to prove in the field, with two conspicuous exceptions: (*a*) the N 5° W fault marking the eastern Nile margin at and north of Gebel Silsila and (*b*) the Rofa Fault (5° E), which delimits the eastern terminus of the limestone cuestas between Gebel el-Barqa and the Sinn el-Kaddab (see Chap. 5).

The northern edge of the Kom Ombo Plain proper is demarcated by an intersecting fault system striking N 65° W (Silsila Fault, Fig. 2–3) and N 45° W. Judging by displacement of the Chalk blocks at Raghayim el-Bid (surface elevation 120–127 m) from intact Chalk exposures some 16 kilometers to the northeast (at 24° 42′ N, 33° 07′ E), the throw of the Silsila Fault is about 150 meters. The Chalk exposures along the Nile banks range in level from 100 to 105 meters, so that there may be additional displacement along the N 45° W fracture on the order of 20 meters. A number of parallel faults are indicated on the northeastern margin of the plain. Along the southern margin of the plain there is a system of less well defined *en échelon* faults striking N 85° W, delineating a hinge line from which the bedrock dips to the north at 10% to 25% (Fig. 2–4). This suggests the Tethyan fault trend of Said (1962: 35). Judging by the sandstone-granite contact at about 750 meters below sea level at Kom Ombo town and by several hundred meters of post-Nubian shales, the structure of the graben center is much more complex than it appears to be at the surface. But unless geophysical work is attempted, the structural details of the plain remain obscured by the Pliocene and Pleistocene cover.

A N 85° W fault appears to delineate the lower course of Wadi Abu Subeira, and Kolbe (1957) has mapped the major faults between this wadi and the iron-ore mines east of Aswan. The two dominant systems in the Nubian Sandstone strike N 40° to 45° W and N 5° to 10° W. Vertical displacements commonly range between 5 and 8 meters, although a maximal throw of at least 40 meters was observed in one place. Mapping and assessment of the faults in this area is facilitated by the distinctive beds of oolitic iron exposed in strongly dissected terrain. The overall impression gained is that, with the exception of the Kom Ombo Graben, structural features *per se* are relatively minor, despite their evident influence on the topography.

Further faults of geomorphic significance undoubtedly exist north of Gebel el-Barqa and in the basins of Wadis Kharit and Shait, but these remain to be studied.

The exact age of the block faulting which created the Kom Ombo Plain is difficult to ascertain. The Middle to Upper Pliocene beds, to be discussed below, were unquestionably deposited into an existing graben. Eocene strata, on the other hand, have been dislocated by the Silsila Fault. It would seem, then, that the graben was formed during the Oligo/Miocene tectogeny as verified from the Cairo area (Knetsch, 1957; Said, 1962: 225 f.), although minor movements may have continued during the Pliocene.

Fig. 2–4. Inclined Nubian Sandstone strata near New Tushka.

Geomorphic Units

The landforms of the Kom Ombo Plain and its environs can be outlined and discussed briefly with respect to the several geomorphic units shown in Fig. 2–5.

THE ETBAI UPLANDS

The rough country of the Eastern Desert is formed by broadly horizontal strata of massive sandstones with some shaley or ferruginous lenses. Several erosional surfaces have been cut into these uplands, the most prominent of which is an extensive, subcontinuous level in 180 to 210 meters that can be followed southwards into Nubia. Designated as the Aswan Pediplain (Lower Nubian Pediplain in Butzer and Hansen, 1965), this surface suggests development as a series of coalescent pediments, related to a fluvial base level at a little under 180 meters (some 90 m above modern floodplain). Vestiges of two higher erosional surfaces in 210 to 220 meters and 230 to 240 meters can be recognized, although their significance is obscure. Lower, distinctly fluvial platforms related to Pleistocene stages of the Nile and its tributaries can be followed around the peripheries of the Kom Ombo Plain and along the margins of the Nile Valley. These platforms are found at elevations of approximately 150 to 160, 140, 130, and 120 meters.

All the erosional surfaces and platforms of the Etbai Upland have been deeply dissected (or were originally cut across badly dissected sandstone

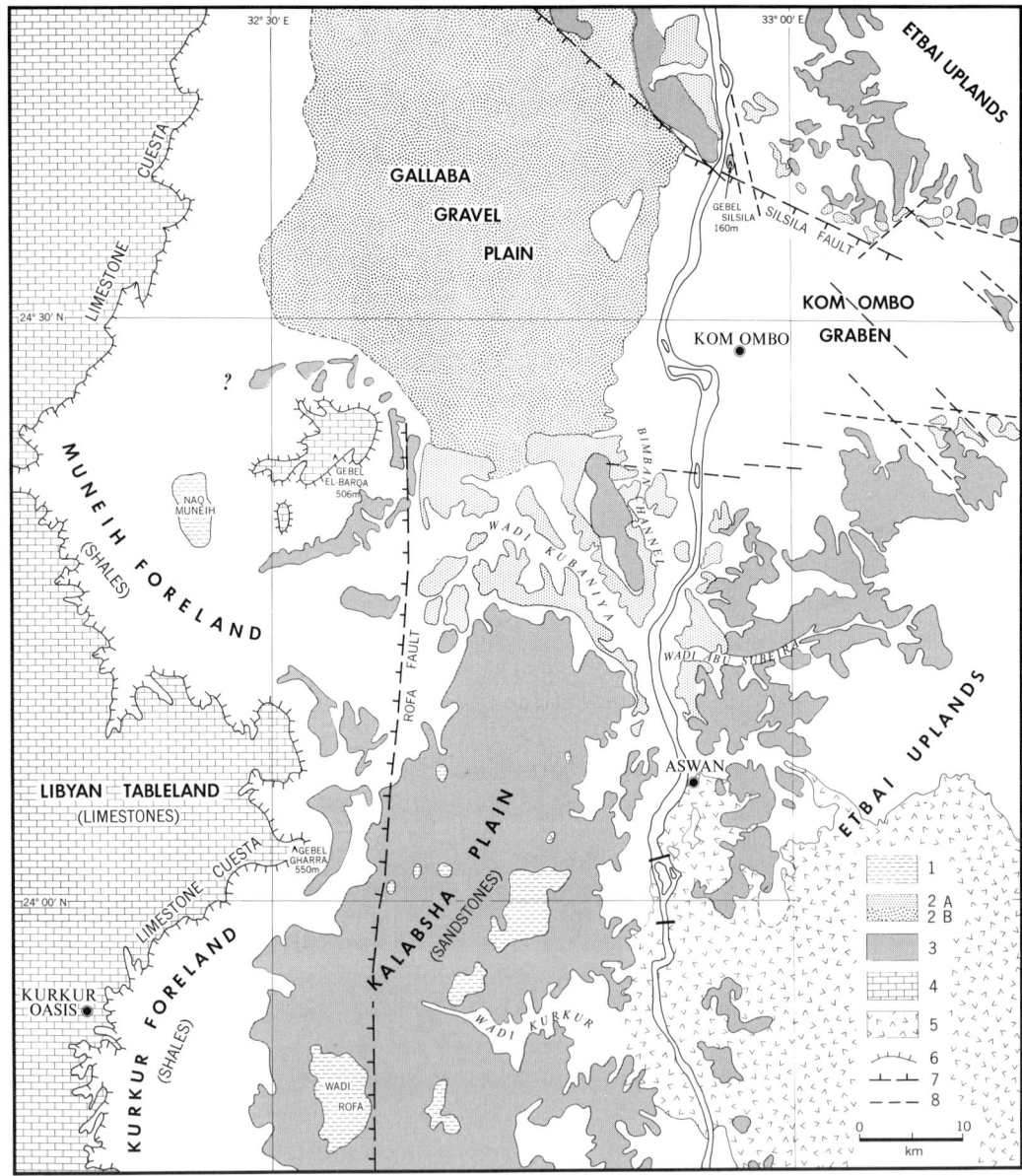

Fig. 2–5. Geomorphology of the Aswan–Kom Ombo area. *1:* Eolian depressions (*dayas*); *2:* fluvial platforms (*A*) and gravels (*B*) of the Gallaba Terrace; *3:* Kalabsha Plain; *4:* Limestones of Libyan Tableland; *5:* igneous and metamorphic rocks of the Basement Complex (in part after Kolbe, 1957); *6:* limestone cuestas; *7:* major faults; *8:* minor fractures (shown in Kom Ombo area only). Map: UW Cartographic Lab. (This map also available at 1:250,000. See p. 524.)

strata). The wadis are steep-sided (45% to almost vertical slopes) and flat-floored. Both the convex and the concave changes of slope are abrupt and angular, giving the landscape a tabular or mesaform aspect from a distance (Fig. 2–6). In detail, however, most of these surfaces are fairly irregular as a result of dissection, denudation, and lithological changes. Rocky hammadas with many low hummocks alternate with smoother flats mantled by colluvium. Generalized slope classes are essentially bimodal, with 40% to 70% of the surface inclined at less than 5%, and 20% to 40% inclined at over 45%. Local relief in five-kilometer squares averages 80 to 130 meters in the northeastern, 65 to 100 meters in the southeastern segment of the Etbai Uplands.

THE KOM OMBO GRABEN

Stretching for 30 to 35 kilometers east of the Nile, this is a depositional counterpart of the Etbai Uplands. The fretted sandstone bluffs terminate

Fig. 2–6. The Etbai Uplands north of the Kom Ombo Plain, looking east across the Silsila Gap from Gebel Silsila. Fluvial platforms at 140 meters and 150 to 160 meters visible in background. Aswan Pediplain on horizon, surmounted by Gebel el-Mizan (248 m).

abruptly along the flat surfaces of late Pleistocene alluvium—studded here and there by remnants of earlier Pleistocene gravel terraces. Local relief in five-kilometer squares varies from less than 5 meters on some of the uniform sediment surfaces, to 50 or 60 meters in the terrace complex of the Burg el-Makhazin. Extremely gentle slopes of less than 1% or 2% account for over 90% of the surface. A similar open plain, the Atmur Nuqra, is found in the Wadi Shait-Natash basin beyond the limestone ridges terminating the eastern end of the Kom Ombo Plain. The Gemini IV photographs suggest that much of the Atmur Nuqra Plain (Figs. 2–3 and 2–8A), presumably of tectonic origin, is an alluvial depression mantled with Pleistocene deposits.

THE KALABSHA PLAIN

West and southwest of Aswan this monotonous sandstone surface forms part of the Lower Nubian Plain (see Chap. 5). In the Kom Ombo area, an irregular fringe of the Kalabsha Plain at 180 meters elevation is still present southwest of Darau and northwest of Gebel Silsila. These moderately dissected upland plains show exceptionally well-developed fluvial platforms related to former courses of the Nile, particularly the level at 150 to 160 meters. Local relief in five-kilometer squares averages 30 to 80 meters, and surfaces are far more regular and extensive than in the Etbai Uplands. Bimodal slope distributions and abrupt, angular breaks of gradient are characteristic.

THE DARB EL-GALLABA GRAVEL PLAIN

This plain represents the depositional counterpart of the Kom Ombo Graben in the Western Desert. Gently undulating gravel surfaces (Fig. 2–7) of early Pleistocene age, veneered with colluvial soils or more recent sands, extend some 30 kilometers to the west of Kom Ombo and then swing northwards for almost 90 kilometers, rejoining the Nile Valley near Esna. Local relief is on the order of 5 to 15 meters. The morphology of the Gallaba Gravel Plain can only be appreciated from the air. The shallow, large-scale undulations are probably in large part due to compaction and sag in areas of deeper or finer-grained alluvium. Deflation may have had local significance. In addition, the wadis draining the limestone escarpments to the west have produced gentle swales that commonly persist for several kilometers out onto the gravel surface. A variety of subdued dune forms—large-scale ripple patterns developed in sand sheets for the most part—are superimposed on this gravel.

Fig. 2–7. Edge of the Gallaba Gravel Plain north of Faris, looking southwest from Gebel Silsila.

The Pliocene Deposits

The indubitable existence of Upper Pliocene deposits under the Kom Ombo Plain was first shown by Sandford and Arkell (1933: 8–10). A number of exposures on both banks of the Nile from south of Muneiha to north of Aklit were mapped on a small scale (*ibid.*: 4), and a characteristic section, lying unconformably on Dakhla Shale and disconformably overlain by Pleistocene gravels, was described. The basal facies alternates between cross-bedded or festoon-bedded micaceous sands and laminated clays, passing up vertically into coarse quartz sands. The uppermost sands are set in a white calcareous matrix containing root drip and other organic impressions and nowhere exceeding 130 meters in elevation. On the basis of their occurence, overall stratigraphic position, and lithology, these beds were correlated with the so-called Pliocene Gulf deposits as recognized in Upper Egypt and in the Fayum-Cairo area (Blanckenhorn, 1901; Sandford and Arkell, 1929: Chap. 3, 1939: Chap. 2; Sandford, 1934: Chaps. 2–3).

The work of the Yale Expedition confirmed most of the fundamental observations of Sandford and Arkell. More detailed study of the Nile bank exposures and discovery of related beds as much as 28 kilometers east of Kom Ombo (Fig. 2–12) now permit identification of a dominantly sandy facies with occasional clay or gravel lenses in the west and of a distinctive shale and evaporite facies in the east (Fig. 2–8A). Both facies appear to be laterally conformable.

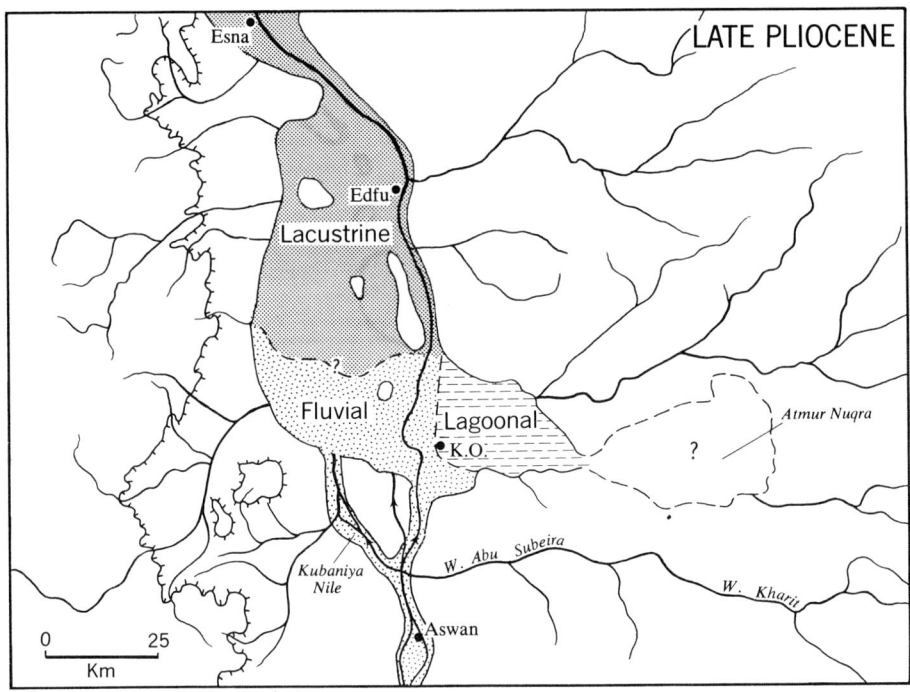

Fig. 2–8A.

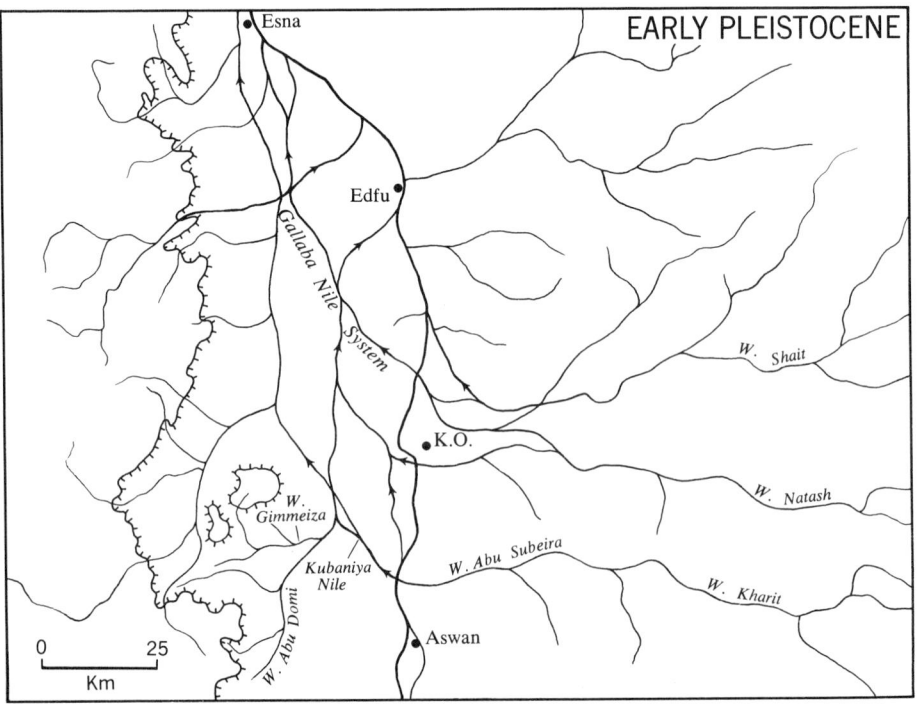

Fig. 2–8B.

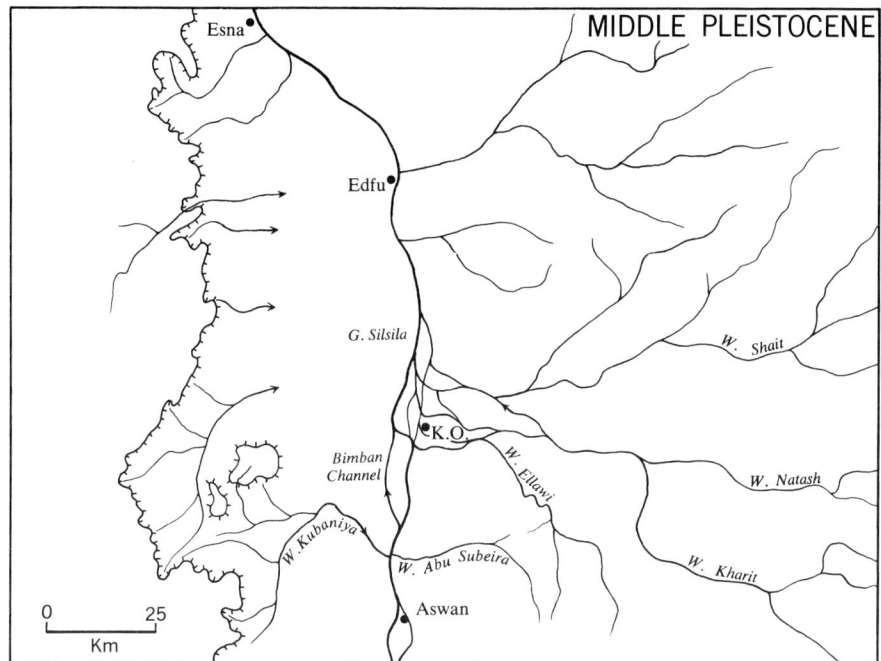

Fig. 2–8C.

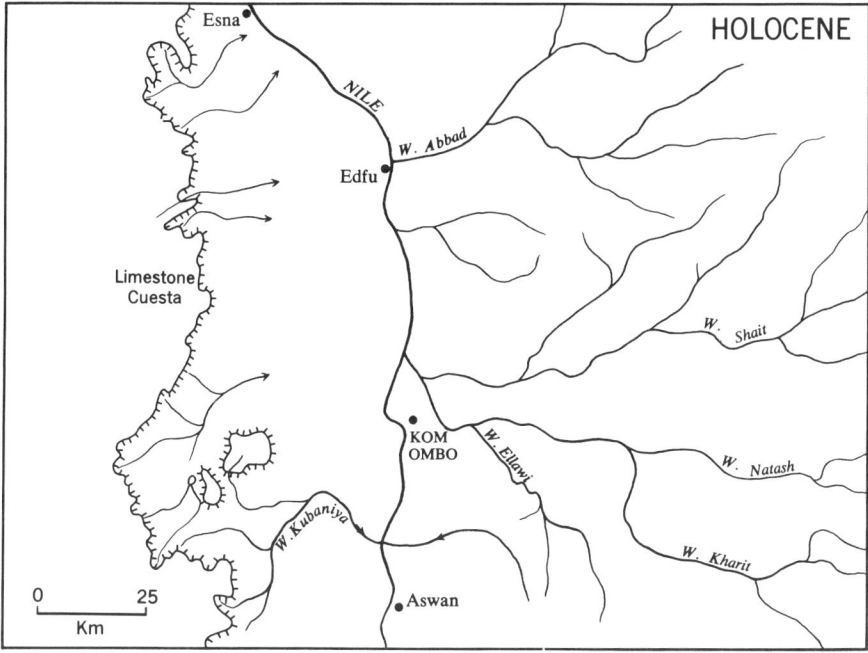

Fig. 2–8D.
Fig. 2–8. Paleogeographic reconstructions of the Kom Ombo region during (A) Late Pliocene, (B) Early Pleistocene, and (C) Middle Pleistocene times. Present hydrography shown in (D). Maps: UW Cartographic Lab.

THE NILE VALLEY FACIES

Between Bayara and Muneiha, the Pliocene deposits are exposed to a thickness of over 10 meters. These deposits are primarily current-bedded, coarse quartz sands (No. 86–I, Table 2–1). Quartz granules are abundant in all beds, and lenses of medium to coarse quartz pebbles mixed with coarse, rounded igneous gravel occur. Brownish-yellow discoloration is due to limonitic veneers on the sand grains. Subcontinuous ferricrete bands (2 cm thick) occur, but are rare. Beds are both topset and foreset, with easterly inclinations of 5% to 35%, primarily SE and NE. Evaporites are confined to veins and pockets and are probably intrusive. Attaining an elevation of 97 meters, these beds are unconformably overlain by 1.5 to 2.5 meters of Pleistocene gravel (dolerite and diorite), dipping 0.5% to the NW.

Another informative exposure can be found between Fatira and Gebel Silsila, where some 6 to 7 meters of homogeneous, coarse sands rest unconformably on the Chalk unit or disconformably on a sequence of uncertain age of alternating pink marls and light-brown silty sands (Figs. 2–9, 2–10, 2–11, 2–13). These typical, well-stratified Pliocene sands attain 97 meters elevation and consist of alternating beds of unconsolidated, light-yellowish-brown beds with root drip (No. 86–I, Table 2–1) and of semicemented, white, marly strata with considerable colloidal silica (No. 86–II, Table 2–1). The dip is 2% to 10% eastwards, ranging from SE to E. Three further units rest disconformably on these sands.

a) 2 m. Basal concretionary strata, consisting primarily of white, semicemented, calcareous concretions 1.5 to 3 centimeters in diameter, embedding coarse quartz sands (No. 87, Table 2–1). Alternating beds show horizontally stratified and vertically set concretions, all probably derived from massive networks of root casts or root drip. Beds dip 5% to 40% N and NE.

b) 1 m. Poorly stratified, coarse quartz sand, dispersed granules, and occasional medium gravel set in a semiconsolidated, calcareous matrix (No. 88, Table 1–1). The basic white to light-gray material is stained with limonite and pyrolusite, suggesting seasonal waterlogging.

c) 8 m. Crudely stratified, well-rounded, coarse-to-cobble gravel with a calcareous matrix (No. 89, Table 2–1) of very-pale-brown to white, coarse sand (quartz, feldspar, hornblende) with limonitic staining. Gravel lithology includes a wide range of igneous and some sedimentary elements, described further in a later section. Like the semiconformable contact between beds *b* and *c*, the gravel strata consistently dip 1% to 2% WNW. The surface of the extensive Fatira gravel exposure, at 99 to 104 meters, dips 0.7% WNW or NW.

The significance of these two examples is, first, that all of the Pleistocene gravel exposures between Muneiha and Fatira dip in a northwes-

Table 2-1. Sediment characteristics of some Pliocene and Pleistocene deposits of the Kom Ombo Plain (see Appendix A)
(Gravel excluded in the case of Pleistocene sediments.)

Sample number	Stratigraphy	Color	Texture	Sorting	CaCO$_3$ (%)	pH
		New Abu Simbel				
19	Paleosol on LT I	(5 YR 6/6)	Silty coarse sand	Poor	5.7	7.75
		New Sebua				
133	Paleosol on LT II	(7.5 YR 7/4)	Silty medium sand	Poor	8.4	7.45
126	Paleosol on MT	(2.5–5 YR 5/6)	Silty coarse sand	Poor	3.0	7.70
129	Upper Pliocene	(10 YR 6/4)	Clayey silt	Moderate	0.9	6.95
		New Shaturma				
121	Paleosol on LT I	(5 YR 6/6)	Silty coarse sand	Poor	0.5	6.65
80	Paleosol on HT II	(2.5 YR 6/6)	Silty coarse sand	Poor	1.0	7.30
124	Upper Pliocene	(10 YR–2.5 Y 6/6)	Medium-sandy silt	Moderate	0.0	6.20
125	Upper Pliocene	(5 Y 4/2)	Clayey silt	Moderate	0.4	6.80
		Fatira				
89	Bed *c*	(10 YR 7/3–6, 2.5 Y 8/2)	Silty coarse sand	Poor	10.8	8.05
88	Bed *b*	(10 YR 7–8/0–2)	Marly coarse sand	Moderate	16.9	7.90
87	Bed *a*	(10 YR 8/0)	Medium sand	Moderate	11.9	8.10
86–II	Upper Pliocene	(10 YR 8/0–2)	Medium-sandy marl	Poor	26.4	8.10
86–I	Upper Pliocene	(10 YR 6–7/3–4)	Medium sand	Moderate	7.1	7.95
90	Dakhla Shale	(2.5–5 Y 7/2)	Silty clay	Moderate	5.1	7.60

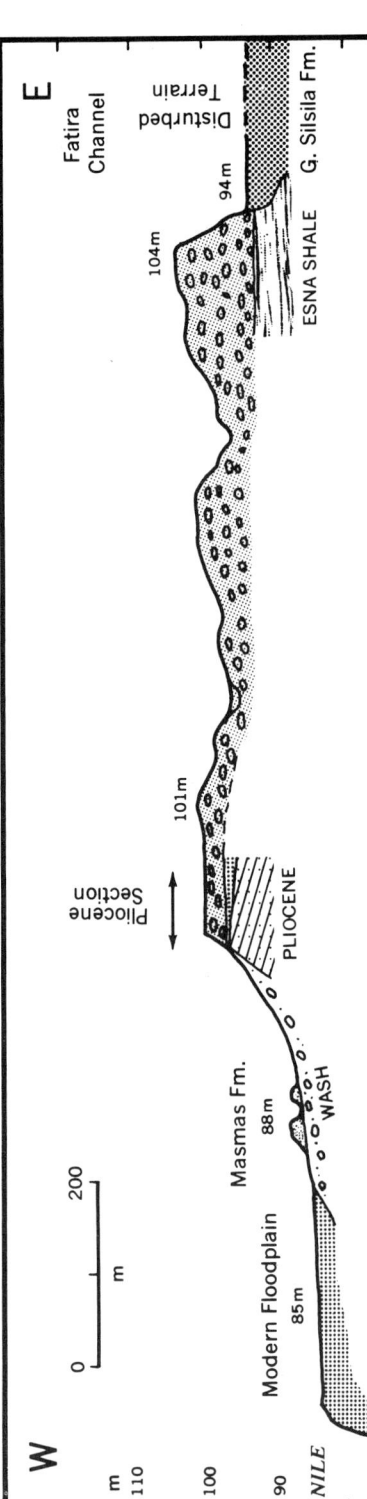

Fig. 2–9. West-East section of the Fatira Terrace. The Masmas and Gebel Silsila Formations are of late Pleistocene age. Map: UW Cartographic Lab.

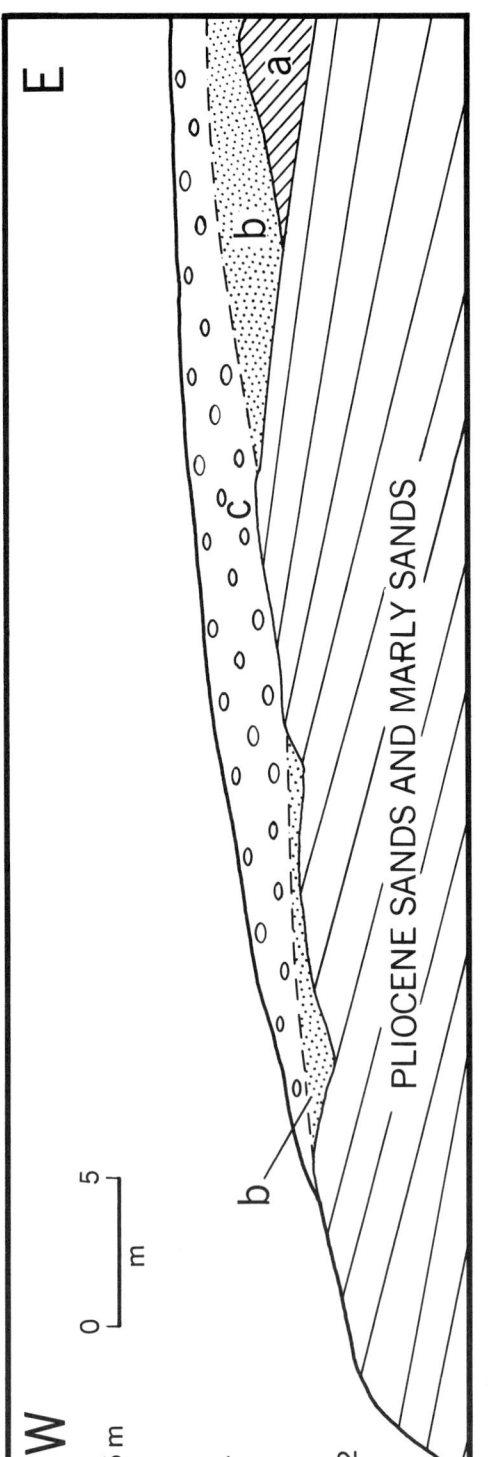

Fig. 2–10. Detail of Pliocene section, Fatira Terrace. Letters refer to sedimentary units described on p. 36. Map: UW Cartographic Lab.

terly direction and show no evidence of downwarping towards the center of the Kom Ombo Graben—as postulated by Sandford and Arkell (1933: 10, 32–33), who erroneously indicate an eastward dip for the Pleistocene gravels. Second, at least some of the topmost concretionary beds which rest disconformably on the typical Pliocene sands strike at right angles to

Fig. 2–11. Pliocene and Pleistocene beds of the Fatira Terrace (see Fig. 2–10). Contact of beds *a* and *b* at base of hammer.

them and must be considerably younger. Third, the fact that the alternating topset and foreset beds of the Pliocene sands variably dip SE, E, and NE can only be explained by original deposition in those directions—whether or not there has been additional eastward tilting at a later date. This interpretation of fluvial or estuarine sedimentation from a westerly source is supported by the occurrence of finer-grained shales and evaporites east of Kom Ombo.

THE KHARIT-SHAIT FACIES

East of Kom Ombo, Pliocene deposits (Fig. 2–12) are first found at the base of the northern Burg el-Makhazin, exposed under Pleistocene gravels in several pits between New Korosko and Sebua. Occupying a vertical elevation range from below 107.8 meters to at least 111.8 meters, the basal facies is an olive-gray shale (No. 125, Table 2–1) passing upwards into an olive-yellow, gypsiferous shale (No. 124, Table 2–1). Light-gray shales and interbedded white gypsum are found under late Pleistocene silts south of Wadi Kharit to 108 meters elevation at New Abu Simbel. An interesting exposure of similar facies is preserved about 150 meters south of Wadi Shait, some 4.5 kilometers east of New Korosko, at *ca.* 112 meters elevation.

a) 220 cm. Well-stratified, light-yellowish-brown (10 YR 6/4), gypsiferous shale, with widespread limonitic staining, occasional laminae of limonite or hematite, and incipient nodules. Rests unconformably on Nubian Sandstone.

b) 10–25 cm. Irregularly stratified, partly laminated, spongy, white evaporite (gypsum with some halite) interbedded with sandy shales as bed *a* and interrupted by ferruginous concretions. Some evaporite has intruded joints and bedding planes at the top of *a*.

c) 125 cm. As bed *a* but with evaporite bands replacing the ferruginous laminae.

These fine-grained beds and evaporites may mark the center of the former sedimentary basin. At the foot of the Chalk bluffs, marking the eastern edge of the Kom Ombo Graben, the Pliocene beds underlie various Pleistocene gravels to a maximum elevation of 130 meters. The facies is coarser, varying from light-gray, silty medium sands to sandy shales, all charged with evaporites.

In general, the Pliocene facies of the Kharit-Shait area is a monotonous, stratified, locally homogeneous shale or sandy shale with a rather high evaporite content. The depositional environment suggested is a shallow, saline body of water, subject to irregular fluctuations in hypersaline

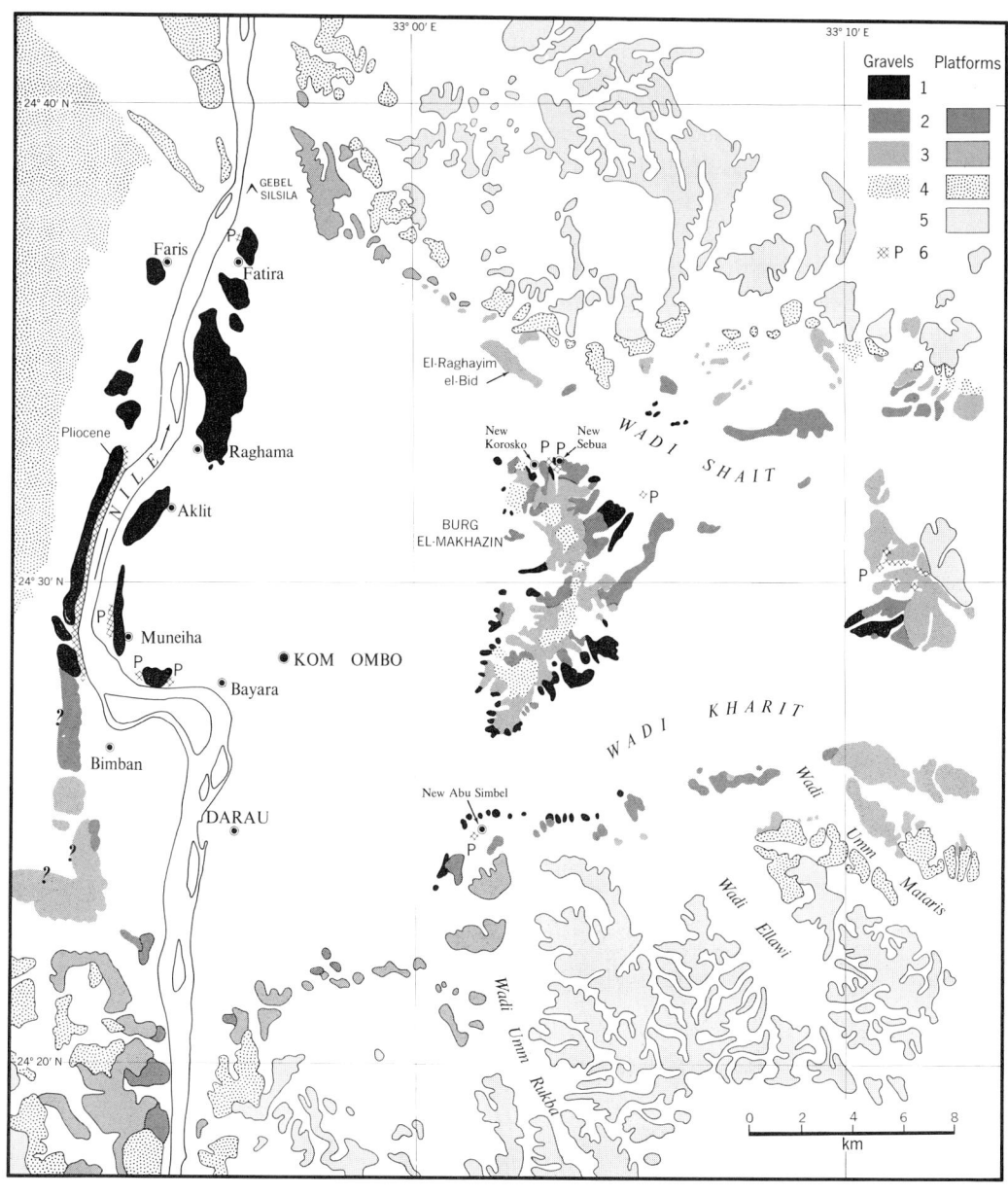

Fig. 2–12. Late Tertiary to Middle Pleistocene geology of the Kom Ombo region. *1:* Low Terrace, *2:* Middle Terrace, *3:* High Terraces, *4:* Gallaba Terrace, *5:* Aswan Pediplain, *6:* Upper Pliocene beds (subsurface only). Map: UW Cartographic Lab. (This map also available at 1:100,000. See p. 524.)

waters at temperatures of 30° to 40° C (see Posnjak, 1940). Although it is possible that the central Kom Ombo Graben harbored hypersaline lagoons during part of the Upper Pliocene, evaporite beds are sporadic and may only be a product of local and temporary conditions.

Carbonates are almost entirely absent in the Kharit-Shait facies and, except for some of the sandy beds of the Nile Valley facies, are generally rare in the Kom Ombo Pliocene. An explanation for this might be a drainage basin analogous to the one of today, which lacks sizeable outcrops of calcareous rocks. It is also probable, however, that any locally derived calcium was precipitated as a sulfate rather than as a carbonate. Of further interest is the limited amount of all but fine-grained detrital sediments on the northern, eastern, and southern peripheries of the former basin. This suggests little or no local runoff of a torrential type.

INTERPRETATION OF THE KOM OMBO PLIOCENE

On the basis of the present study, it seems that, during much of the Upper Pliocene, the Kom Ombo depression was occupied by a saline body of water which was fed by a major stream or streams from the west, with the shoreline in the zone between Gebel Silsila and Darau. Deposits of this period are found from below 87 meters to a maximum elevation of 130 meters. The only vestiges of organic life are root drip or root casts found in some of the estuarine beds. Rather similar deposits have been studied by Chumakov (1965, see Appendix J) at similar elevations in two wadis southeast of Aswan. We believe that all of the Kom Ombo Pliocene can be correlated with the upper part of Chumakov's Middle Series, which probably pertains to the regressive Upper Pliocene (Astian) of the Mediterranean region. Chumakov (1965) also identifies a Lower Series, related to the transgressive Middle Pliocene (Plaisancian) and tapped at the base of the buried, entrenched Nile Valley just upstream of Aswan to 172 meters below modern sea level. Further, carefully controlled, deep borings at Kom Ombo may also reveal a deep Pliocene column capable of direct correlation with the subsurface strata upstream of Aswan.[2]

2. The Upper (Alluvial) Series of Chumakov (1965, also Appendix J), thought to be of Villafranchian age, would pertain to the Basal Pleistocene in the more widely accepted usage. We found no gravels at 160 to 200 meters elevations in the Kom Ombo area, although our earliest Pleistocene alluvium, the Gallaba Terrace, is found to an absolute elevation of 163 meters. Sandford (1934: 48 ff.) recognizes older and higher gravels north of Luxor, while Chumakov (*in litt.*, February 1, 1966) believes that at least a part of the massive conglomerates of Upper Egypt, attributed by Sandford (1934: Chap. 2) to the Upper Pliocene Gulf, belong to the

The relationship of the Kom Ombo salt lake to the Upper Pliocene Gulf of Upper Egypt remains conjectural. The closest deposits downstream are encountered in the Esna-Qena area as limestones and marls of lacustrine type interfingering with or possibly overlain by conglomeratic beds (Sandford and Arkell, 1933: 11–14). It would seem that, due to some accident of geography, the Kom Ombo depression did not have free access to the brackish or freshwater environment of the southernmost Upper Pliocene Gulf. The westerly drainage into the Kom Ombo depression may provide an answer to this hydrographic problem.

Fig 2–8A provides a working hypothesis of the Upper Pliocene paleogeography of southern Upper Egypt. The general extent of Pliocene submergence is inferred from the distribution of Pliocene and Pleistocene deposits, in part following Sandford and Arkell (1933, folding map) with modifications. The distinction of estuarine and lagoonal facies on the Kom Ombo Plain has been discussed already, and a similar facies distinction may have applied to the great depositional basin lying 25 kilometers west of the modern Nile between Kom Ombo and Esna. Major discharge of the Nubian Nile may have been channelled through what is today called Wadi el-Kubaniya. And the greater part of the Wadi Kharit drainage probably still emptied into the Kubaniya Nile via Wadi Abu Subeira, rather than into the Kom Ombo Graben. Consequently, the major drainage into the graben may in fact have come from the southwest. In that case, significant deposition at the mouth of the Kubaniya Nile may well have isolated a lagoon of variable size in the Kom Ombo Graben.[3]

Detailed study of the lithology, heavy minerals, and bedding directions of the Pliocene exposures on both Nile banks between Muneiha and Fatira should resolve these questions.

The Early Pleistocene Proto-Nile System

With the total emergence of the Pliocene Gulf, through-drainage was reestablished on the Kom Ombo Plain. The first alluvial deposits succeeding the Upper Pliocene beds are related to a Proto-Nile system domi-

---

Upper Series. For these reasons it would appear that there is a considerable hiatus in the Kom Ombo area between the Upper Pliocene beds and the Gallaba gravels.

3. A complicating factor is the recent discovery of some evaporite beds in the Middle Series at 80 to 90 meters elevation near Aswan (Chumakov, *in litt.*, February 1, 1966). These deposits were absent in all of Chumakov's other sections and appear to be very localized. They are probably not directly associated with similar evaporites east of Kom Ombo.

nated by the Wadi Shait–Wadi Natash drainage from the Red Sea Hills.[4] This stage is best represented by the massive fill of the Gallaba Gravel Plain (Fig. 2–7), the maximum elevation of which is 163 meters (74 m above modern floodplain[5]). Available information is summarized by Figs. 2–5 and 2–12, and can be illustrated by several transects (Figs. 2–13 to 2–17).

The Gallaba Plain averages 150 to 160 meters elevation along the western bluffs of the Nile Valley, although a lower-level channel at approximately 145 to 150 meters appears to cut through this surface opposite Muneiha. Similarly, the highest gravel surface of the Burg el-Makhazin (Fig. 2–15) is at 150 to 160 meters. Similar gravel terraces are preserved on the northern margins of Wadi Shait (Fig. 2–12) and small vestiges can be found along the southern side of Wadi Kharit (Fig. 2–16).

The Makhazin and Shait gravels are rounded, coarse- to cobble-grade igneous rocks, whereas the Kharit gravel spurs—too small to be shown on Fig. 2–12—consist of subrounded to subangular local materials. Time did not permit a systematic study of the gravels or of the truncated red paleosols. However, a large collection of pebbles from a 151-meter gravel terrace of the southern Makhazin showed a dominance of dolerite and quartz (mainly pegmatitic), together with granodiorite, diorite, gabbro, felsite, olivine basalt, syenite, gneiss, ferruginized sandstone, and ancient conglomerate. The bulk of the material is clearly derived from the Basement Complex in the Shait-Natash headwater zone (Fig. 2–18), with notable underrepresentation of the less resistant schistose rocks.

The lithology of the Wadi Shait, Natash, and Kharit drainage basins has only been mapped on a reconnaissance scale of 1:1,000,000 (*Atlas of*

4. The designation "Proto-Nile" is employed in a number of different ways in the recent literature, and without definition the term is next to meaningless. The definition followed here is that outlined in Chapter 1, namely: an ancestral Nile River confined to the Proto-Nile Basin, north of about the Third Cataract. The existence of such a Proto-Nile, following roughly the present axis of the Nile Valley, is already suggested by the geomorphic evidence in Egyptian Nubia for the early Miocene (see Chap. 5), and this basic pattern was reinstated at the beginning of the Pleistocene.

5. The modern floodplain level used to determine relative elevations is 90 meters for the southern Kom Ombo Plain and Wadi Kharit, 88.5 meters for Gebel Silsila and Wadi Shait. These values represent the highest flood maxima observed prior to the inauguration of the Aswan Dam, i.e., 9 meters above the mean summer low-water stage. The latter was 81.0 meters at Kom Ombo, 79.4 meters at Gebel Silsila (see Willcocks, 1899: 44 f.). All values with plus symbols refer to relative elevations.

Egypt, 1928). There are differences in terms of major outcrops between the basins, and these may assume paleohydrographic importance. So, for example, felsites as well as ultrabasic rocks appear to be confined to the Wadi Shait catchment area, while typical basalts are only found in a small part of the Kharit drainage. Diorites and dolerites are found in all three drainage basins but are widespread only in Wadi Natash. Consequently, the potential implications of lithological indicators are of considerable interest.

The uneven nature of the 150- to 160-meter gravel surfaces could be explained either by subsequent denudation or by several substages at successively lower levels. Examination of the fluvial platforms at 150 to 160 meters suggests the latter alternative. On the north side of Wadi Shait, and again west of Gebel Silsila, the superimposition or close juxtaposition of these gravels and the fluvial platforms can be clearly seen, so that the gravels and platforms shown in Fig. 2–12 are, in all probability, functionally equivalent. The same platform is recorded on the summit of Gebel Silsila (Figs. 2–13, 2–14), where minor erosional benches, independent of lithology, occur at 150, 155, and 158 meters. Although the platforms fringing the northern edge of the Kom Ombo Graben could not be examined in detail, they consistently show similar substages, with the 150-meter level dominant (Fig. 2–6).

The paleogeography of the 150- to 160-meter Gallaba stage (GT) is shown by Figs. 2–5 and 2–8B. The modern Nile was restricted to a minor role, and the major discharge was derived from the wadis of the Eastern Desert: Shait, Natash, and Kharit–Abu Subeira. The lower combined course of Wadis Shait and Natash developed a floodplain 30 kilometers in width. The coarse bed-load deposits suggest a complex, braided stream. The development of benches in Wadi Kubaniya implies a width of 5 kilometers for the Kubaniya Nile, which drained Lower Nubia and the Kharit–Abu Subeira system. Extensive benches in Wadi Abu Domi suggest significant drainage from the Western Desert between Gebel Gharra and Gebel Barqa. Although the fluvial platforms north of Gebel Silsila indicate that this segment of the Nile was significant at the time, the bulk of the water and sediment moved over the Gallaba Plain northwards to Esna. The final 150-meter substage appears, however, to have occupied only a restricted channel on the Gallaba Plain, fed by a minor intake of Shait-Natash waters west of Muneiha, with its major source in the Kubaniya Nile.

In summary, the 150- to 163-meter (+60 to +74 m) Gallaba Terrace marks a period of transition, during which the Pliocene drainage lines in

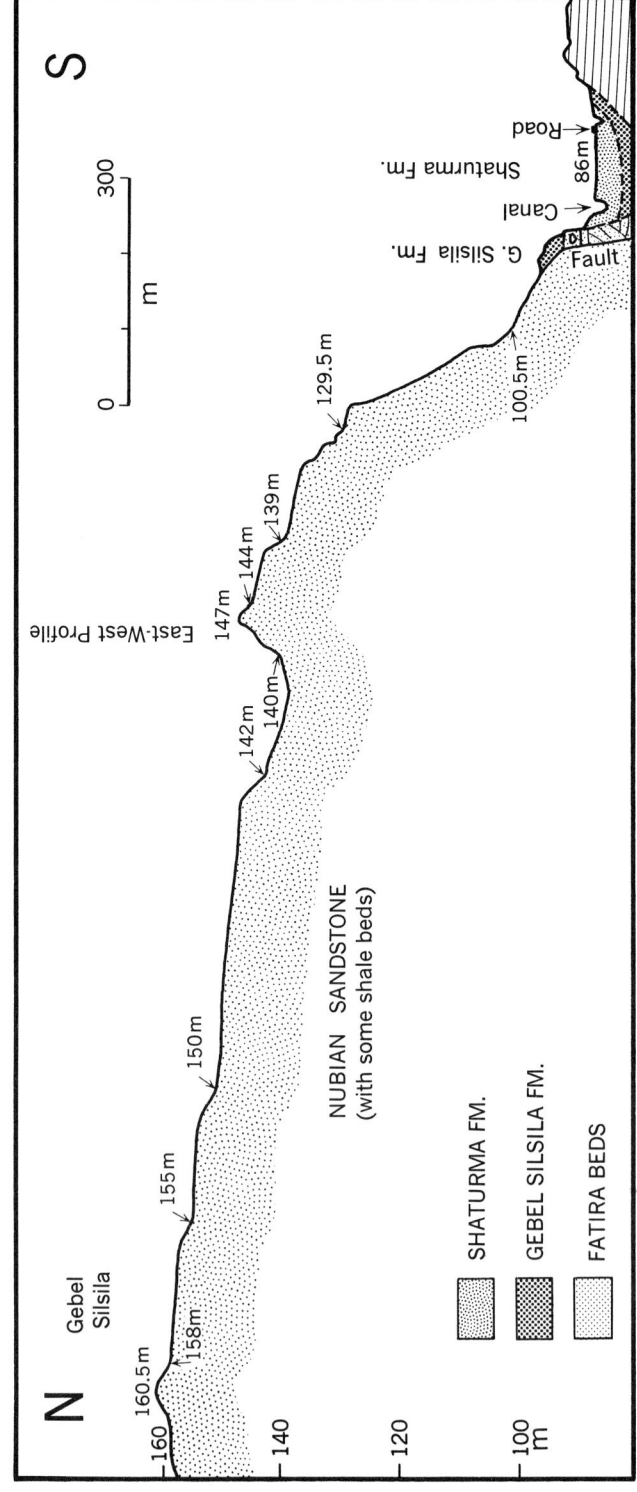

Fig. 2–13A.

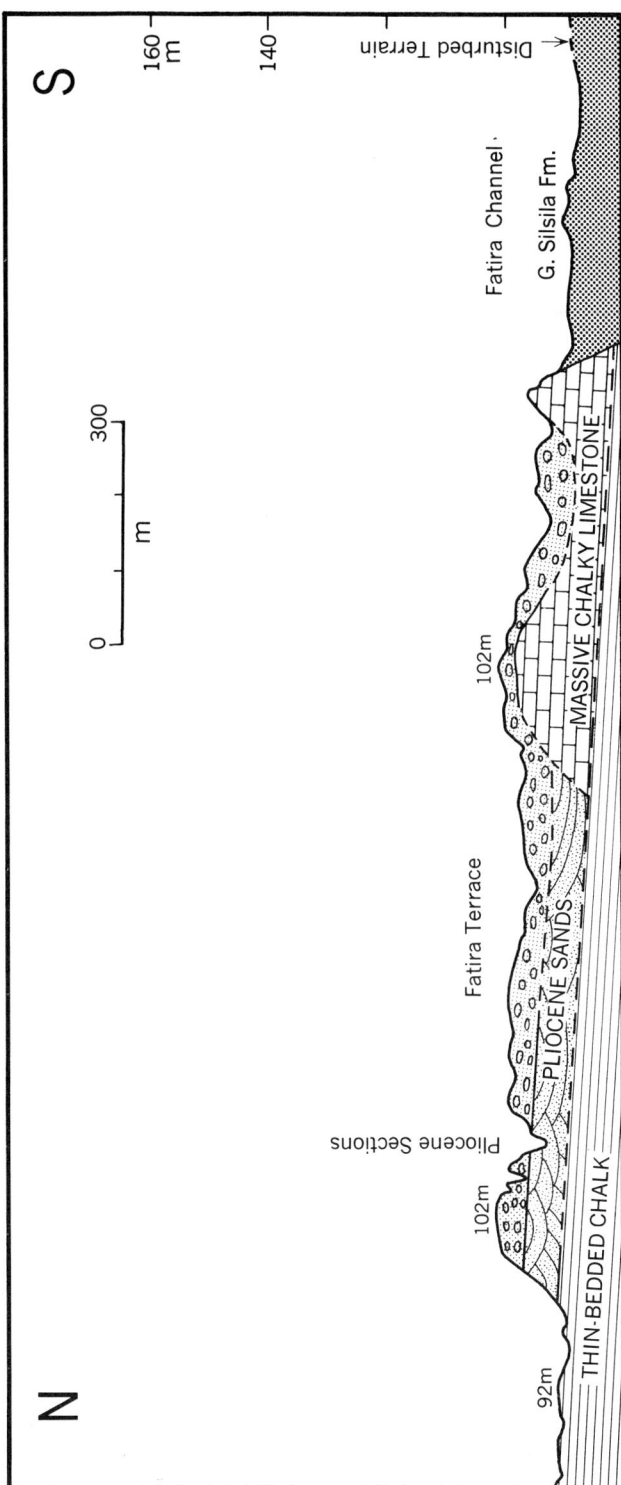

Fig. 2–13B.
Fig. 2–13. North-South section of Gebel Silsila and the Fatira Terrace. The Gebel Silsila and Shaturma Formations are late Pleistocene.
Map: UW Cartographic Lab.

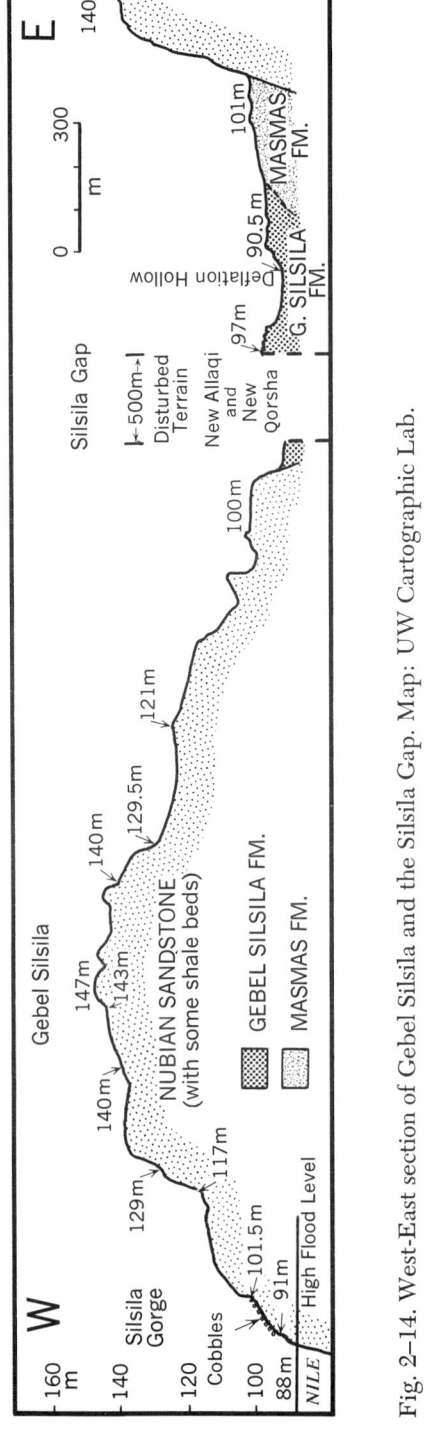

Fig. 2–14. West-East section of Gebel Silsila and the Silsila Gap. Map: UW Cartographic Lab.

Fig. 2–15. Gravels of the southern Burg el-Makhazin (facing north). Gallaba Terrace on horizon and younger gravels in middle background.

the Western Desert were still used extensively but were gradually being superseded by the modern Nile Valley. In fact, the present Nile course between Aswan and Esna may have been constructed during this stage by linking segments of previous overflow channels. The development of extended bedrock benches and accumulation of at least 50 to 75 meters of alluvium under the Gallaba Plain suggests a long period of time. If we assume that there were several phases of nondeposition or dissection, the 150- to 160-meter stage may represent a time interval on the order of $x.10^5$ years. At the height of deposition, the uplands of the Western and Eastern Deserts obviously received frequent, heavy rains permitting seasonal transport of a very coarse bed load by the Proto-Nile. Presumably, a semiarid mediterranean regime was involved. There is no evidence for subsaharan drainage or for summer floods of tropical origin.

The Proto-Nile–Gallaba stage postdates what have long been accepted as Upper Pliocene beds. Simultaneously, it records a radical change in sedimentary environment and facies. On these grounds alone, an early Pleistocene age seems warranted. Presumably, the time interval is partly coeval with that represented by the Villafranchian faunas of Europe.

Age of the Aswan Pediplain

The Aswan Pediplain is one of the most extensive and pervasive geomorphic features in southern Egypt (Figs. 2–7, 2–12). Although the landform characteristics and development are discussed further in Chapter 5, the age of this surface is most susceptible to study in the Kom

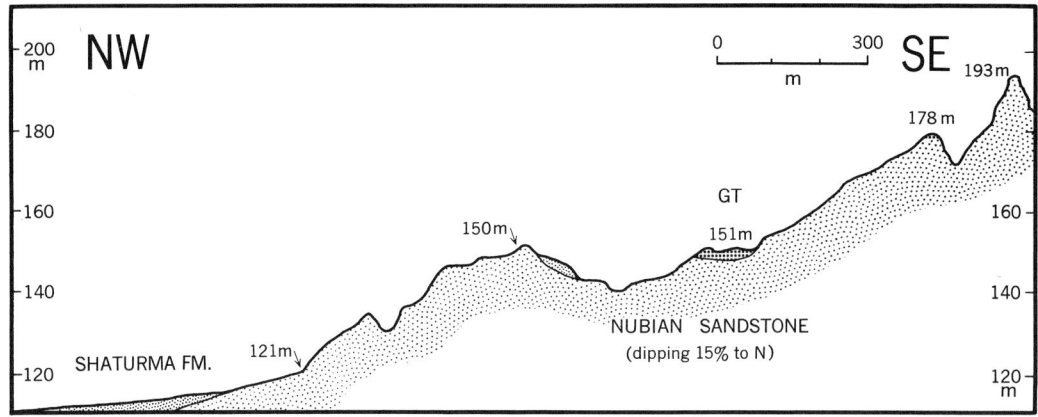

Fig. 2–16. Profile of sandstone uplands at New Farqanda. Map: UW Cartographic Lab.

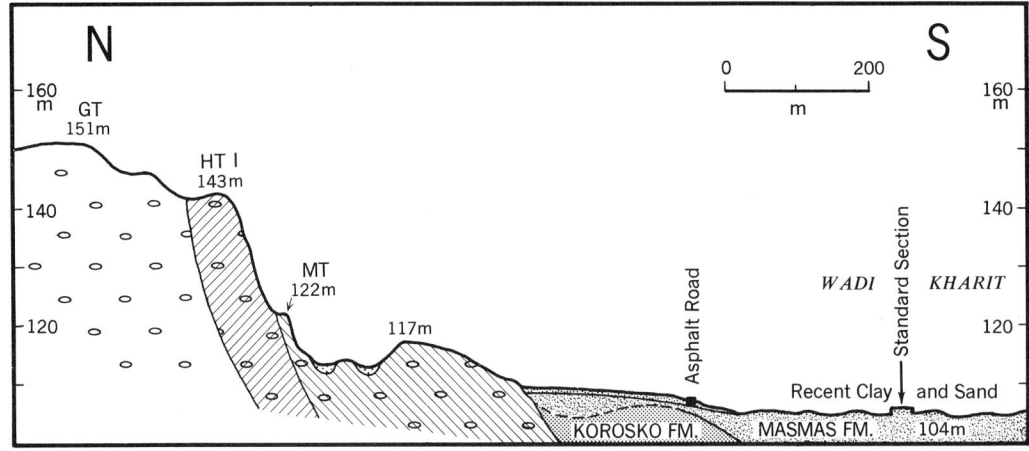

Fig. 2–17. The Pleistocene sequence at New Masmas. Map: UW Cartographic Lab.

Ombo area, in relation to the Oligo/Miocene faulting, the Pliocene beds, and the Pleistocene gravels and platforms.

The 150- to 163-meter (+ 60 to + 74 m) platforms and the Gallaba gravels are appreciably younger than the Aswan Pediplain. At least 30 meters of vertical cutting by the major drainage lines separates the two stages. The pediplain may therefore be very early Pleistocene (as suggested in the preliminary report of Butzer and Hansen, 1965) or it may be Tertiary in age.

Several pieces of indirect evidence at Kom Ombo indicate a Mio/Pliocene age for the pediplain. Numerous tilted blocks along the northern

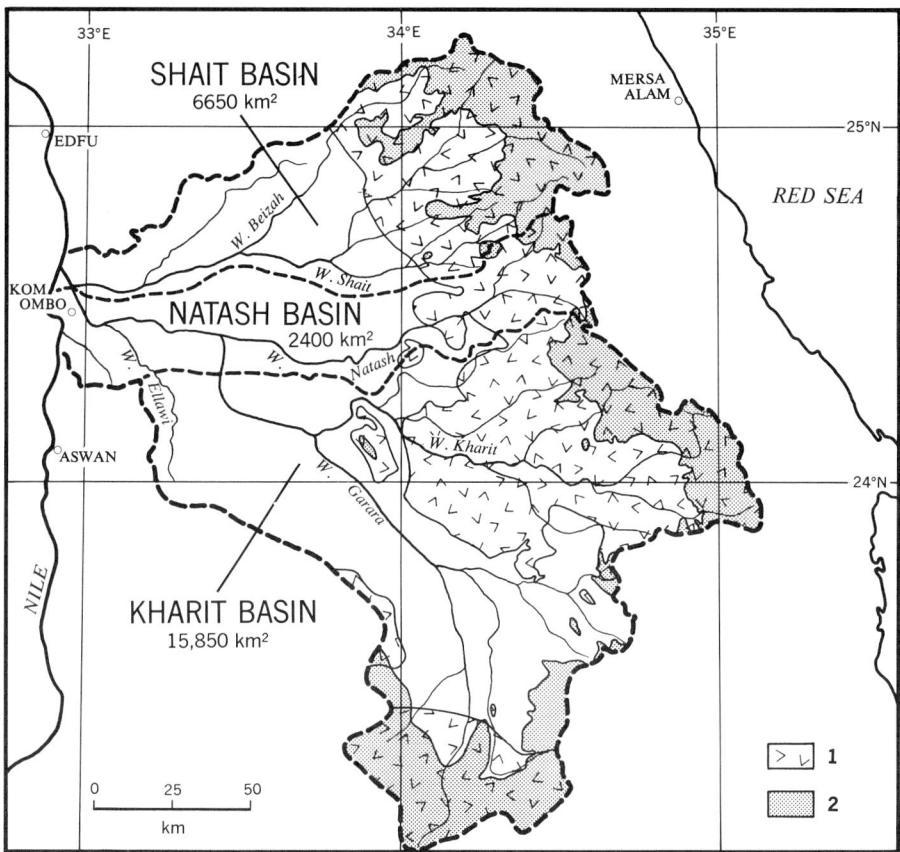

Fig. 2–18. The drainage basins of Wadis Shait, Natash, and Kharit. *1*: Basement Complex, after *Atlas of Egypt* (1928), Sandford and Arkell (1933), and Gemini IV photography; *2*: elevations above 500 meters. Demarcation of the Kharit-Natash divide near Kom Ombo is arbitrary to permit comparison of the formerly independent Natash and Kharit basins. Map: UW Cartographic Lab.

and southern edges of the Kom Ombo Graben have surfaces that appear to represent dislocated remnants of the pediplain. The great, tilted fault block marking the eastern end of the graben is almost certainly a former part of the pediplain that has been tilted northeastwards with an inclination of 10%. Although not quite conclusive, these are indications that fairly appreciable tectonic deformation postdates the development of the pediplain.

Equally pertinent is the fact that the Pliocene beds at the eastern end of the graben are embanked against the limestone fault block but appear to be undisturbed. If this is indeed so, the pediplain would be earlier than the Upper Pliocene Gulf. This is quite possible since the Pliocene deposits of the Kom Ombo–Aswan area are generally found at elevations under 130 meters. On the other hand, the extension of the pediplain into the Atmur Nuqra area speaks for the existence of the Shait-Natash drainage system in more or less modern dimensions. Pending further field work, the Aswan Pediplain must be considered as earlier than the Pliocene Gulf, but later than the primary downfaulting of the Kom Ombo Graben.

The "High" Gravel Terraces

"High" terraces, locally corresponding to modern drainage patterns and antedating a deep, red paleosol, occur as two distinct stages at 140 to 144 meters (+ 51 to + 54 m above Nile floodplain) and at 129 to 133 meters (+ 40 to + 43 m) in lower Wadis Kharit and Shait. Since these terraces were originally mapped as one unit, it has not been possible to differentiate the two stages for Fig. 2–7, although the mutual relations of HT I (140 to 144 m) and HT II (129 to 133 m) can be observed on several profiles (Figs. 2–13 to 2–23). Deposits of both stages are comparatively extensive, and fluvial platforms of HT I are found along the Nile Valley margins between Aswan and Darau, and near Gebel Silsila.

Detailed examination of the Gebel Silsila platforms (Figs. 2–12, 2–13) shows a single bench around most of the mountain at 129 meters, recording HT II, while broad platforms at 139 to 140 meters and 142 to 144 meters suggest the presence of two substages within HT I.

Gravel lithology is dominated by dolerite and quartz, and red surface paleosols are commonly evident at terrace margins, when not obscured by wash. A representative paleosol was studied near New Shaturma on a gravel terrace at 133 meters, 29 meters above the floor of Wadi Shait (Figs. 2–21, 2–22) and forming part of HT II. The petrographic spectrum includes granule-to-coarse-grade quartz and coarse-to-cobble-grade dolerite, diorite, granodiorite, granite, felsite, gneiss, chert, and conglomerate, listed in order of importance. From top to bottom, the profile (Fig. 2–21) can be described as follows:

a) 25 cm. (*B*)*E*. Light-brown (7.5–10 YR 6/4), homogeneous, coarse sandy silt; stratified, with coarse granular structure; highly calcareous. Under a pavement of coarse to cobble gravel with dolerite artifacts which are commonly patinated on exposed faces, superficially rubefied on buried faces. The important Middle Paleolithic surface collection made at this locality is discussed in Chapter 4.

b) 15 cm. (B)Sa. Reddish-yellow (5 YR 5/6), silty coarse sand as matrix to stratified, coarse quartz gravel, with diffuse evaporites and dendritic veins or microconcretions of calcareous salts.

c) 100 cm. $(B)_1$. Light-red (2.5 YR 6/6), silty coarse sand as matrix to stratified, coarse quartz gravel, with very porous, medium to coarse granular structure and less than 1% carbonates (No. 80, Table 2–1). The limited quantity of fine silt and clay suggests later, mechanical eluviation of fines. Kaolinite is the dominant clay mineral.

d) 200 cm. $(B)_2$. Reddish-yellow (5 YR 6/6), silty coarse sand as matrix to stratified, coarse or cobble gravel of quartz and other igneous rock (Table 2–3); medium-to-coarse, blocky structure; frequent concentrations of evaporites. The dolerites, diorites, granites, and gneisses are, in part, badly decomposed.

e) 125 cm. (B)C. Reddish-yellow (7.5 YR 5–6/6), coarse sand as matrix to stratified, coarse quartz and other igneous rock, in part decomposed; loose, single-grain structure.

f) Over 185 cm. C_1. Reddish-yellow (7.5 YR 6–7/6), coarse sand as matrix to stratified, coarse or cobble gravel, with considerable quantities of rotted, pale-yellow (2.5 Y 7/4), igneous rock.

Although the different calibrations of pebbles (Fig. 2–21) are a result of the original sedimentation, the absence of all igneous materials other than quartz in the top 1.5 meters of the profile must be ascribed to more or less complete weathering of the less-durable igneous rocks. This is supported by the fact that the number of recognizable but rotted igneous rocks increases downwards in the profile, as the intensity of weathering decreases. The presence of evaporites both above and below the $(B)_1$-horizon suggests two periods of salt enrichment, separated by a cycle of leaching and clay eluviation. Both periods of evaporite enrichment postdate the rubefaction—as does the (B)E-horizon—which may be due to colluvial washing or eolian deposition. The pavement contains cobble-grade materials, many of which have been reduced in size through fracturing by salt hydration. Similar size material is absent from the top 170 centimeters of the profile. Consequently, the now salt-fractured cobbles of the pavement must be attributed to lateral transport from higher-lying exposures by subsequent sheetfloods.

The local paleogeography of the High Terrace is basically modern in aspect (Fig. 2–8C). The Gallaba Plain was never again inundated by waters from Lower Nubia or the Eastern Desert, and all of the Nile discharge was channelled through the gorges on both sides of Gebel Silsila, following the course of the modern valley. The drainage direction of Wadi Kubaniya was reversed, with the oversize valley serving as an outlet for Wadis Abu Domi and el-Gimmeiza and emptying into the Nile

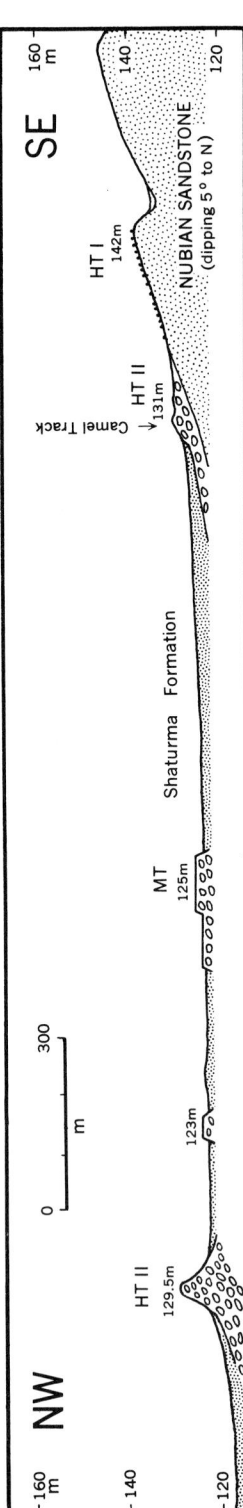

Fig. 2–19. Pleistocene deposits east of the Wadi Ellawi embouchure. Map: UW Cartographic Lab.

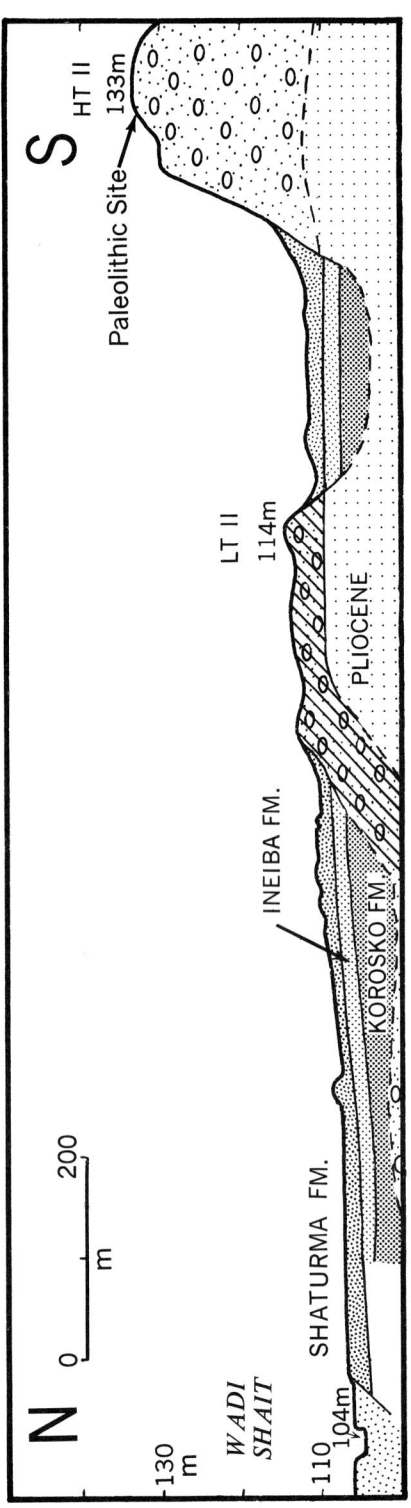

Fig. 2–20. Pleistocene deposits at New Shaturma. The Korosko and Ineiba Formations are late Pleistocene. Map: UW Cartographic Lab.

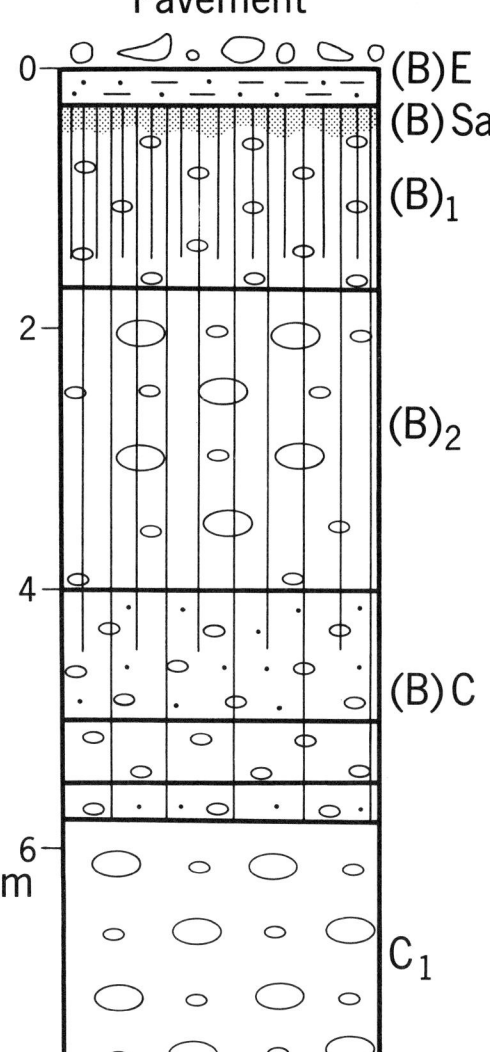

Fig. 2–21. Soil profile of High Terrace II between New Korosko and New Shaturma. Diagram: UW Cartographic Lab.

Valley. Wadi Kharit had been fully linked onto the Wadi Natash system, depriving Wadi Abu Subeira of all but an insignificant part of its catchment area. A branch of the Nile still occupied the Bimban Channel running several kilometers west of the valley, southwest of Darau. And in the Kom Ombo Graben, Wadis Kharit and Shait wove out a pattern of shallow, shifting, anastomatic channels, depositing coarse bed-load sediments.

The final abandonment of the Kubaniya Nile and the desiccation of the

Fig. 2–22. High Terrace gravels between New Korosko and New Shaturma (see Fig. 2–21). The Middle Paleolithic surface collection of New Shaturma (see Chap. 4) comes from this locality.

Gallaba Plain can probably be ascribed to reduced activity of the Eastern Desert wadis and a relative increase in importance of Nile flow. Westerly flow components were clearly reduced after deterioration of Wadi Abu Subeira into a minor, local wadi, whereas northerly components may have been appreciably strengthened by enlargement of the catchment area of the Nubian Nile. With no evidence to the contrary, it seems that heavy winter rains provided most or all of the runoff, and local climate was subarid to semiarid mediterranean in type.

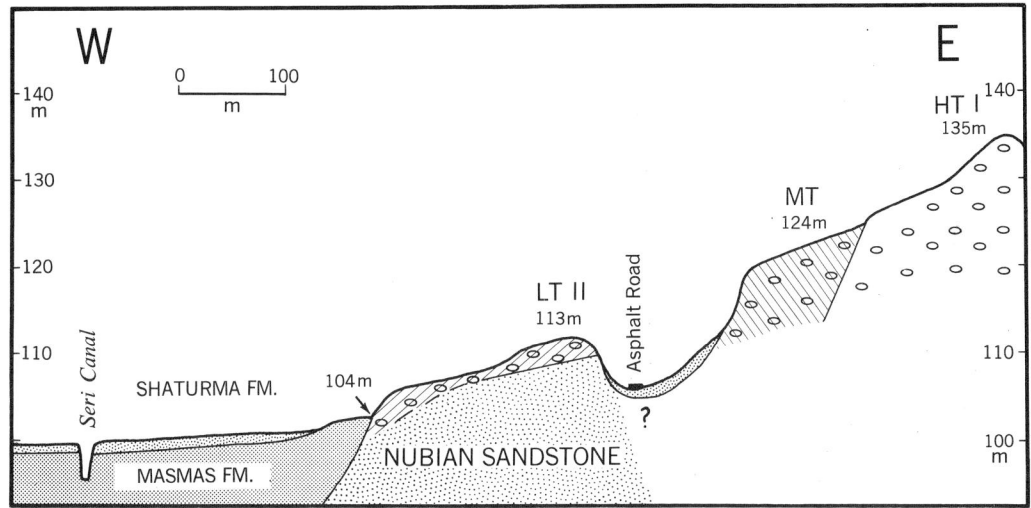

Fig. 2–23. Pleistocene deposits at El-Nasser. Map: UW Cartographic Lab.

The deep, red paleosol on the Gallaba and High Terrace deposits suggests a subsequent period of moderately intensive chemical weathering with at least a fair degree of seasonal moisture. Whether or not more than one such red paleosol developed during this early time range cannot be determined from the field evidence.

Since contemporary fossils, artifacts, or vulcanics are absent from the gravel body of the High Terraces, only relative stratigraphic dating is possible. On the basis of a four-fold Pleistocene classification, the Gallaba Terrace can be considered as Basal Pleistocene, the High Terraces as Lower Pleistocene.

The "Middle" Gravel Terrace and the Fatira Beds

A "Middle" Terrace is developed in the lower valleys of Kharit and Shait (Figs. 2–12, 2–17, 2–19, 2–23, 2–25), with moderately rubefied gravel surfaces at elevations of 121 to 125 meters. Except for a few platforms at similar elevations in the Wadi Umm Rukba embayment, related erosional features are rare on the east bank. Gebel Silsila, however, preserves a 117-meter bench on its western face and a 121-meter platform on its eastern face (Figs. 2–13, 2–14). Similar gravels and fluvial platforms can be seen in the Western Desert north and south of Bimban. These particular exposures were not visited. All in all, the Middle Terrace (MT) suggests a Nile base level of about 122 to 123 meters (+ 34 m).

The lithology of the Middle Terrace is analogous to that of the High Terraces, with a variety of igneous rocks derived from the Basement Complex, dominated by dolerite and quartz. A typical, rather indistinctive paleosol profile can be described from the southern side of Wadi Kharit east of New Tushka (Fig. 2–25), from top to bottom:

a) 10 cm. (*B*)*E*. Brown (10 YR 5/3), sandy silt; stratified, with coarse granular structure; rich in powdery gypsum and carbonates. Fractured, patinated dolerite pavement at surface.

b) 15 cm. (*B*)*Sa*. Light-brown (7.5–10 YR 6–7/4), coarse sandy silt matrix to stratified, coarse, rounded gravel; coarse granular structure; rich in evaporites.

c) Over 10 cm. (*B*). As above, but reddish-yellow (5–7.5 YR 6/6) and with limited salts.

Both the lithology and the paleosols are quite analogous to those of the later Low Terrace complex. The intervening period of downcutting seems to have been of some duration, however, since most of the Middle Terrace deposits were removed prior to renewed alluviation—so, for example, in the Burg el-Makhazin area.

The paleogeography of the Middle Terrace stage was rather similar to that of the present, although stream competence was appreciably greater and the Nile presumably deposited bed-load rather than suspended-load sediments. Such Nile deposits may be represented in the stripped gravel surfaces between Muneiha and Fatira that are shown as Low Terrace features in Fig. 2–12. These truncated gravel terraces are developed on the Chalk Unit or Pliocene outcrops near Muneiha (100 m), Aklit (91 m), Raghama (104 m), and Fatira (104 m) on the east bank. On the basis of excellent exposures studied near Fatira, these gravels can be informally designated as the Fatira beds. Similar gravels at somewhat higher levels (110 to 120 m) are present on the opposite bank of the Nile.

The Fatira exposures are shown by Figs. 2–9, 2–10, and 2–13 and were partly described (No. 89, Table 2–1) during discussion of the local Pliocene beds. About 8 meters of gravel are recorded, to a maximum elevation of 104 meters. The moderately dissected surface dips WNW or NW towards the Nile at 0.7%, while the actual gravel strata dip WNW at 1% to 2%. The lithology of the crudely stratified, unsorted, well-rounded, coarse to cobble gravel (Table 2–3) is dominantly dolerite, diorite, and quartz. Other elements include chert and flint nodules (locally derived?), sandstone, ferruginous sandstone, granite, granodiorite, felsite, basalt, olivine basalt, gneiss, schist, chlorite schist, and quartzite (with quartz veins). Since granites and schists are very rare in the

Makhazin gravels, these elements are derived at least in part from the Aswan area. The facies suggests a fanglomerate, part of several coalescent alluvial fans shot into the Kom Ombo Graben by Wadis Kharit and Shait and by the Nile.

Although no distinct paleosol horizons can be seen on any of the Fatira beds, there is reason to believe that the uppermost 1 to 2 meters of the gravels record the truncated base of a fairly well-developed paleosol. Weathering rinds are common on all rocks other than quartz. In the case of felsites and basalts, such rinds are 0.3 millimeters thick. Flint and chert nodules with typical yellow (10 YR 6/8) cortices may have red (10 R 5–6/6) rinds as much as 1.5 millimeters thick. The feldspars in granites and gneisses are moderately weathered at the core of pebbles 5 centimeters in diameter, while sandstones are completely decomposed to friable, limonite-stained sand. Even the rare mesocrystalline quartz pebbles show a 2-millimeter discoloration rind. In addition, illuvial sesquioxides are frequently found precipitated on the top of rocks. These mainly form a red (2.5 YR 5/8, 5 YR 5/6), hematitic cement. There can be little doubt that several meters of a well-developed (B)-horizon have been stripped off the surface. This indirect argument suggests that the Fatira beds were deposited as part of the Middle Terrace, or even the High Terrace-II, gravels but were subsequently truncated to form a platform corresponding to the Low Terrace. The west-bank gravels—which attain a maximum elevation of 120 meters but generally form a 110- to 112-meter surface—deserve further study in this connection.

An abundant, crude Acheulian industry with coarse flakes was collected from several surface sites on the Fatira beds exposed between Bayara and Muneiha (Muneiha East) by David Boloyan (see Chap. 4). Several collections of coarse, undiagnostic flake tools were made near Fatira. Although the bifacial elements are missing (suggesting that earlier visitors have picked off the more attractive pieces), these implements would fit into the Muneiha East collection. The rolled Acheulian hand axes found *in situ* in the Fatira beds at Bayara and Aklit (Sandford and Arkell, 1933: 32–34, Pl. 21–23) have close counterparts among the unrolled Muneiha East surface collection.[6]

---

6. We found no evidence that the Low Terrace (Sandford's 50 ft. stage) is contained as an "inner lining" in the ancient channel of the Middle Terrace (100 ft. stage) (Sandford and Arkell, 1933: 32). The Fatira beds appear to be uniform along the 18-kilometer stretch of the Nile where they are exposed. We also found no evidence for an eastward dip of any Pleistocene beds in this sector, contrary to the commentary of Sandford and Arkell (1933: 10, 32–33). This point has already been raised in the preceding discussion of the local Pliocene.

The "Low" Gravel Terraces

"Low" terraces are fragmentarily preserved in lower Wadis Kharit and Shait, particularly around the peripheries of the Burg el-Makhazin (Fig. 2–12). Many of the latter deposits represent older wadi gravels that have been reworked into a local fill and graded onto the former wadi floodplains. The surface elevation of this Low Terrace ranges from 107 to 118 meters, and two substages can be recognized at 116 to 118 meters (LT I) and 112 to 114 meters (LT II) (see Figs. 2–20, 2–23, 2–24). The mutual relationships of these substages could not be determined, because of indistinctive facies and a lack of suitable exposures. In addition, there is at least one denudational step, at about 110 meters, cut into LT II. Correlative sediments appear to be lacking in the actual Nile Valley, although the Fatira beds at 100 to 104 meters presumably record erosional levels functionally related to LT II, while the west-bank terraces at 110 to 112 meters may be related to LT I. A Nile floodplain level of about 105 meters ($+15$ m) can be tentatively suggested for LT II and a level of 110 to 112 meters ($+22$ m) for LT I.

Whereas the lithology of the Low Terraces is generally quite analogous to that of the older gravels, the beds along the southern embouchure of Wadi Kharit contain an appreciable sandstone component, in addition to the dominant dolerite and quartz. Generally, calibration is smaller, with medium to coarse pebbles characteristic (Table 2–3). The overall impression of the Low Terrace gravels is that of an arid-zone alluvium, derived in large part from local materials, e.g., reworked older gravel. Much of this activity was due to short-distance transport by local wadis. The large-scale through-drainage of Wadis Kharit and Shait was insignificant compared with that of the earlier alluvial stages.

Although no artifacts were found *in situ* in the gravel body of the Low Terrace,[7] fresh but patinated, coarse Levalloisian artifacts made from dolerite are common on the surface of these gravels or in colluvial wash resting on them. A broken, rather waterworn hand axe was found on a contemporary erosional surface at New Sebua, and occasional unworn bifacial implements were collected from the surface of LT I near New Abu Simbel. This suggests that the Low Terrace antedates the "late Lower Levalloisian" industries as outlined by Caton-Thompson (1946).

---

7. Sandford and Arkell (1933: 44) report rolled Acheulian implements and a few, slightly waterworn "Mousterian" implements from the body of a younger gravel embanked against the Fatira beds north of Muneiha, with an elevation of *ca.* 95 meters.

Fig. 2–24. Polygonally cracked clays deposited by recent spates in Wadi Kharit, near New Masmas.

Paleosols are generally preserved on the surface of the Low Terrace (Nos. 19, 133, and 121, Table 2–1). A composite profile—based on five fairly complete profiles (Pits 22, 40, 44, 45, and 47) studied between New Korosko and New Sebua—can be described from top to bottom:

a) 5 to 30 cm. (B)E. Light-brown (7.5 YR 6–7/4), sandy silt; stratified and with loose, single-grain structure; calcareous; rich in heavy minerals by contrast with the subsoil, suggesting some eolian derivation. This colluvial wash is overlain by a pavement of partly fractured, coarse igneous pebbles.

b) 5 to 15 cm. (B)Sa. Pink (7.5 YR 7/4), sandy silt; coarse granular structure; with microconcretions of calcareous gypsum or other salts.

c) Over 120 cm. (B). (base never seen.) Yellowish-red to reddish-yellow (5 YR 5–6/6), silty coarse sand; coarse, subangular blocky structure. In Pit 44 the clay content is 13% (No. 121, Table 2–1); colloidal silica is present and unpeptized masses of fines can commonly be seen under the microscope. Kaolinite with some montmorillonite constitute the clay minerals present (see Appendix D).

Although none of the pebbles are rotted, they may be stained with hematite. Obviously, the intensity or duration of effective chemical

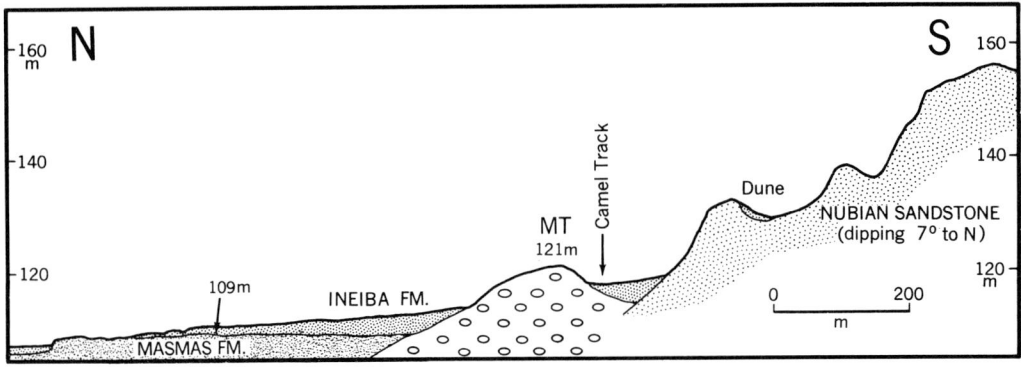

Fig. 2–25. Pleistocene deposits east of New Tushka. Map: UW Cartographic Lab.

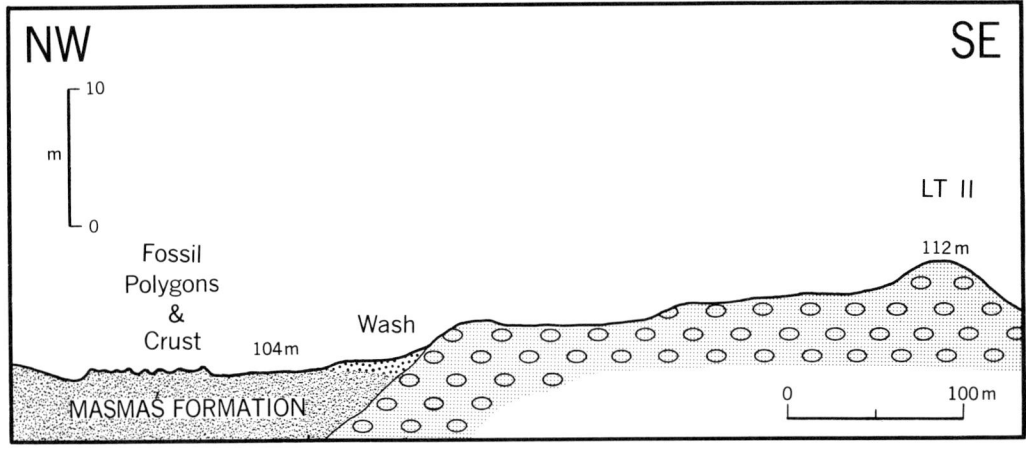

Fig. 2–26. Pleistocene deposits west of New Abu Simbel. Map: Cartographic Lab.

weathering was appreciably less than in the case of the nearby paleosol developed on HT II, yet the clay content of the Low Terrace paleosols is at least three times as great. This, together with the microscopic evidence, underscores that subsequent mechanical eluviation of clays is far less advanced than in the case of the High Terrace paleosols.

The red paleosol developed on the Low Terrace is of some stratigraphic interest. Red soils of similar depth, comparative intensity, or such general occurrence have not subsequently developed in southern Egypt. In northern Egypt such soils last developed on the $+10$ to $+13$ meter Nile gravels that appear to be contemporary with the Tyrrhenian II

(Butzer, 1959a), while red paleosols are absent on all continental deposits contemporary with or younger than the Tyrrhenian II coral reefs on the Red Sea coast near Mersa Alam (see Chap. 8). This indirect evidence suggests that the Low Terrace is Middle Pleistocene rather than Upper Pleistocene.

Comparative Terrace Sequences at Aswan, Edfu, Luxor, and Qena

The Kom Ombo terraces can best be reviewed in the wider perspective of Pleistocene gravels in southern Upper Egypt. Deposits of comparative age are absent between Kubaniya and Aswan, although fluvial platforms recording the Gallaba and High Terraces have been cut into the Nubian Sandstones bordering the valley. South of Aswan, where Ball (1907: Pl. 2) and Sandford and Arkell (1933: 57–58) have mapped the ancient Nile channels and surficial deposits, gravels are again preserved. They occur in the eastern valley followed by the Shellal-Aswan railroad track, near the local watershed at 125 meters (i.e., 29 meters above the floodplain at Shellal or 33 meters above that at Aswan). Clearly this valley is an abandoned course of the Nile, a fact corroborated by recent borings which show that the buried entrenched valley of the Nile is located under this ancient channel, with poorly sorted fluvial sands tapped down to 45 meters *below sea level*, where they rest on Pliocene beds (Voute, 1963). These sands are unlike the gravels of the + 29 to + 33 meter stage and are probably somewhat earlier. The surface gravels, however, qualify for the Middle Terrace.

Turning to the north of Kom Ombo, the Gallaba Plain in the Western Desert; the embouchure of Wadi Abbad, opposite Edfu; and the Nile Valley margins from Luxor to Qena reveal a rich record of Pleistocene sediments. None of these areas have been geologically mapped.

Wadi Abbad deposited great alluvial fans onto the Nile floodplain several times in the course of the Pleistocene. These deposits are completely ignored by Sandford and Arkell (1933) and were subject to a reconnaissance by the authors in 1963. Transects run at Km 7 showed the existence of two low terraces at + 5 meters and + 10 to + 12 meters (relative to wadi floor), an intermediate terrace at + 21 meters, and a high terrace at + 43 meters. The low terraces are fully comparable to the Low Terrace complex of Kom Ombo but are not preserved at the mouth of Wadi Abbad. The two higher terraces can be traced downstream, with the well-preserved intermediate terrace graded to a Nile floodplain of about 115 meters (83 m today). With a relative Nile elevation of +32 meters, this intermediate terrace is probably an equivalent of the Middle Terrace. The higher terrace subdivides near the mouth of the wadi, where it is recorded on the southern side at 125 meters (+42 m) and at 135 to 137 meters (+52 to +54 m), a remarkably close correspondence to HT II and HT I at Kom Ombo. There are interesting lithological differences between the Middle and High Terraces of Wadi Abbad, suggesting appreciable hydrographic changes. The Middle Terrace has about 80% igneous gravel, 20% limestone and chert; the High Terrace has 50% igneous gravel, 50% limestone (badly corroded) and chert. With a comparative abundance of calcareous materials, both the Middle and High Terraces are moderately well-cemented, a contrast with the quite unconsolidated gravels of Kom Ombo.

In the Luxor area, despite the extensive work of Sandford (1934), nothing systematic is known concerning higher gravel terraces. Of considerable stratigraphic interest is the well-known 9- to 10-meter wadi terrace (relative to wadi floor) (Sandford, 1934: 63–67; Butzer, 1959a, 1960a), with its rather scanty *in situ* artifacts of Acheulio-Levalloisian type (Caton-Thompson, 1946). In Wadi Madamud, Wadiyein (opposite Luxor on the west bank), and Wadi Ibeidalla (near Ballas) this steeply inclined local terrace is graded to a Nile floodplain level of approximately 95 meters (75 m today).[8] This suggests correlation of the 9- to 10-meter wadi terrace of the Luxor area with LT I at Kom Ombo.

The broad valley of Wadi Qena is over 10 kilometers wide where it debouches from the limestone hills of the Eastern Desert onto the Nile Valley at Qena. Lower Pleistocene alluvial spreads of the once-mighty stream account for all but 1.5 to 2 kilometers of this expanse (for earlier references see Blanckenhorn, 1921: 164; Sandford, 1929, 1934: 59, 62, 74; Butzer, 1959a, 1961). An irregular gravel plain graded to a Nile floodplain of about 107 meters (73 m today) is the dominant geomorphic feature. About 4 kilometers wide at its apex, some 8 kilometers upstream from the floodplain edge, this great alluvial fan spreads out to a width of 15 kilometers near the floodplain margins. Alternating, water-laid beds of

8. Sandford and Arkell (1933) and Sandford (1934) commonly list their wadi-terrace elevations (relative to wadi floor) interchangeably with Nile-terrace elevations relative to modern Nile floodplain. This assumes that the modern wadi floors are completely graded to the Nile floodplain, which they very seldom are. It also assumes, incorrectly, that present and Pleistocene wadi gradients were generally identical. Consequently, our interpretation of wadi gravels relative to Nile floodplain, based on projecting gradients to the former Nile floodplain perimeter, necessarily produces different values than those given by Sandford and Arkell.

coarse, well-stratified sands and rounded stream cobbles are frequently exposed where local drainage lines have subsequently dissected the ancient wadi channel. This 34-meter terrace has a substage or independent lower stage graded to a floodplain level of 96 meters (+23 m). Ridges of older terraces are commonly preserved a few kilometers north of Qena town, and again on both sides of the valley. These are graded to local base levels of 124 meters (+51 m) and *ca.* 143 meters (+70 m). Spurs of one or more low terraces can be seen within the 1.5- to 2-kilometer-wide channel cut into the 34-meter terrace fan.

Based primarily on reconnaissance in 1963, this lower Wadi Qena sequence indicates the presence of extended gravel terraces graded to successive floodplain levels at about +70, +51, +34, and +23 meters, with one or more lower, less significant stages. The altimetric and geomorphic correspondence with the Gallaba stage, HT I, MT, and LT I at Kom Ombo needs little emphasis.

Although the Pleistocene terraces of Upper Egypt clearly require intensive and systematic study, the broad lines of a recognizable pattern do seem to emerge over an area wider than the immediate environs of the Kom Ombo Plain.

Stratigraphic Conclusions

The geomorphic evolution of the Kom Ombo Plain during the later Tertiary and earlier Pleistocene can only be understood in the context of major tectonic events and changes of climate and hydrography. The sequence of events tentatively outlined in Table 2-2 seems reasonable in view of the available evidence. But considerably more intensive study of gravel lithologies, combined with large-scale mapping, will be necessary before a more definitive framework can be established. Until such a time, it does not seem warranted to define rock-stratigraphic units. Consequently, a local terminology of high, middle, and low terraces will be employed for the Kom Ombo area. Correlations with the Lower Nubian sequence will be suggested in a later chapter.

Contemporary Geomorphic Activity and Soil Development in the Kom Ombo Drainage Basin

Paleoclimatic interpretation of Pleistocene processes can only be relative and qualitative since absolute generalizations are not yet possible either for stream competence or for intensity of soil development. It is therefore

Table 2–2. Geomorphic evolution of the Kom Ombo area during the late Tertiary and early to middle Pleistocene

16.	Chemical weathering with development of a red paleosol.	MIDDLE PLEISTOCENE
15.	Alluviation of the Low Terrace, sandy coarse gravels in large part derived from older alluvia. Two substages (LT I, + 22 m; LT II, + 15 m), separated by fill cutting (vertical differential at least 7 m). Pluvial erosion of uplands; accelerated activity of Eastern Desert wadis.	
14.	Dissection of fill (vertical differential at least 25 m).	
13.	Alluviation of the Middle Terrace (+ 34 m); coarse to cobble gravels derived from the Basement Complex (Red Sea Hills, Aswan). Significant fluvial activity of larger wadis and Nile.	
12.	Dissection of fill (vertical differential at least 25 m).	
11.	Chemical weathering with development of a deep, red paleosol. Moist rainy season with limited geomorphic activity.	
10.	Alluviation of HT II (+ 40 to + 43 m); coarse-to-cobble gravels derived from the Basement Complex. Planation of associated fluvial platforms. Significant fluvial activity of wadis and Nile. Last use of Bimban Channel at flood stage.	LOWER PLEISTOCENE
9.	Dissection of fill (vertical differential 15 to 30 m).	
8.	Alluviation of HT I (+ 51 to + 54 m), as stage *10*. Bimban Channel in use.	
7.	Dissection of fill (at least 45 m) and bedrock, with disintegration of the Wadi Kharit–Abu Subeira drainage, linking of Kharit-Natash drainage, and reversal of Wadi Kubaniya drainage. Enlargement of catchment area of Nubian Nile(?).	
6.	Alluviation of the Gallaba Gravel Plain and related terrace deposits (+ 60 to + 74 m) by the Proto-Nile system. Activity of Nubian Nile dwarfed by major fluvial activity of the great Eastern Desert wadis (Shait, Natash, Abu Subeira–Kharit). Complex period with three or more substages, interrupted by fill dissection.	BASAL PLEISTOCENE
5.	Emergence of the Pliocene Gulf and dissection of Pliocene beds. Uplift in Red Sea Hills(?).	
4.	Pliocene regressive deposits (fluvial, lacustrine, lagoonal) to 130 m elevation. Kom Ombo salt lake fed primarily by Kubaniya Nile(?). Transgressive marine-estuarine beds of Middle Pliocene age are known from Aswan and may occur beneath the Kom Ombo Plain.	UPPER PLIOCENE
3.	Bedrock dissection and minor tectonic deformation, e.g., tilting of fault blocks.	LOWER PLIOCENE

(*Table 2-2*, continued)

2.	Lateral planation with development of Aswan Pediplain. Local base level a little below 180 m.	MIO/ PLIOCENE(?)
1.	Major block faulting, with primary rifting of Kom Ombo Graben. Maximum vertical throw at least 150 m.	OLIGO/ MIOCENE(?)

necessary to outline contemporary geomorphic activity and soil development.

CLIMATOLOGICAL DATA

Although a third-order meteorological station has operated at Kom Ombo since 1930, it has not recorded precipitation data. The 16-year temperature record, published by Hamed (1950), can be considered representative of the greater Kom Ombo region.

The January mean temperature is 59° F (15° C). The July mean is 89° F (31.5° C); the July mean daily maximum, 105° F (40.6° C); and the annual range, 30° F (16.5° C). The smallest mean daily temperature range occurs in December (28° F, 15.5° C), the greatest, in April (35° F, 19.5° C). The mean daily minimum in January is 44° F (7° C), and the absolute minimum (February) is 28.5° F (−2° C). Further north at Luxor (78 m), the 11-year record 1935–45 shows an absolute minimum of 32° F (0° C), while at Aswan (111 m) the 45-year record 1901–45 has an absolute minimum of 35° F (1.7° C). Frost, in other words, is sporadic and extremely rare, while temperature ranges are not excessive under the artificial conditions of a ventilated climatological station at 5 feet above surface. Although uncommon, ice occasionally forms on standing water during cold winter nights with temperature inversions. Ground temperatures in the sun almost daily reach 150° F (65° C) for over 6 months of the year. Consequently, although frost is precluded as an effective agent of mechanical weathering today, thermoclastic weathering, through alternating expansion and contraction, is possible.

Mean monthly wind speeds range between Beaufort 1.0 and 1.5 (approx. 2–3 km/hr.), with March the windiest month. Prevailing wind directions at all times of the year are northwesterly (29.8%) and northerly (26.5%), with all other directions accounting for only 9.3%.

The mean annual precipitation (1926–59) for Aswan is 3 millimeters, with an average of 0.8 days per year with 0.1 millimeters or more precipitation (British Meteorological Office, 1958; S. P. Jackson, 1961). The maximum fall recorded in 24 hours is 5.5 millimeters. Average cloud cover at Kom Ombo is at a maximum in January (1.2 tenths), at a minimum June through September (0.1 tenths). The heaviest and most frequent rainfalls at Aswan have been recorded in April-May and October-November. No rainfall has yet been registered for the months July through September. The sparse, light rains that do affect the area are, therefore, related to meridional circulation of the circumpolar westerlies during the transitional seasons (see El-Fandy, 1946, 1950; Butzer, 1960b) and have no relationship to the summer monsoons of subsaharan Africa. At Luxor (1936–45), April and October are the months with most frequent rain; at Quseir on the Red Sea Coast (1927–45), October through December. At both Luxor and Quseir, June through August is quite rainless. Mean annual precipitation at Luxor is 1 millimeter; at Quseir, 4 millimeters. Unfortunately, no climatic observations of any kind are available from the Red Sea Hills, particularly in the high country over 1,000 meters elevation, where precipitation presumably is somewhat higher.

During 1962–63, one rainfall occurred at Kom Ombo during the passage of a cold front with sandstorm conditions on December 21, 1962. Although only a trace of rain was registered at Aswan, an estimated 1 to 3 millimeters reached the ground at Gebel Silsila. Coarse sands were moistened to a depth of 1 to 2 centimeters, and blowing dust temporarily subsided. No other occurrence of geomorphic significance was noted.

MODERN WADI ACTIVITY

The only watercourses that have been active in recent decades are Wadis Kharit and Shait. In April, 1941, Wadi Kharit flooded extensive tracts of the Kom Ombo Plain, emptying into the Nile. Light rains throughout southeastern Egypt probably climaxed in a series of heavy cloudbursts in the Red Sea Hills, and the flood struck unexpectedly a day or two later. According to the senior engineers of the Wadi Kom Ombo Company, similar spates occur in both Wadis Kharit and Shait about once a decade, and Vignard (1955c) has described such a flood sweeping through Wadi Shait in December, 1923.

The only tangible geomorphic impact of these recent spates are finely laminated, pink (7.5–10 YR 8/4) clayey silts or marls that veneer the wadi beds. Thicknesses vary from several millimeters to a maximum of 25 centimeters (Fig. 2–24). Polygonal systems of dehydration cracks create patterns on the surface of these clay spreads, which rest on up to 40 centimeters of well-stratified, light-yellowish-brown (10 YR 6/4), coarse sand to sandy silt of fluvial origin as well as on loose, coarse eolian sands. Small wadi-bed shrubs in Wadi Shait show no signs of disturbance. Either flood velocities must be moderate or stream flow is laminar rather than torrential. At any rate, transport capacity is almost nil, and deposition has been restricted to suspended-load sediments for some time because these clayey beds have uniform texture and commonly lack coarse laminae.

The coarser bed-load deposits underlying the clay beds suggest periods of flow with greater stream turbulence, local bed erosion, and redeposition. These subrecent alluvial deposits, which are probably only a few centuries old, can be traced into the local wadis, where pebble-grade materials are generally present. So, for example, the inactive bed of Wadi Ellawi consists of loose, light-brown (7.5 YR 6/4), coarse sand with concentrations of angular-to-subangular, fine-to-medium-grade (2 cm maximum) sandstone or ironstone gravel. The subrecent wash of the second-order wadis between New Korosko and New Sebua consists of unconsolidated, well-stratified, light-brown to very-pale-brown (7.5–10 YR 7/4, 10 YR 7/3–4), coarse sand with subangular, medium-to-coarse gravel derived from unconsolidated local older beds (Table 2–3). Silty bands may be interbedded. Maximum thickness of the wadi-floor fill is 130 centimeters, although subsequent dissection has usually reduced this value to about 20 to 50 centimeters. Rare Roman and Islamic potsherds, possibly derived, occur on top of this wash. A final example can be cited from Wadi Shurafa, where up to 4 meters of subrecent wash have been largely dissected and reduced to a bed fill of 1 to 1.5 meters. The material is an unconsolidated, well-stratified or current-bedded, very-pale-brown (10 YR 7/3), coarse sand with strata of subrounded, medium-grade sandstone gravel.

The subrecent character of the coarse wadi-floor deposits is suggested both by their stratigraphic position and by the period of dissection which preceded deposition of the contemporary, well-sorted, and laminated clayey silt. The bed of central Wadi Ellawi is veneered by thin, rippled sand sheets that thicken into dunal forms upstream. And on the southern side of Wadi Kharit all first- and second-order wadis are fossilized by blown sand. Consequently, it can be safely asserted that the minor and intermediate-sized local wadis of the Kom Ombo area were virtually inactive during the present century, although they have occasionally been able to transport bed-load sediments over limited distances by torrential flow in the not too distant past. This period of slightly accelerated wadi activity in Roman or Islamic times was geomorphically confined to the immediate wadi beds, with erosion and quick redeposition of materials from older alluvial beds that fringe the wadis. In response to such localized erosion and deposition, the bed load of a higher-order stream, such as Wadi Kharit, remained almost entirely free of pebble-grade materials, and the nature of these shallow bed-load deposits suggest limited stream competence.

Modern wadi vegetation plays no role in a discussion of erosion and alluviation. Except for the floors of exotic watercourses such as Kharit and Shait, there is a complete lack of vegetation in the Kom Ombo region away from the Nile Valley. Upstream in Wadi Kharit, there are rare, scattered, woody, non-thorny, deciduous perennials, and occasional bushes or low trees of the species *Acacia tortilis* (see Täckholm, 1956). Acacia bushes and tree-size *Acacia tortilis,* up to 4 meters tall, are slightly more abundant along some stretches of Wadi Shait but are still, on the average, well over 100 meters apart. Clusters of deciduous thorn bushes, including *Zilla spinosa,* occur on sandy washes, possibly together with very rare shrubs of *Tamarix nilotica* Bunge. Patches of *Colocynthis vulgaris* Handal, with abundant gourd-like fruits littering the ground, are seen sporadically. Obviously, all of this vegetation is badly degraded through grazing and use for fuel or, in the case of the colocynths, for medicinal purposes. The less disturbed wadis draining eastwards from the Red Sea Hills are more representative of the natural vegetation. Here *Acacia tortilis* parklands, with trees up to 8 meters tall and a discontinuous undergrowth of thorny shrubs, follow the wadi floors. Presumably, the spontaneous vegetation of the larger desert wadis would be a sparse desert savanna (desert scrub according to Kassas and Imam, 1954), physiognomically related to the thorny desert savanna of the southern margins of the Sahara. Beyond this edaphically favored wadi association, the spontaneous desert vegetation remains insignificant.

Although blown sand may accumulate in the lee of rare desert shrubs or scrub, none of the plant types present exhibit significant soil-binding properties. Their effects on rain interception and infiltration are negligible, and their density is insufficient to influence wadi flows. All in all, the geomorphic role of such vegetation can be discounted, even in those remote wadis where vegetation is considerably more abundant today.

MODERN NILE ACTIVITY

The hydrological regime of the Nile River, in contrast to the activity of the Egyptian wadis, is fairly well understood. The mean monthly discharge of the river at Wadi Halfa during the thirty-year period 1912–42 was as follows (in million cubic meters per day) (Hurst, 1944: 46):

J	F	M	A	M	J	J	A	S	O	N	D
117	86	66	55	51	64	164	614	718	479	247	152

Some 80% of the total annual discharge is derived from Ethiopia via the Atbara, Blue Nile, and Sobat rivers, most of this during the flood season (August through October). On the other hand, the major part of the low-water discharge (December through July) is derived from the White Nile Basin and from the East African lakes in particular. So, for example, only 30% of the Nile waters are derived from Ethiopia during the month of April, but 95% during September. Corresponding to the derivation of the Nile waters, the dissolved load varies with the season. The levels of calcium nitrate, calcium phosphate, and colloidal silica are considerably higher during flood stage than during low-water stage (see Ball, 1939: 115–16). Sodium chloride, on the other hand, increases from 8.4 to 26.9 parts per million between flood stage and low-water stage, an increase from 6.1% to 14.1% of the total dissolved load.

The flood silts laid down beyond the natural or artificial levees today range from sandy silt to silty clay in the Aswan–Kom Ombo area. These are stratified and laminated topset beds, prone to cracking when dry (Fig. 2–27) because of the high montmorillonite content. The heavy mineral spectrum of these nilotic sediments varies considerably, presumably as a result of mechanical sorting during deposition. A mean of four recent Upper Egyptian heavy mineral samples analyzed by Khadr (1961) shows 34% pyroxenes (mainly augite), 31% amphiboles (mainly hornblende), 29% epidotes, 11% altered, and 1% titanite. Most of these heavies are derived from Ethiopia today, but they were available in the Basement Complex rocks of Egypt and the northern Sudan and were supplied to the Nile by the major Eastern Desert wadis during the Pleistocene. Although Shukri (1950) believes that most of the augite in Pleis-

Fig. 2–27. Modern floodplain silts near Fatira, exhibiting dehydration cracks and vertisol phenomena.

tocene deposits can be attributed to the Blue Nile–Atbara influx, the same mineral was once supplied in abundance by Wadis Natash and Kharit. Because there are no objective comparative studies and because the Pleistocene pyroxenes are generally in an altered state, we are sceptical as to whether it is possible at the moment to distinguish between the varieties represented in the Red Sea Hills and those derived from the Ethiopian tributaries of the Nile.

At present, the major local constituent of the Nile bed load is quartz sand of eolian origin, either blown directly into the river or eroded from dunes forming the west bank of the Nile in several parts of Nubia. Local wadi influx is quite insignificant, and, although wadi sands may occasionally be transported out onto the floodplain, they almost never reach the Nile banks. Other quartz sands are derived from erosion of coarse-grained igneous rocks along the Nile channel at Aswan and above the Second Cataract (see Appendix C).

RECENT EOLIAN ACTIVITY

Eolian deposits in the Kom Ombo area include small tied dunes of the nebka and the rebdou types (see Tricart and Cailleux, 1960–61, II: 72), irregular dunal forms clogging wadis, and thin sheets with medium-scale ripples (up to 1 m in wavelength), as well as small superimposed dune forms of barchan or transverse type. Such deposits consist primarily of quartz sand but may include heavy minerals and crumbs of fine-grained materials derived from nilotic silts. Dunes and sand sheets occupy only a minute fraction of the surface area, however.

Deflation phenomena range from desert lag and sand-pitted surfaces to yardangs and blowouts. A coarse lag of sandstone detritus, patinated or ferruginized in varying degree, mantles the bedrock uplands. Fine materials, when present, are concentrated in a light-brown (7.5–10 YR 6/4), sandy silt or silty sand wash, presumably of Pleistocene age and up to 50 centimeters thick on older terrace surfaces. Unsorted and only in part stratified, these colluvial mantles usually contain a fair amount of local rubble. In the case of Pleistocene gravel terraces, the lag consists of fractured pebbles; in the case of finer nilotic deposits, it consists of concretions, root drip, and root casts, and, possibly, of cultural materials. The pavements studied were primarily due to winnowing out of loose, fine-grained materials through deflation.

Micropitting by sand blast at about ground level can be frequently observed on consolidated silts or duricrusts, primarily with N 30° W exposure. Far more striking are the yardangs developed in late Pleistocene deposits in the northwestern quadrant of the Kom Ombo Plain. In the Silsila Gap, east of Gebel Silsila, a dozen major yardangs with a local relief of 2 to 10 meters show orientations ranging from N 20° W to N 60° W. The patterns of the ridges and a great interior blowout depression can only be explained by the complex aerodynamics of the Silsila Gap. At the Gebel Silsila archeological sites of the Yale Expedition (see Fig. 3–10), a number of classic yardangs and associated blowouts have a local relief of up to 7 meters. Some of the well-developed yardangs are oriented due north-south; other, less distinct forms strike in the range between N 65° W and N 25° E, primarily as a result of bedding characteristics of the sediments themselves. And at Khor el-Sil, yardangs with a relief of 2.5 meters show orientations varying from N 25° W to N 60° W, with microsculpture and corrasion suggesting that the N 45° W direction is dominant.

In general, then, eolian activity has been significant during the last ten millennia or so, with effective wind directions in the quadrant between due north and N 45° W. With so few characteristic sediments, however, such eolian activity could hardly be expected to leave a well-defined fossil record during the early and middle Pleistocene (see also Appendix C). And the combined effects of several periods of deflation on an older surface could only under special circumstances be disentangled stratigraphically.

PATINATION AND SOIL DEVELOPMENT

A common form of chemical weathering under fairly arid conditions is the development of desert "varnish," a patina of pyrolusite and ferric oxides on rock surfaces. Thickness and intensity in the case of sandstone may vary from a brownish stain to a heavy black rind almost a millimeter thick. Secondary ferruginization may also penetrate many centimeters into the original rock. This feature is well developed in Nubia and is, presumably, a result of impregnation from without (see also Engel and Sharp, 1958), rather than of capillary movement within the rock, carrying dissolved materials to the surface. Meckelein (1959: 44–45, 50) has illustrated similar features from the Fezzan, while Tricart and Cailleux (1960–61, I: 79–82) believe that the thickness of the patina increases with higher precipitation on sandstones in the southern Sahara: a minute film of stain with 100 to 130 millimeters rainfall, a 0.2 to 0.4 millimeter rind with 200 millimeters, and a massive crust 0.5 to 1.0 centimeters thick with 500 millimeters. A friable, white eluviated zone may underlie such rinds.

Although a variable degree of patination is evident on petroglyphs of the Eastern Desert and Nubia (all dating from late prehistoric to Islamic times), contemporary patination proceeds at a very slow rate. It can occasionally be observed that patination is absent on sand-blasted bedrock faces. This demonstrates that corrasion outpaces patination. Certainly none of the indurated, quartzitic rinds or impregnations are of post-Pleistocene vintage. In his study of rock patination in the Aswan area, Passarge (1955) also concludes that major patination must be attributed to moister Pleistocene paleoclimates.

The second characteristic form of chemical weathering is enrichment of soluble salts in the upper soil, with development of diffuse, spongy, or concretionary *Sa*-horizons dominated by chlorides and sulfates with some carbonate. In these characteristic *salt Yerma* soils (Kubiena, 1953: 180–82), salt hydration (see Knetsch, 1960; Knetsch and Refai, 1955) plays a significant role in splitting pebbles. Evaporites often encrust individual

Fig. 2–28. Patinated lag gravels (surface) and the effects of salt-weathering *in situ* (face) in the Fatira beds near Gebel Silsila.

rocks and can be found within natural lines of weakness in the rock. Pebbles in various stages of mechanical breakdown can be observed to depths of 50 to 100 centimeters below the surface in gravel terraces of any age (Fig. 2–28). Although thermoclastic splitting in the zone of surface heating may also be involved, most of the subangular, fractured igneous pebbles that form the local gravel pavements are a product of physical hydration. They result from disintegration of salt-weathered pebbles into two or more pieces after exhumation. This process of disintegration is clearly underway today, since it affects all gravels. The rare surface frosts that can be postulated from the climatological data could certainly not split pebbles at depths of 20 centimeters, let alone 50 to 100 centimeters (see also soil temperatures recorded at Giza, Hamed, 1950: 156). On the other hand, salts can invariably be found within any pebble fracture. This indisputable evidence for salt hydration within the modern Sa-horizon should serve as a warning against the facile interpretation that fractured pebbles in Egyptian Pleistocene deposits were caused by frost.

Rubefaction, although not a typical soil process today, was observed in quartz sands moistened by the light rainfall of December 21, 1962, near Gebel Silsila. The color of coarse sands changed from very pale brown (10 YR 7/3) to light brown (7.5 YR 6/4) in a matter of hours. These sands are 7.1% finer than the 0.06 millimeter sieve and contain 1.3% calcium carbonate (No. 141, Table 3–8). There was no textural or microscopic change, but some of the free iron present among the fines adhering to the quartz grains appears to have oxidized or hydrated. Such highly superficial oxidation and rubefaction may, in part, explain the reddish-yellow (7.5 YR 8/4–6) color of the eolian sands in Wadi Ellawi. It will not, however, account for rubefaction accompanied by clay-mineral formation *in situ*.

WEATHERING OF CLIFF FACES

In his empirical study of cliff weathering in Egypt, Knetsch (1960) recognizes several processes that may produce analogous forms, possibly alternating with one another through time. In as far as applicable to the Nubian Sandstone, these include:

a) Undercutting through fluvial erosion;
b) Release of pressure through horizontal and vertical unloading of adjacent or overlying rock, with development of fissures perpendicular and parallel to new rock faces and with settling of rock masses along older joints and fresh fissures;

c) Geochemical exfoliation or undermining through subflorescence and hydration.[9] These two agencies commonly work in combination in Egypt and lead to granular disintegration or to rock splicing. Weathering of this type is most effective on shade slopes with higher humidities, and in the capillary zone above a seasonal or permanent ground moisture table.

These observations by Knetsch can be applied fairly directly to the wadi walls of the Etbai Uplands. Massive spheroidal exfoliation—a result of unloading—is particularly evident along the Nile Valley margins in the massive, coarse-grained sandstones on the upland surface south of Wadi Abu Subeira. Great blocks of sandstone settle gravitationally along joint plane axes and develop well-rounded edges through exfoliation. In the immediate environs of Kom Ombo, shaley sandstones are frequently intercalated with the more standard facies. In such strata, salt efflorescence and subflorescence are fairly common, and expansion-contraction of hygrophilic clays provide another effective form of weathering. Deterioration of shaley beds and subsequent microslumping or rock falls gradually undermine more resistant strata. In combination with wadi undercutting at long intervals of time, this is the characteristic mode of backwearing.

Waste products are concentrated on the slopes below more resistant rock ledges, particularly at the base of the midslope wherever slumping or rock falls have occurred. Generally speaking, however, detrital accumulations are fairly thin, ranging in caliber from sand grains to boulders, with eolian material occurring locally. Either slope denudation is in equilibrium with weathering, or the rate of weathering is incredibly slow. Presumably, both factors are involved. Midslopes typically range between 45% and vertical (25° to 90°), and average 57% to 106% (29° to 47°). This is sufficiently steep for extensive mass-movement transfer, sporadic washing, local rill-cutting, and occasional deflation. On the other hand, however, rock drawings cut into massive sandstones near Gebel Silsila in late prehistoric times are remarkably well-preserved. Together with the evidence for limited wadi activity discussed above, this all suggests that recent backwearing has been, in fact, very slow.

---

9. *Subflorescence* involves capillary migration to the rock surface, removing cement and other materials from the matrix and forming a weakened subcutaneous zone parallel to the rock face. *Hydration* involves expansion and contraction of efflorescent or subflorescent salts in response to changes of temperature or of humidity.

Despite the differential rate of weathering in different lithofacies, midslopes are remarkably rectilinear unless undercut. Resistant ferruginized or silicified strata form small facets that interrupt, but do not change, the essential profile. The lowermost midslope may be gently concave when not undercut, but this is the exception rather than the rule. Footslopes almost always coincide with pediments, fluvial platforms, wadi floors, or other spreads of alluvium.

Paleoclimatic Interpretation of Pleistocene Features

Turning now to the Pleistocene geomorphic record, we find the alluvial deposits of the Nile, of the local wadis, and of the exotic Eastern Desert wadis on the one hand and evidence of fill or bedrock dissection on the other hand. Eolian deposits are conspicuously absent, although this does not preclude significant Pleistocene eolian activity. The soil record includes a number of red paleosols but no direct evidence for other forms of chemical weathering.

ALLUVIATION

A common aspect of the early to middle Pleistocene alluvial deposits of the Kom Ombo region is that aggradation of the Nile, of the local wadis, and of the great watercourses draining the Red Sea Hills was always contemporary. No sharp facies distinctions have been preserved, and there can be little doubt that all three media of deposition responded to a broadly similar regional climate.

The alluvium of the Nile Valley consists of a coarse bed load supplied by numerous Egyptian affluents, large and small. Major wadis developed great alluvial fans at their confluences with the Nile, and similar bed materials dominated the Nile course downstream to the next major tributary. Consequently, the Nile record speaks for an Egyptian river rather than for an exotic stream such as we know today. Whether or not the Nile received water from south of the Sahara, or experienced summer floods of Ethiopian origin, cannot be deduced from the existing record. The heavy-mineral spectrum, in as far as it is indicative of anything, corresponds fairly closely to the mineralogy of the gravel. Clays and silts within the sandy matrix have either formed *in situ* or are otherwise derived from the Red Sea Hills. The gravel itself is almost exclusively derived from Basement Complex rocks outcropping in the Eastern Desert, at Aswan, at Kalabsha, or above Wadi Halfa. Components related to the sedimentary cover are minor. Yet it is still quite possible that summer floods did take place. The competence and general impact of these floods would have been insignificant, however, compared with the local runoff, and any suspended sediments would have been quickly flushed out. In all probability, the available Nile record pertains exclusively to winter-season runoff, related to a mediterranean-type climate in southern Egypt.

Deposits of the minor, local wadis can only be recognized in the middle Pleistocene record. They suggest that local rains induced sheet-wash and torrential runoff, whereby alluvium was graded to the local base level dictated by the major wadis and the Nile. Where recorded, such local alluvia differ only in degree from the subrecent wadi wash. This suggests alluviation under arid conditions, although local rainfall must have been more frequent and of greater duration or intensity than at present.

The great gravel masses transported by Wadis Shait, Kharit, and Natash are out of all proportion compared with stream competence or capacity today. Several gravel collections were analyzed morphometrically by the modified Lüttig method after Butzer (1964b: 160–64). They are symptomatic of all of these sorted, rounded, coarse-to-cobble gravels (Table 2–3). The degree of rounding of these gravels decreases from well-rounded—in the case of the High Terrace at New Korosko, the Middle Terrace at New Sebua, or the Fatira beds near Gebel Silsila—to subrounded Low Terrace gravels near New Korosko and New Shaturma. As degree of roundness decreases, the coefficient of variation increases from very homogeneous to heterogeneous with increasing detrital components. This corroborates the qualitative observations that local wadi activity was increasingly significant in middle Pleistocene alluviation. Sliding was the dominant transport motion in each case, despite the pebbly nature of the deposits which would intrinsically favor rolling. This suggests that discharge was comparatively high and uniform rather than exceptionally torrential. By contrast, a sample of the subrecent wash (Table 2–3) is subangular, highly detrital, and very inhomogeneous. The hydrological changes effecting this part of Egypt are apparent from these analyses.

In general, the exotic gravel masses of the Kom Ombo region suggest major seasonal rivers with at least several weeks of torrential water flow each year. Contemporary parallels can be found in semiarid parts of the southwestern United States, on alluvial fans debouching from mountain country. The catchment areas of Wadis Kharit-Natash and Shait still retain considerable relief-energy, with median stream gradients of 0.20% to 0.55% (see Fig. 2–18). Present-day rainfall is obviously insufficient to provide the requisite runoff, however, and an annual precipitation of several hundred millimeters in the high country would be necessary to stimulate a similar degree of geomorphic activity today. With such a semiarid climate, sediment yield should also be somewhat higher than at present (see Schumm, 1965).

Seen in retrospect, the early to middle Pleistocene alluvial deposits of the Nile and its Egyptian tributaries clearly suggest a "pluvial" climate, with more frequent rains of greater duration and with seasonal discharge in at least the major wadis. Climate may have varied from arid to subarid at lower elevations, with semiarid conditions prevailing in the high country, above about 500 meters elevation. Rainfall at Kom Ombo may have been on the order of $x.10^1$ millimeters and in the Red Sea Hills, of $x.10^2$ millimeters. There is no reason to doubt that a winter-rainfall regime persisted.

With the exception of some root drip preserved in the Fatira beds, no organic vestiges of any sort have been preserved. It is consequently difficult even to guess the character of the local vegetation. One thing is clear, however: the vegetative mat was incomplete at best, providing only partial protection from the agents of denudation to soil and regolith. This would rule out a complete mat of grassy vegetation or a closed woodland.

Table 2-3. Morphometric gravel analyses
(Lithology: fine-grained igneous rocks other than quartz. For localities see text.)

Stratigraphy	Sample Size	ρ (%)	CV of ρ	Detrital component	E/L	E/l	L (cm)
Subrecent Wash (New Korosko)	75	12.9	135.1	60.0	45.4	65.0	4.24
Low Terrace (New Shaturma)	75	23.7	97.7	32.0	46.7	61.5	3.95
Low Terrace (New Korosko)	75	33.3	73.1	8.0	48.5	64.9	4.57
Fatira Beds (Fatira)	60	65.7	29.9	0.0	42.6	58.2	6.72
Middle Terrace (New Sebua)	75	66.7	29.1	0.0	43.9	59.0	5.15
High Terrace II (New Korosko)	75	72.6	24.0	0.0	47.0	60.3	4.95

DISSECTION

Linear erosion may either involve dissection of *fill* or of *bedrock*. Bedrock incision requires not only considerable relief-energy and time, but also discharge and bedload. Such downcutting appears to have been insignificant in the Kom Ombo area during Pleistocene times and may reflect long periods of fairly moist climate during the late Tertiary. The auxiliary role of chemical weathering in the stream bed may have been appreciable.

Fill dissection, on the other hand, can be performed fairly rapidly in unconsolidated deposits. Downcutting of the Nile and its tributaries some 1000 kilometers from the Mediterranean Sea can be explained only by variations in discharge, and not by movements of base level. Reduction of flood volume would almost automatically lead to remodelling of the floodplain, with redistribution of the bed load within a narrower channel. Consequently, stream incision of a part of the former floodplain may have been most active during the periods of slightly greater, but waning, rainfall immediately following the moister aggradation phase. Such erosional intervals were probably complex, interrupted by brief periods of renewed channel widening and aggradation. Similar periods of dissection presumably interrupted the major aggradation phases repeatedly. And long-term cycles of alluviation and downcutting are only the net effect of innumerable short-term alternations of scour and fill in the wake of each flood (see Leopold, Wolman, and Miller, 1964: Chap. 11).

It has been suggested by some authors that erosion and fill dissection should be at a maximum during the onset of a pluvial stage. On the basis of a study of sediment yield in relation to mean annual precipitation (Langbein and Schumm, 1958), Schumm (1965) suggests that, in an arid climate, an increase in rainfall will increase sediment yield, since the increased runoff would—at least initially—be more effective in promoting erosion than the improved vegetation cover would be in retarding it.[10] In fact, the empirical data indicates that, when annual rainfall is less than 600 millimeters (with a mean temperature of 70° F or 21° C), any increase in precipitation will increase the sediment yield and vice versa. The question is: what geomorphic changes occur in the stream basin? Schumm (1965) believes that dissection and headward erosion of the

---

10. Assuming a drainage basin with a mean annual temperature of 60° F (15.5° C) and a mean annual precipitation of 250 millimeters, a 10° F (5.5° C) decrease in temperature and a 100% increase in precipitation will at least double the annual sediment yield (Schumm, 1965). The runoff will be increased by a factor of at least 20.

tributary valleys will be compensated for by aggradation in the channels of high-order streams. During the waning stages of a pluvial cycle, runoff and sediment yield decrease, while concentration of sediment per unit of runoff increases rapidly. As a result of the increased load, aggradation is postulated for the tributary streams with some scour of the main river channels (Langbein and Schumm, 1958; Schumm, 1965).

The geomorphic evidence in southern Egypt does not support Schumm's hypothesis of erosion upstream and aggradation downstream, in response to an increase in rainfall. Wadi terraces can be followed upstream from the Nile Valley into the wadi headwaters, as is the case for the small wadis west and northwest of Luxor (Butzer, 1959a) or for Wadi Or (Chap. 6). When terraces are missing upstream, this can usually be attributed to subsequent erosion, as can be clearly shown in Wadi Alam-Khariga (Chap. 8). In the case of the major wadis of the Eastern Desert, it is common that ancient alluvial deposits begin in first-order tributary valleys and that these beds are graded to the major wadi terraces, e.g., in Wadi Korosko (Chap. 5). There can be no question that alluviation was a general, regional phenomenon in southern Egypt, leading to aggradation along the length of the wadis. The interfluves were subjected to denudation, and alluvial fill does indeed thicken downstream. But in almost all well-defined wadis of the Eastern Desert, Pleistocene aggradation can be observed along the length of the stream. Presumably, there was a time lag involved, with aggradation beginning downstream and terminating along the entire wadi, but this cannot be demonstrated in the field.

Two explanations can be offered for the remaining discrepancy between theory and empirical observation. First, loose surficial materials of mixed calibration are moderately abundant on upland interfluves, on hillslopes, on pediment surfaces, and in wadi valleys. Transport, other than by wind, is today limited to steeper slopes and to wadi floors. An increase in rainfall sufficient to promote effective, general runoff would transport an enormous amount of sediment into the wadis. Since we are mainly dealing with bed-load materials, transport distances would be short, and an alluvial-colluvial mantle could be expected even at the head of first-order streams. Such veneers would represent material in the course of periodic transfer. Only after rainfall had decreased once again would the alluvium of the uppermost tributaries assume a more permanent aspect. A second reason for general aggradation with increased precipitation involves concentration of sediment per unit of runoff. In a truly arid climate, running waters carry an exceptionally high load at most times, with the ratio of sediment to water often approaching that of a mudflow. Under these circumstances, it is not surprising that transport distances may be short, despite a high sediment turnover through scour and fill during any one flood. On an annual basis, the empirical data (Schumm, 1965: Fig. 3) shows that sediment concentration increases asymptotically with decreasing annual precipitation. Theoretically, by extrapolating the curve for a mean annual temperature of 70° F (21° C), we obtain the absurd result that sediment concentration is one million parts per million with a mean rainfall of 500 millimeters. In other words, in a warm, arid climate, mean annual sediment concentration is uniformly high within a considerable range of precipitation, so that an increase or decrease of rainfall will not affect sediment concentration in a predictable fashion. And, as a corollary, aggradation along the length of a stream valley is theoretically reasonable in response to an increase in precipitation from almost nil to 100 millimeters or more.[11]

PALEOSOLS

The red paleosols of southern Egypt, as developed in noncalcareous gravels, are all of one and the same soil type. This does not preclude

11. Chavaillon (1964: 298–306), in his study of the alluvial terraces of the Saoura Valley in the western Sahara, seems to apply the principles of Langbein and Schumm (1958) and of Schumm (1965) when he equates maximum discharge with erosion of fill and waning discharge with alluviation. The lower Saoura is a high-order stream, however, collecting the drainage waters of a considerable catchment area in the Saharan Atlas. Chavaillon employs the terms erosion and alluviation in a rather loose way. Clearly, short-term scour and fill go hand in hand during all stages of a period of accelerated discharge, with lateral erosion accompanying the original formation of a wider floodplain. And, obviously, exposed terrace gravels necessarily record the last stages of alluviation, prior to scour-fill equilibrium or active dissection. The load carried during a period of pluvial scour exists, however, and has to be deposited, periodically or seasonally, on the current stream bed or floodplain. Here too, it ultimately comes to rest. In other words, whereas pluvial channel erosion and fill dissection are not necessarily synonymous, channel scour and alluviation cannot be chronologically or paleoclimatically separated. As used by Chavaillon, the two concepts are part of the inseparable co-agencies, scour and fill.

Like Chavaillon, we interpret our sediments by a subarid or semiarid climate, but, unlike Chavaillon, we believe that no part of the alluviation-dissection cycle indicates conditions of subhumid or even humid character. We do not believe that cutting restricted channels into a broad, abandoned floodplain requires greater discharge than the floods which had both the competence and capacity to transport and aggrade the original floodplain. Initiation of an aggradation cycle will involve both scour and fill at depth, as well as lateral erosion. But these processes are all part of the same geomorphic trend and cannot be isolated as distinct events.

differences in weathering intensity and profile depth, but the essential features are generally similar. These include: (a) kaolinitic clays (see Appendix D); (b) rubefaction by very modest quantities of hematitic iron; (c) coarse-grained, permeable parent material; and (d) later enrichment of salts or carbonates. Mechanical eluviation of fines is indicated in the older red soils as well.

The soil type represented by these paleosols is difficult to identify with precision. For one thing, the original A-horizons have been eroded. Eluviation of fines has deprived the gravels of much or most of their nonsand matrix, particularly in the case of the better-developed, older soils. Then again, the lack of consolidation led to complete structural disintegration of our samples during shipment. Thin sections could not, therefore, be made, precluding study of microfabrics. Finally, only one or two samples were taken from any profile, so that systematic mechanical and other analyses could not be made of all horizons present. Consequently, our study of these paleosols must be based primarily on field records and so remains inconclusive.

Despite the deep (B)-horizon and intensive discoloration of the High Terrace paleosols, original clay content was certainly less than 25%, and feldspars are still abundant in the matrix of the $(B)_2$ horizon. Obviously, we are not dealing with plastic, fine-textured *Rotlehms* such as Kubiena (1955, 1957) has described from the Hoggar Mountains. In macroaspect, these soils most closely resemble the sandy, moderately weathered *meridional Braunerdes* (Kubiena, 1953: 290–92) of the Mediterranean region. Such soils have a poorly developed $A(B)C$ profile and are limited to noncarbonate bedrock. The iron of the Mediterranean brown woodland soils is dominantly hydrated, however, and discoloration is rather limited. Profiles most similar to the Egyptian soils, also eluviated in their upper parts, have been described from Pleistocene alluvia of the Wadai, between Ennedi and Lake Chad, by Franz (1958, 1966). These reddish (5–7.5 YR 4/4; 5 YR 5–6/8) paleosols have a matrix of sandy silt and contain abundant feldspars. In one such soil, the upper 60 centimeters have subsequently been somewhat eluviated, with development of an illuvial zone at 60 to 80 centimeters. In another case, the top 20 centimeters were eluviated, and the clay content increases from 5% to 30% in the illuvial zone below.

Although further Pleistocene soil profiles will be described in subsequent chapters, it seems that the red paleosols were formed under moister conditions, with moderately significant biochemical weathering. The anhydrous iron may be interpreted as an indication of strong seasonality of rainfall. The rainfall amount or the vegetation type are impossible to assess, although the climate probably was semiarid in character.

All of the paleosols described so far are autochthonous relict soils (Erhart, 1965) and have *in situ* profiles. They are not to be confused with red soil sediments. Such sediments do occur in hollows developed on top of the terraces but are more common in Nubia than in the Kom Ombo area. Red soil sediments were not found incorporated into any of the alluvial gravels. In each case, the soil developed after alluviation had ceased and at a time when geomorphic activity was at a minimum. Whether such a period of biostasis—in Erhart's (1956) sense of the word—immediately followed alluviation cannot be determined.

CONCLUSIONS

The geomorphic and paleosol record of the earlier Pleistocene at Kom Ombo (Table 2-2) suggests at least a half-dozen pluvial periods of varying intensity, duration, and character. In no case is it necessary to assume a radical change in climatic regime or character, but rather a shift from arid to subarid or semiarid. Although the substantive features—such as deposits or soils—all suggest greater moisture, much of the long time interval involved must have been about as dry as the present. The limited geomorphic activity with contemporary climate precludes any significant record. Such hyperarid, morphostatic phases presumably followed or preceded each major dissection of fill, as outlined by Table 2-2.

3

Late Pleistocene and Holocene Sediments of the Kom Ombo Plain

Introduction

Following upon the terrace gravels and red paleosols of the early and middle Pleistocene, true nilotic deposits mark the late Pleistocene and Holocene of the Kom Ombo Plain. The term "nilotic" is used to describe sedimentation by the summer floods of the Nile, including materials of subsaharan origin. Although partly deposited by the Nile River, the earlier gravels were fully conformable with the local wadi deposits, suggesting aggradation during the low-sun season. Distinctive material of Ethiopian provenance appears to be missing. In other words, the contemporary summer-flood regime may only have been inaugurated during the late Pleistocene. Or—said in another way—in response to local climatic factors, Nile aggradation was then dominated for the first time by exotic, summer-flood sedimentation, rather than by winter alluviation.

Three major nilotic units and two wadi formations can be identified in the later prehistoric record of southern Egypt. Since each of these units is common to both the Kom Ombo area and Egyptian Nubia and is susceptible to accurate field mapping, formational status will be assigned to them. In the preliminary reports (Butzer and Hansen, 1965, 1966; Butzer, 1967), lithostratigraphic terms were informally used to describe these units. In the subsequent pages, rock-stratigraphic units (as studied on the Kom Ombo Plain) are formally defined, followed by a description of the type sections. The equivalent designations, from top to bottom, are as follows:

Informal	Formal
Upper Wadi Alluvium	Shaturma Formation
Lower Wadi Alluvium	Ineiba Formation
Younger Channel Silts	Gebel Silsila Formation
Older Floodplain Silts	Masmas Formation
Basal Sands and Marls	Korosko Formation

The Korosko Formation (Basal Sands and Marls)

The late Pleistocene sequence begins with marly-sandy sediments of nilotic origin, unlike any earlier or later deposits preserved in the Nile Valley of southern Egypt. Designated as the Korosko Formation, this lithostratigraphic unit is exposed in four areas along the peripheries of the Kom Ombo Plain (Fig. 3–1): (a) in the lower course of Wadi Ayed, near Silwa Qibli; (b) in a series of pits between New Korosko and New Sebua; (c) in a pit at New Masmas, on the north bank of Wadi Kharit; and (d) in two pits and a surface exposure in Wadi Shurafa, about 500 meters north of New Ballana II. Although the Korosko Formation is preserved under more recent nilotic deposits along the margins of the graben, it is generally not exposed at the surface. With the completion of the Nubian resettlement project, only the Wadi Ayed sections remain available for study.

The upper stratigraphic limit to the Korosko unit is set by the Masmas Formation, which rests disconformably on the dissected surface of the Korosko Formation in three of the above areas. In Wadi Ayed, the Korosko Formation occupies an elevation range from *ca.* 93 meters to 99 meters, resting on bedrock. In Wadi Shurafa, these beds occupy an elevation range of 94.5 to 103 meters, again resting on the Nubian Sandstone. At New Masmas, a 10-centimeter lense at approximately 105.5 meters lies disconformably on reddish-yellow alluvium or colluvium. In the New Korosko–New Sebua area, where the Korosko Formation is recorded in 18 different profiles, the base of the sedimentary unit is nowhere seen, although identical beds are found at the base of wadis cut into the Low Terrace complex (Figs. 3–2 to 3–5). The vertical range here is from below 102.5 meters to 107.5 meters. Although a suitable type section showing the Korosko Formation embanked directly against the Low Terrace edge is lacking, the relationship is unquestionable. Rubefaction and at least 10 meters of fill cutting separate the two alluvial units in

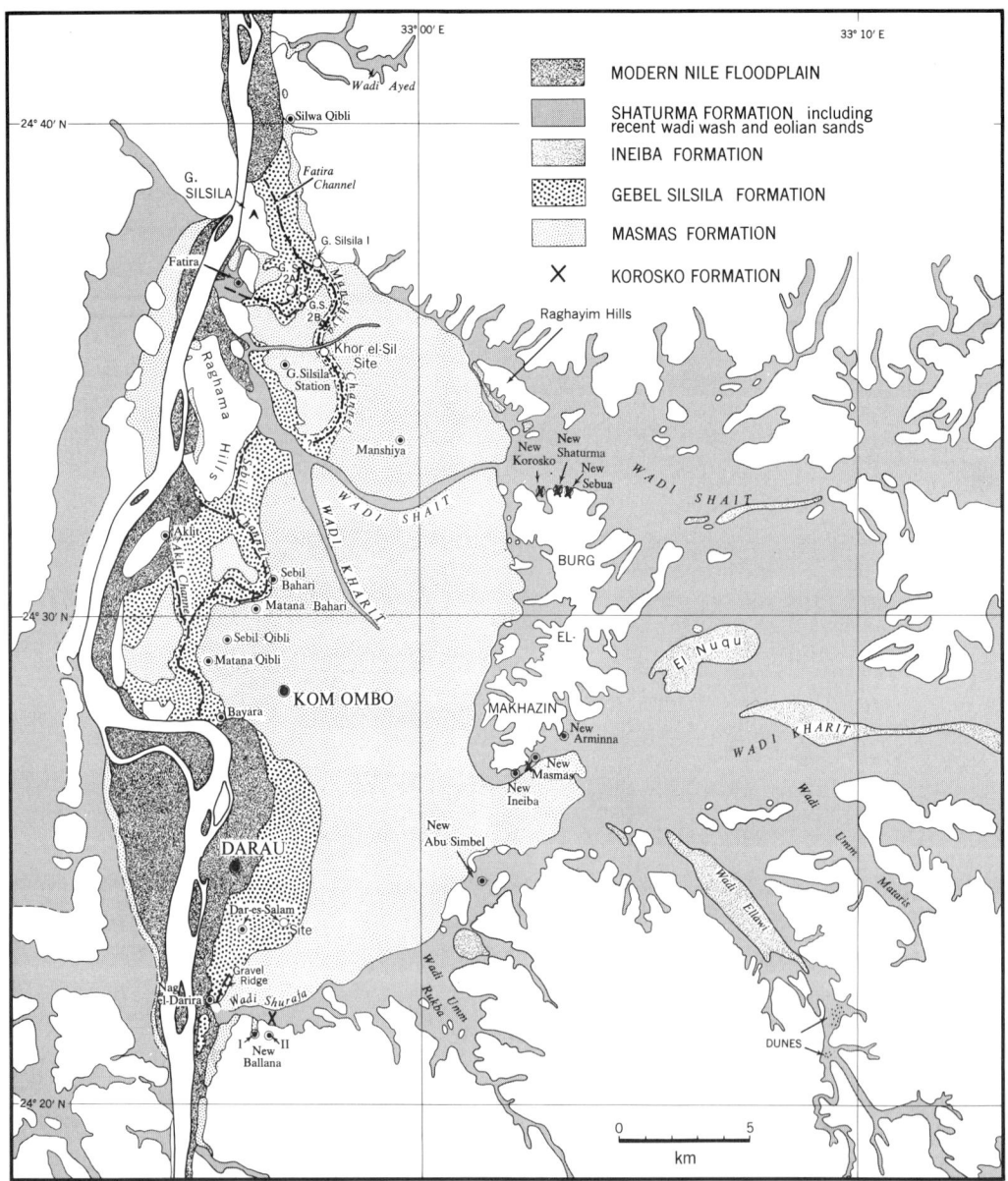

Fig. 3–1. Late Pleistocene and Holocene deposits of the Kom Ombo Plain. Map: UW Cartographic Lab. (This map also available at 1:100,000. See p. 524.)

the Shait area. The total elevation range of the Korosko Formation is from 93 meters to 107.5 meters, approximately 15 meters. The highest deposits are at 19 meters above modern floodplain, although their relative elevation before erosion may well have been over 20 meters.

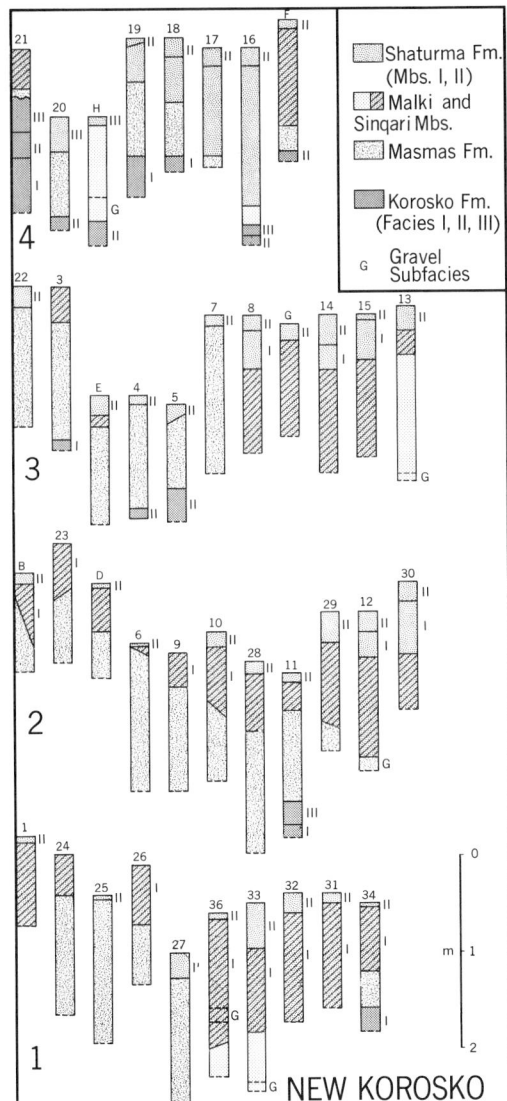

Fig. 3–2. Microstratigraphy at New Korosko. Tops of profiles have been reduced to comparable elevations within each cross section. Section 1 near mouth of wadi; other sections located progressively further downstream. Numbers attached to profiles follow Egyptian engineers' designations. For locations see Fig. 3–3. Graph: UW Cartographic Lab.

The typical facies of the Korosko Formation is a light-gray to light-brownish-gray (2.5 Y 6–7/2), poorly sorted, marly medium sand to sandy marl, with about 15% to 40% $CaCO_3$ and with pH values ranging from 6.7 to 7.9. Dispersed quartz granules are common. Evaporite, mainly halite, is common in small quantities and occurs in diffuse or concentrated form, with soft precipitates up to 3 millimeters in diameter. Concretions of calcareous salts, up to 1.5 centimeters in diameter, are found

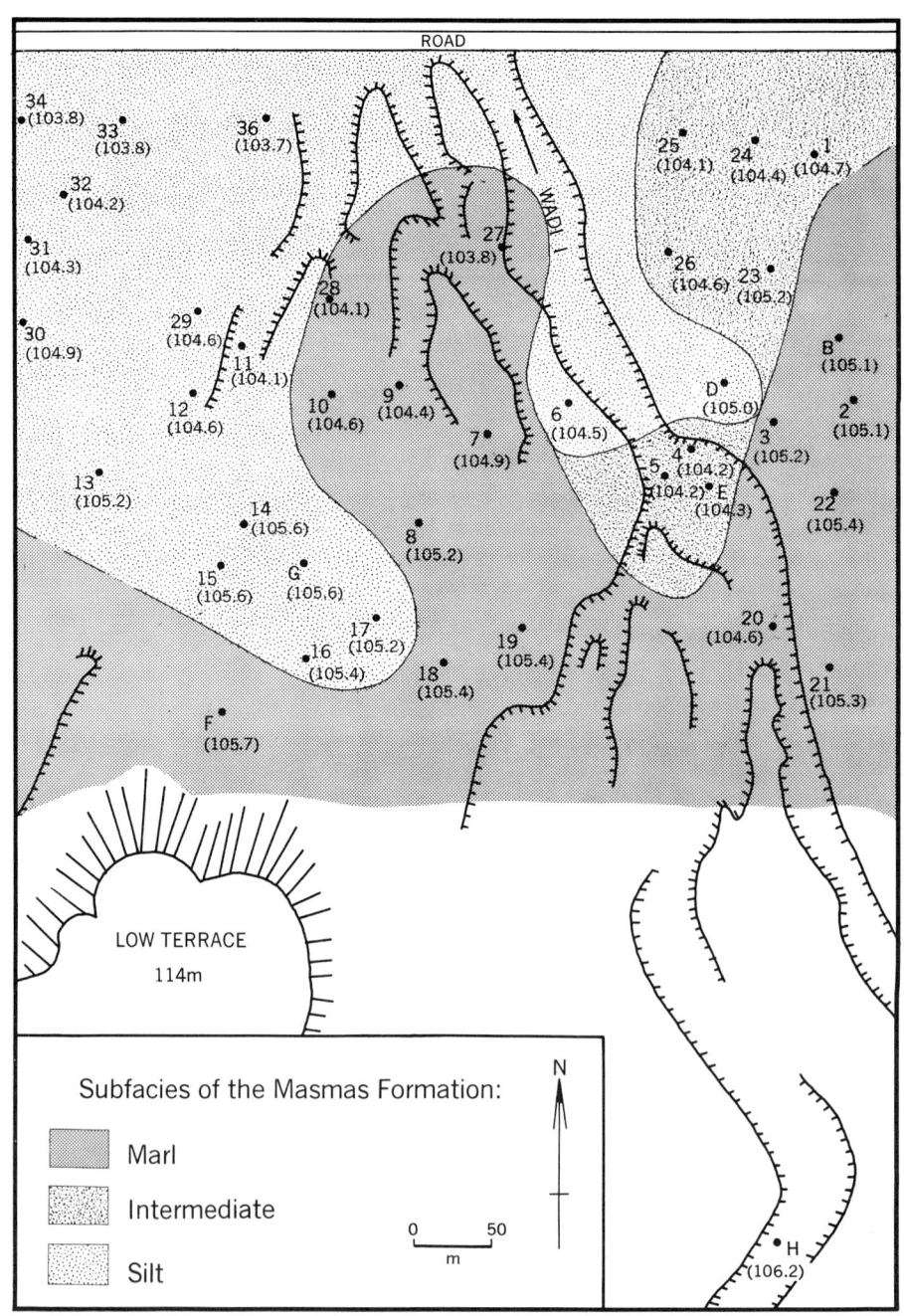

Fig. 3–3. Topography and profile locations at New Korosko. Elevations of tops of profiles given in meters in parentheses. Subfacies of Masmas Formation indicated by overlays. Map: UW Cartographic Lab.

sporadically. The common occurrence of yellow (10 YR 6–8/6), limonitic streaks or stains provides evidence of temporary waterlogging, although local reducing environments are suggested by pyrite found at New Sebua and at New Masmas. Semiconsolidated, the structure is coarse, angular, blocky, with inconspicuous stratification the rule. Impressions of rootlets under 0.5 millimeters in diameter are very common, and such casts may be up to 8 millimeters in diameter and lined with carbonized vegetable matter. Coarse root drip was not observed. Mollusca are fairly abundant, dominated by *Planorbis ehrenbergi* and *Valvata nilotica*, together with shells of *Bulinus truncatus*, *Lymnaea* sp., and *Corbicula fluminalis* (see Appendix G).

The more or less homogeneous sandy marls that characterize this formation at Kom Ombo occur with two variant facies in the Wadi Shait area (Figs. 3–2 and 3–4). The first of these, found at the base of the standard facies, contains dispersed pebbles or local gravel concentrations and a high sand component (Nos. 13, 69, 74, Table 3–3; Nos. 77, 81, 84, Table 3–8). These marly sands with gravel indicate local wadi activity. Three lots of gravel from New Korosko were analyzed morphometrically [Korosko Formation (II) in Table 3–1]. These are a very inhomogeneous, subangular, coarse detritus, moved primarily by sliding. The lithology reflects that of the local Pleistocene wadi alluvia, and most of the gravel was derived from angular lag. Transport distances were short, so that pebble edges were barely rounded off. Sheetflooding, with the gravel sliding on a sandy base, was probably responsible. The nature of the deposits suggests that some of the gravel was washed into standing waters, while other detritus was deposited subaerially, including the incorporation of aggregates of eroded marls into the matrix.

Table 3–1. Morphometric gravel analyses
(Lithology: fine-grained igneous rocks other than quartz.)

Location of sample[a]	Formation	Sample size	ρ (%)	CV of ρ	Detrital component	E/L (%)	E/l (%)	L (cm)
20–A	Shaturma (II)	75	12.9	135.1	60.0	45.4	65.0	4.24
36–C	Ineiba (Sinqari)	100	29.0	97.9	27.0	42.5	55.1	3.59
26–C	Ineiba (Malki)	75	18.2	120.5	46.6	47.1	65.0	2.99
20–C	Korosko (II)	50	14.8	121.8	56.0	47.4	63.5	5.17
5–C	Korosko (II)	75	14.6	135.6	57.3	49.6	66.4	4.33
21–A	Korosko (I)	75	11.6	149.1	73.3	42.8	56.6	4.07
18–D	Korosko (I)	75	19.1	133.6	50.6	44.1	61.1	4.93

[a] Localities refer to pit numbers and horizons at New Korosko and New Sebua.

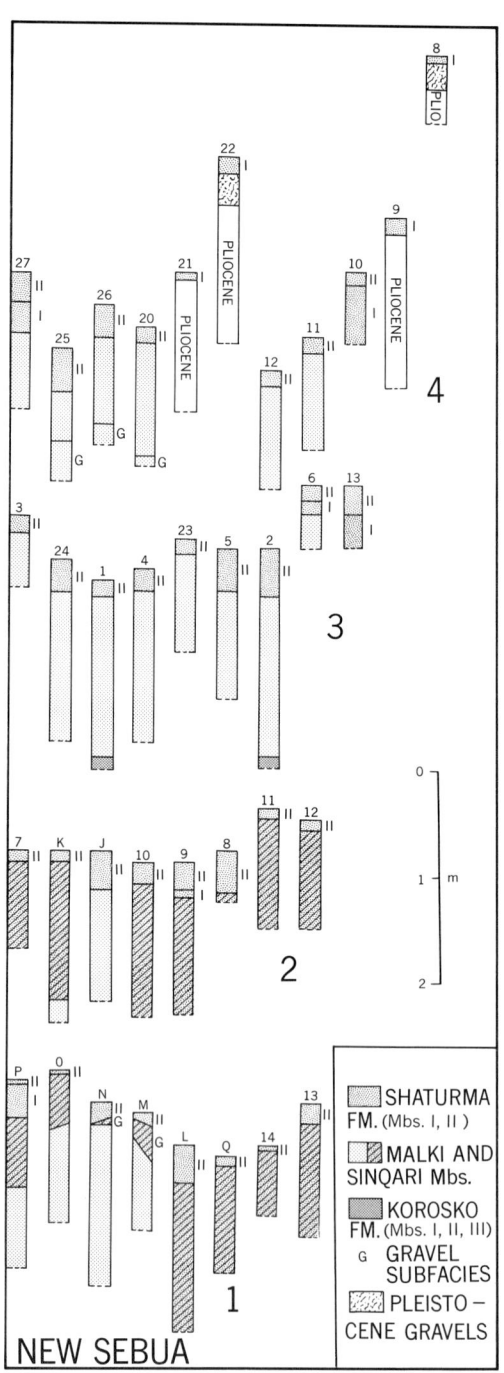

Fig. 3–4A.

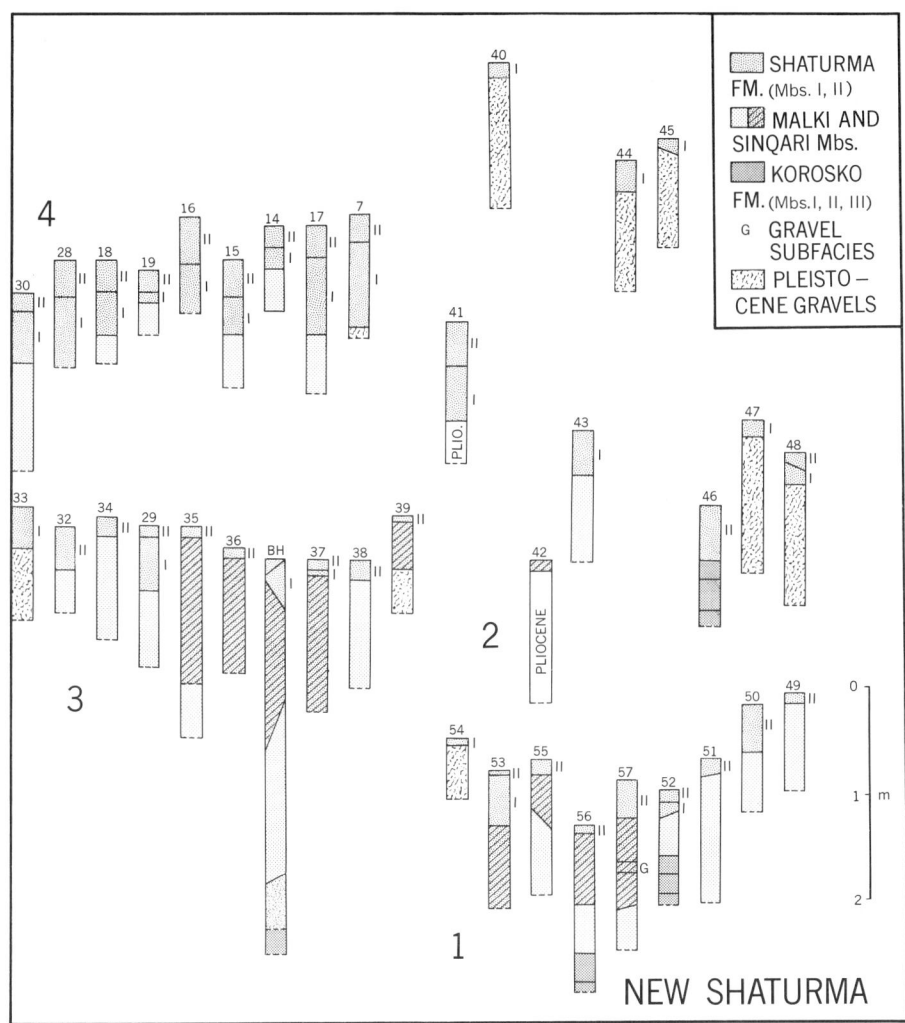

Fig. 3–4B.
Fig. 3–4. Microstratigraphy at New Sebua (*A*) and New Shaturma (*B*). Sections *1* and *3* near wadi mouths. For locations see Fig. 3–5. Graph: UW Cartographic Lab.

The second subfacies is a pale-brown (10 YR 6/3), sandy gravel, with abundant carbonates (10% to 15%) but no salts (No. 76, Table 3–3; No. 47, Table 3–8). Sorting is poor, stratification is moderate, and structure is medium-to-coarse, subangular blocky. Although apparently a little better rounded, the gravel is identical in type [Korosko (I) in Table 3–1] to that of the gravelly marl. This subfacies is, in part, laterally conformable with

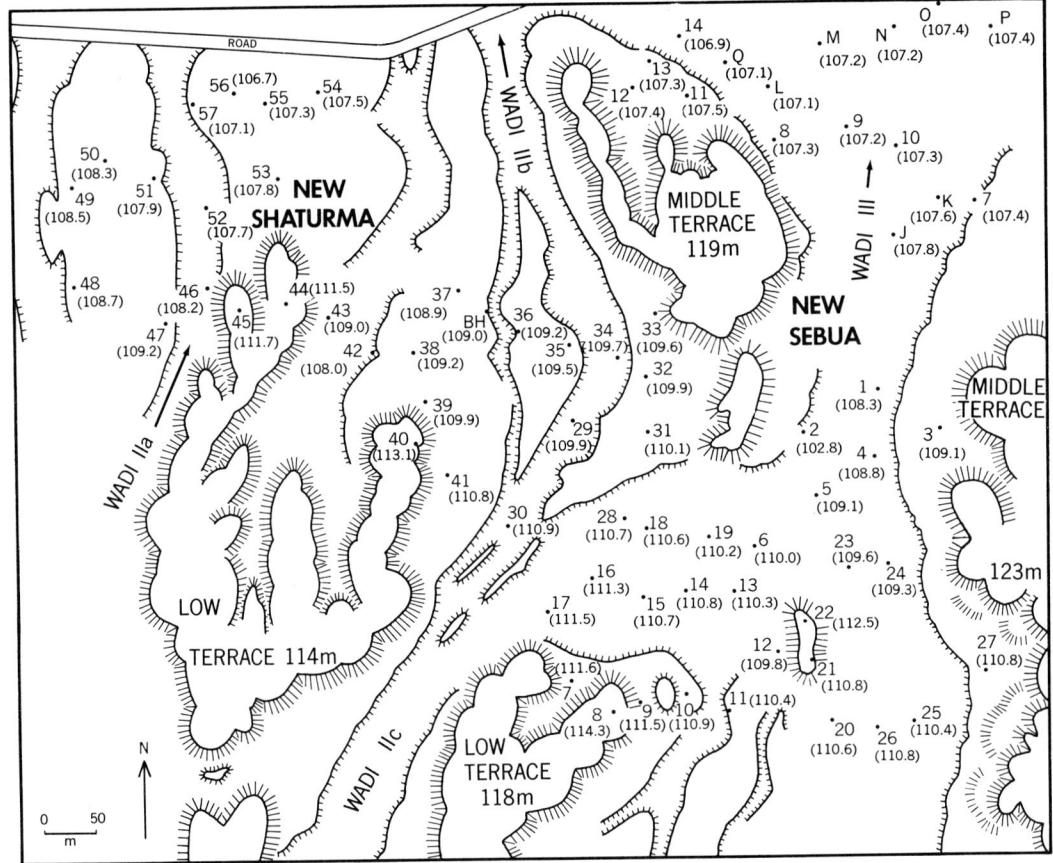

Fig. 3–5. Topography and profile locations at New Shaturma and New Sebua. Elevations of tops of profiles given in meters in parentheses. Map: UW Cartographic Lab.

the gravelly marls. Some of these sandy gravels underlie the gravelly marls more or less disconformably, however. These earlier gravels have stratigraphic interest, since they indicate alluviation underway in the local wadis, prior to establishment of a semi-aquatic environment. Local fluvial activity abated during the earlier phases of the marl deposition and then ceased completely. Despite their local character, these subfacies of the Wadi Shait sections show that climatic changes of some significance occurred during the deposition of the Korosko Formation. Detailed sections will be further described below.

The overall composition of the Korosko Formation suggests derivation from both local and distant sources. The homogeneous character of the beds throughout the Kom Ombo Plain and Lower Nubia, despite lateral subfacies, suggests deposition by the Nile with little *direct* participation of the Egyptian wadis. Yet the lateral subfacies in Wadi Shait and also in Nubia indicate accelerated wadi activity during the earlier part of this sedimentary interval. It would seem, therefore, that the deposits were related to a summer-flood regime of subsaharan origin, at a time of year when the Egyptian wadis were dormant. The bulk of the quartz sands and all of the gravel must have been locally derived, either from eolian or fluvial sources (see Appendix C). The carbonates and salts, and most of the silts and clays, were not of Egyptian origin, however. These were introduced to Egypt by the waters of the Blue Nile and the White Nile. Montmorillonite and kaolinite are the major clay minerals represented, although kaolinite is only a minor element among recent nilotic clays (see discussion in Appendix D). The heavy-mineral spectrum is similar to, but not quite identical with, that of recent nilotic silts derived primarily from Ethiopia (see Appendix B).

The Korosko Formation suggests either a lacustrine or a mixed fluvial-semiaquatic environment of deposition. Bedding is exclusively topset and is generally indistinct. The high carbonate content suggests precipitation in standing waters, primarily through inorganic agencies. The presence of salts in limited quantities and only localized evidence of pyrite also speak for a moderately well aerated depositional environment, for the most part in periodic communication with the Nile flood waters. Unfortunately, the beds studied are all found in relatively sheltered embayments and do not pertain to the center of the former floodplain. In Nubia, the Korosko Formation is also confined to wadi embouchures and to other embayments along the Nile Valley margins. In only one locality (Khor Hamra, see Chap. 6) have backset river-margin strata been preserved.

The molluscan fauna, dominated by the genera *Bulinus, Planorbis*, and *Valvata*, is ecologically consistent with the facies character. All of these gastropods are abundant in the muddy, and often foul, drainage ditches of contemporary Egypt. Harris (1965), in his significant study of modern molluscan ecology on the floodplains of Iraq, provides pertinent quantitative data concerning *Planorbis ehrenbergi* and *Bulinus truncatus*. Both species have very similar preferences and tolerances and appear to be present at all times of the year. They occur in areas of quiet water (current velocities less than 20 m/min.), with less than 0.13 milliequivalents per liter (meq/liter) nitrate ions and with a very slightly saline (sodium ions 0.3–1.6 meq/liter) content. Neither species thrives when total dissolved salts exceed 400 parts per million. The subordinate genera *Lymnaea* and *Corbicula,* also present in Iraq, display much broader limits of tolerance.

On the basis of the abundant *Planorbis, Valvata,* and *Bulinus* shells found in the Shait Pits, standing but possibly stagnant waters with a limited quantity of solubles can be inferred. Large quantities of decaying organic matter were either absent or subject to rapid oxidation. The presence of planorbids in the pyrite beds suggests that the ferrous sulfide was a result of temporary reducing conditions (possibly in the groundwater zone after sedimentation), rather than of decomposition of animal matter. All this indicates abundant water with a nonacidic, dominantly oxidizing environment. Any lakes or ponds that did form appear either to have maintained communication with the Nile or to have infiltrated rapidly into the moderately coarse-textured subsoil. Since only the less soluble part of the chemical load was deposited, waters were not allowed to evaporate completely and did not, therefore, precipitate their more soluble salts. Complete evaporation could have been precluded by rapid return drainage of lakes as the floods receded or by rapid percolation of ponds formed in closed depressions. Nothing speaks for extensive swamp conditions or for true lacustrine environments except in direct association with seasonal inundations of the floodplain.

The conclusion that the Korosko Formation was essentially fluvial in origin is further supported by the longitudinal gradient of related deposits (at + 34 m near the Sudanese border and + 20 m near Gebel Silsila) of 1.33 per 10,000, compared with a modern gradient of only 0.94 per 10,000. General or even widespread lacustrine conditions are excluded by such a steepened gradient. We suggest, instead, that the preserved sediments reflect temporary or sporadic subaqueous environments along the valley margins. Presumably coarser, well-stratified, cross-bedded deposits were laid down in closer proximity of the major Nile channels. The steeper, higher-velocity river must have transported a comparatively coarse load, since sands and granule gravel were deposited well beyond the river bed. This could be explained by extensive sand splays developed on an undulating, presumably braided floodplain. Lacustrine situations would develop in the numerous swales or irregular depressions at the flood maximum, after which the seasonal lakes would drain back to the river or, in the case of shallow ponds, infiltrate rapidly into the moderately permeable base. This type of environment could best account for the mixture of sand, chalk, silt, and clay, as well as for the minor role played by the soluble salts.

Precise dating is not available for the Korosko Formation on the Kom Ombo Plain. A radiocarbon assay on marls from Wadi Or in Nubia gave an age of 25,250 B.C. ± 950 (I–2061), while a date from the overlying Masmas Formation at Kom Ombo gave 16,350 B.C. ± 310 (I–2060).

Mammalian remains are absent. The mollusca are of some interest since *Corbicula fluminalis vara* is extinct, while *Lymnaea* is now absent from Egypt (Gardner, 1932a: 108). Archeological sites were not found in these sediments, although a few isolated artifacts were found at New Korosko. These included a waterworn, proto-Levalloisian core in dolerite (Pit 21), presumably washed in from older surface sites nearby, as well as a mint, felsite blade segment with patination on one face only (Pit 18) (see Chap. 4). In relation to the local stratigraphic sequence, and as corroborated by the Nubian evidence, the Korosko Formation may be considered mid-Upper Pleistocene.

The Masmas Formation (Older Floodplain Silts)

Deposition of the Korosko Formation was followed by an extended period of Nile and wadi downcutting to below modern floodplain level. The total absence of contemporary deposits or paleosols precludes paleoenvironmental interpretation for this erosional interval.

Renewed sedimentation was characterized by horizontal, sandy to clayey silt of fluvial origin, deposited over an extensive floodplain by summer floods analogous to those of today. These deposits of the Masmas Formation are exposed over 200 square kilometers of the Kom Ombo Plain (Fig. 3–1) rising from 87 meters elevation in beds preserved west of the Fatira Terrace (Fig. 2–9) to 107 meters elevation in lower Wadi Kharit. The total vertical thickness exceeds the 20 meters recorded, since the Masmas Formation clearly goes well below floodplain level in Wadi Shurafa. Two borings at distances of 2 to 4 kilometers southeast of Kom Ombo struck 33 meters of clays and fine micaceous sands, resting on coarse sands and gravels (Attia, 1954: 11–13). The surface deposits at 100 meters elevation pertain to the Masmas Formation, while the coarse sands are found at only 5 to 6 meters below surface at Kom Ombo itself and at points further west, where the Fatira beds are exposed. It would seem that the sands and gravels represent middle Pleistocene beds, while the nilotic silts and clays pertain in large part or entirely to the Masmas Formation. The contact of the silts and gravels (Fatira beds?) is at 67 meters, indicating that the Masmas Formation may well have a vertical thickness exceeding 43 meters.

The upper stratigraphic limit to the Masmas unit is set by the Gebel Silsila Formation between Darau and Gebel Silsila and by the Ineiba Formation in the embouchures of Wadis Kharit and Shait (Figs. 3–2 to 3–5). The lower stratigraphic limit is given by the Korosko Formation in

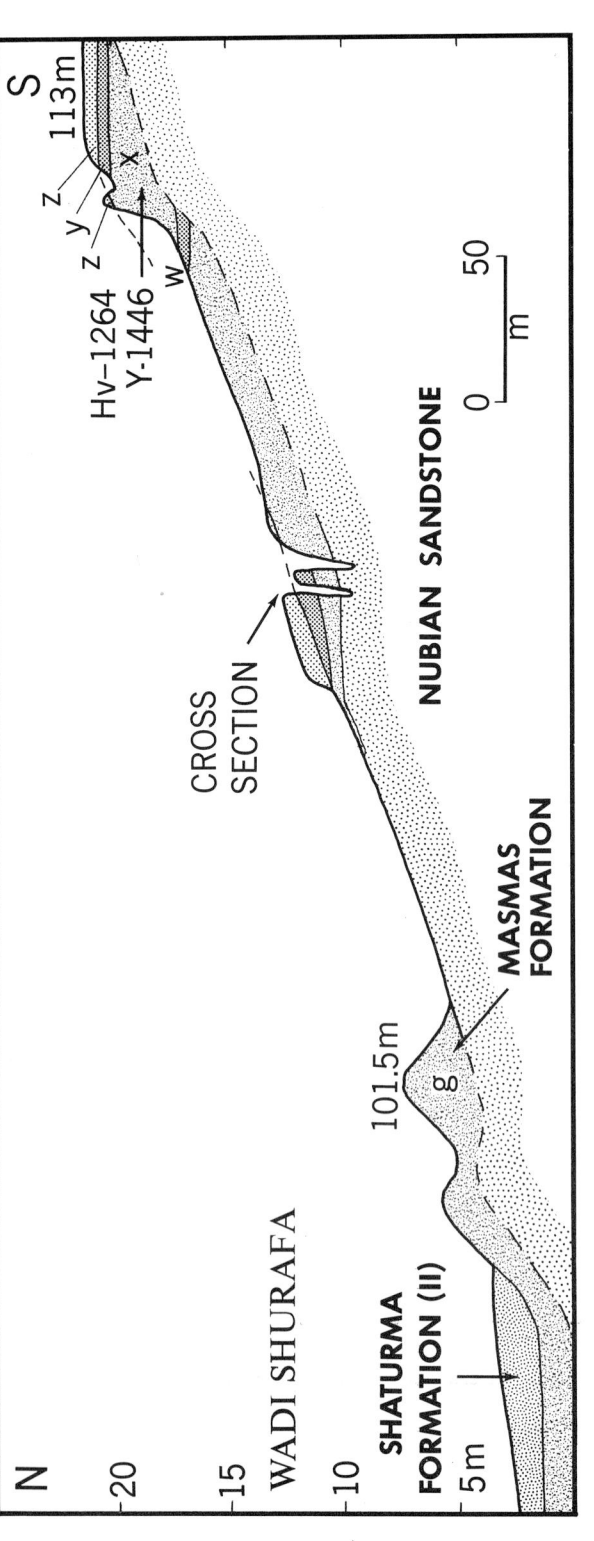

Fig. 3–6. The Masmas Formation as exposed in tributary wadi at New Ballana I. Flood silts of terminal Pleistocene age (upper bed x) indicated by radiocarbon code numbers (Hv–1264, Y–1377). For descriptions of beds w to z see p. 130. Map: UW Cartographic Lab.

Wadi Shurafa and at New Korosko. Sections illustrating these disconformable contacts are further described below. Although the best profiles have been destroyed during the Nubian resettlement project, the Masmas Formation remains exposed on the west bank of the Nile and can still be studied by artificial sections beyond the edge of the cultivated lands in Wadis Kharit and Shait.

The typical Masmas facies consists of thick, horizontal beds of internally uniform, alternating clayey silt and sandy silt. The clayey silt beds are dark grayish-brown (10 YR 4/2), with a very coarse, angular blocky structure, commonly with prismatic tendencies or with slickensides. Stratification is inconspicuous, partly because of repeated swelling and cracking following wetting and drying. Crack networks of different sizes may be filled by younger, coarse-grained sediments. Clay content characteristically is between 20% and 30%, varying from 10% to 55%, while coarse sands (over 0.2 mm) are usually absent. Sorting is moderate. Carbonates range from 0.5% to 10.0%, while pH values typically lie between 7.3 and 7.8. Calcareous salt (halite or gypsum) concretions 0.3 to 4.0 millimeters in diameter are common, and diffuse evaporites are often evident on ped faces. Pyrolusite stains or dendrites and, less commonly, limonitic staining suggest some seasonal waterlogging. Rootlet hollows up to 5 millimeters in diameter with traces of carbonized vegetable matter or evaporites are common. Local proliferations of *Corbicula fluminalis vara* occur in such clayey silts along the southern margins of the Kom Ombo Plain.

The sandy silt is brown (10 YR 5/3), with a coarse, subangular blocky structure. Stratification is the rule, while undulating or cross-bedded laminations are common, suggesting considerable water velocity. Sorting is poor because these beds are less homogeneous. Carbonates also range between 0.5% and 10.0%, but pH values are more variable, lying between 6.7 and 8.4. Concretions are absent, although diffuse evaporites and empty rootlet zones are typical; pyrolusite or limonitic stains are rare.

These alternating textural grades of flood silts form beds that range in thickness from 10 centimeters to over 3.5 meters and can often be traced horizontally for several hundred meters. They reflect deposition on alluvial flats or in backswamp areas. However, an excellent cut through over 1 kilometer of related channel and levee beds was visible in 1962–63 along a new ditch running west to east about 350 meters north of the Khor el-Sil. One 200-meter cross section was recorded in detail (Fig. 3–7) and several others were sketched in rough or photographed (Fig. 3–8). The deposits in question relate to several abandoned channels once

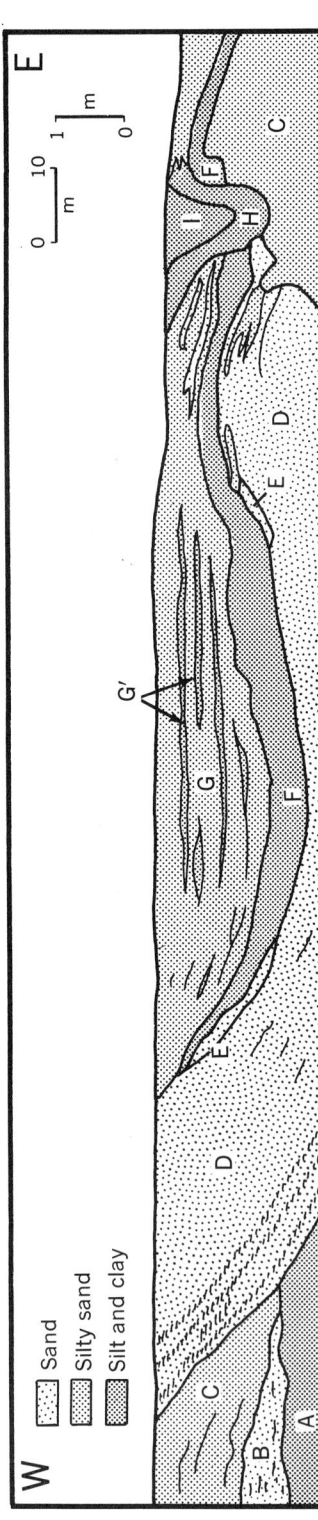

Fig. 3-7. Cross section of the Masmas Formation northwest of Gebel Silsila Station (1 km east of railway, 350 m north of the Khor el-Sil). Beds A and B antedate the major channel (beds D and E), while bed C records the related levee. Bed F was deposited in the cutoff prior to inruption of a minor overflow channel which deposited units H and I, with associated levee beds (G) and alluvial flat deposits (bed G'). These sections, intersected northeast of Silsila Station, provide fundamental information concerning the depositional environment of the Masmas beds. Map: UW Cartographic Lab.

running approximately parallel to the modern Nile, passing east of the Cairo-Aswan railroad track into the Silsila Gap. The major channels averaged 150 to 200 meters in width and have a filling of poorly stratified, coarse sands. After abandonment, a compact clayey silt was deposited in the cutoff meanders. The levees include complex backset and topset beds, varying in texture from medium sand to sandy silt and dipping 5% to 15% towards the channel. Sorted laminae of dark, heavy mineral concentrations are usual in coarser beds. Coarse root drip, absent in the other facies, is common among the levee beds. The deposits of the alluvial flats are identical to those already described as typical for the Masmas Formation. Fig. 3–7 and Table 3–2 illustrate the facies variation.

A regional variant of the Masmas Formation is found in parts of Wadis Kharit and Shait. This is a marly facies with 15% to 30% calcium carbonate, varying from light brown to grayish brown in color (10 YR–2.5 Y 5–6/2–3) (No. 72, Table 3–3; No. 83, Table 3–8). Other properties are analogous to those of the typical flood silts. The conformable, lateral variations of facies are clear in the New Korosko embayment (Fig.

Fig. 3–8. Different facies of the Masmas Formation northeast of Gebel Silsila Station (750 m east of railway, 350 m north of Khor el-Sil). An abandoned channel subsequently filled with dark, clayey beds (left) is flanked by sandy beds of channel-margin type (right).

3–3). Temporary ponds with carbonate precipitation are suggested for the floodplain margins. More problematical are thin, localized lenses with ashy to vesicular structure, with a high sulfide content and with proliferations of planorbid shells (No. 299, Table 3–3; No. 332, Table 3–8). These will be discussed further in connection with the detailed sections. Most of the mollusca collected from the Masmas unit were found in such marly or ashy beds. The total assemblage is similar to that of the Korosko Formation, including *Planorbis ehrenbergi* and *Valvata nilotica* as dominant forms, together with *Bulinus truncatus,* and rare *Corbicula fluminalis* or *Cleopatra bulimoides* (see Appendix G). On a very local scale, aquatic snails may also be found in more calcareous beds of the standard facies.

In general, the sand fraction of the Masmas Formation is concentrated in the fine to medium grades, less than 0.2 millimeters in diameter (see also Appendix C). These are unquestionably nilotic in origin, corresponding closely to the sands present in the modern Nile. Coarse- to granule-grade sands only show up locally, in what are the axes of modern wadis, and then usually in association with dispersed, subangular pebbles. This is the case in sections that expose the Masmas Formation below the bed of Wadi Shurafa and in 6 of 27 pits at New Korosko (Fig. 3–2). It is remarkable that only three general but shallow lenses of wadi wash interrupt a 13-meter vertical profile of the Masmas Formation exposed beneath Wadi Shurafa (Figs. 3–9 and 3–10). This implies that local wadi activity was minimal during the nilotic aggradation and was certainly no greater than at the present time. That part of the sand fraction which was obtained north of the Sudanese border was probably derived in large

Table 3-2. Sediment characteristics of different facies of the Masmas Formation
(See Fig. 3–7.)

Sample number	Bed	Color	Texture	Sorting	$CaCO_3$ (%)	pH
39	I	(10 YR 4.5/2.5)	Clayey silt	Moderate	1.0	7.6
40	H	(10 YR 5.5/3)	Clayey silt	Moderate	4.9	7.8
38	G'	(10 YR 5-6/3)	Medium-sandy silt	Moderate	8.2	7.7
35	G	(10 YR 6/3)	Silty medium sand	Moderate	0.5	8.0
34	F	(10 YR 4/1.5)	Silty clay	Moderate	0.0	7.3
37	E	(7.5 YR 5/5)	Coarse sand	Moderate	0.3	7.8
36	D	(10 YR 6/3)	Coarse sand	Moderate	0.8	7.8
43	C (West)	(10 YR 5/3)	Silty medium sand	Poor	3.3	8.0
41	C (East)	(10 YR 4.5/3)	Clayey medium sand	Poor	0.8	7.7
42	A	(10 YR 5/2)	Coarse-sandy silt	Poor	0.9	7.6

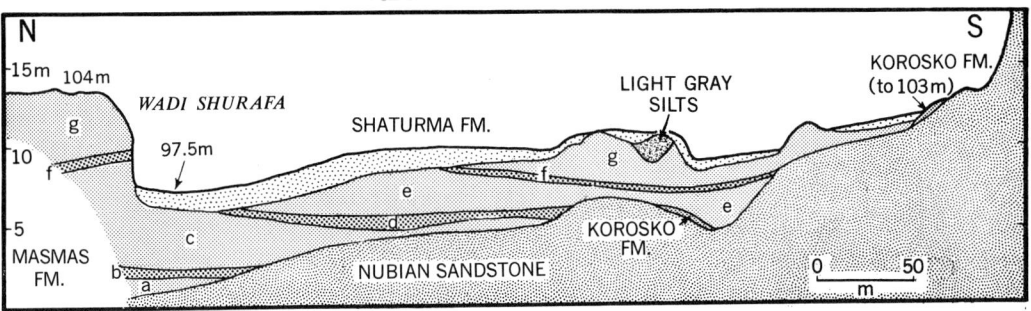

Fig. 3–9. The Masmas Formation in Wadi Shurafa (1.6 km east of railway). Beds *a* to *g* are described on p. 129. Map: UW Cartographic Lab.

part from eolian sources, materials blown directly into the river, or from riverbank dunes eroded by the Nile.

All in all, the Masmas Formation represents a series of fluvial deposits quite similar to those laid down by modern Nile floods. The heavy mineral spectrum is basically analogous to that of the modern silts (See Appendix B), and the clay minerals are dominated by montmorillonite, although the proportions of kaolinite are greater than in modern flood silts (see Apprendix D). Sandy silts are also interbedded with clayey silts in modern deposits but are infrequent in the present basin lands. The bedding of the Pleistocene sandy silts reflects violent flooding following rapid levee-breaching. The crevasses responsible for such deposits, although almost unknown in Egypt even before inauguration of the Aswan Dam, must have been fairly common on these late Pleistocene floodplains. Aggradation of a much broader floodplain (12 to 22 km, compared with 1 to 6 km today) also implies a somewhat more vigorous flood regime than in recent times. The highest Masmas deposits near the Sudanese border are at + 33 meters, so that the average gradient downstream to Kom Ombo was 1.28 in 10,000, compared with 0.94 today. It would probably be correct to infer that the summer floods were higher (in relation to low-water stage), swifter, and of greater duration. This, in turn, seems to suggest a climatic change in the Ethiopian highlands.

Fully satisfactory dating has not yet been established for the Masmas Formation. Mammalian remains—except for a single bone found *in situ* near Gebel Silsila 2B—appear to be absent, and the molluscan fauna has no stratigraphic interest. Archeological remains seem to be lacking, both on the surface and *in situ*. A radiocarbon determination was obtained from a clayey marl at New Korosko—16,350 B.C. ± 310 (I–2060)—al-

Fig. 3–10. Nile silts of Masmas Formation underlying wadi alluvium of Shaturma Formation in Wadi Shurafa (see Fig. 3–9). Note tendency to prismatic structure and slickensides in the nilotic beds.

though a second date from a sandy marl at New Shaturma—15,150 B.C. ± 400 (I–2178)—is a little too young (see Appendix E).

Today, the major, undissected Masmas surfaces at Kom Ombo dip gently westwards, maintaining a general level of 98 to 105 meters. The lack of relief on these featureless plains indicates but limited dissection of the original floodplain surface at 105 to 107 meters. Nonetheless, some planation is evident, possibly in response to a local base level of perhaps 95 to 98 meters. In a broad way, this interval of denudation is probably associated with the subsequent Gebel Silsila stage, which coincided with a period of accelerated wadi activity. At about this time or shortly thereafter, vertisol phenomena developed in exposed sediments. Although syngenetic crack networks occur in all clayey strata of the formation, there are some indisputable epigenetic crack networks that frequently penetrate to depths of over 1.5 meters, exhibiting polygonal structures at the surface.

Two different types of cracked soils can be illustrated from New Korosko and from New Abu Simbel. At New Korosko the sediments are fine-grained but rich in carbonates and comparatively poor in salts. The entire vertical depth of sediment is here fissured by both a fine and a coarse geometric network which was subsequently filled with sediment. The cracks of the fine network are only several millimeters wide and may be vertical or horizontal. They break up the primary sediment into blocky aggregates up to a few centimeters in diameter. The coarser cracks occur with a vertical plan only, averaging between 1.0 and 4.5 centimeters in width and thinning out a little at depth. The intrusive sediments found within these former cracks are primarily those of the Ineiba Formation, although they are inhomogeneous and quite variable. They range from a brown (10 YR 5/3), marly silt to light-yellowish-brown (10 YR 6/4) sand with fine- to medium-grade gravel.

Excellent exposures of polygonal networks and salt duricrusts were present about 1.5 kilometers northwest of New Abu Simbel until October, 1962. The local facies consists of dark-grayish-brown (10 YR 4/2), medium-sandy silt with considerable salt and carbonate (No. 23, Table 3–8). Two classes of polygons occur together: "major" polygons 75 to 250 centimeters in diameter, with 3- to 5- centimeter-wide crack fillings, and "minor" networks a few centimeters in diameter, with individual cracks 0.5 to 3 centimeters in width. The major cracks are filled with brownish (10 YR 5–6/2–3), silty sand, cemented with carbonates (35%) and water-soluble salts (10,000 parts per million), and bedded vertically, parallel to the cleavage faces. Gypsum concretions, 0.5 millimeters in diameter, were deposited normal to these cracks, within a range of up to 2 centimeters. The major cracks provide tough, high rims rising 10 to 30 centimeters above the lower-lying polygon centers (Fig. 3–11). A gypsum duricrust originally fossilized this irregular surface (Fig. 3–12), impregnating the upper 5 to 8 centimeters of crack fillings and the top 3 to 5 centimeters of the sandy silt. The minor crack networks intersect the polygon centers and are filled with brown (10 YR 5/3), sandy silt. The base of these two networks was not seen but is at least 120 centimeters below surface. Slickenside structures are also apparent at depth.

Although many or most of the minor crack networks may be syngenetic, the major polygons are clearly epigenetic. Their origin can be attributed to alternating swelling and cracking of dense sediments, rich in montmorillonitic clays. The annual flood regime may have provided

Fig. 3–11. Polygonal crack network developed in Masmas Formation near New Abu Simbel and fossilized by carbonates and gypsum.

Fig. 3–12. Deflated, irregular gypsum crust formed over Masmas silts near New Abu Simbel, originally formed as an *Sa*-horizon, probably precipitated by rare Nile flood waters of exceptional height *ca*. 12,000 years ago.

periodic waters adequate to saturate the soil. Except for the gypsum crust, there is no evidence of discoloration, and humification of the topsoil is lacking, although limonitic mottling may occur in the subsurface. These hydromorphic paleosols correspond closely to the Mazaquert type of vertisol (see G. D. Smith *et al.*, 1960: 126–28, 132).

The gypsum surface crust is a younger feature, judging by minor cracks filled with gypsum that pervade both the original silt and the calcareous fillings of the major polygonal cracks. This crust almost certainly developed during a period of periodic or sporadic very high Nile floods some 12,000 years ago, when this part of the plain was occasionally inundated by Nile waters rich in Blue Nile solubles.

The Gebel Silsila Formation (Younger Channel Silts)

The last Pleistocene nilotic unit of the Kom Ombo Plain consists of complex fluvial beds deposited by several Nile arms and overflow channels. The Gebel Silsila Formation occupies an undulating plain, some 1 to 3 kilometers wide, along the western fringes of the Masmas Formation. Areal extent is approximately 75 square kilometers (Fig. 3–1). From south of Darau to north of Gebel Silsila, this formation can generally be observed to dip below the modern floodplain. These deposits attain a maximum elevation of 101.5 meters in the Silsila Gap (Fig. 2–14), 100 meters in the Gebel Silsila 2 area (Figs. 3–13 and 3–23), 102.5 meters at Sebil (Fig. 3–31), and 102 meters south of Darau, and on the western face of Gebel Silsila there is a fresh fluvial bench with unpatinated, well-rounded sandstone cobbles to 101.5 meters (Figs. 2–14, 3–16). Vertical thickness consequently exceeds 14 meters.

The upper stratigraphic limit to the Gebel Silsila unit is given by the Shaturma Formation, deposited in channels that dissect the Gebel Silsila beds. Unfortunately, no contacts with the Ineiba Formation were observed in the Kom Ombo area. The lower stratigraphic limit to the Gebel Silsilan is generally given by the dissected margins of the Masmas Formation. A prime area of Nubian resettlement, the Gebel Silsila exposures have been graded and placed under cultivation everywhere except for a few localities south of Darau.

Highly variable texture, with rapid vertical and horizontal changes of facies, are a field characteristic of the Gebel Silsila Formation. Textural classes range from coarse sand to clayey silt, although silty sand, sandy silt, and clayey silt are most common. Medium-grade pebbles are abundant in bed-load deposits (Fig. 3–14). Stratification is very conspicuous in almost all beds. The sands, pale to very pale brown (10 YR 6–8/3) in color, are moderately to well sorted and generally inhomogeneous in character. Structure varies from loose, single grain in the sands to medium-to-coarse, subangular or angular blocky in the silty beds. Few fine, horizontal deposits pertaining to alluvial flats or backswamps have been preserved. The complex alternations of backset, foreset, and topset strata suggest instead multiple channel shifts and a sedimentary environment close to the low-water channel and its immediate banks.

The Gebel Silsila unit includes a considerably higher proportion of sands than the Masmas Formation, while clay content (10% to 25% in beds other than sands) is considerably lower. The clays exhibit a rather intense montmorillonite peak on the X-ray diffractograms, with kaolinite yielding first and second-order peaks of very low intensity (Appendix D). pH values average between 7.1 and 8.2, while carbonates range from 0% to 50%, being richest in former surface or subsurface crusts or concretionary zones. Lime-salt concretions of such origin, ranging from

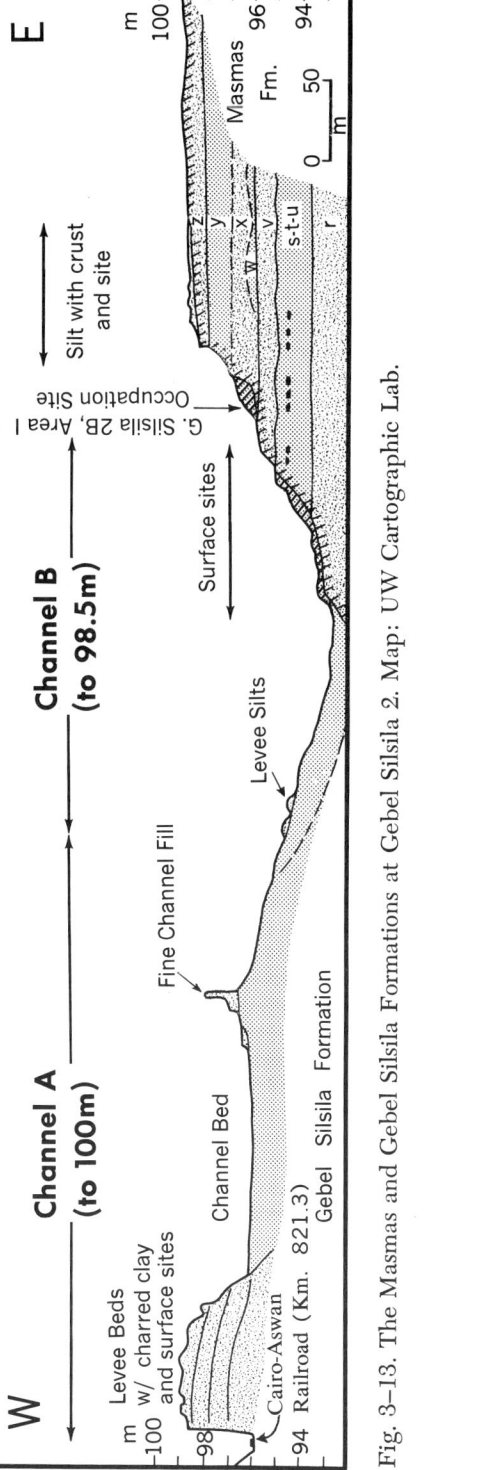

Fig. 3–13. The Masmas and Gebel Silsila Formations at Gebel Silsila 2. Map: UW Cartographic Lab.

Fig. 3–14. Inclined bed-load deposits of the Gebel Silsila Formation south of Darau.

0.5 to 8 centimeters in diameter, may be locally eroded and redeposited in the form of concretionary gravels. Salt microconcretions, generally 0.5 to 4 millimeters in diameter, are moderately common in silty beds. Lime-salt duricrusts are also evident in some fine beds at or near the surface. Such soil phenomena seldom exceed 3 centimeters in thickness. Occasional limonite and pyrolusite staining suggest temporary waterlogging.

Except for the quartz sands, ultimately derived in large part from Egypt and the northern Sudan (see Appendix C), the sand fraction of the Gebel Silsila Formation generally includes abundant mica and heavy minerals (hornblende, amphiboles, magnetite, and pyroxenes) (see Appendix B). The fresh state of the micas, which may be as much as 3 millimeters in diameter, suggests that the clay minerals are derived. The ferromagnesian minerals are probably derived in good part from the Ethiopian basalts, although the olivine-basanite lava fields of the Bajuda Steppe in the central Sudan (Putzer, 1958) must also have contributed a share (Fig. 3–15).

Pebble lithology is more informative. A count of 100 pebbles comprising two samples taken in gravel pits south of Darau (Fig. 3–14) showed

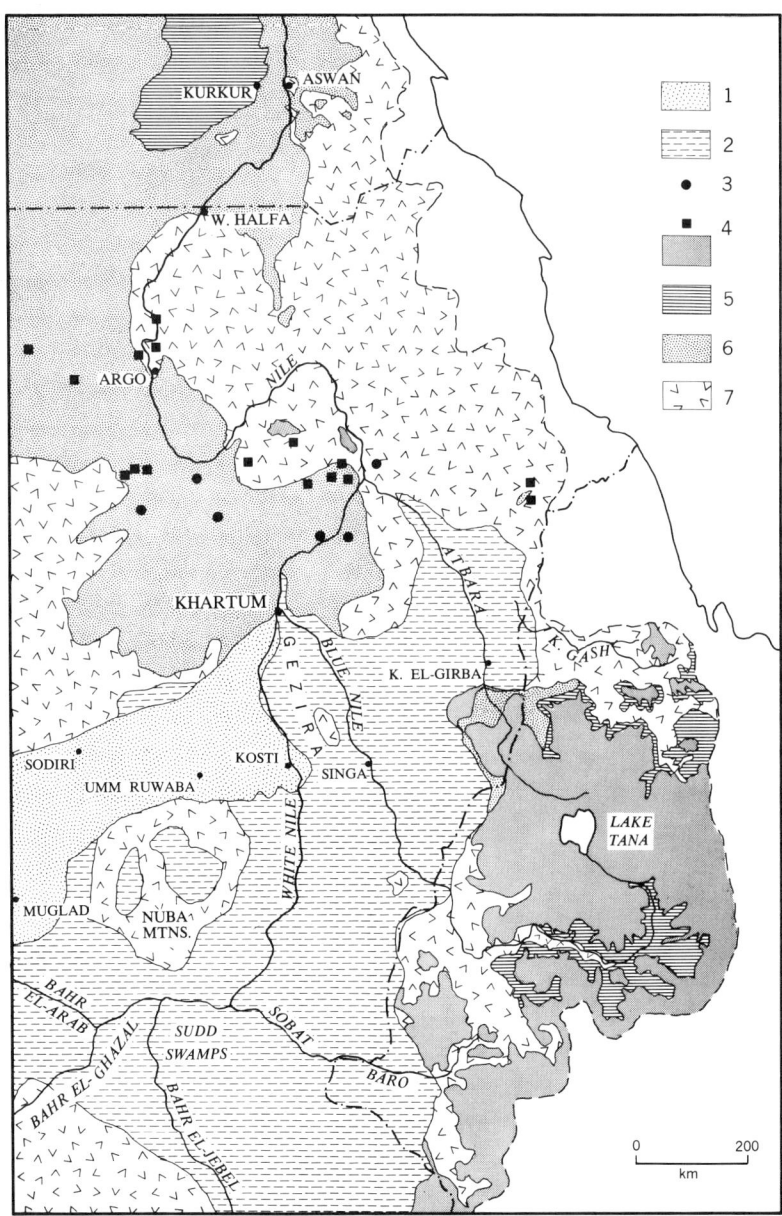

Fig. 3–15. Surface lithology of a part of the Nile Basin: *1:* Pleistocene eolian deposits; *2:* Pleistocene alluvial deposits; *3:* Hudi Chert; *4:* Tertiary vulcanics; *5:* Mesozoic and Tertiary limestones; *6:* Mesozoic sandstones; *7:* Basement Complex. After 1:4,000,000 Geological Map of the Sudan (Khartum, 1963), and 1:5,000,000 Geological Map of Africa (Paris, 1963). Map: UW Cartographic Lab.

72% mesocrystalline and macrocrystalline quartz; 12% granite; 5% chert or cherty flint; 2% each for feldspar, schist, agate, sandstone, and ironstone; and 1% felsite. Together with a part of the quartz, the granite, aplitic feldspar, and schist are almost certainly derived from the Basement Complex in the Aswan-Kalabsha area. The ironstone and sandstone are local. Some of the chert may be from the Basement Complex, although the more chalcedonic varieties must be attributed to the Hudi Chert of the central Sudan (Fig. 3–15) (Andrew and Karkanis, 1945; de Heinzelin and Paepe, 1965). The banded agate and isolated pebbles of white chalcedony, carnelian, and jasper found in the same quarry are also probably derived from the Hudi Chert, although a more local origin in the Egyptian Basement Complex cannot be entirely excluded, particularly for the brown jasper (see Lucas and Harris, 1962: 386–87, 391–92, 397–98). The abundance of chalcedonic rocks among Gebel Silsila pebbles along the length of the Nile Valley in southern Egypt and the northern Sudan (de Heinzelin and Paepe, 1965) speaks for a southerly provenance in the Hudi Chert, however.

The only special subfacies of the Gebel Silsila Formation that deserves mention is the sandy gravel that forms distinctive ridges south of Darau. To be further discussed below, these channel-bed gravels (Fig. 3–14) suggest considerably greater stream competence. Morphometric study of 100 quartz pebbles indicated transport by both sliding and rolling (E/L, value 51.6%, E/l value 67.8%), despite the sandy nature of the deposits. Average length of pebbles over 1 centimeter long is 1.66 centimeters, with a standard deviation of 0.5. Since pebbles are virtually absent from the modern Nile bed, increased stream competence with stronger turbulence is suggested (Fig. 3–16).

Organic vestiges from the Gebel Silsilan include abundant root drip and occasional root fragments pertaining to shrubs or small trees. These are generally confined to levee beds. Hollow rootlets zones, sometimes coated with carbonized organic matter, are common in silty beds. The molluscan fauna is quite distinct from that of the older nilotic formations. The most characteristic and abundant mollusc is *Unio willcocksi*, found in association with either the bivalve *Corbicula fluminalis* or the gastropod *Cleopatra bulimoides*—but rarely in association with both at the same time. Rolled fragments of the Nile oyster (*Etheria elliptica*) are occasionally found in association with *Cleopatra* (see Appendix G). Except in some beds exposed at Khor el-Sil, *Planorbis*, *Valvata*, and *Bulinus* seem largely to be absent, a fact that can be readily explained by rapidly moving waters in channel proximity. Mammalian faunas, generally found in stratified human-occupation sites, are rather abundant. The

Fig. 3–16. Cobble bed load and fluted sandstone on west face of Gebel Silsila at 101.5 meters.

1962–63 collections of the Yale Expedition will be published in detail by C. A. Reed. In the meanwhile, the older faunal collections of Vignard and Sandford from the same deposits near Sebil Bahari can be enumerated, as identified by Gaillard (1934: 3 ff.) and Sandford (1934: 86):

> *Bubalis* (*Alcelaphus*) *buselaphus* Pallary (Hartebeest)
> *Gazella dorcas isabella* Gray (Gazelle)
> *Bos primigenius* Bojanus (Aurochs)
> *Bos brachyceros* Owen (Probably females of *B. primigenius*, according to Reed, 1965b).
> *Bubalus* (*Homoioceras*) *vignardi* Gaillard (Extinct buffalo)
> *Equus asinus* L. (Nubian wild ass)
> *Equus caballus* L. (Either *E. asinus* or a zebra, but not horse according to Reed, 1965b).
> *Hippopotamus amphibius major* Owen (Hippo)
> *Hyaena crocuta* Zimmermann (Spotted hyena)
> *Crocodilus niloticus* Laurenti (Crocodile)
> *Testudo* (*Trionyx*) sp. (Tortoise)

Struthio sp. (Ostrich)
Synodontis schall (Fish)
Nodularia coelatura (Fish).

Though incomplete, this faunal list is already of considerable ecological interest since—as could be expected between riverbank and galeria woodland—it includes aquatic, woodland, and steppe species (Butzer, 1959*b*: 22).

The courses of these abandoned channels are dotted by innumerable Late Paleolithic archeological sites. Vignard (1923) had already defined three successive Sebilian industries employing a form of Levalloisian technique, and P. E. L. Smith (1964*a*, *b*, 1966) has indicated the presence of at least four further Late Paleolithic industries on the Kom Ombo Plain. All of these lithic assemblages are closely circumscribed in their occurrence by the distribution of the Gebel Silsila Formation. Most of the sites have been deflated and constitute a lag deposit today. Waterworn artifacts occur in almost all exposures of these sediments in the Kom Ombo region. A number of undisturbed occupation floors, geologically *in situ*, were found, however. Gebel Silsila 2B, Area I, with radiocarbon dates of 11,120 and 11,610 B.C. (Stuiver *et al.*, 1967), is a classic example of such a site (Reed *et al.*, n.d.). It includes fragments of a rather massive human frontal bone, pertaining to an adult male of basically modern type (Reed, 1965*a*).

The age of the Gebel Silsila Formation at Kom Ombo is fairly well established through radiocarbon dating (Stuiver *et al.*, 1967; P. E. L. Smith, 1964*b* and *in litt.*, March 15, 1966; Crane and Griffin, 1965, 1966):

15,050 B.C. ± 600	(I–1297)	Khor el-Sil, loc. 3, non-Sebilian, Levalloisian-type industry
14,050 B.C. ± 300	(M–1551)	Gebel Silsila 1, Sebekian
13,360 B.C. ± 200	(Y–1376)	Gebel Silsila 2B, Area II
13,250 B.C. ± 700	(M–1642)	Gebel Silsila 1, Sebekian
12,290 B.C. ± 370	(I–1298)	Gebel Silsila 1, Sebekian
12,150 B.C. ± 450	(I–1299)	Gebel Silsila 1, Sebekian
11,900 B.C. ± 200	(Y–1806)	Gebel Silsila 2B, low channel B bed
11,660 B.C. ± 600	(M–1641)	Gebel Silsila 1, Sebekian
11,610 B.C. ± 120	(Y–1447)	Gebel Silsila 2B, Area I
11,120 B.C. ± 120	(Y–1375)	Gebel Silsila 2B, Area I
10,450 B.C. ± 400	(I–1300)	Sebil Qibli, late Sebilian level

The time range of approximately 15,000 to 10,000 B.C. appears consistent with both the geological evidence and the radiocarbon chronology established in Nubia. It is of interest that the faunas include two totally extinct species, the freshwater clam *Unio willcocksi* (see Gardner, 1932a: 51) and the extinct buffalo *Homoioceras vignardi* (see Bate, 1951: 18). The remaining mammalian species were still present in historical times (see Butzer, 1959b: 36 ff., 54 ff.).

In conclusion, the Gebel Silsila Formation marks a third, and final, Upper Pleistocene aggradation stage (to $+13$ m). Two or more major channels, presumably braided, wound their way over a floodplain about twice as wide as that of today, with one branch of the Nile flowing through the Silsila Gap. The highest contemporary deposits in southern Egyptian Nubia are at 22 meters, so that the gradient to Kom Ombo is 1.17 in 10,000 (compared with 0.94 today). Bed-load deposits are as significant as suspended-load deposits on the Kom Ombo Plain. Gravel, including exotic pebbles from the central Sudan, suggests greater stream velocities, as does the almost total absence of quiet-water gastropods. The fine silt and clay fraction as well as the dissolved load can probably be attributed to the subsaharan drainage basin of the Nile. Although direct field relationships could not be established with the local wadi formations of Wadis Kharit and Shait (see below), indirect geomorphic considerations and some limited radiocarbon determinations indicate that a major period of wadi alluviation coincided with these nilotic deposits. All in all, it would seem that a moister local climate can be correlated with this period of rather vigorous summer floods of southerly origin.

More problematical is the sporadic evidence of occasional or temporary high Nile floods of exceptional amplitude, apparently terminating the Gebel Silsila stage on the Kom Ombo Plain. At New Ballana I (see below), between 1 and 3 meters of dark silts were deposited to 112 meters (22 m above floodplain) *ca.* 12,000 years ago (9770 B.C. $\pm$ 195: Hv–1264; 10,050 B.C. $\pm$ 120: Y–1446). Similarly, surface proliferations of *Corbicula fluminalis vara* found upon the Masmas Formation at New Abu Simbel (*ca.* 101 meters) gave an age of 10,070 B.C. $\pm$ 205 (Hv–1265). The shells are presumably equivalent to shell middens with later Sebilian industries described from the foothill zone of the Burg el-Makhazin by Vignard (1955c), and certain veneers of nilotic deposits, similar to those of the Masmas Formation, can be cited from the lower courses of Wadis Kharit and Shait. Although the evidence remains inconclusive, a short-term stage at this level would help explain a variety of geomorphic and pedological phenomena. And similar, radiocarbon-dated indications have been uncovered by several expeditions working in the Wadi Halfa area (see Chap. 6).

If it proves possible to demonstrate this brief period of exceptionally high floods, its paleoclimatic implications should be of unique interest. Deposits are limited to veneers around the peripheries of the Kom Ombo Plain, and high channel beds clearly did not exist. Equally obvious is the lack of fluvial denudation or dissection of the older Gebel Silsila stage deposits. In other words, the floodplain had no opportunity to adjust to this supposed high flood level. The flood waters also appear to have had little energy to erode. This would suggest no more than periodic or even sporadic floods of exceptionally great amplitude, perhaps 20 meters or so, compared with about 9 meters today.[1] After a few centuries at most, these abnormal conditions ceased, and there is no further record of nilotic deposition at Kom Ombo prior to alluviation of the contemporary floodplain.

The Ineiba Formation (Lower Wadi Alluvium)

Whereas the Gebel Silsila Formation directly overlies the Masmas Formation along the nileward edge of the Kom Ombo Plain, another alluvial unit plays the same stratigraphic role in the embouchures of Wadis Kharit and Shait. This is the Ineiba Formation, a widespread wadi accumulation masked by the more recent deposits of the Shaturma Formation (Figs. 3–1, 3–17). It is well exposed in the beds of Wadis Kharit and El-lawi and along the southern bank of Wadi Shait. The Ineiba Formation is subdivided into two members, the Malki and Sinqari, distinguished both stratigraphically and lithologically. The older, Malki subunit is contemporary with the Gebel Silsila Formation as represented on the Kom Ombo Plain. The younger, Sinqari subunit has no nilotic equivalent at

---

1. This geomorphologic interpretation is not contradicted by the biological evidence. Jean de Heinzelin (*in litt.*, March 9, 1967) examined the *Corbicula* shells from New Ballana. They vary in length from 9 to 16 millimeters and some of the largest seem to have a growth interruption. On the basis of available information on *Corbicula* growth rates, he suggests that these are 1- and 2-year old specimens, similar to those found on the modern floodplain at Debba in Upper Nubia. Since *Corbiculae* do not appear to survive complete desiccation of their microhabitat, it would seem that floodplain colonization must take place by mature individuals. It also seems probable that these molluscs are capable of transport by low-energy stream waters (B. W. Sparks, *in litt.*, March 7, 1967). Consequently, the presence of *Corbiculae* at New Abu Simbel or New Ballana does not imply growth *in situ* and therefore does not require the availability of perennial waters.

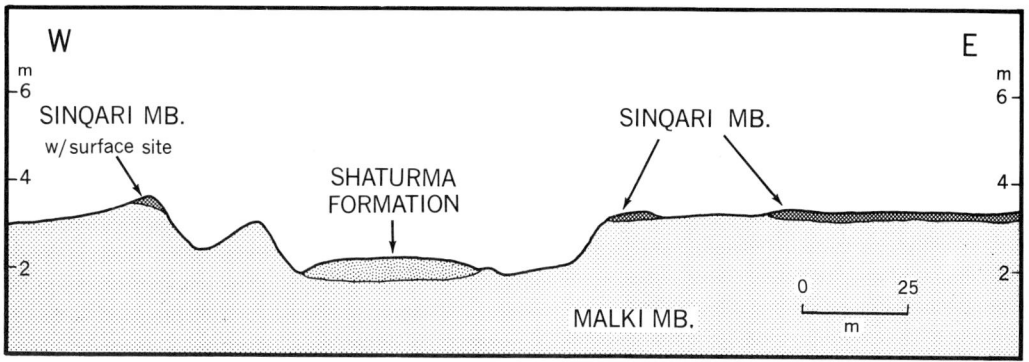

Fig. 3–17. The Ineiba and Shaturma Formations near the mouth of Wadi Ellawi. Map: UW Cartographic Lab.

Kom Ombo, although another member of the Gebel Silsilan occupies the same stratigraphic position in Nubia.

THE MALKI MEMBER

The Malki subunit can locally be observed resting disconformably on the Masmas and Korosko Formations in Wadis Kharit and Shait (Figs. 3–2 to 3–5). Wherever preserved from subsequent dissection, the surface exhibits a well-developed vertisol and, in most areas, is overlain disconformably by the Sinqari Member. The lowest elevation at which the Malki subunit was recorded is 101.9 meters at New Korosko, and there is reason to believe local base is at about 101 meters. Unfortunately, similar deposits are absent closer to the Nile and cannot be recognized in the minor wadis which abut directly against the Nile floodplain north or south of the Kom Ombo Graben.

The facies of the Malki Member shows little lateral or vertical differentiation. Beds tend to be massive and homogeneous, so that stratification is limited even though fine laminations are typical. Slickensides may disturb the bedding. Structure is coarse-to-very-coarse, angular blocky, with prismatic tendencies. Generally speaking, these beds consist of light-brown (7.5–10 YR 6/4), clayey silt (No. 15, Table 3–3; Nos. 54, 57, 62, 75, 85, 122, 127, Table 3–8). Despite their impermeability, oxidation staining and other evidence of pseudo-gley conditions are completely absent. Diffuse salts are present, as well as less common salt veins, soft microconcretions (0.5 to 4.0 mm), or partly calcareous concretions up to 5 millimeters in diameter. In fact, the deposits of El Nuqu are saturated with salts to such a degree as to suggest a sebkha environment. Carbonates are high, varying from 7.5% to 30%, while pH values are within a range 7.2–7.9. Fine rootlet zones or impressions, in part with carbonized organic matter, are quite common. They suggest fine herbaceous vegetation rather than sporadic xerophytes. Land snails are conspicuous by their total absence, although aquatic snails were found within basal strata in two instances, certainly derived from the underlying Korosko and Masmas Formations.

The Malki beds consist of suspended sediments deposited on alluvial flats by the exotic waters of Wadis Kharit and Shait. They now form 4- to 6-meter wadi terraces. Total thickness near the mouth of Wadi Shait exceeds 9 meters (101.9 to *ca.* 111 m), near the mouth of Wadi Kharit, 4 meters (107 to 111 m). The average longitudinal gradient of these deposits in Wadi Kharit is 0.2% at the eastern end of the Kom Ombo Graben, 0.4% in Wadi Ellawi. Inclination of individual beds is obviously greater, although topset beds are the rule. Syngenetic or subsequent precipitation of salts and carbonates suggests widespread, slightly saline flats, possibly in the backwater zones of the great wadis from the Red Sea Hills. Clay mineral X-ray diffractograms (Appendix D) are dominated by montmorillonite and kaolinite. First- and second-order kaolinite peaks are noticeably more intense than those of the nilotic deposits, suggesting relatively greater importance in the Malki subunit.

By exception, a basal gravelly bed exceeding 60 centimeters in thickness can sometimes be recognized under the typical Malki facies at New Korosko and New Sebua. The coarse clastic components are quite identical to the locally derived wash of the Korosko Formation at the same sites (Nos. 60, 68, Table 3–8; see also Table 3–1). Clay mineral diffractograms, however, are identical to those of the typical Malki facies (Appendix D). These gravels suggest accelerated local wadi activity, in part contemporary with floods of Wadi Shait (sandy silt to clayey silt matrices), in part out of phase (silty sand matrices). After this initial phase of local wadi activity, accelerated discharge and alluviation were limited to the floodplains of the great exotic wadis. A modest but protracted pluvial phase can be suggested for the highlands.

Mammalian bones and mollusca appear to be absent from the Malki Member and archeological vestiges are limited to a few scattered, patinated, and slightly waterworn artifacts such as the two small chert flakes of Late Paleolithic type found at New Korosko (Pit 16). A radiocarbon determination of 15,450 B.C. ± 300 (I–2179) was obtained from a clayey marl at New Sebua representing the lower part of the stratigraphic column (see Appendix E). Thus, this beginning of wadi alluviation must have been synchronous with the earliest beds of the Gebel Silsila Formation. The terminus to the Malki beds is given by a conspicuous vertisol. Polygonal crack networks at New Korosko and New Sebua extend at least 140 centimeters into these deposits, with cracks attaining a width of 4 centimeters. The fillings consist of light-yellowish-brown (10 YR 6/4), silty sands of the overlying Sinqari or Shaturma units. The slickensides present may have similar implications, although pseudo-gley phenomena are absent. The bulk of these vertisol phenomena are clearly epigenetic and suggest a Mazaquert paleosol, prone to periodic saturation with salt-charged waters to considerable depth. They are presumably related to sporadic or periodic high Nile floods, spilling back into the lower wadis to 112 meters elevation. Consequently, the C^{14} date of 10,070 B.C. ± 205 (Hv–1265) obtained from surface *Corbiculae* at New Abu Simbel probably provides a terminal date for the Malki Member.

THE SINQARI MEMBER

The surface of the Malki beds was both stripped and dissected to a variable extent prior to deposition of the Sinqari Member (Figs. 3–2, 3–4). In Wadi Shait, a broad channel was cut through the Malki beds along the length of the lower wadi, and dissection at New Korosko reached 102.3 meters, suggesting that the local floor of Wadi Shait was excavated to below 99 meters. At least 12 meters of fill cutting separate the two members at this locality. On the other hand, there was little dissection in Wadi Kharit, and no more than 4 or 5 meters of fill cutting are evident at New Arminna.

Resting disconformably on the Malki beds as well as on the older formations, the Sinqari Member consists of well-stratified, often laminated or current-bedded, unconsolidated alluvium, ranging from light brown to light yellowish brown in color (7.5–10 YR 6/4). Structure is coarse, subangular blocky. Moderately sorted, the texture varies from silty medium sands to coarse sands (Nos. 16, 71, 300, Table 3–3; Nos. 22, 29–II, 58, 78, 398, Table 3–8). These sands may be interbedded with lenses of coarse sandstone or ironstone gravel, or with thin beds (0.3 to 5.0 cm) of medium-sandy silt or clay silt, dominated by kaolinitic clays (Appendix D). The gravel (Table 3–1) averages in the subrounded category and is rather flattish in shape, suggesting a dominance of sliding motions during transport. The comparatively high index of rounding reflects a component of derived, Pleistocene pebbles. Salts are present in small quantities, occasionally in soft aggregates less than 0.5 millimeters in diameter. In the Sinqari Member and throughout all of the Shaturma beds, carbonates lie under 10%, with the highest values in fine-textured materials. pH values generally range from 7.4 to 7.7.

Coarse, vertical or horizontal root drip and root impressions, often a meter or more in length, are relatively frequent (Fig. 3–18) in coarse-textured beds, while fine vertical root casts and minute rootlet zones with carbonized vegetable matter may be evident in finer-grained deposits. Occasional spheroidal hollows, about 6 to 8 centimeters in diameter, may be casts of the gourd-like fruits of *Colocynthis vulgaris*. The land snail *Zootecus insularis* is commonly found in small numbers or local proliferations *in situ*, normally in association with *Pupoides coenopictus* (see Appendix G). Since land snails are absent from the Malki beds, such shells are contemporary with the Sinqari Member.

In the New Korosko-New Sebua area, the Sinqari beds have been recorded in an elevation range of 102 to 110 meters. Because they were graded onto a higher Shait floodplain, a total thickness of at least 8 meters, and probably over 11 meters, is suggested. No local archeological, mammalian, or radiocarbon materials are available for dating purposes. These beds are evidently younger than 10,000 B.C. and the phase of very

Fig. 3–18. The rubefied sandy facies of the Ineiba Formation exposed in the bed of Wadi Kharit 12 kilometers upstream of New Arminna. Root drip is well developed, and *Colocynthis* casts are evident.

high Nile floods inferred for that time, however. On the other hand, stratigraphically equivalent beds in Nubia (Wadi Or, Chap. 6) have a radiocarbon date of 6940 B.C. ± 160 (Y–1377, Stuiver *et al.*, 1967). This suggests a terminal Pleistocene to early Holocene age. Unlike the Malki beds, the Sinqari subunit records local as well as exotic wadi activity, with some evidence of local sheet wash. Together with the snail fauna and root drip at the type area, this suggests a period of winter rains and fairly abundant wadi vegetation in the Kom Ombo area. In view of these suitable ecological conditions, it is curious that there is no evidence whatever for contemporary settlement in the alluvial area.

At a later date, dissection removed a good part of these sediments from the wadi embouchures. Subsequently, a zone of at least 50 centimeters, and possibly as much as 100 centimeters, in depth was rubefied in the process of soil development. This rubefaction is recorded by distinct color shifts from 10 YR 6/4 to 5 or 7.5 YR 6/4, and there is little question that hues redder than 10 YR are a result of pedogenesis. Similarly, the conspicuous kaolinite peak of the X-ray diffractograms distinguishes these soil horizons from the parent material (Appendix D). No other morphological changes can be cited from such soils, which are similar in type, although not in intensity, to the red paleosols discussed in Chapter 2. Nonetheless, greater chemical weathering is a prerequisite, suggesting a time of frequent, gentle rains, with a high percentage of infiltration and limited surface runoff. This reddish desert soil appears to date from the mid-Holocene. Although an exact age cannot be determined, it has considerable stratigraphic significance, serving as a *terminus post quem* for later wadi alluvia on the Kom Ombo Plain and in Nubia.

The Shaturma Formation (*Upper Wadi Alluvium*)

The youngest surficial deposits of Wadis Kharit and Shait, consisting of screes, sheetflood deposits, and broad spreads of subrecent alluvium, are here defined as the Shaturma Formation. Two members are recorded, designated simply as I and II. Beds are generally rather shallow, suggesting either that sedimentation was very slow or that only brief periods of net aggradation are involved.

MEMBER I

Dissection of the Sinqari beds to a base elevation of 104 meters (or lower) preceded alluviation of the earliest Shaturma beds in the lower course of Wadi Shait. This indicates some 6 meters of fill cutting. Again, there is little evidence in Wadi Kharit of dissection, which amounts to only 1 or 1.5 meters at the mouth of Wadi Ellawi and a scant 50 centimeters upstream of New Arminna.

Local thicknesses of the earlier Shaturma beds vary between 1.5 and 3 meters, and there is no direct evidence for general gradation to a common base level. Member I is a well-stratified, unconsolidated, light-brown (7.5–10 YR 5.5/4) alluvium which, in contrast to the Sinqari beds, lacks evidence of current bedding. Structure is coarse, subangular-to-angular blocky. Inhomogeneous and poorly sorted, the texture varies from a silty sand to sandy silt, with local bands of medium gravel (No. 79, Table 3–3; 128, Table 3–8). Salts are more common than in the Sinqari beds, with soft aggregates to 0.5 millimeters in diameter and rare lime-salt microconcretions (2 mm). Organic vestiges are limited to some rootlet impressions. True root drip and land snails are completely absent, suggesting considerably less wadi-floor vegetation.

A coarse, detrital alluvium corresponding to Member I is widespread in most of the wadis of the Kom Ombo Plain, averaging 30 to 200 centimeters in thickness. It includes a considerable admixture of subangular to subrounded gravel (mainly ironstone and sandstone) except in the vicinity of Pleistocene gravels, where igneous components are abundant. This wash rests on top of the Sinqari beds along the channels of the major wadis, and grades upslope into a coarse colluvium, generally 30 to 40 centimeters thick. Land snails are absent. Although no archeological vestiges were observed *in situ,* milling stones and chert flakes and blades, as well as potsherds of black-exterior ware, lie on the surface. Some of this material may have been deflated from within the colluvium. But at least one undisturbed flaking site was observed on top of Member I wash in Wadi Kharit, 1.5 kilometers east of New Tushka. In addition to handstones, milling stones, and a few blades, a big dolerite nucleus with its detached, primary waste flakes in close functional association was collected here by Martin Baumhoff. At the mouth of Wadi Ellawi (Fig. 3–16), a similar site—with a dolerite core, several coarse waste flakes, a denticulated blade (11 cm long) and a broken handstone—rests on top of Shaturma beds. It seems that late prehistoric occupation in Wadi Kharit clearly postdates Member I, although such occupation may be related to the terminal part of the moister, aggradation phase.

In general, Member I comprises wadi wash and slope colluvia derived from Pleistocene terraces or from the lag on the sandstone cliffs. In addition to accelerated activity by both the local and the Red Sea Hill wadis, these deposits imply rather effective sheetflooding. The stripping of surface lag as well as of fresh, unpatinated, and unfractured gravel from the body of many Pleistocene terraces can only be explained by heavy, torrential rains and violent surface runoff. Although direct isotopic dates are lacking (see Chap. 4), Member I appears to record the mid-Holocene "subpluvial" recorded by various lines of evidence in Upper and Middle Egypt from the late prehistoric and protohistoric period (see Butzer, 1959b: 48 ff., 54 ff.).

MEMBER II (SUBRECENT WASH)

At both New Korosko and New Sebua, Member I was almost entirely eroded and the remnants of the Sinqari beds dissected, again prior to renewed aggradation. Local fill cutting averaged about 1 to 1.5 meters and subsequent alluviation was limited to about 1.5 to 2.5 meters. The deposits are well-stratified, loose, coarse sands with dispersed medium gravel (No. 6, Table 3–3). Thin lenses of silty sand or sandy silt occur locally. Colors vary from very pale brown (10 YR 7/3–4) to light yellowish brown (10 YR 6/4). Salt aggregates are absent, as are all traces of organic life. Dissected to a depth of 1.5 or 2 meters today, the modern channel bed contains a variety of derived archeological materials, the youngest of which are Roman and Islamic potsherds.

Correlatives of Member II in Wadi Kharit and elsewhere have been mentioned in Chapter 2, during discussion of modern wadi activity. Local rainstorms were more frequent than at present, although the local climate remained hyperarid. It is particularly unfortunate that archeological dating is unavailable for the Shaturma Formation. Potentially, these two comparatively recent alluvial cycles could provide valuable information on climatic fluctuations during the last few millennia. Member II was almost certainly contemporary with the early Islamic alluviation recorded in Middle Egypt (see Butzer, 1959b: 67 ff.). And in the floor of Wadi Qena there is a massive, suballuvial clayey marl sequence with several hearth zones, one of which yielded a radiocarbon date of 1090 A.D. ± 115 (I–2561). This section, which is pertinent to an understanding of historical changes of climate in Egypt, will be described elsewhere (Virginia Burton and K. W. Butzer, in preparation).

Description of the Type Sections

Selection of type sites and sections for rock-stratigraphic units in the Nile Valley is hampered by a scarcity of cultural toponyms for potential sites.

To overcome this difficulty in the Kom Ombo area, the names of resettlement villages in Nubia have been chosen for those four units which are present at both places in an identical stratigraphic context. Consequently, where suitable type sections are available, the terms Korosko, Masmas, Ineiba, and Shaturma can be applied to the same formations on the Kom Ombo Plain and in Egyptian Nubia. By exception, the Gebel Silsila Formation is named after the railroad station north of Kom Ombo. In this particular case, the lithological counterparts in Nubia cover a greater time range and include subunits absent north of Aswan.

The type section for the Korosko Formation is provided by Pit 21 at New Korosko (Figs. 3–2, 3–3), surface elevation 105.3 meters. From base to top, the following beds are exposed (see Table 3–3):

a) Over 55 cm, base not exposed. Crudely stratified, coarse, subangular gravel (Table 3–1) with a matrix of light-yellowish-brown, silty coarse sand. The pebbles, of heterogeneous rounding and lithology, are mainly dolerite and quartz. Although the base is not seen, Pit 20 nearby suggests a total thickness of over 1.5 m, necessarily embanked against older Pleistocene gravels laterally and at depth.

Table 3–3. Sediment characteristics at the type sites of the Korosko, Masmas, Ineiba, and Shaturma Formations

Sample number	Bed	Color	Texture[a] (matrix)	Sorting	$CaCO_3$ (%)	pH
			New Shaturma			
6	f	(7.5–10 YR 6/4)	Coarse sand	Moderate	6.3	8.05
79	e	(10 YR 6/4)	Medium-sandy silt	Poor	8.9	7.35
16	d	(10 YR 6/4)	Silty medium sand	Poor	6.6	7.35
15	c	(7.5–10 YR 6/4)	Clayey silt	Moderate	15.1	7.30
14	b	(2.5 Y 6/2)	Marly sandy silt	Poor	16.6	7.75
13	a	(5 Y 7/2)	Medium-sandy silt	Poor	11.5	7.95
			New Ineiba/New Masmas			
301	g	(7.5 YR 7/4)	Silty coarse sand	Poor	7.3	7.7
300	f	(5–7.5 YR 6/4)	Silty medium sand	Poor	1.4	7.5
299	e	(2.5 Y 6/2 to 10 YR 5–6/2)	Clayey silt	Poor	18.4	7.1
298	d	(10 YR 3–4/2)	Medium-sandy silt	Poor	9.8	7.0
322	c	(10 YR 7/2–3)	Marly silty sand	Poor	16.7	6.65
397	b	(5–7.5 YR 6/6)	(Coarse-sandy silt)	—	—	—
323	a	(7.5 YR 6/5)	(Coarse-sandy silt)	—	—	—
			New Korosko			
71	f	(10 YR 6/4)	Clayey silt	Moderate	10.2	7.7
72	e	(2.5 Y 5.5/2)	Marly sandy silt	Poor	23.9	7.5
70	d	(2.5–5 Y 7/2)	Medium-sandy marl	Moderate	31.4	7.6
69	c	(2.5 Y 7/2)	Marly sandy silt	Poor	20.7	7.6
74	b	(2.5 Y 7/2)	Marly sandy silt	Poor	23.3	7.5
76	a	(10 YR 6/4, 7/3)	Silty coarse sand	Poor	11.4	7.5

[a] Textures in parentheses not determined by quantitative method.

b) 25 cm. Light-gray, marly sandy silt with dispersed subangular pebbles and coarse, columnar-to-blocky structure. Rare oxidation stains; soft, salt aggregates (to 0.5 mm) and lime-salt concretions (to 1.5 cm). Soluble salts, 295 parts per million. Very fine rootlet impressions (to 0.5 mm diameter). A waterworn, coarse, proto-Levalloisian core (identified by Leslie G. Freeman) was found in this bed.

c) 15 cm. Light-gray, marly sandy silt with coarse, angular blocky structure. Soft salt aggregates (to 2 mm) and lime-salt concretions (to 1 cm). Rootlet impressions up to 4 mm diameter, with carbonized vegetable matter. Rare *Planorbis*, *Valvata*, and *Bulinus* shells (see Appendix I).

d) 20 cm. Light-gray, medium sandy marl with coarse, angular blocky structure. Soft salt aggregates (to 3 mm) and very fine rootlet structures. Soluble salts, 275 parts per million. Rare *Planorbis*, *Valvata*, *Bulinus*, and *Corbicula* shells. Disconformity.

e) 10 cm. Light-brownish-gray, marly sandy silt with medium, angular blocky structure. Lime-salt concretions (2 mm) and innumerable fine rootlet structures with traces of carbonized vegetable matter. Highly abundant *Planorbis* and *Valvata*, occasional *Bulinus*, and rare *Corbicula*. Disconformity.

f) 40 cm. Light-yellowish-brown, alternating bands of laminated clayey silt and well-stratified, medium sandy silt, with coarse, angular blocky structure. Soft salt aggregates (less than 0.5 mm). Soluble salts, 750 parts per million. Under a lag of dolerite cobbles.

Beds *a* through *d* record the Korosko Formation, with *a* representing the sandy gravel subfacies and *b* the gravelly marl subfacies. Bed *e* represents the Masmas Formation, *f*, the Shaturma Formation.

The type section for both the Masmas and the Ineiba Formations[2] must be chosen from two exposures (Fig. 2–18, 3–19), conveniently located between the coalescent villages of New Masmas and New Ineiba. Units *a* through *c* are described as recorded in two adjacent pits formerly located beside the asphalt road; units *d* through *g*, as recorded in the bed of Wadi Kharit, 300 meters due south. Units *d* through *f* are common to both exposures. From base to top (see Table 3–3):

a) Over 70 cm, base not exposed. Crudely stratified, fine quartz gravel, with a matrix of reddish-yellow, coarse-sandy silt. Occasional pockets of light-

---

2. The best type site for both the Malki and Sinqari Members of the Ineiba Formation is that of Pit BH (see below) in the composite resettlement village of New Shaturma, Malki, and Sinqari. Due to the scarcity of good type sections, it was found necessary to employ the name Shaturma for a younger formation, selecting a less suitable type site for the total Ineiba Formation at New Masmas–New Ineiba. This last section lacks clear evidence of the Malki Member, although it may record a terminal nilotic equivalent. For the moment, the Malki and Sinqari units are still regarded as members of the Ineiba Formation, although various considerations may ultimately favor giving formational status to each.

Fig. 3–19. Part of type section of Masmas and Ineiba Formations at New Masmas.

gray (5 Y 7/2) color are marly in appearance. Limonite precipitates occur on many granules.

b) 25 cm. Crudely stratified, subangular, coarse gravel, with a matrix of reddish-yellow, coarse-sandy silt.

c) 10 cm. Light-gray, stratified, marly silty sand, with coarse, subangular blocky structure. Disconformity.

d) Over 130 cm, attaining over 3 m in other exposures. Dark grayish-brown, medium-sandy silt with some laminae of coarse sand and of clayey silt, suggesting mixed wadi and nilotic sedimentation. Coarse, subangular-to-angular blocky structure, with prismatic tendencies. Local limonite and pyrolusite stains suggest seasonal waterlogging. Lime-salt concretions (2–7 mm) common. Soluble salts, 2050 parts per million. Wadi admixture more prominent in upper 60 cm, and the topmost 20 to 30 cm comprise derived silt aggregates set in current-bedded wash. Rare *Planorbis, Valvata,* and *Bulinus* shells (see Appendix G), together with terrestrial snails in one pocket of mixed wadi facies. Minor disconformity.

e) 5 to 20 cm. Light brownish-gray to grayish-brown, clayey silt with gypsum and much organic matter. Medium, subangular blocky to platy structure. Section cuts through several semicontinuous saucer-shaped pan sediments, averaging 1.5–3.0 m in width near base, less than 1.0 m near top of bed *e*. The pan materials are marly in facies, while 1–2 mm laminae of light-brown

(7.5 YR 6/4), sandy silts indicate intermittent wadi flow. Rootlet zones with sand fillings and silt casts, up to 1 cm diameter, are frequent, and a rich aquatic snail fauna (*Planorbis, Valvata, Bulinus*) is present. Primarily a backswamp deposit, bed *e* records evaporation pans as well as wadi wash. Soluble salts, 1850 parts per million. Subsequently, a vertisol developed, with polygonal crack networks penetrating well into bed *d*, filled with brown (10 YR 5/3), medium-sandy silt. Disconformity.

f) 10 to 40 cm. Light-brown, silty medium sand with lenses of silty coarse sand or silt. Well-stratified and current-bedded, with coarse, subangular blocky structure. Lower 15 cm contain abundant derived aggregates of beds *d* and *e*, up to 1 cm diameter, as well as a few planorbids. Abundant terrestrial snails and derived aquatic gastropods (Appendix G.) This bed has been superficially rubefied. Disconformity.

g) 2 cm. Stratified, medium gravel, consisting mainly of subrounded calcareous concretions and sandstone, in a matrix of loose, pink, silty coarse sand. The derived concretions are stained by pyrolusite and are rich in colloidal silica. They are probably derived from a lag developed on bed *d*. The surface bed *g* is obviously deflated and is littered with archeological materials: diminutive chert and agate flakes and occasional bladelets, rare dolerite flakes, several handstones of quartzitic sandstone or coarse ironstone as well as a few sherds of black-topped pottery. Although disturbed, the concentration of artifacts suggests a site of Predynastic or Early Dynastic age (see Chap. 4).

Beds *d* and *e* record the Masmas Formation overlying the Korosko Formation (bed *c*), while *a* and *b* may represent the coarser subfacies of the Korosko. Unit *e* probably marks a terminal facies of the Masmas Formation, although it may also pertain to the period of high floods *ca.* 12,000 years ago. Bed *f* is a typical channel facies of the Sinqari Member, Ineiba Formation, while *g* can be traced discontinuously upslope into 30 or 40 centimeters of coarse colluvial wash typical of Member I of the Shaturma Formation.

The type section of the Gebel Silsila Formation—at Gebel Silsila 2B, Area I—is located about 2 kilometers north of the railroad station of the same name, or 650 meters east of Km 819.5 of the Cairo-Aswan railroad (Figs. 3–13, 3–23, 3–26, and 3–27). The relevant sections will be described in further detail below.

Finally, the type section of the Shaturma Formation, as well as of the Ineiba Formation, was originally exposed in the Pit BH under the village of New Shaturma (Fig. 3–4), surface elevation 109.0 meters. From base to top (see Table 3–3):

a) Over 40 cm, base not seen. Light-gray, medium-sandy silt, with dispersed pebbles. Medium, subangular blocky structure. Considerable waterlogging suggested by lenses and flecks of yellow (10 YR 7/6) limonite.

b) 45 to 55 cm. Light brownish-gray, marly sandy silt with dispersed pebbles and laminae of pink marl. Coarse, subangular blocky structure. Carbonized vegetable matter in fine rootlet zones. *Planorbis, Valvata,* and *Bulinus* present (see Appendix G). Disconformity.

c) 130 to 175 cm. Light-brown, clayey silt with abundant dispersed salt, forming evaporite films on ped faces. Soluble salts, 2950 parts per million. Coarse, angular blocky structure with slickensides. Fine rootlet zones. Disconformity.

d) 80 to 160 cm. Light yellowish-brown, silty medium sand. Well-stratified, with coarse, subangular blocky structure. Disconformity.

e) 45 cm. Light yellowish-brown, medium-sandy silt. Stratified and laminated, with medium, angular blocky structure. Lime-salt evaporites in occasional rootlet hollows. Soluble salts, 1500 parts per million. Disconformity.

f) 5 to 15 cm. Light yellowish-brown, coarse sand. Stratified and loose.

Beds *a* and *b* record the Korosko and Masmas Formations, while *c* and *d* respectively characterize the Malki and Sinqari Members of the Ineiba Formation. Beds *d, e,* and *f* represent Members I and II of the Shaturma Formation.

Deposits and Stratigraphy South of Darau

A series of informative exposures at the southern exit of the Kom Ombo Graben was studied in the wadi just south of the hamlet El-Shurafa ("Wadi Shurafa"), and along the railway between Wadi Shurafa and Darau. Both the Masmas and Gebel Silsila Formations are well developed in this area.

Six deep pits in Wadi Shurafa (located about 1.6 kilometers east of the railroad) plus a natural section of 5 to 7 meters in height (exposed along the northern edge of the wadi) were used to reconstruct a sequence of events during the middle phase of the Masmas Formation (Fig. 3–9). This middle phase occupies an elevation range of *ca.* 91 to 104 meters. Two facies alternate continually: (*1*) brown, medium-sandy silt to clayey silt, with some dispersed, subangular ironstone detritus; poorly stratified, locally with slickensides; and (*2*) pale-brown, silty medium sand or sandy silt, well-stratified or current-bedded, locally with dispersed detritus. Both sediments contain lime-salt concretions, root drip, or hollow rootlet zones. But the massive-bedded, amorphous, compact silts (Fig. 3–10) are strictly of nilotic origin, whereas the thin-bedded, well-stratified sands are wadi deposits including much derived silt. From top to bottom, the local sedimentary suite is as follows:

g) 4.2 m, clayey silt to silty medium sand (10 YR 5/2–3);
f) 0.5 m, silty sand (10 YR 6/3);
e) 3.5 m, clayey silt (10 YR 5/2);
d) 1.1 m, silty sand (10 YR 6/3);
c) 3.6 m, medium-sandy silt (10 YR 5/3);
b) 0.7 m, silty sand (10 YR 6/3);
a) 1.0 m, medium-sandy silt (10 YR 5/3).

Locally, the various sediments rest against the Nubian Sandstone or the eroded remnants of white (5 Y 8/1, 10 YR 8/2), marly silty sands of the Korosko Formation (Fig. 3–9). Beds of the Gebel Silsila Formation are disconformably embanked against Masmas silts at the mouth of Wadi Shurafa.

The same impression of periodically renewed wadi activity of moderate proportions is conveyed by a part of the upper phase of the Masmas Formation (ca. 104 to 110 m), exposed in the minor tributary wadi followed by the road up to New Ballana I. At least 4 meters of bed g (No. 329, Table 3–8) are exposed at the base of the steep wadi (Fig. 3–6). Then, after a break, deposits are again preserved in the middle of the wadi (Fig. 3–20). Overlying 1.8 meters of coarse-sandy gravel are 2 meters of interdigited grayish-brown (10 YR 5/2), clayey silt (No. 325, Table 3–8) and pink (7.5 YR 7/4) sands with medium-grade, subangular

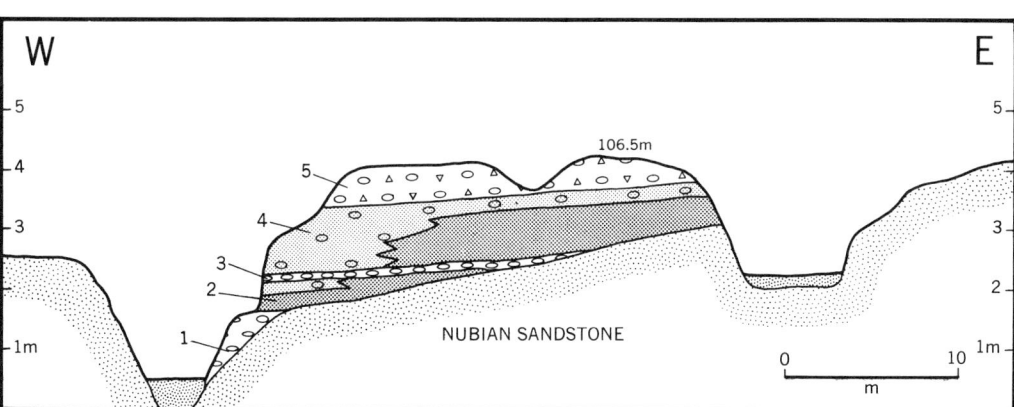

Fig. 3–20. Intercalated gravelly sand and clayey silt of the Masmas Formation near New Ballana I. 1, 3: Late Pleistocene rolled gravels; 2, 4: Clayey silts of Masmas Formation intercalated with wadi sand and gravel; 5: Holocene wadi detritus. Map: UW Cartographic Lab.

sandstone pebbles. These in turn are disconformably overlain by later sandstone detritus.

Near the head of this tributary wadi, the following section is exposed, from top to bottom:

z) 75 cm. Very-pale-brown (10 YR 7/4), coarse sand with lenses of medium-to-coarse, subangular sandstone gravel; well-stratified, single-grain, loose. The 1% dip of the beds increases sharply to 6.5% at the edge of the exposure. Identical beds sporadically mantle nilotic deposits further downvalley.

y) 40 cm. Brown (10 YR 5/3), silty medium sand with dispersed medium detritus; well-stratified, single-grain, loose. Abundant mica and some *Corbicula* shells are derived from the underlying unit.

x) 300 cm. Dark grayish-brown (10 YR 3-4/2), silty clay (No. 328, Table 3-8); angular, coarse blocky structure with slickensides. Limonite and pyrolusite staining widespread. Frequent salt precipitates (1 to 4 mm), and 290 parts per million soluble salts. This highest exposure of nilotic silts attains 112 m. A proliferation of *Corbicula fluminalis vara*, weathering out of the top meter or so of this sediment, gave radiocarbon determinations of 9770 B.C. ± 195 (Hv-1264) and 10,050 B.C. ± 120 (Y-1446, Stuiver et al., 1968). These dates preclude correlation of the silts in question with the Masmas Formation, as they are several millennia younger than the spectrum of C^{14} dates obtained from the younger Gebel Silsila Formation at lower elevations on the Kom Ombo Plain. Presumably, we are dealing with local surficial silts of later age. The clay-mineral spectrum (Appendix D) suggests close affinities with the Gebel Silsila Formation. The exact stratigraphic relationships of the shell-bearing upper part of bed x to the older deposits downhill cannot be reconstructed in the field.

w) 30 cm. Brown (10 YR 5/3-4), silty medium sand; well-stratified, with coarse, subangular blocky structure.

Bed w conforms with the derived wash recorded from the Wadi Shurafa section or with the interdigited alluvia 100 meters further downslope. The essentially uniform character of the middle and upper phases of the Masmas Formation seems evident from these sections. Some six phases of slightly accelerated wadi activity, responsible for sandy beds averaging 0.5 to 1.0 meter in thickness, interrupt a total profile of about 21 meters, but the bulk of the nilotic deposits are remarkably free from wadi intrusions. Later Holocene alluviation—some 4 meters of well-stratified, coarse sands and subrounded, medium gravel (Member II of the Shaturma Formation) (Fig. 3-10)—is much more impressive. This suggests that during deposition of the middle and upper parts of the Masmas Formation local climate was at least as dry as that of today.

In one pit of the Wadi Shurafa section (Fig. 3-9), younger deposits of uncertain affinity rest disconformably on the Masmas Formation and under Member II of the Shaturma Formation. Up to 2 meters of foreset-bedded, light-gray (10 YR 7/2), clayey silt is overlain by 50 centimeters of poorly stratified, marly sandy silt of the same color. With a maximum elevation of 101 meters, these curious silts may be a lateral facies variant of the Gebel Silsila Formation.

The Gebel Silsila Formation in the area south of Darau is mainly represented by ridges of sandy gravels. Sandford and Arkell (1933: 4, 49–50) discuss these "shingle banks" and consider them to be a lag of nilotic deposits, reworked at a later substage by the Nile as a "lateral beach" on the convex side of a wide, wind-swept river bend. Closer examination of the ridge between the tombs of Sidi Hammuda and Sidi Mahmud, just south of the Shurafa embouchure, showed deposits of the Gebel Silsila unit embanked disconformably against the Masmas Formation. The younger beds consist of very-pale-brown (10 YR 8/3), well-stratified, steeply inclined backsets of coarse sands with dispersed pebbles and rich in mica and hornblende. They include lenses of pale-brown (10 YR 6/3), medium-sandy silt. Abundant, waterworn Late Paleolithic artifacts litter the surface. These are typical channel-margin backsets, marking the right bank of the Nile (96 m), although without affinity to "lateral beaches."

A little further north—to the east of the railroad, between Km 846 and 848—are other sinuous ridges with a local relief of 7 meters and a maximum elevation of 102 meters. They run south to north, downstream of a large sandstone residual (Fig. 3-1), maintaining an asymmetrical shape, dipping as much as 20% to the east and 7% to the west (Fig. 3-21). The dominant facies is a medium-grade gravel with a matrix of coarse quartz sand, abundant *Cleopatra* shells, and waterworn early Late Paleolithic artifacts. These ridges can be interpreted as a complex gravel bar aggraded within the main Nile channel in the lee of the bedrock outcrop. Shearing effects on stream velocity produced several complexes of recurving, inclined beds, with shallow backsets dipping to the SW, steeper foresets to the NE. Significantly, however, these sandy gravels are laterally interdigited with horizontal beds of brown (10 YR 5/3), clayey silt. Other, finer-grained lateral beds have been deflated or eroded by surface washing. In other words, these are true channel deposits, exceptionally coarse and complex by virtue of their position in a shallow, uneven stream bed.

Deposits and Stratigraphy of the Gebel Silsila Area

The Masmas Formation in the northern reaches of the Kom Ombo Plain forms a surface planed off at 98 to 100 meters (Fig. 3–13). The strata exposed to the east of the occupation site Gebel Silsila 2B, Area I, are typical of these horizontal floodplain sediments (Table 3–4, Figs. 3–26, 3–27), and these sections complement the information provided by the abandoned stream beds near Gebel Silsila Station (Table 3–2, Fig. 3–7). The clayey silt is typical of alluvial flats or backswamps; the sandy beds show turbulent lamination, suggesting rapid surges of water onto the alluvial flats in the wake of crevasses.

More spectacular is the development of the Gebel Silsila Formation in this area. For much of the late Upper Pleistocene, a Nile channel bifurcated from the main river near Fatira, swinging eastwards and then northwards through the Silsila Gap (Figs. 2–14, 3–22). At least two distinct stages can be recognized, one at 100 to 101.5 meters ("Channel A"), the other at 98.5 to 99 meters ("Channel B") (Fig. 3–23).

Channel A (+13 m) was approximately 150 to 200 meters wide. Homogeneous, very-pale-brown (10 YR 7/3), coarse sands with rare quartz pebbles (6–12 mm) mark the actual channel beds, with backset-bedded levee deposits interdigited laterally (Fig. 3–24). The levee facies

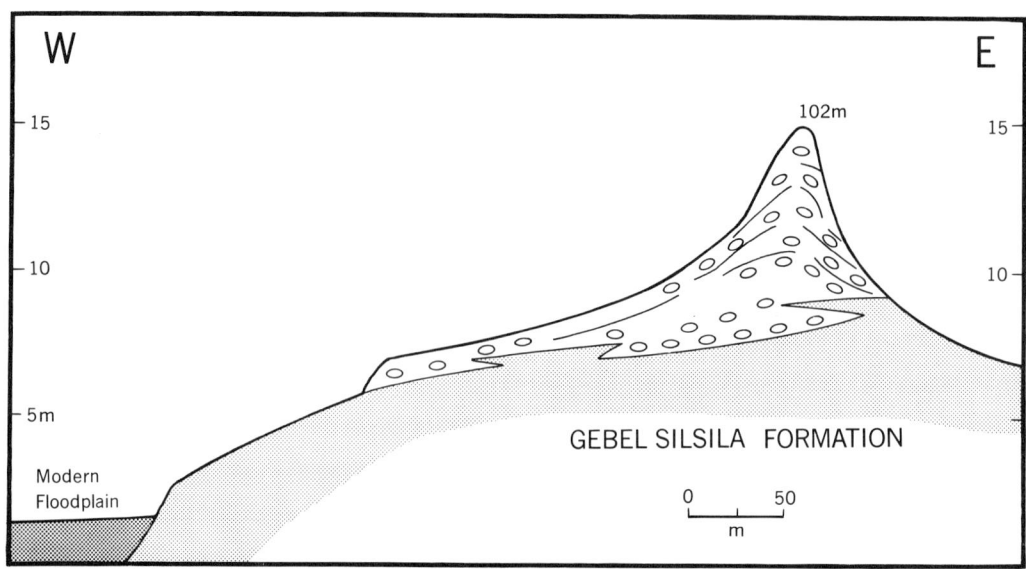

Fig. 3–21. Channel gravel bar of the Gebel Silsila Formation south of Darau at Nag el-Darira. Map: UW Cartographic Lab.

Table 3-4. Sediment characteristics of the Masmas Formation at Gebel Silsila 2B

Sample number	Maximum thickness (meters)	Bed	Color	Texture[a]	Sorting	CaCO₃ (%)	pH
94	0.55	z	(10 YR 5/2)	Clayey silt	Moderate	4.6	7.6
92	1.20	y	(10 YR 6/3)	(Medium-sandy silt)	—	—	—
95	0.70	x	(10 YR 5/3)	Clayey silt	Moderate	3.2	8.1
93	0.30	w	(10 YR 4/3)	(Medium-sandy silt)	—	—	—
101	0.50	v	(10 YR 4-5/2)	Clayey silt	Moderate	5.0-6.5	7.3-7.6
102, 229	0.40	u	(10 YR 4.5-5.5/3)	(Silt to clayey silt)	—	—	—
98	0.10	t	(10 YR 6/3)	Medium-sandy silt	Poor	8.5	7.8
97	0.90	s	(10 YR 5/3)	Clayey silt	Moderate	7.4	7.6
96	over 0.70	r	(10 YR 4/3)	Clayey silt	Moderate	4.7	7.6

[a] Textures in parentheses not determined by quantitative methods.

alternates rapidly between pale-brown (10 YR 6/3), medium-sandy silt (No. 144, Table 3-8) and silty medium or coarse sand. Backsets may dip as steeply as 55%. The finer deposits are studded with vertical or horizontal root drip, in part preserving calcified root tissue or wood up to 12 millimeters wide. Carbonized vegetable matter can also be seen in hollow rootlet zones. Part of the root drip forms a spongy mat up to 10 centimeters thick, recalling a grass sod. Whereas almost no bone or shell was found *in situ* in these levee deposits, a fair amount of fragmentary bone and large quantities of artifacts are found in the surface lag. *Unio* and *Cleopatra* shells abound in the channel beds, and a radiocarbon determination of 11,900 B.C. ± 200 (Y-1806) was obtained from *Etheria elliptica*

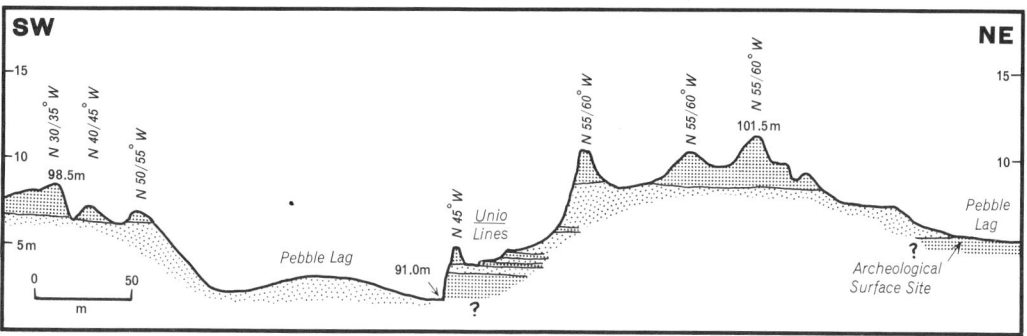

Fig. 3-22. Silt and sand strata of the Gebel Silsila Formation exposed in the Silsila Gap. All relief features are due to deflation; strike of major yardangs is indicated in degrees. Section approximately 100 meters south of Fig. 2-14. Map: UW Cartographic Lab.

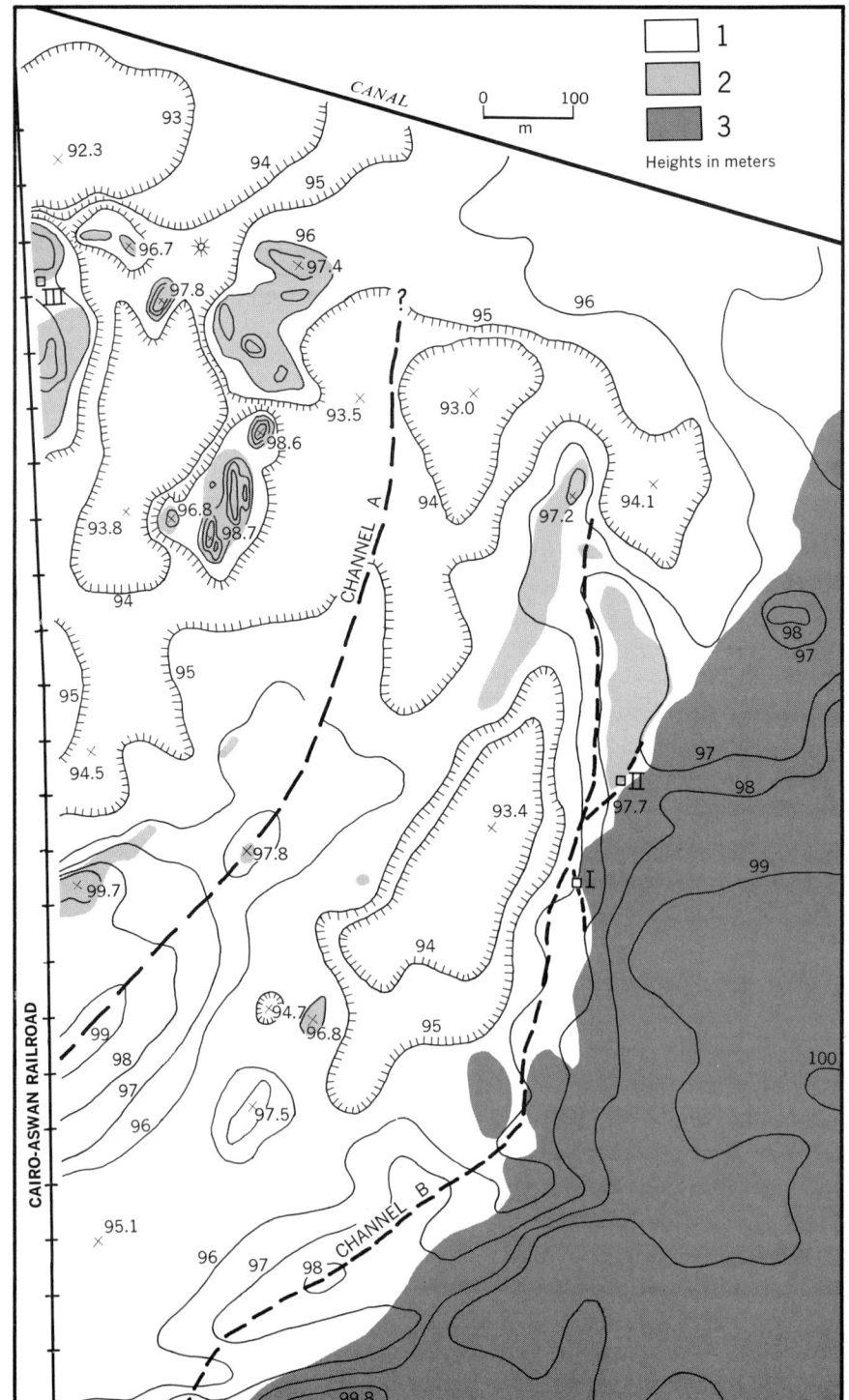

Fig. 3–23 (*facing page*). Geology and topography of the Gebel Silsila 2 area, showing orientation of the Fatira Channels A and B. *1:* Sand facies of Gebel Silsila Formation; *2:* Silt facies of Gebel Silsila Formation; *3:* Masmas Formation. *I:* Site 2B, Area I; *II:* Site 2B, Area II; *III:* Site 2A, location of duricrust with artifacts, adjacent to square S-4, E-1. Topography corrected from an unpublished 1:2500 Egyptian survey map. See Fig. 3–1 for general location. Map: UW Cartographic Lab.

shell in a position of growth at the south end of the site Gebel Silsila 2A (Charles A. Reed, *in litt.*, January 27, 1967).

Most of the Channel A sediments have been badly scoured by wind action and have been reduced to streamlined mounds or ridges that form typical yardangs (Figs. 3–24, 3–25). Erosion of the loose, coarser beds has been prominent, tending to preserve the finer levee deposits. Local relief at the Gebel Silsila 2A site varies from 2 to 6 meters; in the Silsila Gap, local relief increases to 10 meters.

A brief interval of vertical incision preceded a second Nile arm with a thalweg some 900 to 1500 meters east of that of the first channel. The new stream bed was much narrower, measuring 75 to 150 meters from bank to bank. The highest sediments of Channel B probably attained 99 meters (+ 10.5 m), and they are almost all confined to coarse channel sands and to east-bank levee beds skirting a low erosional step cut into the Masmas Formation. Reverse bedding of levee remnants between the two channel thalwegs precludes confusion of the two. Downcutting to below 96 meters separated these substages.

Channel B meandered and braided into a number of temporary channels during its existence. The occupation site Gebel Silsila 2B (Area I), which forms the type site of the formation of that name, was occupied along the concave side of a meander bend. Originally, a small undercut "cliff" of 80 to 100 centimeters was incised into the Masmas Formation, followed by deposition of backset and occasional foreset levee beds in relation to a local thalweg oriented at about N 70° W. A little later, this meander bend was abandoned and was replaced by a stream thalweg oriented N 10° W just west of Area I and N 40° E at Area II. By the time of occupation of the second site at Area II, this last stream position was being abandoned, and the local channel was reduced to an overflow runnel only 7 to 8 meters wide.

The microstratigraphy of Area I is summarized in Figs. 3–26 and 3–27 and in Table 3–5. Embanked against or overlying the Masmas Formation (No. 229, Table 3–4), the following strata are exposed from bottom to top:

a) 0.5 to 4 cm. White lime-salt crust, embedding a silty coarse sand, derived concretions, occasional *Unio* shells and some artifacts or flaking waste, resting disconformably on the Masmas. Soluble salts, 2750 parts per million.

b) 90 cm. Very-pale-brown, coarse sand with some silty lenses; loose, single-grain structure, well-stratified and cross-bedded N 20° E.

c) 50 cm. Pale-brown, coarse sand with backset lenses of derived concretions and of brown, silty coarse sand with concretionary pebbles. One of these lenses (Level 1) includes a concentration of *Unio* and *Corbicula* shells and archeological materials, to be discussed in Chapter 4.

d) 2 to 6 cm. Pale-brown, silty coarse sand matrix for calcareous nodules, concretions, and consolidated silt pebbles, all derived from older beds upstream. Quartz granules and pebbles to 1.5 cm in length are also present. This heterogeneous backset bed dips 10–40% to N 10–20° E and is consolidated with a lime-salt cement. At least a part of the abundant *Unio*, common *Corbicula*, and rare *Cleopatra* shell is natural, but all of the mammalian bone is part of the principal occupation floor (Level 2). A long period of subaerial exposure is suggested by root drip penetrating from bed *d* into beds *c* and *b*, as well as by concentration of secondary carbonates and other evaporites. Soluble salts range from less than 500 to over 7,000 parts per million, primarily sodium chloride. Radiocarbon dates of 11,120 B.C. (charcoal) and 11,610 B.C. (shell) were obtained from materials resting on top of bed *d*.

Fig. 3–24. Yardangs of Gebel Silsila deposits preserved along western bank of former Channel A at Gebel Silsila 2A, looking north to the Silsila Gap. Backset and foreset beds are conspicuous, with deflation selectively removing sandy deposits.

e) 160 cm. Pale- to very-pale-brown sands, ranging from silty medium-to-coarse sands at base, to coarse sands in upper profile; loose, single-grain structure; basal part includes derived, light-gray to light-olive-gray (2.5–5 Y 6/2) concretions, abundant mica, and rolled bone fragments. Well-stratified beds range from gentle backsets and topsets to complex cross beds with lenses of derived concretions. The finer basal unit is intensively stained to a strong brown (7.5 YR 5/6) color by limonite, probably because of waterlogging over an impermeable base. At least one disconformity interrupts this unit.

f) 10 cm. Pale-brown, coarse-sandy silt with derived concretions and pebbles, as unit *d;* backset N 15° E; coarse, subangular blocky; rich in evaporites and subsurface root drip. Concentrations of *Unio* and *Corbicula* shell and artifacts qualify this as another archeological level (3).

g) 80 cm. Pale-brown beds changing from medium-sandy silt at base to clayey silt at top, as direction of bedding veers from N 20° E to N 65° E, thus recording abandonment of the local meander bend. Structure changes from coarse subangular to angular blocky. Rolled hippopotamus bone and fragment

Fig. 3–25. General view of deflated Gebel Silsila 2A area, looking south along western bank of Channel A.

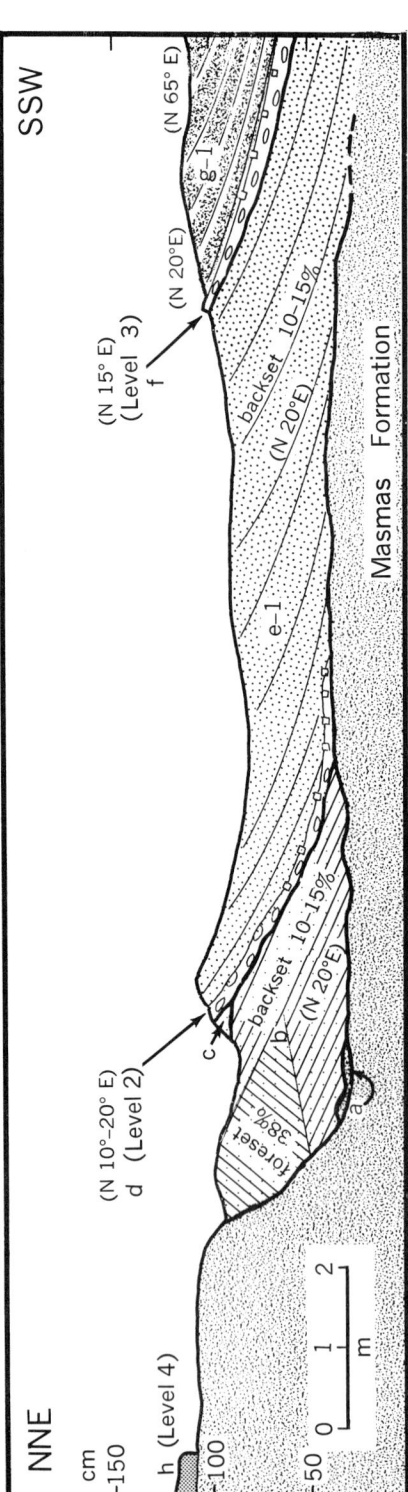

Fig. 3-26. North-South profile through Gebel Silsila 2B, Area I. Map: UW Cartographic Lab.

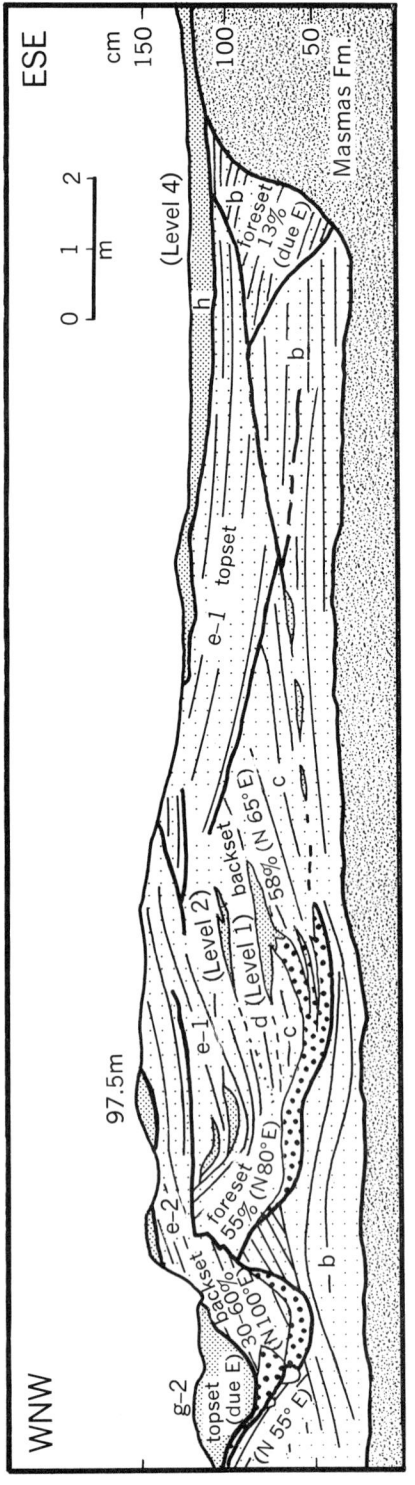

Fig. 3-27. West-East profile through Gebel Silsila 2B, Area I. Map: UW Cartographic Lab.

of catfish skull. A local 5 cm duricrust suggests subsequent weathering of the surface under arid conditions. Disconformity.

h) 15 cm. Brown clayey silt; stratified, single-grain, loose; with small, unidentifiable snail fragments and archeological Level 4. Resting disconformably on the Masmas Formation and bed *e*, this highly calcareous silt is of nilotic origin but may subsequently have been reworked as a colluvium.

Area II is located 120 meters north of Area I (Fig. 3–23), and the 2 meters of deposit are concentrated in a shallow channel cut into the Masmas Formation with an orientation of N 40° E. From base to top (Table 3–6), the following strata are exposed:

a) Over 90 cm. Alternating brown, clayey silt and pale-brown, silty medium sand. The sands are well stratified with single-grain or very coarse granular structure; the silts are less conspicuously stratified, with medium, subangular blocky structure; minor vertical root drip is present in both facies. Sterile. Bedded against the Masmas beds at inclinations up to 30%, the highest ex-

Table 3–5. Sediment characteristics at occupation site Gebel Silsila 2B, Area I

Sample number	Bed	Color	Texture	Sorting	CaCO₃ (%)	pH
99	h	(10 YR 5/3)	Clayey silt	Moderate	29.8	8.4
231	g–2	(10 YR 6/2–3)	Clayey silt	Moderate	1.8	7.9
388	g–1	(10 YR 5–6/3)	Clayey silt	Moderate	5.0	7.8
387	f	(10 YR 5.5/3)	Coarse-sandy silt	Poor	43.0	7.3
226	e–2	(10 YR 7–8/3)	Coarse sand	Moderate	0.6	8.5
227, 230	e–1	(10 YR 6/3)	Silty medium or coarse sand	Poor	0–0.6	7.6–7.8
233, 396	d	(10 YR 6/3)	Silty coarse sand to coarse-sandy silt	Poor	11.2–16.0	7.4–8.3
228	c	(10 YR 5–6/3)	Silty coarse sand	Poor	3.5	8.0
234	b	(10 YR 7/3)	Coarse sand	Good	0.3	7.9
232	a	(10 YR 8/1)	Silty coarse sand	Moderate	32.2	7.7

Table 3–6. Sediment characteristics at occupation site Gebel Silsila 2B, Area II

Sample number	Bed	Color	Texture[a]	Sorting	CaCO₃ (%)	pH
392	ε	(10 YR 6/3)	Clayey silt	Moderate	7.8	7.5
391	∂	(10 YR 6/3)	Silty medium sand	Moderate	1.2	8.2
390	γ	(10 YR 5/3)	Medium-sandy silt	Poor	7.0	8.1
389	β	(10 YR 5–6/3)	(Medium-sandy silt)	—	—	—
398	α	(10 YR 5–6/3)	(Silty medium sand to clayey silt)	—	—	—

[a] Textures in parentheses not determined by quantitative methods.

posures, somewhat impregnated with salts, attain 98 m elevation. Maximum width of bed a is 17 m. Disconformity.

β) 5–10 cm. Pale-brown, medium-to-coarse sand matrix to an agglomeration of derived nodules and silt pebbles, with a major archeological level. Radiocarbon determination of 13,360 B.C. based on charcoal from the occupation floor. Beds β through ε form part of a 7 to 8 m wide overflow channel resting on a and running N 40° E. Strata dip 5% to 10% from either side to the stream axis.

γ) 30 cm. Brown, medium-sandy silt matrix to derived pebbles of clayey silt. Occurs as a localized wedge, interrupted by a 5 cm lense of pale-brown, coarse sand. Sterile.

δ) 40 cm. Pale-brown, silty medium sand; well-stratified and current-bedded; medium, subangular blocky structure, embedding abundant archeological materials.

ε) 30 cm. Pale-brown, clayey silt; well-stratified; fine-to-medium, angular blocky structure. This sterile bed marks the final stage of alluviation with flood silts in the overflow channel.

Although the radiocarbon determinations would suggest that Area II is older than Area I, the geology speaks for the opposite conclusion: (*1*) All of the Area II deposits are related to a stream bed or banks oriented N 40° E, i.e., the east-bank backsets dip N 130° E. Only the upper beds of unit *g* at Area I approximate this orientation. Since a gradual channel shift can be observed in bed *g*, it seems reasonable that units *a* to *f* record an earlier major channel and that, after abandonment of the local meander bend, discharge was eventually confined to a small overflow channel with a different orientation (units α to ε). Such deterioration of the Fatira Channel preceded the end of aggradation. (*2*) Although obscured by a veneer of sand, bed *g* can be discontinuously traced northward and, in terms of facies and orientation, seems to be equivalent to a part of bed *a*. (*3*) Units β to ε have no duricrusts or zones of carbonate enrichment as do beds *a* to *g*. Instead, the last crustification at Area I is recorded in bed *g*; at Area II, in unit α. (*4*) The Area II strata attain a higher elevation (98.0 m) than the deposits of Area I (97.5 m). (*5*) The microlithic artifacts found in units β and δ compare well with those in unit *h* (Baumhoff, *in litt.*, February 7, 1966). Although no direct stratigraphic link was established in the field, all of the field evidence strongly favors a younger age for Area II. The time difference may be only a century or two, however.

The sands exposed in the Gebel Silsila area are almost always rich in micas and hornblende of nilotic derivation. Yet, on casual inspection, many of the cross-bedded sands suggest an eolian origin. Sorting in the medium and coarse sand grades is never as pronounced as in true eolian deposits (see Harris, 1957, 1958), however, and an appreciable fine component under 0.02 millimeters diameter tends to rule out an eolian origin. Only one of twenty hydrometer samples (No. 234, Table 3–5) tested from the local Gebel Silsila Formation has the textural spectrum of an eolian deposit, with 85.5% of the sample between 0.2 and 0.6 millimeters and with 99.5% between 0.06 and 2.0 millimeters. But even this cross-bedded, well-sorted coarse sand is an integral part of a strictly fluvial sequence of channel-margin deposits. In general, directions of bedding apparent in the late Pleistocene deposits near Gebel Silsila cannot be interpreted from modern wind directions, although these beds can be readily related to different channel positions. The question of eolian reworking of the fluvial beds prior to final deposition is more difficult to answer. There is certainly no evidence of this on a macroscopic scale, although in detail there probably are many sandy beds of mixed origin.

The lenses of derived concretions or silt pebbles are of some interest. The derived nodules or rolled concretions consist of light-brownish-gray (10 YR 6–7/2) silt, indurated with calcite and some salt. Many are stained or deeply patinated to a dark-brown (10 YR 3/3) color by ferric oxides and pyrolusite. Similar staining is common to rootlet zones evident inside these pseudo-pebbles. Dominant size is 2 to 5 centimeters, although some are as much as 8 centimeters long. The second type of pseudo-pebble consists of consolidated sandy silt or clayey silt, varying from pale brown to brown (10 YR 5–6/3) in color, impregnated with carbonates and a trace of salt. Rounded and ellipsoidal in shape, these silt pebbles fall into the same coarse grade-size as the derived concretions. But, whereas a part of the concretions appear to come from the Masmas Formation, the silt pebbles are entirely derived from the Gebel Silsilan and can often be attributed to the next bed below.

The nature of the archeological strata and the overall geographical setting of the sites at Gebel Silsila 2A and 2B will be discussed in Chapter 4.

Channel Stratigraphy at the Khor el-Sil Sites

The archeological sites of Khor el-Sil, excavated jointly by the National Museum of Canada and Yale expeditions in the autumn of 1962, are located about 1.5 kilometers east of Km 821.3 of the railroad (Fig. 3–1). Deposits of the Gebel Silsila Formation are well developed as part of the abandoned Manshiya Channel—named after el-Manshiya, the nearest hamlet. This channel was cut into the Masmas beds and meandered northwards towards the Silsila Gap. The southern starting point for this channel is obscured by cultivation.

The geomorphology of Khor el-Sil is outlined in Fig. 3–28. Homogeneous, well-sorted, pale-brown (10 YR 6/3), medium or coarse sands, with abundant mica and hornblende, form the dominant local facies of the Gebel Silsila Formation (No. 25, Table 3–8). Crossbeds are common, with dips exceeding 55% in places. The complex nature of the bedding suggests a channel environment. Broadly horizontal, extensive bands of sandy silt (No. 30, Table 3–8), varying from 3 centimeters to 50 centimeters in thickness, are common in the upper part of the sands. Sedimentation terminated after a massive 1 to 3 meter layer of pale-brown (10 YR 6/3), medium-sandy silt (Nos. 26–27, 31–33, Table 3–8) was deposited on top of the sand complex in relation to a channel moving in a westerly direction (N 85° W). The massive silts are part of a minor overflow channel and its floodplain. The highest deposits attain 98 meters, but deflation has been very active. Dune-shaped yardangs, with a local relief of 2 meters, stud this desolate plain. Remnants of former subsoil crusts, exposed through deflation, often cap resistant mounds.

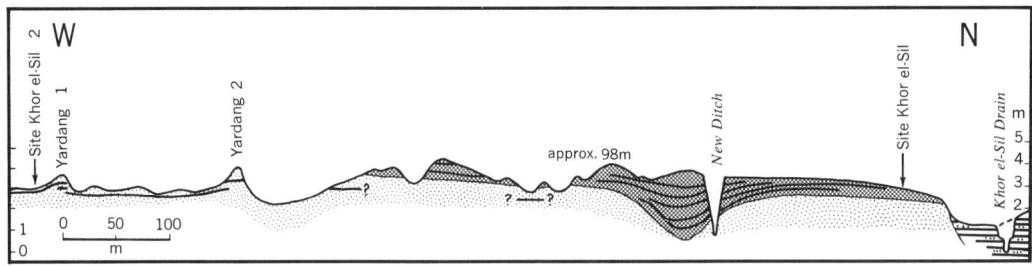

Fig. 3–28. The Gebel Silsila Formation at the Khor el-Sil archeological sites. Section describes a quarter circle N–S–SW–W. The Manshiya silt channel strikes N 85° W. Map: UW Cartographic Lab.

An approximate date for the beds of the Manshiya Channel is given by a radiocarbon assay on shell from one of the archeological sites: 15,050 B.C. ± 600 (I–1297, P. E. L. Smith, *in litt.*, March 15, 1966). This date suggests that the fossil Manshiya Channel may be broadly contemporary with the basal deposits of Channel A at Gebel Silsila 2. Abundance of *Cleopatra* shells in both channels supports this suggestion, since this aquatic gastropod is rare in the younger, Channel B deposits.

The modern irrigation drain that intersects the northern periphery of the site cuts a filled wadi (Fig. 3–29) that once flowed westwards across

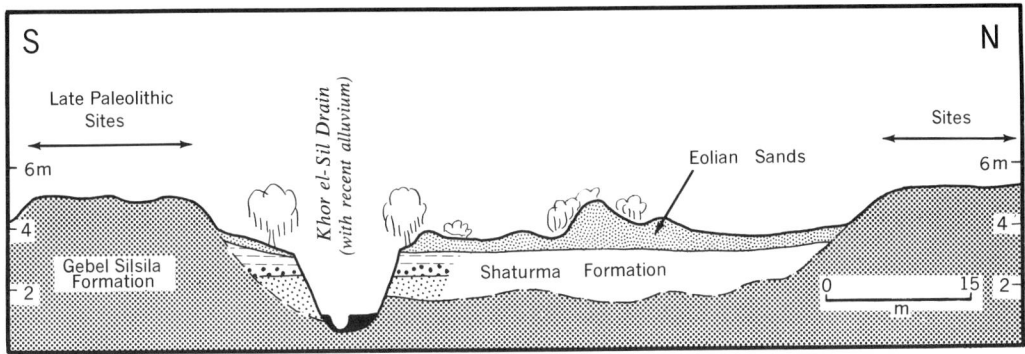

Fig. 3–29. Section of the Khor el-Sil alluvia (*ca.* 1.5 km east of railway). Map: UW Cartographic Lab.

the Kom Ombo Plain towards the confluence of the Nile and Wadi Shait. This fossil wadi cuts through the Masmas and the Gebel Silsila Formations and is filled in with beds recalling the Shaturma Formation. This alluvial fill averages 2 meters in thickness and is masked by modern dunes with barchan, parabolic, and nebkha forms. A typical section is exposed from base to top (Fig. 3–30):

a) 75 cm. Very-pale-brown (10 YR 8/4), coarse sand with derived concretions or pebbles of sandy silt (2–20 mm diameter). Alternating topsets and foresets.
b) 3–4 cm. Lense of fine gravel, mainly concretions with some quartz granules.
c) 35 cm. Brown (10 YR 5/3), medium-sandy silt; current-bedded.
d) 15 cm. Pale-brown (10 YR 6/3), silty medium sand with derived concretions and quartz granules; stratified.

The concretions of beds *a* through *d* are identical with surface lag found on the Gebel Silsila Formation. This suggests that a long period of deflation, as well as dissection by the former wadi, preceded alluviation. Similar sections are exposed along the Khor el-Sil drain beneath the railway track and again near the highway. Almost certainly the sterile, current-bedded wadi wash of beds *a* to *d* represents the Shaturma Formation.

Channel Stratigraphy of the Sebil Area

Abandoned Nile channels were rather conspicuous among the Gebel Silsilan deposits in the Sebil area before the spring of 1963. Vignard

Fig. 3–30. Alluvial fill, probably a part of the Shaturma Formation, cut by the irrigation drain at Khor el-Sil (see Fig. 3–29).

(1923: 4, map 2), in his classic study of the Late Paleolithic cultures, interprets these features as distributaries of a Wadi Kharit delta. Although Sandford and Arkell (1933: 51–52) pass over this obvious error, they implicitly accept Vignard's topographic levels (1923: Fig. 1), without an explanation of the former geography.

Several brief surveys and mapping by the authors suggest that two or three minor channels of the Nile could once be traced through this rather complex topography (Fig. 3–1). They all pertain to the Gebel Silsila Formation. The highest of these left the present Nile channel about 1 kilometer west of Bayara, curving well west of Matana Qibli and continuing north to a point 1.5 kilometers northwest of Sebil Qibli. From here, the channel braided as it swung east to the Sebilian type-site at Sebil Bahari, then followed a meandering course northward to Wadi Shait, approximately 1.5 kilometers west of the railroad. A lower channel broke away from this circuitous route west of Sebil Qibli, heading directly toward Aklit.

The higher channel complex is illustrated by Fig. 3–31. Two braided channels, each some 200 to 250 meters wide, are accompanied by two benches with persistent levels of 100 to 101 meters and 98 to 99 meters, separated by a distinct step. These can be interpreted as flood benches —in part erosional, in part alluvial—set within an older body of silts to 102 meters maximum elevation (Fig. 3–32). Both benches were accompanied by longitudinal shell ridges and marked by archeological sites. The action of swiftly flowing water in these channels is apparent from the steep, foreset-bedded sand banks lodged at the rear of former channel islets. Since local relief averages only 4 meters, these sinuous depressions probably served only as seasonal overflow channels. On the other hand, the shallowness of these channels may also be due to infilling during the stages of abandonment. In modern times, they are partly masked by wind-blown sands and are locally subject to severe deflation. In addition, seepage water from higher-lying canals fills the lowest depressions, promoting growth of vegetation. In terms of elevation and development, these two overflow stages compare closely with Channels A and B at Gebel Silsila 2, and there seems ample reason to correlate the two substages.

After abandonment of the higher channel complex, flood-season discharge continued to flow through the southern segment, between Bayara, Sebil Qibli, and Aklit. A low flood bench at about 97 meters (+ 7 m) developed at this time. After functioning briefly, this Channel C was also abandoned, and modern hydrographic patterns were established.

Three substages can consequently be suggested for the time interval represented by the Gebel Silsila Formation:

Channel A Phase (to 101 or 102 m; 12 m over modern floodplain). Age very approximately 15,000–12,500 B.C. Recorded at Gebel Silsila 2A, Sebil, Darau, and possibly at Khor el-Sil.

Channel B Phase (to 98 or 99 m; 9 m above floodplain). Age approximately 12,000–11,000 B.C. Recorded at Gebel Silsila 1 and 2B, and at Sebil and Darau.

Channel C Phase (to 97 m; 7 m above floodplain). Guess age possibly as late as 10,000 B.C. Recorded between Bayara and Sebil Qibli.[3]

Brief periods of downcutting separate these intervals of alluviation.

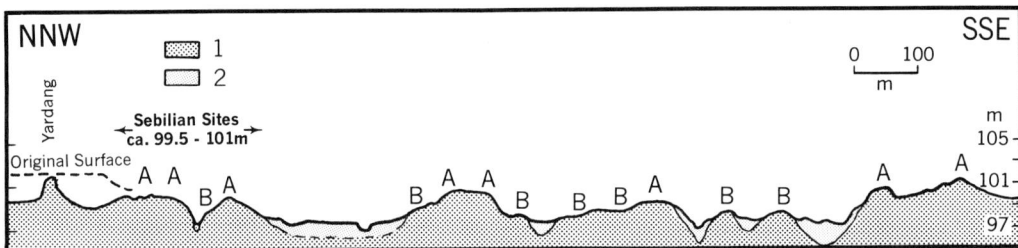

Fig. 3–31. The Sebil Channels and related flood benches A (100 to 101 m) and B (98 to 99 m) between Sebil Qibli and Matana Bahari. *1:* Silts of the Gebel Silsila Formation; *2:* Eolian sand filling channel beds. Map: UW Cartographic Lab.

Late Pleistocene Nilotic Sediments North of the Kom Ombo Plain

Nilotic sediments of Late Pleistocene age have been reported from many parts of Upper and Lower Egypt and from the Fayum Depression. Of considerable interest are exposures described by Sandford and Arkell (1933: 45–46) from the west bank of the Nile, 11 to 13 kilometers south of Edfu. These deposits include a massive basal unit of sand, sandy marl,

---

3. Sites in this area were excavated by P. E. L. Smith in 1963 and their geology was studied by Robert J. Fulton. Hopefully, they will provide further geomorphic and chronological information on this terminal channel. A radiocarbon determination of 10,450 B.C. ± 400 from shell (I–1300, P. E. L. Smith, *in litt.*, March 15, 1966) from a terminal Sebilian site may be relevant.

Fig. 3–32. Abandoned late Pleistocene channel near Sebil Bahari. The breach in the old levee (foreground), cut by recent irrigation overflow, exposes channel-bank strata (hidden in shadows).

and marl with *Unio willcocksi, Corbicula* ssp., *Cleopatra bulimoides, Hippopotamus amphibius, Equus* sp., *Bos* sp., *Crocodilus* sp., and fish bones. "Mousterian" implements were also found in this sediment, which strongly recalls the Korosko Formation. Higher in the same sequence are hard clays, dark-brown earthy muds, and marls to 18 meters above floodplain, suggesting the Masmas Formation. These exposures merit reinvestigation.

Typical sediments of the Masmas Formation are found to + 15 meters between Edfu and Esna at Hierakonpolis (Butzer, 1960a), and Sandford and Arkell (1933: 46–47) indicate the presence of similar deposits just south of Luxor. At Armant, these authors believe they have found a 1-meter lenticle of cross-bedded eolian sand within massive silts. Throughout this area, nothing has been reported that resembles the Gebel Silsila Formation, although a low wadi gravel in Wadiyein, west of Luxor, contains a few artifacts that suggest the Middle Sebilian (Butzer, 1959a).

Further north, after a considerable break, sands and well-rolled medium gravel with abundant agate are found at 5 meters or more *below* modern floodplain level in Middle Egypt at Qau, Sheikh Timai, and El-Fashn (Sandford, 1934: 86–90, 95). These recall the Gebel Silsila Formation and are correlated with the 22-meter (above sea level) shoreline in the Fayum Depression (*ibid.;* 92–94). Contemporary beach deposits include a Late Paleolithic industry (Sandford and Arkell, 1929: 57–61, 65; Caton-Thompson, 1946), At Qau, a rich fauna with

> *Bubalis* sp.
> *Gazella dorcas isabella* Gray
> *Bos* sp.
> *Equus* sp.
> *Hippopotamus amphibius* L.
> *Sus* sp.
> *Crocodilus niloticus* Laurenti
> *Crocodilus* nov. sp.
> *Testudo* (*Trionyx*) sp.

has been reported by Sandford (1934: 86) and suggests close affinities with that from the Gebel Silsila Formation at Sebil. Although none of these correlations are proven, it seems likely that this particular formation is indeed represented by suballuvial gravels in Middle Egypt and by beach deposits in the Fayum.

A posssible correlative of the Masmas Formation are the fine gravels and silts to 8 meters above floodplain between Sedment and the Delta (Sandford and Arkell, 1939: 54–56, 79–80), and the equivalent 34-meter beach deposits of the Fayum (Sandford and Arkell, 1929: 37–53; Sandford, 1934: 92–94; Caton-Thompson, 1946). These deposits have an Upper Levalloisian industry. Reinvestigation of the Fayum shorelines and radiocarbon assays on the shells embedded in related deposits should permit a fairly secure Late Pleistocene chronology for all of Egypt.

Stratigraphic and Paleoclimatic Conclusions

The sequence of Late Pleistocene events, climatic fluctuations, and geomorphic change outlined in the previous sections is synthesized in Table 3–7.

The existence of three distinct nilotic sedimentary units in the late Pleistocene record of the Kom Ombo area is apparent from both the lithological and the stratigraphic evidence. In addition to the sections and facies characteristics already described in detail, the X-ray diffractograms (Appendix D) show very distinctive and internally consistent clay fractions for each sedimentary unit. The Gebel Silsila Formation shows a

Table 3-7. Geomorphic evolution of the Kom Ombo area
during the late Pleistocene and Holocene

30.	Dissection of fill in minor wadis (vertical differential 2 to 3 m) and clay accumulation in Red Sea Hills wadis. In general, very limited wadi activity.	HISTORICAL
29.	Alluviation of Shaturma Formation by wadis: Member II. 2.5 m thickness. Some winter rains, no organic vestiges. *Ca.* 1000–1200 A.D. Dissection (about 1.5 m).	
	Member I. 3 m thickness, grading into colluvial screes, suggesting heavy, torrential rains. Fourth millennium B.C. First alluviation of modern Nile floodplain probably contemporary.	HOLOCENE
28.	Dissection of wadi fill (total cutting since phase *26* exceeds 6 m). Limited wadi activity.	
27.	Biochemical weathering with formation of red paleosol. Frequent, gentle rains and greater vegetation. A period of minor dissection separates phases *26* and *27*.	
26.	Alluviation of Sinqari Member, Ineiba Formation, by major wadis (thickness over 11 m). Accelerated wadi activity in Red Sea Hills and greater local rains, with abundant wadi activity *ca.* 9500–6500 B.C.	
25.	Dissection by Nile (amplitude uncertain) and wadis (vertical differential at least 12 m). Limited wadi activity.	LATE WÜRM
24.	Brief stage of sporadic or periodic Nile floods of exceptionally great amplitude, attaining maximum elevation of 112 m *ca.* 10,000 B.C. Formation of Mazaquert paleosol on Malki deposits and secondary gypsum impregnation of Masmas silts.	
23b.	Alluviation of Malki Member, Ineiba Formation, by major wadis (thickness over 9 m). Contemporary with phase *23a*. Accelerated wadi activity in Red Sea Hills; some local rains recorded by basal gravel stratum.	
23a.	Alluviation of Gebel Silsila Formation by Nile (thickness well over 14 m). Influx of exotic gravel in bed load; greater Nile competence. Channel C (97 m) *ca.* 10,500 B.C. Dissection by Nile. Channel B (98 to 99 m) *ca.* 12,000–11,000 B.C. Dissection by Nile (over 6 m). Channel A (100 to 102 m) *ca.* 15,000–12,500 B.C.	
22.	Dissection by Nile (to below modern floodplain, vertical differential at least 20 m) and wadis. Lower Nile floods and limited wadi activity.	EARLY TO MIDDLE WÜRM
21.	Development of hydromorphic paleosol on floodplain (a vertisol of Mazaquert type).	
20.	Alluviation of Masmas Formation by Nile (thickness at least 23 m, possibly over 43 m). Flood regime similar to that of today; minimal wadi activity in general, interrupted by at least six periods of limited wadi flow. Maximum floodplain elevation 110 m.	
19.	Dissection by Nile (to below modern floodplain; vertical differential at least 19 m, possibly over 40 m) and wadis (well over 5 m). Lower Nile floods and limited wadi activity.	

(*Table 3-7*, continued)

18.	Alluviation of Korosko Formation by Nile and wadis (thickness at least 15 m). Greater Nile velocity. Maximum floodplain elevation over 108 m. Valley margin: 　Facies III: Marl. Limited wadi activity. 　Facies II: Gravelly marl. Some wadi activity. 　Facies I: Sandy gravel. Accelerated wadi activity.	EARLY TO MIDDLE WÜRM
17.	Dissection by Nile (to below modern floodplain, vertical differential at least 12 m) and wadis (over 10 m cutting in fill). Lower Nile floods and limited wadi activity.	EARLY UPPER PLEISTOCENE

very intense montmorillonite peak and a very minor secondary peak for kaolinite. In the Korosko Formation, the kaolinite peak is very prominent and is at least as intense as the montmorillonite peak. The situation is intermediate in the case of the Masmas Formation, with the montmorillonite peak most prominent and intense, but with the kaolinite peak still significant. The heavy mineral analyses (Appendix B) also seem to indicate certain differences among the heavy mineral suites pertaining to each sedimentary unit.

The composition of each of the three nilotic units can be ascribed to summer floods of subsaharan origin. In each case, fairly rapid aggradation of an expanded floodplain to elevations substantially higher than today suggests greater and more persistent floods than are now usual. The coarse sands of the Korosko unit and the gravel lenses of the Gebel Silsila unit further suggest greater stream competence. In fact, the presence of exotic gravels derived from the Hudi Chert suggests accelerated wadi activity in Upper Nubia as well as greater summer-flood velocities for the Gebel Silsila stage. And each of the three nilotic alluvia presumably record pluvial substages in the summer rainfall belt of Ethiopia or the southern Sudan.

Apart from the several unsatisfactory stratigraphic details already discussed above, two more general problems remain to be emphasized. (*1*) The terminal stages of the Gebel Silsila Formation at Kom Ombo remain enigmatic, a point of concern since there are later alluvial substages of identical character in Lower Nubia. Publication of the local stratigraphy

Table 3-8. Sediment characteristics of miscellaneous samples from the Kom Ombo Plain

Sample number	Location	Color	Texture[a]	Sorting	CaCO₃ (%)	pH
		Shaturma Formation (Member I)				
128	New Shaturma (Pit 28)	(7.5–10 YR 5/4–6)	Silty coarse sand	Poor	8.5	7.45
		Ineiba Formation (Sinqari Member)				
22	Wadi Ellawi	(5 YR 5–6/4)	Silty medium sand	Moderate	3.0	7.85
29-II	Upper Kharit	(5–7.5 YR 5–6/4)	Coarse sand	Good	0.7	7.95
58	New Korosko (Pit 36)	(7.5–10 YR 6/4)	Coarse sand	Good	4.0	8.20
78	New Korosko (Pit 12)	(10 YR 6/4)	(Silty medium sand)	—	3.7	7.70
398	Upper Kharit	(5 YR 6/4)	Medium-sandy silt	Poor	4.5	7.90
		Ineiba Formation (Malki Member)				
54	New Korosko (Pit 36)	(10 YR 5.5/3)	(Clayey marl)	—	28.5	7.20
57	New Korosko (Pit 33)	(10 YR 5.5/3)	(Clayey marl)	—	28.5	7.80
62	New Sebua (Pit 20)	(7.5–10 YR 6/4)	(Clayey silt)	—	14.7	7.40
75	New Korosko (Pit H)	(7.5–10 YR 6/4)	(Clayey silt)	—	10.6	7.35
85	New Shaturma (Pit 52)	(10 YR 6/4)	Marly clayey silt	Poor	18.0	7.30
122	New Sebua (Pit 1)	(10 YR 6/4)	(Clayey silt)	—	7.7	7.60
127	New Sebua (Pit 26)	(7.5–10 YR 6/4)	(Clayey silt)	—	8.4	7.15
		Ineiba Formation (Basal gravel of Malki Member)				
60	New Korosko (Pit 13)	(7.5–10 YR 5–6/4)	(Coarse-sandy silt) (matrix)	—	2.0	7.85
68	New Korosko (Pit H)	(7.5–10 YR 6/4)	(Coarse-sandy silt) (matrix)	—	6.8	7.40
		Gebel Silsila Formation				
33	Khor el-Sil 1/2 (b)	(10 YR 6/3)	Medium-sandy silt	Moderate	1.1	7.60
25	Khor el-Sil 1/2 (a)	(10 YR 6/3)	Medium sand	Good	1.1	8.20
27	Khor el-Sil 2 (a)	(10 YR 6/3–4)	Medium-sandy silt	Moderate	0.4	8.00
28	Khor el-Sil 2 (b)	(10 YR 6/4)	Silty sand	Poor	52.0	—
30	Khor el-Sil 6	(10 YR 6/3–4)	Medium-sandy silt	Poor	5.1	7.20
26	Khor el-Sil 1 (c)	(10 YR 6/2)	Clayey silt	Poor	14.0	7.90
31	Khor el-Sil 1 (b)	(10 YR 6/3)	Clayey silt	Poor	9.8	7.30
32	Khor el-Sil 1 (a)	(10 YR 6/4)	Clayey silt	Poor	4.3	7.10
63-I	Silsila Gap	(10 YR 6/3)	Clayey silt	Moderate	4.0	7.70
141	Gebel Silsila 2A	(7.5 YR 6/4)	Coarse sand	Moderate	1.3	7.80
142	Gebel Silsila 2A	(10 YR 6/3)	Silty medium sand	Moderate	5.8	6.65
144	Gebel Silsila 2A	(10 YR 6/3)	Medium-sandy silt	Poor	2.8	7.80
325	New Ballana I	(10 YR 5/2)	Clayey silt	Poor	4.5	7.30
		Masmas Formation				
23	New Abu Simbel	(10 YR 4/2)	Sandy silt	Poor	33.8	7.40
83	New Shaturma (Pit 52)	(10 YR 6/3)	Marly medium sand	Moderate	20.6	7.75
326	Wadi Shurafa	(5 Y 5/1)	Silt	Good	1.3	8.40
328	New Ballana I	(10 YR 3–4/2)	Silty clay	Moderate	0.8	7.40
329	New Ballana I	(10 YR 5/3)	Silty medium sand	Poor	0.4	7.80
332	El Nuqu	(10 YR 5/1)	Marly silty clay	Moderate	24.3	7.75
		Korosko Formation (Unit III)				
56	New Korosko (Pit 11)	(10 YR 6/2)	(Medium-sandy marl)	—	29.0	7.50
		Korosko Formation (Unit II)				
77	New Korosko (Pit H)	(2.5 Y 7/2)	Marly silty sand	Poor	16.4	7.50
81-I	New Shaturma (Pit 52)	(2.5 Y 6/2)	Marl	Poor	35.4	7.40
81-II	New Shaturma (Pit 52)	(2.5 Y 7/2)	Marly silty sand	Moderate	17.8	7.65
84	New Shaturma (Pit 52)	(2.5 Y 5.5/2)	Marly coarse sand	Poor	19.0	7.20
		Korosko Formation (Unit I)				
47	New Korosko (Pit 34)	(10 YR 6/3)	Silty medium sand	Poor	15.2	7.50

[a] Textures in parentheses not determined by quantitative methods.

between Bayara and Sebil Qibli by P. E. L. Smith and Robert J. Fulton may clarify a part of this terminal Pleistocene time range. (2) The origins of the contemporary floodplain are completely obscure in the Kom Ombo region. The modern Nile muds are considerably richer in silt and clay and contain less sand than the Gebel Silsila Formation. Micas also appear to be less abundant at present. Borings under the floodplain of Upper Egypt (see Attia, 1954) generally show late Holocene muds resting disconformably on fine, micaceous sands with abundant heavy minerals. The latter almost certainly pertain to the Gebel Silsila Formation. In the Edfu-Esna area, the general thickness of the recent alluvium is 5 to 11 meters (estimate based on 15 bores selected from Attia, 1954: 14–27). Of a total thickness of 5 meters at Hierakonpolis, midway between Edfu and Esna, some 3 meters had already been deposited by 2250 B.C., when the foundations of a Sixth Dynasty temple were laid. Assuming deposition of 10 or even 20 centimeters per century, late Holocene siltation at Hierakonpolis certainly had begun by the beginning of the Predynastic period. Consequently, earliest alluviation of the modern floodplain in Upper Egypt may be contemporary with Member I of the Shaturma Formation. But, as will be outlined in Chapter 6, the picture is far more complicated in Nubia, where the modern floodplain is younger than 3000 B.C.

Seen in overview, the late Pleistocene deposits of the Kom Ombo area are amazingly complex and are, as a consequence, rather informative. Each section contributes this or that detail towards a fuller understanding of past environments. Yet the totality of geomorphic change has been small since the early Pleistocene. Denudation of older gravels, dissection of wadis, and backwearing of slopes have taken place, but at rather slow rates. Chemical weathering, other than salt hydration, has been almost insignificant from the geomorphic point of view. The amplitude of climatic change has also been small. The only periods of significant local runoff can be associated with the first part of the Korosko Formation and with Member I of the Shaturma Formation. But even then, local climate was arid, with 50 to 100 millimeters of rainfall at most. The other periods of accelerated wadi activity were only a little moister than today, perhaps 10 or 20 millimeters average precipitation. Despite the persistent aridity of local climate since the early Pleistocene, these minor fluctuations of climate certainly had significant ecological implications. And the changing flood regime of the Nile must have affected the riverine zone profoundly, providing ample grazing, game, and water during stages of floodplain alluviation and restricting this same lifeline during intervals of downcutting.

4

Geographical Factors Conditioning Prehistoric Settlement of the Kom Ombo Plain

Introduction

The sequence of Pleistocene and Holocene events outlined in the preceding chapters has obvious significance for an understanding of prehistoric settlement on the Kom Ombo Plain. For one thing, climatic changes in a harsh desert setting are pertinent to any discussion of prehistoric man. In addition, many of the later prehistoric sites at Kom Ombo are found in geological context, permitting detailed analysis of the local habitat. For these reasons, it seems purposeful to focus the geomorphological evidence on the prehistoric geography of early settlement. In the following discussion of regional environments and local settings, we have summarized pertinent aspects of the material culture of prehistoric man at Kom Ombo on the basis of preliminary archeological reports from the different expeditions. This tentative archeological framework is then reviewed in the light of the paleoenvironmental evidence. Needless to say, full ecological interpretation of Stone Age sites and cultures cannot precede publication of the detailed archeological reports and must be undertaken by the archeologists themselves. But a sketch of the physical landscape, with a cultural perspective, can most effectively state the earth scientists' contribution to an interdisciplinary paleoecological study.

Early Paleolithic Industries of Southern Egypt

The oldest human industries of southern Egypt are recorded by stone tools. These are found on the surface as sporadic finds or concentrated sites, as well as within geological formations. These oldest assemblages are dominated by bifaces or "hand axes," with some cores, tools made from detached flakes, occasional choppers, and hammerstones. Waste flakes, the products of stone-working, may be abundant at sites employed as "working floors." Locally available rock forms the raw material. Organic vestiges have not been recovered from any Early Paleolithic sites.

Several groups of sites can be briefly mentioned as representative of the Early Paleolithic industries. An authoritative study of a dozen major sites in the desert east of Wadi Halfa and at Jebel Brinikol (Fig. 4–1) has been made by Jean and Geneviève Guichard (1965: 70–84). Some of these tool concentrations were collected from pediments near the foot of residual hills or "gebels"; others were found on top of the butte and mesaform gebels themselves. Although the majority are surface sites, with a possible admixture of tools from different periods, several stratified sites were excavated at moderate depths within colluvial deposits containing *Zootecus* shells. It appears that these stratified sites were disturbed and bedded by running water, whereas most of the gebel-top sites represent undisturbed working floors. There is no evidence of butchering activities or of habitation. Ferruginous sandstone is the dominant raw material, with some bifaces fashioned from thin ironstone slabs. Various types of bifaces are typical of these assemblages, and only cleavers and trihedral picks are conspicuously rare. A special "Nubian" biface is common in the younger sites of the Early Paleolithic group. Partly dictated by the slablike raw material, these Nubian bifaces are rather flat, symmetrical, and carefully trimmed, with the lower base describing a circle and with the point sharply biconvex. Flake tools, struck in good part from prepared, Levalloisian cores, are uncommon and are largely confined to side-scrapers. Occasional hammerstones and choppers complete the inventory. The Guichards indicate that these industries are broadly similar to the range of early-to-final Acheulian in subsaharan Africa. The Nubian bifaces, however, are peculiar to the sandstone reaches of the Nile Valley.

An Upper Acheulian occupation site (Arkin-8) has been excavated by Chmielewski (1965: 153–55) from the upper part of a wadi alluvium, on the western edge of the Nile Valley near Arkin. About 70% of the artifacts were neither waterworn nor wind polished. Quartz pebbles were the primary raw material, a factor largely responsible for the distinct character of this assemblage compared with those of the Guichards (1965). Nubian bifaces appear to be lacking, and the Levalloisian technique of core preparation is almost entirely absent. Instead, chopping tools are common. Abundant oval, amygdaloid, and cordiform bifaces are

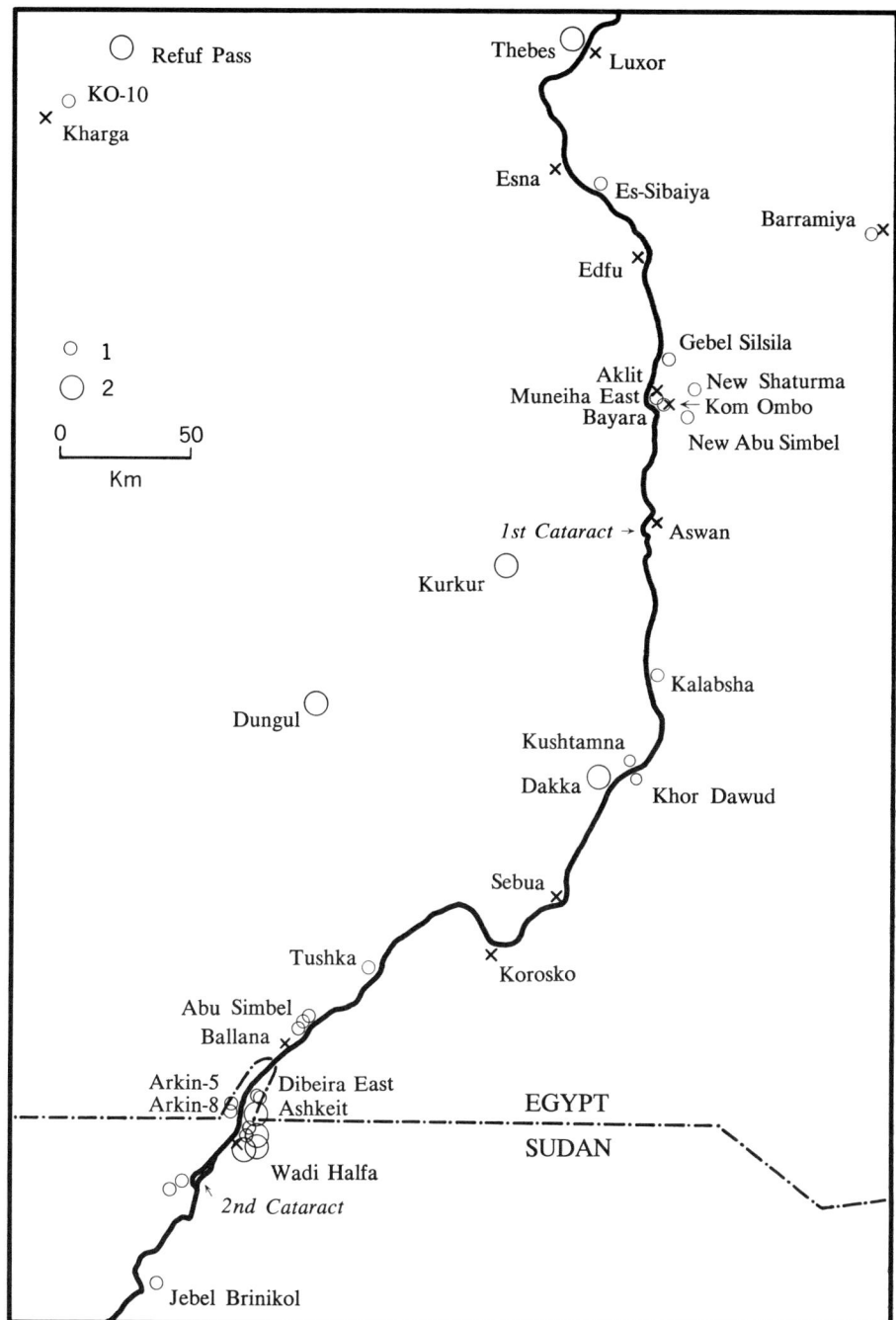

Fig. 4–1. Major Early and Middle Paleolithic sites in southern Egypt. Single (*1*) and multiple (*2*) sites distinguished. Location of modern place names indicated by X. Map: UW Cartographic Lab.

typical of both the Arkin and the Wadi Halfa assemblages. The red (2.5 YR 6/4) wadi alluvium in question is consolidated by carbonates and, by some tenuous indirect stratigraphy, is correlated with an undescribed nilotic silt a kilometer to the east.

North of Edfu, Sandford and Arkell (1933: 27–28) reported a rich Acheulian site on the landward margin of a + 30 meter gravel (Middle Terrace?) on the east bank of the Nile at es-Sibaiya Station. The materials may be contemporary with the gravels. Rather crude bifaces, coarse flakes, and hammerstones, mainly in brown chert, characterize the assemblage. Specific affinities with the Wadi Halfa and Arkin sites are few.

The last significant Early Paleolithic collections from southern Egypt include three stratified sites in the Kharga area, excavated by Caton-Thompson (1952: 25–26, 57–73, 95–98, Pl. 10–56) and E. W. Gardner. At the moment, in fact, the longest published Acheulian series in Egypt comes from mound spring KO–10, in the Kharga Depression. Made in a pale chert, the Khargan bifaces exhibit few specific affinities with the Nubian collections except for the almost total absence of cleavers. A good proportion of the flakes were struck from prepared cores. In addition to this Upper Acheulian industry, Caton-Thompson (1952: 26–27, 99–103, Pl. 57–60) found a so-called Acheulio-Levalloisian site geologically *in situ* at Refuf Pass on the Kharga escarpment. Although younger in the geological sequence, this collection is rather similar to the Upper Acheulian.

Ignoring the scattered surface finds or isolated "stratified" hand axes, these four groups of Early Paleolithic sites show a considerable range of typological differentiation, despite a close similarity in basic tool classes. Although temporal and geographical differences may be involved, the peculiarities of different raw materials may well have been paramount. So, for example, the Guichards (1965) emphasize the similarity of their collections to the Acheulian of Khor Abu Anga, near Khartum, which also employs ferruginous sandstone as primary raw material.

Unfortunately, none of the Early Paleolithic sites is precisely dated. Although correlations between the Kharga and Kurkur Oases (Chap. 7) tentatively suggest a Middle Pleistocene time range for the Upper Acheulian and "Acheulio-Levalloisian," the Arkin site may possibly date from the early stages of the Upper Pleistocene. Nowhere in the Nile Valley have artifacts been found in the High Terrace or its probable equivalents (see Chaps. 2 and 5; Sandford and Arkell, 1929, 1933, 1939; Sandford, 1934). Consequently, Early Paleolithic occupation in southern Egypt appears to postdate alluviation of the High Terrace complex and to predate aggradation of the Masmas Formation. It is unlikely that human occupation spans the whole of this considerable interval of time. Far more intensive and extensive archeological work will be necessary before either the temporal sequence or the geographical distribution of Early Paleolithic settlement can be properly understood.

Early Paleolithic Settlement in the Kom Ombo Area

Bifacial artifacts, patinated and occasionally wind-polished or waterworn, are found scattered on several of the Pleistocene gravel terraces of the Kom Ombo region. Geologically stratified sites were not observed, however, although earlier writers have reported scattered implements from some of the gravels. Vignard (1923: 4; 1954) obtained an Acheulian biface from the Fatira beds at Muneiha, while Sandford and Arkell (1933: 32–33, Pl. 21–23) refer to waterworn or rolled bifaces of "Chellean" type, found within or weathering out of various exposures of the same beds.

A complex surface site of Early Paleolithic type was collected from the Fatira beds between Bayara and Muneiha by David S. Boloyan. The artifacts, still unstudied, include bifaces, cores, flakes, and possibly some hammerstones. Dolerite is the primary raw material, to the total excusion of sandstone and its derivatives. It is therefore not surprising that Nubian bifaces are absent. All of the artifacts are wind polished or patinated, and some are slightly waterworn. Being situated at elevations of 96 to 100 meters, these surface tool concentrations were washed over by the Nile on at least several occasions during the late Pleistocene, so that specific associations could hardly have been preserved intact. But the collections themselves are of considerable typological interest. These sites at Muneiha East may be contemporary with the terminal phases of the Fatira beds, since Vignard and Sandford found analogous bifaces *in situ* within the body of the gravel. Since "Mousterian" implements were found in abundance under the Masmas Formation near Bayara (Vignard, 1923: 3, 8), the Early Paleolithic collections from Muneiha East can hardly be younger than the early Upper Pleistocene. They may represent workshops of prehistoric groups that once frequented the adjacent Nile floodplain, possibly during the Low Terrace stage.

These scant comments summarize the extent of our present knowledge of the local Early Paleolithic. Richer sites certainly exist but remain to be discovered and collected. In the meanwhile, however, we know next to nothing about possible local antecedents of the Middle and Late Paleolithic industries of the Kom Ombo region.

Middle Paleolithic Industries of Southern Egypt

The most common prehistoric remains in southern Egypt are flakes and flake tools of Middle Paleolithic type. These are found scattered through the desert hills and plains on both sides of the Nile Valley and even occur between the dunes of the great Libyan sand seas. Locally, wherever suitable materials outcrop at the surface, great concentrations of tools and flaking debris litter the surface for kilometers. This is the case where chert-bearing limestones or conglomerates are exposed along the Nile Valley edges, as at Thebes (Luxor), or where basal chert conglomerates occur in the Nubian Sandstone, as at Barramiya (Fig. 4–1). Elsewhere in the sandstone country, primary or secondary forms of ferruginous sandstone or quartzitic ironstone were sought out and were found chiefly on the gebel tops. Finally, the Pleistocene gravel terraces themselves provided suitable sites for countless flaking workshops.

The typology and classification of the Egyptian flake industries is currently in a state of flux. Caton-Thompson (1946) outlined a sequence of Middle Paleolithic industries of Levalloisian technique, based primarily on the cultural and geological stratigraphy established at Kharga. Intensive statistical study of large samples of Middle Paleolithic assemblages in the Wadi Halfa area by Jean and Geneviève Guichard (1965), Chmielewski (1965), and Wendorf, Shiner, and Marks (1965) suggests that the older evolutionary scheme is untenable. Instead, at least three distinct cultural traditions are apparent, and final publication of these studies will mark a milestone in our understanding of this fascinating culture phase.

The Guichards recognize a "classical" Mousterian tradition and a Nubian Middle Paleolithic, the latter subdivided into two stages.

The classical Mousterian (Guichard and Guichard, 1965: 84–86) is recorded from three major sites, all found on gebel-top surfaces east of Wadi Halfa. Despite a great quantity of flakes, finished tools are few and consist largely of Levalloisian flakes and a variety of cores. A few sidescrapers and denticulates are also present. These sites are interpreted as workshops. Similarities and differences are immediately apparent between the "classical" Mousterian of the Guichards (1965) and the "typical" Mousterian industries found in colluvial deposits on the pediments west and east of the Nile, north of Wadi Halfa (Wendorf, Shiner, and Marks, 1965: xxii). Comparison of the industries must await publication of findings from more recent excavations.

The Nubian Middle Paleolithic (Guichard and Guichard, 1965: 86–111) is based on 12 major surface sites from gebel tops east of Wadi

Halfa and at Abu Simbel, in some cases found on the surface of reddish-yellow colluvia with *Zootecus*. Three tool types characterize this industry: bifacially retouched foliate points with sharp and rounded bases; convex-edged "Nubian" side-scrapers; and two peculiar types of "Nubian" core. Except for the ubiquitous Levalloisian flakes and cores, bifaces occur in 10 of the 12 collections, and side-scrapers account for 22% to 60% of the total implements in each. Some of the bifacial leaf-shaped objects recall the Sangoan and Lupemban of subsaharan Africa, others the Aterian of northwestern Africa. The Guichards (1965: 98) report six windbreaks from one site. Built of crude rock slabs, these structures are two to four meters in length and up to 70 centimeters high. All open to the south. The working floors and these crude, unmortared constructions may be contemporary.

West of Dibeira (Arkin-5), Chmielewski (1965: 156–58) has described a habitation site with an industry similar to the Nubian Middle Paleolithic, labelled as "Sangoan." Ferruginous sandstone is the primary raw material here also. Arkin-5 was found within a reddish-yellow (5 YR 6/6) medium sand of colluvial origin, below a sand-swept and lag-covered pediment surface. Slabs of ironstone were mined from beneath the colluvium to be used for toolmaking. Some of the slabs were set obliquely against the walls of an artificial pit, possibly making a habitation site.

A third class of Middle Paleolithic assemblage is recognized by Wendorf, Shiner, and Marks (1965: xxii–xxiii) at two sites east and north of Wadi Halfa. This is a Denticulate Mousterian, with few side-scrapers but with 35% to 50% denticulates.

The temporal sequence of the Middle Paleolithic industries near Wadi Halfa is difficult to establish in default of superimposed stratigraphy. On the basis of different degrees of wind polish on tools, the Guichards (1965: 110) suggest that the classical Mousterian is older than, or contemporary with, Nubian Middle Paleolithic I. At the moment, the very general impression obtains that, although workshops were situated at distances of 15 kilometers and more from the present Nile, true habitation sites may have been confined to the actual valley margins. There is little reason to assume any significant, protracted Middle Paleolithic occupance of the desert hill country away from the Nile and the major wadis. Instead, mining and flaking were probably carried out on a periodic or temporary basis, possibly during the course of hunting forays into the desert wadis. The question of environmental change will be considered more specifically for the Kom Ombo area.

A *terminus ante quem* appears to be available for the Middle Paleolithic near Wadi Halfa. More evolved industries of Late Paleolithic type occur within and under the Masmas (Khor Musa) Formation, and the Middle Paleolithic sites are distinctly earlier than this sedimentary unit (Wendorf, Shiner, and Marks, 1965: xvi, xxxv). Consequently, the local Middle Paleolithic appears to be contemporary with, or earlier than, the Korosko Formation.

The Russian Academy of Sciences collected a number of surface sites in the Dakka-Kushtamna area, all of them found among lag littering sandstone outcrops within a kilometer or so of the Nile (Lyubin, 1964). Three reasonably homogeneous collections were made at Khor Dawud (I and II) on the east bank and at Dakka (III) on the west. They consist of cores, flakes, flake blades, points, and some small bifaces, which are compared with the Upper Levalloisian of Kharga. Of the total of 143 artifacts at these three sites, 88% were made in ferruginized sandstone or ironstone, 8% in a greenstone, and 4% in chert. The proportions are the same at each site. The deliberate selection of ironstone in an area with abundant fine-grained igneous pebbles is noteworthy.

Middle Paleolithic Settlement in the Kom Ombo Area

The distribution of Middle Paleolithic artifacts in the Kom Ombo region corresponds closely with exposures of Pleistocene gravels. Artifacts of all kinds are rare on the sandstone uplands away from the gravels and fluvial benches. This implies that the distribution of Middle Paleolithic artifacts is primarily restricted by the availability of suitable raw material for toolmaking. The gravels are dominated by dolerite and other fine-grained igneous rocks, whereas only quartz and ironstone or ferruginous sandstone are found in the local Nubian Sandstone. Artifacts were primarily made from dolerite and, less commonly, from felsite. Quartz and ironstone were seldom employed. Deliberate preferences for certain raw materials are indicated.

Although Vignard (1923: 3; 1954) and Sandford and Arkell (1933: 44) have reported Mousterian-type implements from beds underlying late Pleistocene silts between Bayara and Aklit, the exact nature or stratigraphy of these finds is not quite clear. Middle Paleolithic surface sites were collected from the Low Terrace surface near New Abu Simbel by P. E. L. Smith and Martin A. Baumhoff. These included crude Levalloisian flakes and cores, some scattered bifaces, and a single pick. Associations were unclear since the density of artifacts per unit of surface area was low. A surface site on Gebel Silsila, mentioned by Vignard (1923: 3, 8, 67; 1955a), was only collected in small part and was not published. We failed to locate this or similar sites during our survey of Gebel Silsila in 1962.

One interesting surface collection, a major workshop site, was made on a High Terrace spur near New Shaturma (Fig. 2–20) by P. E. L. Smith and others in 1962. Although the artifacts will be published by Smith, some provisional comments can be made here on the basis of a small, separate collection we made from the southern periphery of the site. These artifacts, not necessarily representative of the site as a whole, were kindly examined by Leslie G. Freeman. There are a number of Levalloisian flakes and cores, including a blade core and a point core; a regularized core; a chopping tool; a few bifaces—some with thin bases; a blade; a (possibly intentionally) broken bifacial leaf-shaped piece; denticulate-edged tools; notches; a side-scraper; and a quantity of cortical flakes. This collection might well represent a local, Egyptian, facies of the Middle Paleolithic.

Dolerite and rare pieces of felsite form the raw material. Exposed faces of all artifacts are sand polished or patinated, while the underparts are frequently veneered by a superficial, reddish weathering rind. Such pieces were probably once embedded in a reddish soil or wash. The artifacts lie scattered among a colluvial lag, but none of several dozen pieces examined showed signs of water wear. It is, therefore, rather probable that the workshop postdates the colluvium, which in turn is younger than the deep, red paleosol developed on the gravels (see Chap. 2). In fact, weathering rinds up to 2 millimeters in thickness are evident on the unworked portions of hand-axe butts, cores, and the external faces of flakes. It is, therefore, not unreasonable to assume that the pebbles were fashioned into artifacts after development of the red paleosol. Subsequent reddening of flaked pebble surfaces has been quite superficial. This indicates that occupation was no earlier than the Middle Terrace. On the other hand, some waterworn artifacts, apparently derived from this site, are found in the Korosko Formation at nearby New Korosko. These include a proto-Levalloisian core and a simple blade segment (Chap. 3), both of which may well belong to the New Shaturma assemblage. Possibly, therefore, Middle Paleolithic settlement in the Kom Ombo area was contemporary with the Korosko aggradation.

Although the collecting of the New Shaturma site was not "controlled," it provides a good example of a Middle Paleolithic flaking floor, recording one or more periods of use. Probably most of the artifacts are discards or are otherwise unfinished. If the suggested correlation with the Korosko Formation is correct, vegetation and game would have been abundant on the floor of Wadi Shait, a few hundred yards away. It is here that the contemporary camp sites and living floors were to be found. They have long been swept away in the course of repeated stream cut and fill.

In retrospect, the Early and Middle Paleolithic inhabitants of the Kom Ombo area have left an incomplete and biased record of their occupation. Of all their implements, only stone tools survive, and most of their stone artifacts, in as far as preserved, consist of waste flakes and imperfect or unfinished implements from working sites. Habitation or butchering sites cannot be verified around the Kom Ombo Plain. They may have once been located on the floors of Wadis Kharit and Shait, along the peripheries of the Nile floodplain, or on the banks of the Nile itself, but these particular deposits have either been eroded or buried by younger sediments. The popular notion that Paleolithic habitation was confined to or concentrated on hilltops is contradicted by the workshop inventory of the gebel sites as well as by the habitation sites uncovered in the Nile Valley near Wadi Halfa.

Seen within an ecological perspective, Early and Middle Paleolithic occupation of the deserts, no matter how transient, suggests strongly that water, vegetation, and game were available in both the major and minor wadis. Lacking sound correlations between sites and paleoclimatic events, it is impossible to prove this point. But it is improbable that prehistoric man would have ventured far from water, even if only to obtain raw materials for toolmaking. This argument is borne out by the Late Paleolithic groups who confined themselves to the Nile floodplain and its immediate environs at a time when the Egyptian climate was a little moister than today, although arid nonetheless. During the longer pluvial periods of the Middle and early Upper Pleistocene, the major wadis harbored stagnant pools of water long after the rainy season. And in most wadi beds, ground water could probably be tapped by superficial digging. A thorn savanna presumably lined the wadis—large or small—providing grazing for a variety of animals as well as food and fuel for man. Whereas settlement was probably concentrated near permanent sources of surface water in the major wadis and in the Nile Valley, temporary or seasonal settlement may have been found wherever game was abundant. Without organic refuse preserved from habitation sites, however, it remains impossible to reconstruct the settlement patterns or economic systems of these early populations.

Late Paleolithic Industries of Southern Egypt

The Late Paleolithic of the Kom Ombo Plain is characterized by a sequence of Sebilian industries as well as by a number of non-Sebilian traditions whose mutual relationships are as yet poorly understood. The classical Sebilian of the Kom Ombo Plain is widely regarded as a direct offshoot of Middle Paleolithic traditions in southern Egypt. Each stage of the Sebilian exhibits further technological innovations, with the abandonment of older tool types and the invention of new forms.

The first stage in the indigenous evolution at Kom Ombo is recorded by the Lower Sebilian, an industry collected from one good site and several mixed sites by Vignard (1923: 5–13; also Caton-Thompson, 1946). Found west of the Sebil Channel near Sebil Bahari, this site is at about 102 meters and may have been exposed by deflation from deposits of the Channel-A stage. Levalloisian cores and flakes are common, but a special, short-platformed core was used to detach subtriangular backed flakes, trimmed by a steep retouch. These backed points, with or without retouch, form the most characteristic item. Blades are very rare. A large anvil stone and several large rock fragments, thought to belong to broken hammerstones, were found. The raw materials include dolerite, "porphyry," ferruginous sandstone, and quartz. Baked clay, in place, provides evidence for hearths, while a fair number of badly smashed, mineralized bone fragments give some indication of the butchering practices used.

The Middle Sebilian sites (Vignard, 1923: 13–29), of which at least 58 were identified, are strung out along the surface of the Channel-A flood platform at an elevation of 100 to 101 meters. Some sites were buried in sand (Vignard, 1923: 26) and may originally have been in geological context. Small Levalloisian cores are common, in part subsequently retouched and used as side-scrapers. Levalloisian flakes were replaced by backed points, often with basal truncation or steeply retouched on one or more margins. Many of the flakes and points so produced have subgeometric shapes, approximating atypical triangles, trapezoids, or lunates. Some blade cores occur as well as 4- to 6-centimeter-long blades, retouched in part and systematically snapped into shorter segments. These, together with the first handstones and millingstones, are attributed to the second stage of the Middle Sebilian, a subdivision made on tenuous grounds, considering that we are dealing with surface sites. A notable reduction in average size was achieved, both by flaking techniques and by deliberate snapping of flakes or blades. Chert and chalcedony form the major raw material, displacing dolerite and quartz.

These Middle Sebilian sites must, in part, have formed sizeable middens with volumes of several cubic meters, and large enough to be exploited for fertilizer (*sebakh*) by modern Egyptians. Shell was an important constituent, primarily that of *Unio* and *Corbicula*. Abundant fauna, with all marrow bones systematically broken in similar places, included the herbivores and fish identified in the faunal list of Gaillard (1934) (see Chap. 3). Masses of rock fragments (dolerite, "porphyry," quartz, sandstone) were carried in from nearby gravel terraces, to be used as hammerstones or, possibly, as "pot-boilers" to heat water—Vignard (1923: 26) suggests that red-hot rocks were taken from the fire and dropped into water held in leather bags. In addition to ash, there was much "baked" clay of red or black color. This is natural silt or clay, partly fired in contact with hearths—either *in situ* underneath the fire or in the form of earth clumps used to shield the flames or cover up glowing embers. The grinding stones were often smeared with ferric oxide and several chunks of brownish or reddish oxides were found among the sites. Although Vignard labels these as "limonite," they were probably derived from the hematite ores found in Wadi Abu Subeira or east of Aswan, 35 or more kilometers upstream. A single piece of white coral may indicate trade connections with the Red Sea littoral.

The Upper Sebilian sites (Vignard, 1923: 29–63) are located on a low bench or on the floor of the Sebil Channel at elevations of 98 to 99 meters. They appear to belong to Channel B. Diminutive forms are characteristic, and the industry is predominantly submicrolithic in character. Small points were retouched to serve as end-scrapers or arrow tips; others were retouched on two or more sides; still others form triangles, trapezoids, or lunates. There are great numbers of blades and bladelets, 1 to 7 centimeters in length and 0.3 to 2 centimeters in width, designed for various purposes. Many of the small geometric flakes and particularly the microburins, obtained from the flaking waste of innumerable triangles and trapezoids, were probably mounted in wood or bone to be used as arrows, spearheads, or harpoons. Microlithic scrapers and borers are also present. As in the case of the Middle Sebilian, there are grinding stones, some smeared with oxide; bone fragments; shell, some deliberately broken; baked clay from hearths; and low middens. Pottery is entirely absent. In one of the 33 sites collected, Vignard (1923: 56) found a semicircle of baked clay measuring 18 meters in length, 2 to 3 meters in width, and 30 to 35 centimeters thick. He suggests that it delimited one side of a living site.

As early as 1941, Huzayyin (p. 262) emphasized that the Sebilian industries at Kom Ombo were differentiated on the basis of surface sites and that typological revision might be necessary. Caton-Thompson (1946: 108, note 7) was equally reluctant to accept Vignard's detailed subdivisions and intermediate stages for similar reasons. P. E. L. Smith (1964a, b; 1966), who collected a number of sites in the Sebil area in 1963, confirms the basic scheme of sequential industries but believes that modifications will be required, particularly in the delineation of the Upper Sebilian. On the whole, however, these reservations are confined to typological details. The presence of both habitation and flaking sites, with evidence of complex human activity, is obvious, and the functional relationship of these occupation floors to successive flood benches of the Sebil Channel (Fig. 3–31) was clearly apparent during the course of our field work.[1]

After publication of his Sebil monograph, Vignard (1955a) discovered an industry of Aurignacian affinities in 1922 near the southern bank of Wadi Shait, about 7 kilometers southeast of Gebel Silsila Station.[2] P. E. L. Smith (1966) excavated an identical site at Gebel Silsila 1, characterized by many scrapers and burins. It is now called the Menchian. Two other, non-Sebilian industries were excavated and first defined from Gebel Silsila 1: (a) the Sebekian, a blade industry with some burins and

1. Vignard (1955b) also reports briefly on a sequence of Sebilian I, II, and III industries collected in the area almost due east of Kom Ombo near the Cassel Canal, which he euphemistically designates as "Burg el-Makhazin." These localities are only described in the vaguest of terms, and they were all lost to cultivation shortly after Vignard's visits in 1922 and 1923. Most distinctive appear to have been the Middle Sebilian collections made from among small shell heaps (primarily *Unio*), aligned along small, sanded-up channels descending from the Burg el-Makhazin. Possibly, we are dealing here with equivalents of the Ineiba Formation. Since there are next to no archeological materials in the Ineiba beds where exposed today, however, it would seem more reasonable to associate these "Burg el-Makhazin" sites with the temporary stage of exceptionally high Nile floods about 12,000 years ago. The presence of *Unio* would, in fact, suggest the proximity of now-obscured Nile channels. Whatever the ecological setting may have been, dolerite was the raw material employed at these Middle Sebilian sites, in contradistinction to the flint, chert, etc., used at Sebil and Gebel Silsila.

Vignard (1955b) also mentions Lower and Upper Sebilian sites from the footslopes of the Burg el-Makhazin. The description of these Lower Sebilian materials is not very convincing, and we are inclined to question the existence in this area of a Sebilian sequence that parallels that at Sebil proper.

2. Here again it would be vital to have an understanding of the physical setting of this "Wadi Shait" site as well as the nearby site of Menchia. Despite three notes by Vignard (1955a, b, c) on the sites from the eastern part of the Kom Ombo Plain, the exact locations all remain uncertain. Unfortunately, since they have all been destroyed, it seems that we will never know their ecological significance.

end-scrapers, but lacking microburins and geometric forms (radiocarbon dates 14,050 B.C., 12,290 B.C., and 12,150 B.C.); chert is the major raw material; (b) the Silsilian, which underlies the Sebekian, is a true microlithic industry with burins, microburins, bladelets, and grinding stones; exotic pebbles form the raw material. Finally, a last, as yet unnamed, non-Sebilian industry utilizing Levalloisian techniques was collected from two small occupation sites of Khor el-Sil (radiocarbon date, 15,050 B.C.). The archeological inventory here includes burins, grinding stones, and worked bone.

The complexity of industrial traditions in the Late Paleolithic of the Kom Ombo Plain is a little surprising, particularly when the sites are almost all found in relation to channels of the Gebel Silsila Formation and in similar ecological settings. Obviously, only a few of these industries are strictly contemporary, but the presence of a more or less continuous, indigenous Sebilian tradition alongside of four or more distinct, and possibly intrusive, groups in a time span of five millennia is nevertheless interesting. The physical setting and stratigraphy of several of these sites will be outlined in the subsequent sections.

North of the Kom Ombo Plain, surface sites of Middle and Upper Sebilian affinities were traced along the western margins of the valley as far downstream as Esna (Sandford and Arkell, 1933: 45–47; also Caton-Thompson, 1946). Upstream, the first and second raising of the Aswan Dam seem to have submerged many of the Late Paleolithic cultures. Sandford and Arkell (1933) mention no sites between Darau and Sebua, and, although the stretch between Aswan and Korosko was rapidly surveyed by the Combined Prehistoric and Yale expeditions in 1964–65, it does not seem to have yielded much Late Paleolithic material. The Russian expedition of 1961–62 reports two interesting Late Paleolithic sites, however, including a collection of 1500 artifacts and chips weathering out of nilotic gravelly sands (120 to 126 meters) at Khor Nabruq, opposite the mouth of Wadi Allaqi (Vinogradov, 1964). Flint and chalcedony form the raw material, and 80% of the material consists of flaking debris. Typologically the artifacts compare in a general fashion with the Upper Sebilian, although geometric forms are uncommon. Vinogradov (1964) emphasizes that this collection is not waterworn and that it probably represents a workshop. A much smaller surface concentration of similar typology was collected from similar sands of the Gebel Silsila Formation a little north of Kushtamna.

Upstream of Korosko, richer and more abundant sites appear once again. Four late Paleolithic industries have been identified in the

Ballana–Wadi Halfa area (Waechter, 1965; Wendorf, Shiner, and Marks, 1965). These include (*a*) the Khormusan, utilizing a highly evolved Levalloisian technique, with a dominance of denticulates and burins; (*b*) the "Sebilian," with certain clear similarities to the Middle Sebilian of Vignard (1923); (*c*) the Halfan, with backed flakes and backed microblades but dominated by a special type of retouched or unretouched flake; these Halfan flakes also occur at Khor el-Sil; and (*d*) the Qadan, including cortex-backed scrapers, lunates, retouched points, burins, and retouched flakes, mainly of microlithic character. The Khormusan has two radiocarbon determinations of 20,750 and 15,850 B.C. and extends through the entire Masmas (Khor Musa) Formation into the base of the Gebel Silsila (Sahaba) Formation. The "Sebilian" extends through the entire Gebel Silsila Formation, and its latest known expression has a C^{14} determination of 9000 B.C. The type site of the Halfan, at the base of the Gebel Silsila Formation, has a radiocarbon date of 14,550 B.C. Finally, the Qadan, which postdates the classical development of the Gebel Silsila Formation, has C^{14} determinations of 9950 and 4400 B.C. Of possible relation to the Qadan are the fine-barbed bone points excavated by Wendt (1966) from an abri ("Catfish Cave") east of Ibrim and which predate a level with a C^{14} determination of 5100 B.C.

Working west of Wadi Halfa, the University of Colorado expedition has devised a scheme of five industrial complexes (Wheat and Irwin, 1965), some of which can be readily equated with those of Wendorf, Shiner, and Marks (1965). The Buhen is a Levalloisian facies using white quartz, chert, and ferricrete sandstone. It includes a large number of burin types and is probably analogous to the Khormusan. The Halfan industry corresponds to the industry of the same name as identified by the Combined Prehistoric Expedition. The Dabarosan, using a double-ended core technique, includes blades, denticulates, scrapers, perforators, and burins. It has a radiocarbon date of 16,140 B.C. (Hewes *et al.*, 1964; Hewes, 1964). An unnamed industry, with curvate-backed bladelets and geometric microliths follows. Finally, the Wadi industry has broad similarities with the Qadan.

Information on the industries of the Ballana–Wadi Halfa area is complemented by a number of paleontological discoveries. A cemetery with 39 mineralized skeletons was excavated west of Wadi Halfa by the Colorado expedition. The human remains, while obviously *Homo sapiens sapiens*, have several strikingly robust, primitive features (Hewes *et al.*, 1964; Armelagos *et al.*, 1965). This cemetery is associated with the Wadi complex of Wheat and Irwin (1965) and is thought to date from about

5000 B.C. A massive human mandible, suggestive of the same population, is associated with a Halfan site nearby (Armelagos, 1964). Another burial, with two skeletons of similar type, was uncovered north of Wadi Halfa (Solecki, 1963)—unfortunately, without direct cultural associations. More recently, the Combined Prehistoric Expedition found two further cemeteries, again with populations of similar morphology, at Dibeira and at Tushka. The former is associated with the Qadan, the latter with the Halfan (Wendorf, 1965). Significantly, the human remains at Gebel Silsila 2B also pertain to a similar physical type (Reed, 1965a).

Although the industries of the Ballana–Wadi Halfa area are part of an independent complex of industries, there are obvious analogies with the Late Paleolithic of the Kom Ombo Plain. The same shift from Levalloisian to microlithic techniques at both localities is accompanied early in the sequence by a change from local raw materials to nilotic pebbles. Skeletal materials from broadly morphologically similar human populations are widespread in both areas. Although some industries were widely dispersed, different traditions, local or intrusive, appear to have persisted side by side with them. A balanced overview of these rather complicated Late Paleolithic assemblages of southern Egypt will only be possible after publication of final reports by all of the archeological missions.

Physical Setting of Occupation Sites at Gebel Silsila 2A

The archeological site Gebel Silsila 2A occupies an area of over 250,000 square meters on the western bank of former Channel A, adjacent to Km 819 of the railroad (Fig. 3–23). Rich but heterogeneous industries litter the surface and were collected during November and December of 1962 by Martin Baumhoff and Heinz Walter, using 30 by 60 meter squares (see Reed et al., n.d.).

Although some artifacts are found on top of the yardangs (see Chap. 3), concentrations are limited to the surface lag in the intervening depressions. Nowhere could artifacts or bone be seen in the abundant natural exposures on yardang faces. In one case, however, approximately 200 artifacts were found within or weathering out of a large slab of halite cemented duricrust, some 4 to 5 centimeters thick. The material is a pale-brown (10 YR 6/3) silty medium sand with 6% carbonates, a pH of 6.4 to 6.7, and 10,000 parts per 1,000,000 water-soluble salts (No. 142,

Table 3–8). This slab was originally part of a pedogenetic lime-salt crust, developed in the terminal levee deposits of Channel A. It contains a dozen small fragments of mineralized, sand-blasted bone. After final abandonment of the Fatira Channel, deflation removed softer sandy beds and gradually undermined the crust. Found at an elevation of 96.5 meters today, the slab has been lowered at least 3 meters from its original position. The only similar crust preserved in place was found at 99 meters elevation on a yardang top 700 meters due south. In that case, no artifacts were found *in situ,* although chunks of fire-reddened clay were embedded in the crust.

The artifacts from the duricrust slab form part of a workshop and consist of flaking waste or rejects. Two atypical triangles (2.4 and 2.6 cm long); a small, retouched point (2.6 cm); and a number of flakes, detached from cores after removal of blades and microbladelets, all suggest Upper Sebilian or other microlithic affinities, however. The minute size of the flaking waste supports the argument for a submicrolithic industry. The raw material is a pale-brown chert, and preserved pieces of cortex infer nodules originally 4 to 6 centimeters in diameter. Several segments of cortex are deeply weathered and show hematite stains, suggesting the Fatira beds as the source of the chert pebbles. The surface artifacts from the adjacent, collected square (S–4, E–1) may amplify the technological picture.

This small collection is hardly representative of the rather variable industries found in the grid rectangles. But it points toward the uppermost strata, above 98 meters, as the probable source of the deflated artifacts. It is difficult to decide today whether there originally was much vertical stratigraphy at Gebel Silsila 2A. Preliminary assessment of the artifact assemblages indicates that several industries, none apparently closely related to the Sebilian sequence, are present (Reed *et al., n.d.*).

Although Channel A is older than Channel B, it is possible that surface sites related to Channel B, perhaps even mantled by flood silts, originally rested on top of the site-A complex. The duricrust with artifacts may be just such an example. Although typological and radiocarbon analyses may sort out many of the inherent difficulties, the internal stratigraphy of the major artifact concentrations may prove impossible to resolve.

The ecological setting of the 2A sites can only be inferred from their riverside location in an area of complex riverbank and levee beds (see Chap. 3). What little bone is preserved is generally fragmentary, mineralized, and wind-abraded. It may also have been waterworn originally. Identification of some of the bone seems possible. Although the levee beds show frequent root drip and occasional pieces of calcified wood, most of the casts and fillings pertain to lower plants. Evidence concerning the nature of the occupation sites must be awaited from the archeological reports.

Presumably, vegetative growth was optimal during the postflood period, between October and January, while Channel A was reduced to an insignificant rivulet. Abundant fish could be taken in shallow pools of water with little effort. Semiaquatic, woodland, and steppe animals could be stalked in the riverine thickets or as they approached the stream at night to water. By midwinter, the Fatira Channel may have dried out so that the seasonal settlements were shifted to the banks of the Nile proper or were abandoned in favor of a more mobile hunting subsistence.

Physical Setting of Gebel Silsila 2B, Area I

Excavation and surface collecting at Gebel Silsila 2B (Fig. 3–23) by Baumhoff (1965) and Walter were begun in December, 1962, and terminated the following March. About 9,600 square meters were systematically collected by 10-meter squares within and on both sides of the former Channel B (see Chap. 3), while two stratified east-bank sites were excavated. Good site preservation permits a far better interpretation of the local site settings. Most of the occupation levels found *in situ* were distinctly localized on inclined, backset levee beds. Much of the cultural material is also found among the surface lag of the old channel itself (Fig. 3–23). This suggests that the channel was abandoned by the Nile during the low-water season for at least a part of the Channel-B period. The abundance of shell and catfish bones and the presence of both aquatic and terrestrial animals suggests a highly riverine setting. This would be compatible with a flood arm active during the high-water season (late July through October), followed by reduction to a small watercourse or to a string of subcontinuous pools for the remainder of the year.

The setting and microstratigraphy of Area I has already been outlined (Chap. 3) and illustrated (Figs. 3–26, 3–27, Table 3–5). The nature of the archeological levels still remains to be discussed.

LEVEL 1

The principal archeological feature of Level 1, bed *c*, includes a scattering of artifacts, made from exotic, multicolored nilotic pebbles, and the human frontal bone (Reed, 1965a) alluded to already. The bed itself consists of a subcontinuous lense of pseudo-pebbles with a silty sand matrix. To obtain an idea of the concentration and type of material within and on top of this lense, we analyzed two random 25-centimeter squares. The first contained:

> 22 rolled concretions and silt pseudo-pebbles
>   3 single *Unio* shells
>   3 single *Corbicula* shells
>   1 artifact.

The second included:

> 206 pseudo-pebbles
>   2 coarse pebbles of light-gray shale
>   9 single *Unio* shells
>   7 single *Corbicula* shells
>   3 subrounded pebbles of fire-reddened, charred silt
>   2 fragments of rolled mammalian bone.

On a more general basis, between 25% and 50% of the bivalve shells are found as contiguous, matching pairs. These are probably the proportions generally found in natural deposits. The rolled bone and baked silt pseudo-pebbles strongly suggest derivation for all of the cultural materials, and, except for the artifacts and the skull fragment, this is a natural deposit. In other words, we are not dealing here with an occupation floor but with archeological materials derived from an eroded, older site upstream.

LEVEL 2

Bed *d*, the major archeological level excavated by the Yale expedition, is a consolidated, backset bed with pseudo-pebbles and medium quartz gravel, set in a matrix of silty sand. The impregnation with secondary carbonates and other evaporites is notable (Table 3–5), as is the coarse vertical and horizontal root drip contemporary with the former exposure of this surface. One such root cast was 4 to 5 centimeters thick and over 110 centimeters long. A protracted period of subaerial weathering is suggested, during which solubles were concentrated at or just below the surface by upward capillary movement of soil water. Seasonal resolution does not seem to have been effective, which suggests a lack of seasonal flooding at the time. At a later date, bed *d* was rapidly buried by the fluvial sands of bed *e*.

The archeological materials are embedded in, or more commonly rest on, bed *d* (Fig. 4–2). They include large quantities of worked chert and flint, suggesting a toolmaking tradition intermediate between the Middle and the Upper Sebilian (Reed et al., n.d.; Baumhoff, 1965). In addition to fragments of innumerable grinding stones, there is a great mass of fractured rock scattered at random over this surface. Fragments of mammalian and catfish bone are abundant, some with deliberate fractures or evidence of artificial working.

A first problem is the nature and origin of the fractured rock. A meter square (C–2) was removed, and all material over 2 centimeters in length, other than fashioned implements, was identified and counted out by the authors. The results are given in Table 4–1. In addition, there were 30 single shells of *Unio* and 8 of *Corbicula*. All of the rock, except for 6 quartz pebbles, was fractured, whereas the pseudo-pebbles were 100% intact. This dichotomy can best be explained by human intervention. Equally significant is the fact that crude rock fragments were entirely absent in Level 1 and in all nonarcheological beds. This material was well beyond the competence of the Silsila Channel and was not freshly rolled. It could all have been obtained from the Fatira terrace, about 2 kilometers away. And it was certainly fractured deliberately or through use at the site. Vignard (1923: 26) records similar rock debris from his Middle and Upper Sebilian middens.

The second question is whether the archeological materials are undisturbed and *in situ*, or whether they were derived and stream bedded—or at least subsequently reworked by fluvial agencies. To resolve this problem, we constructed an orientation diagram. Since the fractured rocks form part of the archeological inventory and since their major axes are ideally suited for compass orientation study, a total of 100 were measured at random from 4 different squares. Dividing these into 9 quadrants of 20° each, the following distribution patterns are obtained:

N 5° – 25° E	12
N 25° – 45° E	16
N 45° – 65° E	14
N 65° – 85° E	7
N 85° –105° E	9
N 5° E– 15° W	12
N 15° – 35° W	7
N 35° – 55° W	15
N 55° – 75° W	8

Fig. 4–2. The occupation floor (Level 2) of Gebel Silsila 2B, Area I. Note intact handstone, chert artifact, and fragments of bone and shell amid rock debris.

Obviously there is no alignment whatever to the normal dip of the beds (N 10°–20° E) or parallel with the former direction of stream flow (N 60°–80° W). The standard deviation is only 4.0, and there can be no question that this is a random scatter, in which distribution reflects human activities rather than stream bedding. In other words, Level 2 is an occupation floor *in situ*, in both the geological and the archeological sense. Deposition of bed *e* does not seem to have disturbed the site in any measurable fashion. The southeastern edge of the site has been washed away, however, and a line of cobbles marking this periphery is oriented with the stream channel.

Table 4–1. Coarse aggregates from Square C–2, Level 2, Gebel Silsila 2B, Area I

Material	Total number	2–6 cm	Over 6 cm	Fractured	Intact
Pseudo-pebbles (concretions, silt, etc.)	1,263	1,260	3	—	1,263
Sandstone	133	101	32	133	—
Quartz and quartzite	92	92	—	86	6
Cherty limestone	4	4	—	4	—
Dolerite	108	99	9	108	—
Total	1,600	1,556	44	331	1,269

The abundance of fractured bone and of grinding stones, together with the comparatively limited proportions of flaking waste, indicate that Level 2 is a living site. The grinding stones were probably used to grind pigment for body paint, since more than 80% of these fragments show traces of reddish stains under the microscope (Baumhoff, 1965). Body paint—but green in color—was a rather significant cultural trait of the Predynastic Egyptians (Baumgartel, 1965), so that, conceivably, these grinding stones were in no way related to food preparation.

LEVEL 3

A secondary archeological horizon is associated with bed *f*. The stratum itself is a subcontinuous surface of pseudo-pebbles with rare quartz. It is consolidated by carbonates (43% $CaCO_3$), which penetrate a little into the top of bed *e* where there is some root drip. The pseudo-pebbles are locally concentrated and appear to have been disturbed by fluvial washing and local stripping. Subaerial weathering appears to postdate this superficial erosion.

A scatter of artifacts and bone fragments occurs on top of this surface, together with abundant, stratified *Unio* and *Corbicula* shells (25% to 50% contiguous). All of the bone has been somewhat rolled, however, while subrounded to rounded pseudo-pebbles of baked, reddish-yellow silt occur throughout. Although most of the artifacts are quite fresh, some are slightly but distinctly waterworn. Two 25-centimeter squares were again examined in total. The first contained:

>   64 pseudo-pebbles (2 to 5 cm long)
>   1 quartz pebble (1.7 cm)
>   4 single *Unio* shells
>   2 single *Corbicula* shells
>   3 artifacts
>   2 dolerite fragments (2.8 and 3.9 cm long).

The second included:

>   111 pseudo-pebbles
>   5 single *Unio* shells
>   1 *Corbicula* shell
>   1 artifact
>   2 sandstone fragments (7.0 and 9.4 cm long).

Except for the scattered rock fragments, this horizon is quite similar to Level 1 and almost certainly does not represent an occupation floor *in situ*. Everything can be readily explained by derivation from an eroded occupation site a little upstream. Such a site need not be younger than Level 2.

LEVEL 4

A great number of microlithic artifacts occur in bed h, which is locally superimposed on g. A surface collection was made of these artifacts, which litter the edge of the Masmas Formation as far as Area II—where similar artifacts, also made in multicolored, nilotic pebbles, occur in two archeological horizons. Several hearths, baked into clayey beds of the Masmas unit, occur at the base of g, about 70 meters southeast of Area I. Since the silt comprising bed g is loose and badly disturbed, it is impossible to ascertain the nature of these occupation sites or, for that matter, whether we are dealing with a true flood silt or a colluvium.

INTERPRETATION

The Area I sites present an excellent type case that includes different examples of Late Paleolithic occupation. Level 2 may have been the most typical, a living site on the sloping bank of a seasonal overflow channel. Comparable physical settings can still be seen along the Nile in November and December, when the river exposes its banks, foot by foot, veneered by a fresh mantle of clean sand. Where low islands or major sandbanks slow down the waters at flood stage, silts may be accreted for several months under slough-like conditions. Tamarisk scrub, sedge, grass, and patches of halfa emerge on the riverbanks, giving way to acacias or palms on the berm of the levees. The margins of the late Pleistocene Fatira Channel must have once been similar, with stretches of sandy and of muddy banks, partly vegetated by scrub and bunch grasses and surmounted by a fringe of woodland or savanna within the range of the groundwater table. Abandoned channel stretches held pools of quiet water where flocks of birds, including Egyptian goose, several kinds of duck, heron, and stork (see Reed, 1965b), presumably congregated in great numbers. The arrival of a group of hippopotamus or crocodile may have caused commotion among the riverbank settlers when wild cattle, buffalo, and warthog (see Reed, 1965b, 1966) browsed in the riverine woodlands. Further out in the desert savanna, Barbary sheep, hartebeest, isabella gazelle, and wild ass grazed, on the alert for preying felids (see Reed, 1965b, 1966).

Many sites comparable to Level 2 were occupied for a season or more and were then abandoned. Sometimes a location used for several seasons was found swept away after the flood, as the channel shifted back and forth. Occupation refuse and possibly a number of burials were eroded in this manner and were reworked among the deposits of the stream bed or laid down again on the banks. This may have been the origin of Levels 1 and 3.

Finally, some prehistoric groups appear to have occupied the levees, the highest ground in stream proximity. Here they might be able to stay out the flood without shifting their campsites. Perhaps there were added advantages to local hunting at that time of the year. Larger animals were easily bemired on the muddy alluvial flats, ready preys to the huntsman with a microlith-tipped arrow or javelin. The location of Level 4 can be most readily explained in this fashion.

On frequent occasions, rocks were carried in from the Pleistocene gravels at Fatira or elsewhere, the flint nodules to be worked, the harder rocks to be used as hammerstones or to smash animal bone in order to get at the marrow. Some groups preferred the exotic pebbles exposed in the main course of the Nile at low water, colorful red jasper and carnelian, milky white chalcedony, brightly banded agate—semiprecious stones even to this day and suitable raw materials for prehistoric flint working. Red ochre from the hematite exposures found in wadi valleys a day or two's travel further upstream was occasionally collected or traded.

Physical Setting of Gebel Silsila 2B, Area II

The Area II site (see Chap. 3) was partly excavated by David S. Boloyan in February, 1963. The archeological materials are confined to beds β and δ, which are locally superimposed and elsewhere separated by a lense of sandy silt (γ). Occupation was more or less continuous, however, and there appear to be no significant differences in the cultural inventory. Fresh microlithic artifacts, rather similar to those found in bed h of Area I, occur in great masses among the pseudo-pebbles of bed β. In part they are stratified, in part, chaotically bedded, as if they had been stamped into wet mud underfoot. Fractured pieces of dolerite and quartz and occasional, fragmentary *Unio* shell are also present. The same features occur in the silty sand of bed δ, together with fragments of fish and mammalian bone.

Since the strata in question belong to a small, subsidiary overflow channel some 7 to 8 meters in width, the site was probably used after the flood peak had passed. The bedding of the artifacts suggests that the sediments were still wet at the time of occupation. This fits in well with the overall geomorphic interpretation of the 2B sites.

Physical Setting of Occupation Sites at Khor El-Sil

The Late Paleolithic sites on the Manshiya Channel at Khor el-Sil were collected or excavated in late October and early November of 1962 by P. E. L. Smith, in part in conjunction with Martin A. Baumhoff. All of these sites are related to the Gebel Silsila Formation and were originally stratified within the beds of a minor Nile channel once flowing 1.5 to 2 kilometers east of the railway, joining the Fatira Channel at the northern end of the plain.

The one site with stratigraphy at depth, Khor el-Sil 2 (Fig. 3–28), illustrates the nature of these occupation floors. The lag surface included root-cast fragments (up to 10 cm long, in part spongy), wind-scoured concretions (under 1 cm diameter), fine to medium quartz pebbles, and broken *Cleopatra* shells. Abundant chert artifacts (see P. E. L. Smith, 1966), chunks of fire-reddened clay, fragmentary animal and fish bone, and "intrusive," coarse concretions of patinated chert can be included in the archeological inventory. The strata, from top to bottom (Fig. 4-3), include:

a) 10 to 20 cm. Pale-brown (10 YR 6/3–4), medium-sandy silt with rare dispersed concretions and rich in mica and heavy minerals; well stratified; loose, single-grain structure (No. 27, Table 3-8). Artifacts are found near the top of this bed and again at the base, although the bulk of these flood silts are quite sterile.

b) 3 to 5 cm. Subsurface crust consisting primarily of same material as bed *a*, partly consolidated by salt and calcium carbonate. A subcontinuous lense of coarse sand, up to 2 cm thick, with some quartz pebbles, often forms the upper part of this crust. Salt and carbonate enrichment is greatest at the base of such sand lenses, with as much as 52% $CaCO_3$. A composite sample of the crust yielded a texture of silty sand (No. 28, Table 3-8). Structure is medium, subangular blocky, favouring breakdown of the crust into concretions similar in shape to those of the surface lag. This is the mixed *Ca/Sa*-horizon of a crust Yerma soil (Kubiena, 1953: 184–85) or orthic Calcorthid (G. D. Smith *et al.*, 1960: 157–58). A fair number of artifacts were embedded within the crust or just on top of it, together with *Unio*, fish, and bone fragments. A piece of charred sandstone, foreign to the site, was found at the base of the crust, penetrating into bed *c*.

c) 40 to 50 cm. Very-pale-brown (10 YR 6–7/3), coarse sand with laminae of mica and heavy minerals; horizontal, well stratified; loose, single-grain structure. Sterile except for a piece of mammalian long bone near the top.

d) Over 30 cm. Alternating foresets of brown (10 YR 5–6/3), medium-sandy silt (averaging 2 cm thick) and very-pale-brown (10 YR 6–7/3), coarse sands, dipping 15% to 45% to NNW. These channel beds are sterile.

The recognition of the *Ca/Sa*-horizon is of value in interpreting similar duricrusts at Gebel Silsila 2. Duripans of this type were commonly observed at Khor el-Sil, always near or along the contact between beds of different texture, at variable depths but always within 40 centimeters of the surface. These are typical arid-zone soil phenomena, reflecting downward leaching as well as upward capillarity. Where destroyed, however, these duripans do not form again under present climatic conditions, and, where deflated, they are gradually undermined. It is almost certain that these are fossil features that require considerably more rainfall to form, possibly on the order of 100 millimeters per year.

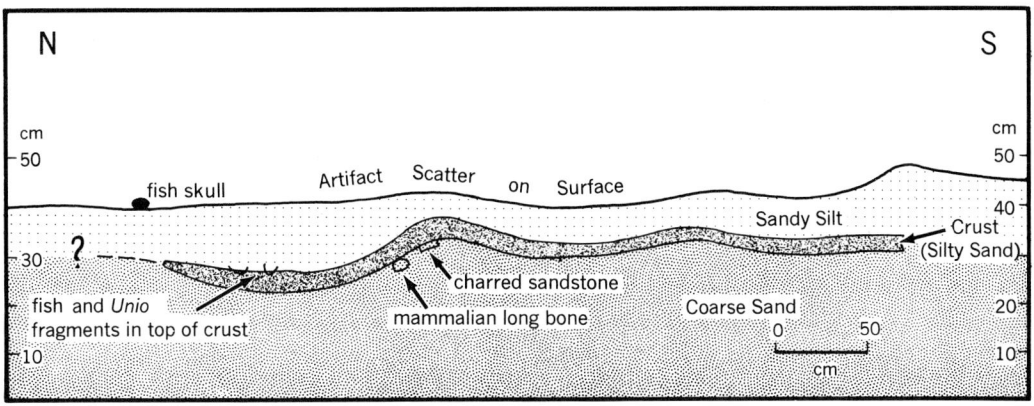

Fig. 4–3. Microstratigraphy of the Late Paleolithic site Khor el-Sil 2. Artifacts occur cemented to upper part of crust and as a surface lag. Map: UW Cartographic Lab.

The second point of interest in this profile is the zone of "clean" silt separating analogous concretions and artifacts in the lag above and in the crust below. One possible explanation is to assume two archeological levels. The deposits are all water-laid, and the artifacts show some evidence of fluvial disturbance. Thus, there might be two successive occupations at the same locality, separated by a rather short time interval. Typologically, however, the surface and subsurface artifacts are identical (P. E. L. Smith, *in litt.*, March 15, 1966), and the argument for an upward migration of coarser materials to the surface, leaving an intermediate zone free of artifacts, concretions, and pebbles, seems a more reasonable one. This desert process has been advocated by Springer (1958), Buol (1965), and Cooke (1965), who explain it by alternate wetting and drying of the soil. The structural deterioration of the sandy silt of bed *a* to a loose, powdery horizon, readily compacted under foot, supports this pedogenetic hypothesis. Other cases at Khor el-Sil of duripan concretions found in exposed lag, up to 15 centimeters above the subsurface crust from which they were derived, support this argument. In conclusion, we believe it likely that only one archeological level is present at Khor el-Sil 2. It would originally have been found in the lower part of bed *a* but would have migrated upward through gradual displacement in a powdery, desert topsoil. Only those artifacts adhering to the duripan may be still strictly *in situ*.

At Khor el-Sil 1 (Fig. 3–28), archeological materials of mixed typology (see P. E. L. Smith, 1966) are essentially confined to the surface lag and include windworn or waterworn bone fragments, sandstone rubble, coarse chert nodules, and a block of shell conglomerate of unknown origin. Among the natural residue are concretions, root cast fragments, pieces of abraded *Unio* shell, and quartz pebbles. A powdery zone of 5 to 10 centimeters of homogeneous, clayey silt (No. 26, Table 3–8) overlies a soft, gypsum horizon (No. 31, Table 3–8) up to 5 centimeters thick.

The site of Khor el-Sil 3, situated a little northwest of locality 1, is analogous. The powdery horizon is 10 to 20 centimeters deep, the *Sa*-horizon up to 3 centimeters thick. *Unio* shells from this site gave a radiocarbon date of 15,050 B.C. ± 600 (I–1297; P. E. L. Smith, *in litt.*, March 15, 1966).

The Khor el-Sil sites 1 to 3, as well as less significant surface sites at Yardang 1 (Fig. 3–28) and near the Khor el-Sil drain (Fig. 3–29), all seem to be related to broadly contemporary, flood-silt horizons, resting on sterile riverbed sands. They may have been situated in relation to the silt-filled channel of Fig. 3–28, marking a terminal stage of the Manshiya Channel.

Another surface collection representing a distinctive industry, Khor el-Sil IV, was made by P. E. L. Smith (1966) a little south of this area in the spring of 1963. Its physical setting has been studied by R. J. Fulton.

Physical Setting of Occupation Sites Along the Sebil Channel

The nature of the occupation sites of the classic Sebil area and their relationship to the Sebil Channel (see Chap. 3) has already been summarized from the original account of Vignard (1923). A few words may be said, however, about the record as visible during 1962–63.

No middens of any size were seen between Sebil Qibli and Bahari. These had either been collected by Vignard or destroyed by *sebakh* diggers. Sandford and Arkell (1933: 51) also fail to reconfirm the observations of Vignard. Between Sebil Qibli and Matana Bahari, a long ridge of *Unio* shell, with intermingled *Cleopatra* and *Corbicula*, obviously marks the western streambank on top of the 100- to 101-meter flood bench (Fig. 3–31). Occasional waterworn artifacts among the shells emphasize that much of this was a natural deposit. Kill sites with animal bones and implement concentrations occur sporadically on or near this same ridge, however. C. A. Reed identified *Bos, Equus asinus,* and antelope or gazelle in association with blades and grinding stones at one such site. Much of the area is obviously deflated; elsewhere, a layer of loose, coarse sand rests on the eroded surface of backset and topset river-margin beds (Fig. 3–32). Quite possibly some of Vignard's sites were natural accumulations of derived materials rather than living sites *in situ*. But there was once, at least before the grading machines destroyed this Pleistocene museum, a wealth of true occupation sites in the area, and it seems very probable that many of these sites were originally buried by flood silts. They have been exposed by deflation, in much the same way as Gebel Silsila 2A.

Physical Setting of the Dar Es-Salam Site

A surface site was collected in March, 1963, by Martin Baumhoff about 1.6 kilometers due east of Dar es-Salam. A microlithic assemblage with catfish bone and some fragments of mammalian bone appears to have been denuded out of the Gebel Silsila Formation at this locality. These materials were exposed on the southwest slope of a yardang and presumably come from a former occupation site. Although the irrigation projects nearby obscure the paleogeography of ancient Nile channels, a meander arm or overflow channel once curved well east of Darau, rejoining the main Nile near Bayara. The topography of the Dar es-Salam area closely resembles that of Gebel Silsila 2A.

Other Late Paleolithic sites were discovered by P. E. L. Smith a little further north.

Late Paleolithic Geography of the Kom Ombo Plain

The Late Paleolithic cultures of the Kom Ombo Plain were contemporary with a geological formation and can be fairly accurately dated within the time range 15,000 to 10,000 B.C., plus or minus 500 years. This well-defined temporal unit has a fairly substantial geomorphic record. Although analysis of the biological and archeological evidence is still incomplete, the available data do allow a tentative reconstruction of Late Paleolithic geography.

REGIONAL ENVIRONMENT

Late Paleolithic climate of the Kom Ombo Plain is best documented in the geomorphic record. Fossil dunes or other eolian deposits appear to be absent, and there is nothing to prove significant deflation or sand blast at this time. Deposition was limited to fluvial agencies. A vigorous flood regime led to aggradation of silts, sands, and gravels within the confines of the Nile floodplain. The wadis of the Red Sea Hill country, as recorded by the contemporary Ineiba Formation, were also comparatively active. Surface denudation by rain wash, or cut and fill by local wadis, remained unimportant. There is no evidence of frost-weathering or of significant chemical decomposition. The soil record is limited to several salt or carbonate horizons pertaining to desert paleosols (crust Yermas or Calcorthids) that developed on several occasions during and after deposition of the Gebel Silsila Formation. They imply some local rainfall, sufficient to leach the top 10 to 15 centimeters of the soil. Paleobotanical remains other than root casts are so few that they contribute very little to the environmental picture.

From this we can infer that climate was perhaps a little moister than today, but certainly still "arid" in type. Rainfall was quite insufficient to generate any effective local morphogenesis. Winter frosts were unimportant. The mean daily minimum temperature of the coldest month—today, 44° F (6.8° C) in January—must have remained above 32° F (0° C). A possible indication of cooler temperatures comes from the absence of certain typical Ethiopian faunal elements such as the elephant, rhinoceros, and giraffe (Reed, 1965b), all of which were present in southern Egypt in late prehistoric times (Butzer, 1959b: 78 ff.).

The hydrography of the Nile and of the exotic wadis from the Red Sea Hills differed appreciably from that of today. Summer flood amplitudes were greater, together with stream velocity, turbulence, and competence, and floods periodically swept down the great wadis to the Nile, temporarily inundating a large part of the lower country. In other words, the very modest shift in precipitation values experienced in the Kom Ombo area was probably accentuated in the highlands.

The effects of a more vigorous Ethiopian flood regime in the riverine zone included more extensive and probably longer lasting inundations, rapid shifts of stream channel, a tendency to stream braiding, and a higher groundwater table. In the lower courses of Wadis Kharit and Shait, temporary playas were created along the valley margins, while pools of water persisted in depressions of the wadi channel for weeks or months. In this way, the groundwater supply under the wadi bed was periodically replenished.

Inferences on the distribution of vegetation can only be made by analogy. A fringing woodland or tree-savanna certainly occupied the Nile floodplain, while a thorn-savanna may be assumed in the major wadis. Away from these flood zones, edaphic sources of moisture were available under the floors of the minor wadis, probably supporting a semidesert vegetation of low scrub or bush, together with a range of desert shrubs. A similar edaphic formation might be expected on the Kom Ombo Plain bordering the alluvial flats of the Nile. As today, groundwaters derived from the summer inundation seeped a kilometer or two into the Pleistocene silts, moistening low-lying ground for as much as two or three months. Under natural conditions, these would be ideal localities for colonization by certain types of vegetation. Finally, desert wasteland lay beyond the edaphically favored environments, probably dotted by such scattered desert shrubs as are found today in the coastal deserts west of the Nile Delta.

An attempt to reconstruct the late Pleistocene environments is made in Fig. 4-4. The approximate location of the major and minor channels, as well as of the floodplain of the Nile, can be inferred directly from surficial deposits. Numerous islands must have studded the main Nile, but their location cannot be determined. The extent of the alluvial land was greater than the surfaces now occupied by the Gebel Silsila Formation, since many shallow deposits once overlying the Masmas have been eroded. Reconstruction of the floodplains of Wadis Kharit and Shait is primarily based on the inferred original distribution of the Ineiba Formation—either as masked by later wash or as denuded in proximity of the Nile Valley. The remainder of the Masmas silt plain and the Fatira beds, now at elevations of 99 to 102 meters, must have experienced considerable seepage during the flood and early postflood season. This area consequently enjoyed a very high water-table for many months of the year. Finally, the larger local wadis are schematically shown as potential areas of concentrated soil moisture.

Assuming that this environmental reconstruction is approximately correct, the exceptionally favorable resource base of the Kom Ombo Plain becomes apparent. Further upstream or downstream the riverine zone was little more than a narrow lifeline. Many rather small wadis opened onto a restricted valley. By contrast, a greater variety of riverine environments was available on the Kom Ombo Plain. Average floodplain width was over 5 kilometers, compared with less than 1 kilometer upvalley or downvalley. Similar broad floodplains in Wadis Kharit and Shait provided excellent pasturage for large herds of game animals, and the groundwater vegetation on the older silts augmented these grazing resources.

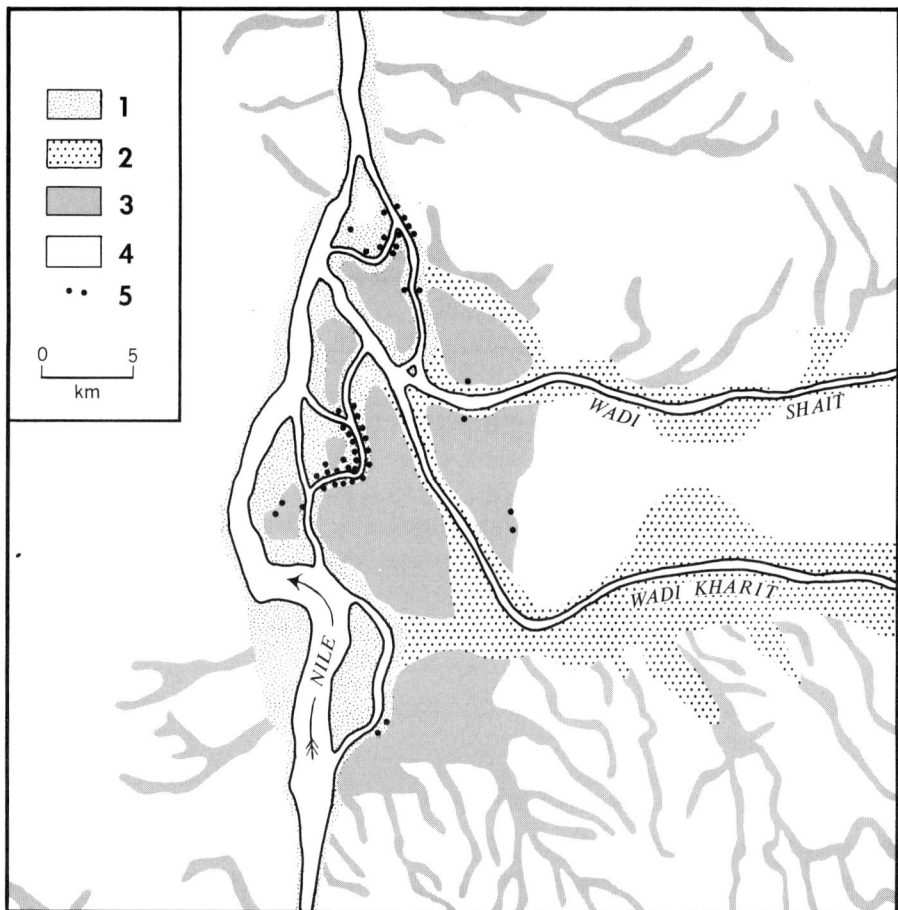

Fig. 4–4. Reconstruction of the Kom Ombo Plain at the time of Late Paleolithic occupation (*ca.* 15,000 to 10,000 B.C.). *1:* Seasonally inundated Nile floodplain, probably with woodland or tree-savanna; *2:* periodically inundated wadi floorplains, probably with thorn-savanna; *3:* edaphically favored zones with high water-table in minor wadis and adjacent to Nile floodplain (99 to 102 m), probably with semi-desert shrub; *4:* desert; *5:* major Late Paleolithic occupation sites. Map: UW Cartographic Lab.

ECONOMIC AREA

An estimate of the regional food resource base can be attempted with reference to Fig. 4–4. The approximate area of the Nile floodplain, excluding the major channels, was 100 square kilometers. That of the wadi valleys and groundwater zones amounted to some 270 square kilometers.

Altogether, the *de facto* economic area can be assessed at between 350 and 400 square kilometers, depending both on the margin of error inherent in Fig. 4-4 and on the extent to which man and animals ranged up into the Red Sea Hill wadis.

Within this economic area there is paleontological evidence for several species of large game, including hartebeest, gazelle, Barbary sheep, wild cattle, buffalo, wild ass, warthog, and hippopotamus (Gaillard, 1934; Reed, 1965*b*, 1966). Of these animals, the hartebeest, gazelle, Barbary sheep, and wild ass probably favored the wadi environments, whereas the other genera were largely restricted to the Nile floodplain. The absence of the oryx, addax, and ibex from the riverine occupation sites (Reed, 1965*b*) may be a result of their predilection for more open environments. All three of these species were almost certainly present in the Eastern Desert at this time (Butzer, 1959*b*: 78 ff.). Unfortunately, too little is known about the ecological requirements of any of these species, precluding an estimate of the local biomass.

Among the strictly aquatic food resources were the freshwater clam (*Unio* sp.), the Nile oyster (*Etheria elliptica*), a soft-shelled turtle (*Trionyx* sp.) (Sandford and Arkell, 1933: 86), and a variety of fishes, mainly the great catfish (*Clarias* sp.) still common to the Nile today (Reed, 1965*b*). *Bulinus* and *Planorbis*, those molluscan genera which today play intermediate host to the human blood fluke and liver fluke, appear to have been very scarce (see Appendix I).

The absence of pollen or identifiable macrobotanical remains from the contemporary formations and the archeological sites does not permit any conclusions on possible vegetable foods. Since thermal conditions were somewhat different, this also precludes discussion of vegetation analogies other than on a physiognomic basis—as already done in the preceding section.

Among the other natural resources for prehistoric man, there may be mentioned the raw materials for toolmaking, natural ochre, and fuel. The first of these raw materials was locally available from older Pleistocene gravels or from the Nile bed itself. Ochre could have been obtained from Wadi Abu Subeira or from the Aswan area, unless deeply weathered red soils were also improvised for paintmaking. Finally, wood was plentiful along the Nile and the major wadis and could be supplemented with dry grass, thorns, or brush as a fuel source. Vignard (1923: 56) attributed the red-tinted fired clay so frequent in the Sebilian hearths to the use of herbaceous fuel, while the blackened clays may have been charred with carbon from combustion of woody or leguminous fuels.

Any assessment of human population levels remains within the competence of the prehistorian but must await complete analysis of the archeological record. For one thing, a thorough functional interpretation of the tool inventory must be made. Then again, the number of individuals of each faunal species at the sites needs to be determined, together with a study of the butchering techniques and practices. The lack of statistical studies made of contiguous, single or broken mollusca from large samples at any of the sites will limit an evaluation of the role of *Unio* and possibly of *Corbicula* in the human diet. Interesting information on nutritional patterns has already come to light from studies of tooth wear among the Qadan skeletons found near Wadi Halfa. An advanced state of tooth wear indicates a rather gritty diet (Armelagos *et al.*, 1965), suggesting that meat did not provide the major part of the diet of this particular group.

Despite the many serious uncertainties involved, a very crude estimate of the potential population level can be attempted by analogy with specialized food-gatherers known from modern ethnological study. In the case of the Yuki and Maidu of California in contact times, population density was on the order of 1 person per 2.2 square kilometers (Baumhoff, 1963). Applying this value to the Kom Ombo Plain, which had an economic area of 350 to 400 square kilometers, a population of 160 to 180 people could be envisaged in late Paleolithic times. Obviously this is only a provisional working hypothesis that may require drastic revision. But it emphasizes the insignificance of man in the physical landscape. Even allowing for destruction of the overwhelming majority of late Paleolithic sites by repeated stream erosion or by human activity during the twentieth century, the number of sites is small when averaged out over five millennia. The evidence would be compatible, however, with a population at no time exceeding 200 persons, possibly subdivided into 4 to 6 groups. With such extenuated settlement among a maze of minor channels, the coexistence of several distinct cultural traditions seems reasonable.

LOCAL SETTINGS OF OCCUPATION SITES

Despite variations in detail, all of the occupation sites described above were found in similar situations within or on top of the Gebel Silsila Formation. In each case, a riverbank or levee location assured a supply of water and an access to aquatic or semiaquatic food resources. In each case too, the faunal remains substantiate that the sites were located near the water supply of terrestrial game species. At the season of occupation, the riverbank sites were located on dry ground, while the levee sites may have been dry for most or all the year. These riverside locations seem to have been chosen without respect to exposure, and there is no evidence of shelters. Whether or not the river channels had any particular significance for communications is impossible to determine.

In overview, the bearers of the Sebilian and other Late Paleolithic cultural traditions confined their habitations to the riverine zone of the Nile. Within that environment, they seem to have selected the riverbanks, levees, and adjacent alluvial flats for many or most of their activities. This picture of areal concentration of Late Paleolithic settlement in the same ecological zone seems to hold true throughout southern Egypt. By contrast, there is little or no evidence for habitation away from the Nile. So, for example, no occupation sites have been found in the Ineiba Formation in Wadis Kharit and Shait, despite the evidence for moderately favorable ecological conditions. Localization of settlement seems, then, to have been culturally determined.

Neolithic Industries of Southern Egypt

A cultural hiatus of perhaps five millennia follows the Upper Sebilian and allied Late Paleolithic industries in most of the Egyptian Nile Valley (see Caton-Thompson, 1946; Arkell and Ucko, 1965). North of the First Cataract, there are only a few industries, including the Helwan microlithic, that might pertain to this time interval. Unfortunately, their age and stratigraphy are obscure, however. Thus, the first appearance of the Fayum "A" Neolithic, about 4650 B.C.,[3] has long been considered as part of an intrusive agricultural colonization with cultural or ethnic affinities to western Asia. The Fayum "A" people settled around the shores of a prehistoric lake in the Fayum Depression, leaving a rich archeological record of arrowheads, bone harpoons, stone adzes, sickle blades, pottery, storage pits for grain, and living shelters (Caton-Thompson and Gardner, 1934: 12 ff., 22 ff., 88–89). Above all, domesticated emmer wheat and barley (Helbaek, 1955) have been found, together with rough linen woven out of flax. Domesticated animals, although probably present,

---

3. Radiocarbon determination of 4630 B.C. ± 180, correcting the original C–550 date by a factor of 1.03 in accordance with the currently accepted C^{14} half-life of 5,730 years. This correction will be made throughout this section in dealing with the Neolithic and Predynastic cultures, because of the difficulty of otherwise reconciling the radiometric and historical chronologies. Ideally, a further correction factor should be applied to compensate for fluctuations of atmospheric radiocarbon in response to cosmic radiation flux. Such a correction formula has been proposed by Stuiver and Suess (1966), but it is based on several tenuous assumptions and is, in addition, applicable only to a time range not exceeding 6,000 years.

were never adequately studied (Reed, 1959). Clearly the Fayum "A" marks a radical break with the Late Paleolithic in terms of both technology and subsistence.

More recent work in the Wadi Halfa area has opened a new perspective on late prehistoric cultural development in the Nile Valley. Although many of the domesticable plants and animals were not native to Africa and had to be introduced from western Asia, there is at least some reason to believe that the earliest farming communities in Egypt exhibited a number of indigenous cultural traits. And the temporal break between the microlithic industries of the last hunting-fishing-gathering groups and the first evidence of agricultural subsistence appears to be fortuitous, accidental, or local.

The first intimations of "late," pre-pottery sites in the Wadi Halfa area were provided by Myers' excavations at Abka (Myers, 1958, 1960). More recent work (Wendorf, Shiner, and Marks, 1965; Wheat and Irwin, 1965; Shiner, 1965; Chmielewski, 1965) closes this apparent hiatus. The Qadan sequence, which terminated about 4600 B.C. (new radiocarbon half-life value), is partly coeval with another Epi-Paleolithic industry, the Arkinian, with radiocarbon determinations of 7440 and 5750 B.C. The two Arkinian sites were found near Dibeira, stratified in nilotic silts (Arkinian unit) which form a younger unit of the Gebel Silsila Formation.[4] The archeological inventory includes backed blades, scrapers, arrowheads, and beads made of ostrich shell. The Shamarki Complex, with one radiocarbon value of 3820 B.C. (new half-life value), may provide a link between the Arkinian and the typical Neolithic. Sites of considerable size are located on top of the Arkinian or in the lower part of the subsequent silt formation[5] (Qadrus unit) near Dibeira West. Geometric and microlithic tools predominate, and ostrich eggshell beads are present. One terminal site of this complex includes adzes, borers, tanged points, and potsherds. Finally, there are abundant surface sites with plain pottery similar to some of that from the Khartum Neolithic, but with a toolmaking tradition different from that of the foregoing Arkinian or Shamarkian complexes. The artifactual assemblage includes crude scrapers and retouched flakes together with well-made borers and groovers.

The apparent absence of Epi-Paleolithic sites north of about Korosko may reflect the geomorphic history of the Nile Valley in early Holocene times (see Chap. 6). Nilotic deposits postdating the Gebel Silsila Forma-

---

4. Defined as the Arminna Member in Chapter 6.
5. Defined as the Kibdi Member of the Gebel Silsila Formation in Chapter 6.

tion, as represented at Kom Ombo, and predating the modern floodplain are either absent or suballuvial in northern Nubia and Upper Egypt. Since such deposits are documented and dated in the Wadi Halfa–Amada area (described in Chap. 6), they have either been eroded or buried by more recent alluvium further north. This might explain the absence of contemporary sites in the more constricted parts of the Nile Valley. Inadequate surveying and collecting may account for the hiatus elsewhere.

The distribution of Predynastic cemeteries and living sites in southern Egypt has been repeatedly reviewed in recent years (Kaiser, 1957; Arkell and Ucko, 1965; Baumgartel, 1965; Trigger, 1965: 66–83, 198–200), and the pertinent data is summarized by Fig. 4–5. Amratian (Nagada I) settlement is recorded at Hierakonpolis and Kubaniya, north of the First Cataract, and at Shellal and Bahan to the south (Kaiser, 1957). Nubian settlements of Gerzean (Nagada II) and later age are concentrated between Kubaniya and Seiyala but occur sporadically further upstream. Early A-Group sites (Early Nubian Ib of Trigger, 1965: 72–73), contemporary with the Semainean (Nagada III) and predating the First Dynasty conquest of Nubia *ca.* 3100 B.C. (Edwards, 1964), are distributed between the Second Cataract and Kubaniya. Information on settlement patterns is adequately summarized by Trigger (1965: 67–83) and is exemplified by the A-Group living sites excavated at Seiyala by the Austrian expedition (Bietak and Engelmayer, 1963). The A-Group settlement of Nubia still falls largely within the last period of high silt alluviation (Kibdi or Qadrus unit) according to Säve-Söderbergh (1964) and Wendorf, Shiner, and Marks (1965).

Late Prehistoric Settlement of the Kom Ombo Plain

Repeated surveys and incidental collecting since the 1890's have failed to discover vestiges of Predynastic or Old Kingdom settlement on the Nile floodplain or along the immediate desert margins between Gebel Silsila and Wadi Abu Subeira. The oldest conventional archeological finds have been Eighteenth Dynasty (1567–1320 B.C.) temples, shrines, and rock tombs at Gebel Silsila and Ombos, as well as a number of stelae at Bimban (Fig. 2–2). Rock pictures and petroglyphs are slightly more instructive. Prehistoric rock drawings are known from both banks of the Nile near Gebel Silsila, and there is an Early Dynastic inscription (*ca.* 3100–2700 B.C.) in the quarries at Gebel Silsila East (Porter and Moss, 1937: 220). An Old Kingdom inscription has been recorded from Nag el-Shibeika (Kagug) (Porter and Moss, 1937: 219) and, further north, a

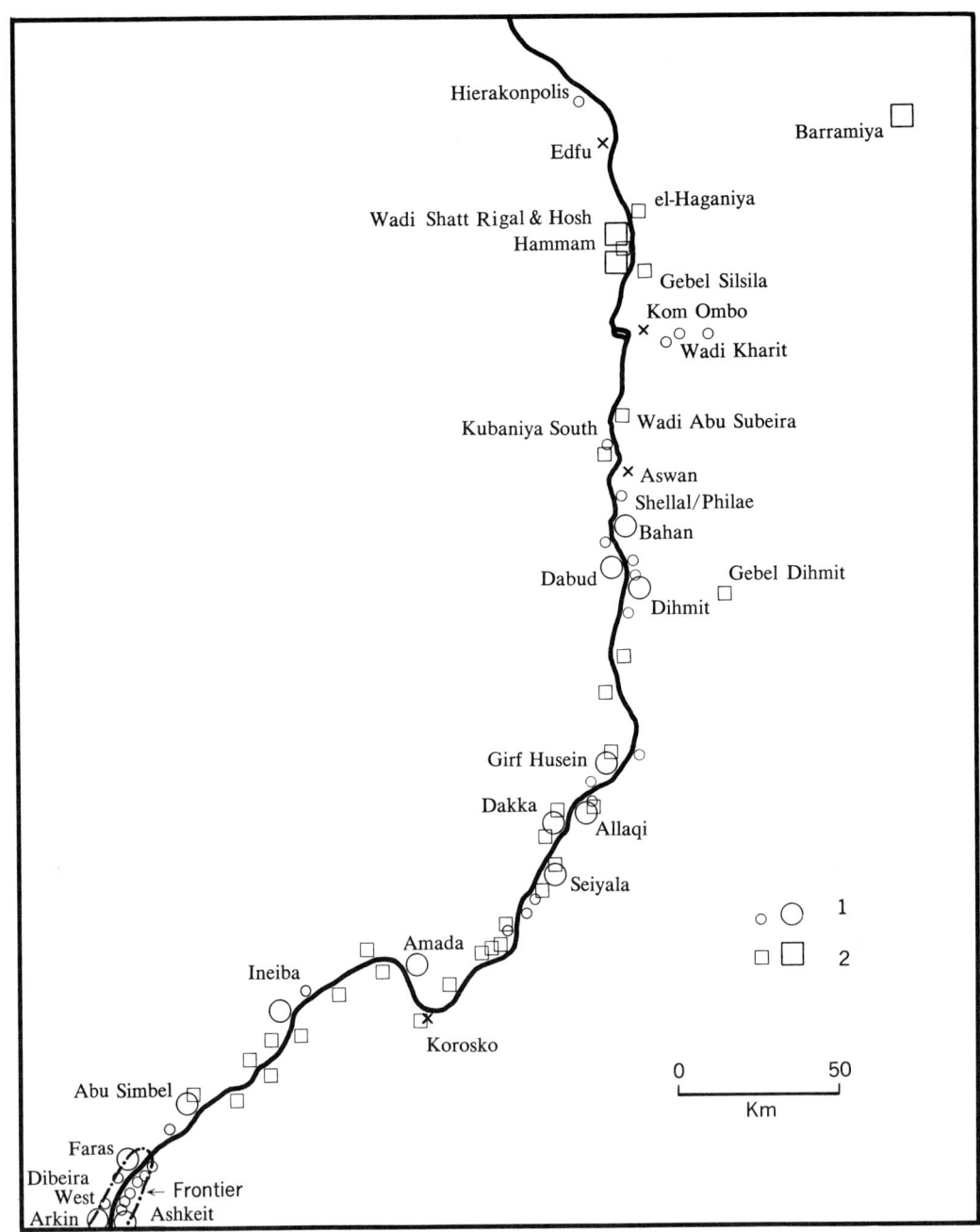

Fig. 4–5. Predynastic sites in southern Egypt. *1:* Habitation sites and cemeteries (Gerzean, Amratian, Early A-Group), after Kaiser, 1957; Säve-Söderbergh, 1964; Trigger, 1965; Wendorf, Shiner, and Marks, 1965. *2:* Early rock drawings, after Winkler, 1938–39; Dunbar, 1941; Resch, 1963a; and others. Multiple sites indicated by larger symbols. Map: UW Cartographic Lab.

number of significant prehistoric and First Intermediate Period (2181–1991 B.C.) graffiti have been published from Wadi Shatt Rigal (Winkler, 1938–39; Porter and Moss, 1937: 206 ff.). In effect, however, the Nile Valley at Kom Ombo appears to be an archeological void between Late Paleolithic and Eighteenth Dynasty times.

Evidence of late prehistoric occupation was found in both Wadis Kharit and Shait by the Canadian and Yale expeditions. Two surface sites, at New Masmas and east of New Tushka (Fig. 2–1), have been mentioned already (Chap. 3). Both sites have grinding stones, and the first included potsherds of black-topped red ware. Similar potsherds were collected by members of the Yale expedition from a surface scatter near el-Raghayim, near the mouth of Wadi Shait.

More significant, for paleoecological purposes, were a number of hearths found 11 kilometers east of New Arminna in Wadi Kharit. These sites, collected by us in November, 1962, occur on top of the Sinqari Member as exposed in the center of the wadi bed. Stream denudation has left small heaps of fire-reddened clay that rise 10 centimeters or so above the wadi surface. In three cases, hearths of carbon powder and fragmentary charcoal were found under or next to the burnt clay, sunken to a depth of 5 to 8 centimeters into the semiconsolidated Sinqari beds. Scatters of fractured rock and poorly worked chert blades and flakes were found around three such sites, black-exterior potsherds around one. At one site, a number of pieces of *Acacia* bark (kindly identified by B. F. Kukachka) were found embedded in reddish sands within 50 centimeters of the charcoal hearth. This bark ranged in size from small fragments to pieces as much as 8 centimeters long. The abundance of minute debris rules out sedimentation by fluvial agencies. The overlying sands suggested the backdirt from the sunken hearth, although this could not be fully demonstrated. In addition, a block of spongy, reddish, sandy silt, including abundant fossil-stem and leaf casts or fragments (to be studied by Madeleine Van Campo), was found near the same site. This may have been derived from the Sinqari beds. The small collection of lithic materials was examined by P. E. L. Smith, who believes they are of post-Paleolithic workmanship, although not particularly diagnostic (*in litt.*, October 17, 1964).

Black-topped or black-exterior red ware was common in Predynastic Egypt but is already absent during the Semainean or Nagada III (Kaiser, 1957). Ware of this general type persisted in Lower Nubia as well as among the upper Nubian mercenaries in Egypt as late as the Eighteenth Dynasty, however (Emery, 1965: 165). Since the occupants of Wadis

Kharit and Shait do not seem to have left a record in the nearby Nile Valley, we are probably dealing with desert folk, possibly of Nubian affinity. This necessarily implies better natural resources and a moister climate in the desert wadis. The desert fauna of Egypt was severely decimated between the First and Fourth Dynasties (*ca.* 2900–2600 B.C.) and modern desert conditions prevailed in Egypt between the end of the Fifth Dynasty (*ca.* 2350 B.C.) and the mid-first millennium B.C. (Butzer, 1959*b*: 64 ff., 67 ff). It would seem, therefore, that this period of occupance in the desert wadis predates the Fourth Dynasty.[6]

In order to obtain a date that might throw some light on the age of Member I of the Shaturma Formation, as well as on the late prehistoric settlement of Wadi Kharit, some pieces of bark from the one sunken hearth were radiocarbon dated (Appendix E). The age is 2850 B.C. ± 100 (I–2567).[7] On the basis of the indirect association between bark and hearth—and again between the Upper Kharit hearths and the late prehistoric archeological vestiges of the area—this would seem to indicate occupation during A-Group or Gerzean times. Similarly, it would suggest that Member I dates from the fourth millennium B.C.

Because a Neolithic-Predynastic archeological record is absent in the Nile Valley near Kom Ombo, further information concerning late prehistoric settlement must be gained from the rock drawings of the area. Winkler (1938–39), who systematically studied the rock art of southern Upper Egypt, recognized several prehistoric groups. The "Earliest Hunters," who may predate the local introduction of agriculture (Resch, 1963*a*), drew pictures of elephant, giraffe, ibex, antelope, and dog among hunting scenes inscribed on the west bank of the Nile, north of Hosh, and in Wadi Shatt Rigal (Sites 35 and 36 of Winkler, 1938: 31–32; 1939: 31). Further giraffes were drawn between Hosh and el-Hammam (Site 48, Winkler, 1939: 31–32). At Gebel Silsila West, the Earliest Hunters drew hunting scenes with giraffe, ibex, hippopotamus, and ostrich (Sites 49–52, Winkler, 1939: 6, 30–32). Finally, in a small west-bank wadi 5 kilometers northwest of Aswan, there are engravings of giraffe, antelope, gazelle, crocodile, and ostrich. The unpublished drawings of wild cattle and other game, found just above Gebel Silsila 1 (P. E. L. Smith, 1964*a*),

---

6. The Canadian expedition systematically collected silts in lower Wadi Kharit during 1962 (P. E. L. Smith, 1964*a*), and the detailed reports of that mission should answer many of the questions concerning the subsistence problems of this settlement.

7. New half-life value. Employing the correction factor of Stuiver and Suess (1966), the true age would be 3475 B.C. ± 100.

seem to be contemporary. All in all, these drawings of the Earliest Hunters may well be the product of Epi-Paleolithic groups resident in or near the Nile Valley. The presence of elephant and giraffe among the fauna seems to imply that the drawings are somewhat younger than the Late Paleolithic occupation of the Kom Ombo Plain.

Winkler's second prehistoric people were designated as "Autochthonous Mountain Dwellers" and were characterized as cattle pastoralists, whose dress included the Libyan sheath (Winkler, 1939: 18–20). Resch (1963a, 1964) considers these people as contemporary with or equivalent to the Badarian and Amratian cultures of the Nile Valley. At Hosh they sketched hunting scenes with elephant, rhinoceros, ibex, gazelle, and ostrich, as well as a man leading a tethered gerenuk gazelle (Sites 35 and 36, Winkler, 1938: 29–30; 1939: 19). Cattle are conspicuously absent. The elephant, rhino, and gerenuk (*Lithocranius walleri*) were all locally extinct in Egypt by the beginning of the historical era (Butzer, 1959b: 54 ff.), so that there can be little question that these drawings are Predynastic.

The third and last group, designated as "Early Nile Dwellers" (Winkler, 1938: 30–31), is probably identical with the Gerzean population of Egypt (Resch, 1963a, 1964). Their rock drawings at Hosh (Site 35, Winkler, 1939: 17) include Nile boats. The other site of Winkler, northwest of Aswan (No. 53, 1939: 17–18), shows giraffe, ostrich, sheep (domesticated or Barbary), and greyhounds. Another site was examined by C. A. Reed and Butzer in March, 1963, at the mouth of Wadi Abu Subeira, above the quarries on the north face. The older, grooved or pecked drawings here show numbers of boats and cattle, at least one animal of which appeared to be domesticated. These graffiti may be Gerzean or Dynastic.

In sum, the prehistoric rock drawings of the Nile Valley between Edfu and Aswan do not verify the existence of agricultural or cattle-herding populations until late Predynastic times, and then only south of Kom Ombo. All the other drawings suggest that widely scattered hunting groups inhabited the area as late as the Gerzean, or at least that hunting still was a significant economic trait. The absence of Predynastic cemeteries, habitation sites, or pottery for a 100-kilometer stretch between Hierakonpolis and Kubaniya South—with the possible exception of Wadi Kharit—relates well to the one-sided and spotty evidence of the rock art.

Two hypotheses can be offered for the Predynastic settlement gap apparent to the north of Kom Ombo. Either natural conditions were unfavorable or cultural factors were involved.

The seasonally inundated valley lands today consist of three geomorphic types. A flat floodplain, without natural levees, commonly accompanies the Nile channel south of Edfu through most of Nubia. This type of floodplain seldom exceeds a kilometer in width. The only example of an irrigation basin is the Hod Bimban, found in the widest section of the valley at Darau (Fig. 4–6). The natural levees of the Hod Bimban rise an average of 2 meters above the lowest parts of the alluvial basin, a pattern that is general on the convex floodplain north of Edfu (Butzer, 1959b: 27 ff.). Mansuriya Island constitutes a third geomorphic unit, partly protected by artificial levees but otherwise similar to the flat floodplain type. In terms of agricultural colonization, the Nile floodplain at Kom Ombo is no less suitable than areas further to the north or south which were settled as much as two or three millennia earlier.

Ethnic or political factors may have been responsible for the apparent break in continuous agricultural settlement along the Nile Valley in Predynastic times. It is probably significant that Gebel Silsila marked the political and linguistic frontier between Egypt and Nubia at that time (Säve-Söderbergh, 1941: 7; Edwards, 1964: 3). The earliest record of an Egyptian campaign against the Nubians belongs to Aha, one of the first pharaohs of the First Dynasty (ca. 3100–2900 B.C.). The conquest of the area between Gebel Silsila and the First Cataract and the establishment of the Elephantine Nome may date from this period (Edwards, 1964: 18). The Nubian archeological record north of Aswan significantly breaks off at about the same time (Trigger, 1965: 73–74). And Aha's successor, Djer (Zer), was able to inscribe a battle scene, commemorating victory over the Nubians, as far south as Wadi Halfa (Emery, 1965: 125). Sporadic border raids or larger campaigns may well have preceded Egyptian conquest of the Elephantine Nome by several centuries. The absence of permanent settlement in the no-man's-land between Edfu and Kubaniya would seem logical in this light.

There is some reason to believe that the Nubian population north of the First Cataract was either deported to Egypt or driven southward after the conquest (Trigger, 1965: 74), a practice that had been established by the time of Sneferu (ca. 2613–2587 B.C.) (Emery, 1965: 127; Trigger, 1965: 78–79). Systematic Egyptian colonization of the buffer zone may have been neglected until the reigns of Hatshepsut (1503–1482 B.C.) and Thutmosis III (1482–1450 B.C.), when serious exploitation of the sandstone quarries at Gebel Silsila was begun.

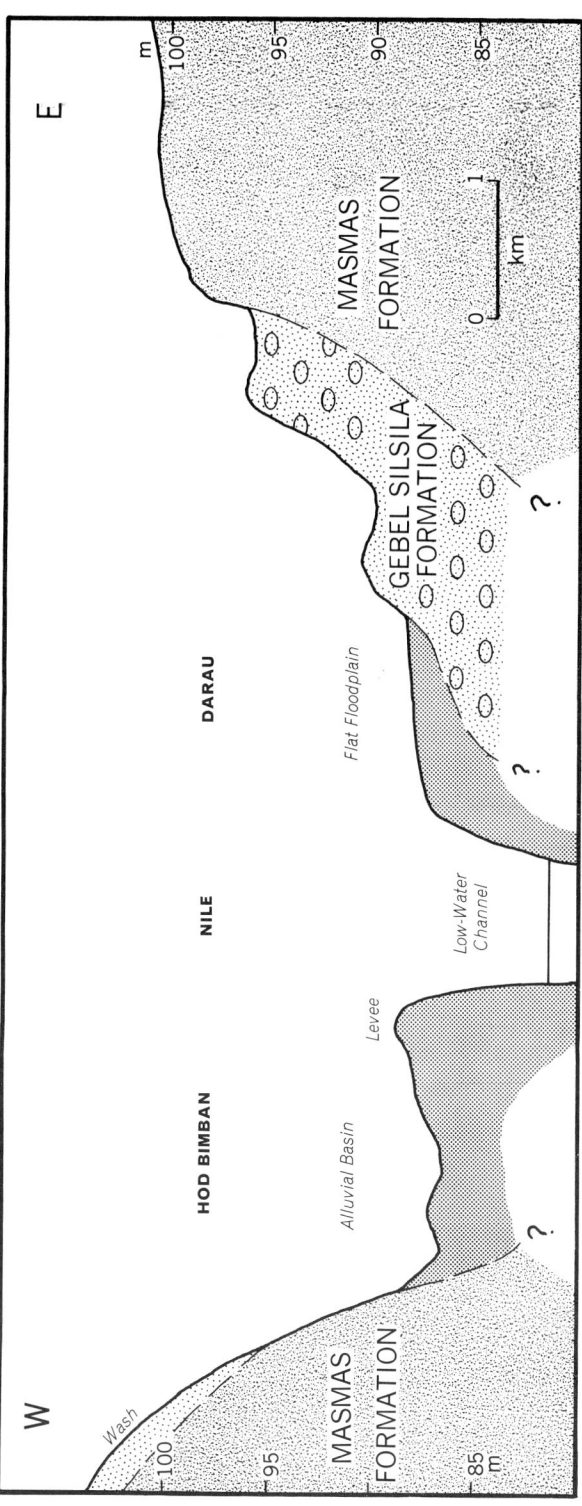

Fig. 4-6. Cross section of the Nile floodplain and late Pleistocene deposits north of Darau. Map: UW Cartographic Lab.

5

Geomorphic Evolution of Egyptian Nubia

Introduction

Nubia, a part of the Nile Valley, is a land of river, rock, and sand that winds its way upstream of Aswan into the distance of the Sudan. Abrupt cliffs of dark igneous rock close in on the river in some sectors, while sandy desert plains, studded with conical or tabletop hills, open broad vistas elsewhere. The scenery varies in response to subtle differences of lithology or structure, with a brilliant sky offset by the starkness of the terrain. A fringe of acacias, dum palms, and gardens traces the riverbanks past each cluster of mud houses. Building styles change with the physical landscape—mud-brick ruins or imposing temples blending with the villages. Five millennia of historical tradition are recorded here in the border marches of Egypt—years of precarious and changing existence in an environment blessed with few resources other than its people.

Today, as the waters rise slowly behind the High Dam, Nubia has become barren and depopulated in order to rescue Egypt from the dilemma of overpopulation. In similar fashion, Egypt has dictated its whims to its less fortunate southerly neighbor through most of the last five millennia. The prehistoric inhabitants of Lower Nubia were first subjugated through a series of expeditions by the pharaohs of the first three dynasties, and Sneferu (*ca.* 2613–2589 B.C.) appears to have deported as many as 7,000 people in order to pacify the area. During the Fourth and Fifth Dynasties (*ca.* 2613–2345 B.C.), Egypt ruled Lower Nubia with an iron hand, and the population sank to an estimated low of 2,600 people (Trigger, 1965: 160). By the end of the Sixth Dynasty (*ca.* 2181 B.C.), independence had been regained, and the indigenous population, known as the C-Group culture, began to increase once more. The Twelfth Dynasty monarchs (1991–1786 B.C.) reconquered Nubia and built a series of imposing mud-brick fortresses (Emery, 1965: 143 ff.) along both banks of the Nile between Dakka and Semna, above the Second Cataract. Gold mining in Wadis Allaqi and Gabgaba, as well as in the Batn el-Hagar, appears to have been the primary form of exploitation. Commercial enterprises were also undertaken in Upper Nubia, based on an entrepôt at Kerma.

During the Second Intermediate Period (1786–1567 B.C.), Lower Nubia regained its independence, possibly as part of the Kingdom of Kush which was centered upon Upper Nubia. The first pharaohs of the Eighteenth Dynasty (1567–1320 B.C.) reestablished Egyptian control and pushed the frontier into the central Sudan. Mining operations were resumed and expanded, and a fair degree of prosperity was attained as Nubia was acculturated to Egyptian civilization. The lever-operated *shaduf* was introduced, permitting irrigation of arable land not reached by the annual floods. By the close of the Eighteenth Dynasty, the population had risen to 17,500 (Trigger, 1965: 160 ff.). Although the major monuments of Abu Simbel, el-Diwan (ed-Derr), Sebua, Girf Husein, and Beit el-Wali date from the reign of Ramses II (1301–1234 B.C.), his wasteful policies, coupled with a rapid decline of annual flood levels, began to affect population levels (Trigger, 1965: 112–14). With the collapse of the Egyptian administration, the natural crises of low inundations ultimately led by 1000 B.C. to the almost total abandonment of Lower Nubia (Trigger, 1965: 160 ff.).

Resettlement of Lower Nubia *ca.* 300 B.C. was accomplished by a new population, the Meroites, who apparently spoke an Eastern Sudanic language (Trigger, 1966). The waterwheel or *sagiya* was introduced at about the same time, permitting large-scale irrigation. Politically, the Ptolemies (323–30 B.C.) and their Roman successors controlled the northernmost part of Nubia between Aswan and the mouth of Wadi Allaqi. The remainder of the country belonged to the Kingdom of Meroe, which had its capital in Upper Nubia. On the whole, relationships between Egypt and Meroe were peaceful, and commercial and cultural contacts flourished (Emery, 1965: 225–26; Trigger, 1965: 120 ff.). A number of temples were built in northernmost Nubia during these centuries of prosperity, including Philae, Dabud, Qirtassi, Tafa, Kalabsha, Dandur, Dakka, and Muharraqa. Population between the First and Second Cataracts rose to a high of 60,000 inhabitants (Trigger, 1965: 160 ff.).

After 253 A.D., Nubia was reduced to chaos by the breakdown of the Kingdom of Meroe and the inroads of the Blemyes, nomads from the Eastern Desert. At about the same time, the Nubian language was introduced by tribes immigrating from further south (Trigger, 1966). Rome abandoned northern Nubia in 297 A.D., and a gradual fusion of the different ethnic groups led to the emergence of a Christian Nubian kingdom after about 540 A.D. During the subsequent centuries, a Christian rural civilization in Nubia withstood the first onslaught of the Arabs and attained a position of political power in the ninth century, controlling parts of Upper Egypt for several centuries. Arab beduin of the notorious Banu Hilal infiltrated into Nubia during the eleventh century, and the devastating Egyptian campaign of 1172 A.D. dealt a mortal blow to the Nubian kingdom. The Christian populations withdrew to a few larger, fortified centers south of the present border, maintaining their identity until the last Christian king was defeated in 1323 A.D.

After the Islamic night had descended on Nubia in the late Middle Ages, there is no record until 1737 when Norden, the first European traveller, penetrated from Aswan to Tumas. The economy of Lower Nubia during the early nineteenth century has been outlined by Christophe (1963), based on the observations of Burckhardt and other early travellers. The misery of the local population, reduced to serfdom by the rapacious governors, probably exemplifies five centuries of merciless exploitation that were not terminated until administrative reforms were made after 1858.

The inauguration of the Aswan Dam in 1902 brought a new era to Nubia. A seasonal lake at 106 meters elevation was created, inundating much of the floodplain for the 100-kilometer stretch north of Dakka. Plans to raise the dam level in 1912 led to a rash of archeological salvage work after 1907. The new reservoir level at 112 meters inundated most of the Nile Valley as far south as Sebua. A second archeological survey was begun in 1929, preceding the major raising of the Aswan Reservoir to 121 meters in 1934. Most of the arable land of Egyptian Nubia was submerged, and the first evacuation of people to Upper Egypt was begun. Loss of agricultural lands was partially compensated for by large-scale irrigation projects on high-lying Pleistocene silts at Dakka, el-Allaqi, Ineiba, and Arminna. Everywhere villages were rebuilt on higher ground.

During the first three decades following completion of the old Aswan Dam in 1934, the reservoir area underwent a fairly well-defined annual cycle. The waters were mainly released during May and June, at the climax of the low Nile season, to permit spring and early-summer agriculture in Egypt. From late July to mid-September or early October, the flood surge was allowed to pass through the dam sluices. In Nubia this served to expose parts of the old floodplain, permitting a short crop season. But the accelerated deposition of silts in the stagnant or slowly moving waters of the reservoir completely masked all of the archeology and most of the geomorphology below about 105 to 110 meters. The accretion of floodplain mud between 1912 and 1963 can, in fact, be estimated at 3 to 5 meters near Seiyala.

The High Dam project, one of the greatest irrigation schemes of this century, began to go into effect late in 1964, and early in 1966 the reservoir level had already been raised to 134 meters. By January 1, 1969, the level will be 164 meters and a maximum of 182 meters will ultimately be attained. In subsequent years, the annual cycle of fluctuation will range between about 165 and 175 meters, with the minimum coming prior to the flood season in May and June, the maximum in early winter. This annual cycle will be superimposed upon a long-term cycle of one 182-meter maximum stand every thirty years and one minimum level of 147 meters each century. The longer term cycles are partly designed to permit silt accretion along the margins of the reservoir. In this fashion, new arable lands will be created where many of the Nubian *emigrés* are ultimately expected to find new homes. The more significant of these new riverine oases will be located southwest of Kalabsha, around the lower course of Wadi Allaqi, and between Abu Simbel and the Sudan border.

The implications of the High Dam for Egypt are, of course, momentous. Whereas the old dam had a total capacity of 5,000,000,000 cubic meters of water, the High Dam will have a capacity of over 125,-000,000,000 cubic meters. This new irrigation capacity will permit an increase by as much as 40% in the cultivated land of Egypt, and many areas of Upper Egypt will be able to add a crop season to their annual schedule. By the time the hydroelectric resources of the High Dam are fully exploited, an annual production of over 10,000,000,000 kilowatt hours per year is expected.

The implications for Nubia are less favorable, at least for the next decade. The 48,000 inhabitants of Egyptian Nubia were evacuated to the Kom Ombo Plain during 1963–64, following earlier resettlement of some 7,000 people to the Esna area. The evacuation to Kom Ombo entailed considerable economic hardship and increased mortality, and it will be some years before the agricultural productivity of the new lands lives up to official expectations. Although the Nubians themselves will undoubtedly adjust and maintain their identity until a return to Nubia is possible, the practical consequences of the High Dam for the scientific record of Egypt will be almost irrevocable. Practically all of the late Pleistocene deposits and archeology have been permanently submerged and lost at the time of writing, insofar as they have not been documented or removed. By the time the 6500-square-kilometer reservoir is full in 1975, all of the Pleistocene record of Lower Nubia will be under water. The subsequent chapters and the accompanying map sheets of the geomorphology and surficial deposits must therefore attempt to record as completely as possible the Pleistocene of Egyptian Nubia.

Early Geological Work

The first geologist to work in Nubia was Joseph Russegger (1844, v. 2, pt. 3: 137–89), who spent considerable time between the First and Second Cataracts during the years 1836–38. Russegger concentrated his attention on the bedrock formations and first defined the Nubian Sandstone. More systematic observations on the Pleistocene date back to Leith Adams (1864), who spent two months in Nubia in 1862–63 and whose molluscan collections were subsequently determined by Woodward. Adams describes ancient nilotic deposits up to as much as 35 or 40 meters above floodplain level at el-Diwan, Wadi el-Arab, Dakka, Qirtassi, Dabud, and Shellal. *Unio willcocksi, Corbicula fluminalis, Etheria elliptica, Cleopatra bulimoides,* and *Bulinus truncatus* were collected from these varied sites, while Falconer identified two hippopotamus molars found *in situ* near Kalabsha Temple. Complex, dissected alluvia of nilotic and wadi origin—some with calcareous nodules—are also mentioned from the wadi embouchures. The permanent value of Adams' account lies in his record of rich late Pleistocene deposits between Dakka and Shellal, exposures already lost to the waters of the Aswan Reservoir by the time of the Oriental Institute Survey by K. S. Sandford and W. J. Arkell (1933).

At the close of the nineteenth century, Sir Henry Lyons (1894) again drew attention to Pleistocene nilotic deposits in the Wadi Halfa area, while Sir William Willcocks (1894) illustrated some of the geomorphic forms in the sandstone of Lower Nubia. These nineteenth-century observations were synthesized by Max Blanckenhorn (1901), who considered the high nilotic silts of Lower Nubia and Kom Ombo pertinent to an early Pleistocene pluvial phase—solely on the basis of their relative elevations. Of more lasting value was John Ball's 1:20,000 mapping of the First Cataract during the winter of 1899–1900 (Ball, 1907) and his observations at the Semna Cataract, above Wadi Halfa, in 1902 (Ball, 1903).

Sandford and Arkell surveyed the Pleistocene of Lower Nubia early in 1930, travelling by felucca from Aswan to Wadi Halfa and working on foot in the desert tracts adjacent to the Nile. A sequence of high Pleistocene "terraces" was set up by Sandford and Arkell (1933: 18–21, 25–27, 29–31, Plates 4a, 5a, 6a), including the following stages: 90 meters, 73 meters, 46 meters, 30 meters, 24 meters, and 15 meters. The 90-meter stage was identified at Abu Simbel and el Riqa, with gravel veneers reported from the first locality. The 73-meter stage was mentioned from Abu Simbel, Kasr Ibrim, and el-Riqa; the 46-meter stage, at Abu Simbel, at Tushka West, between Ineiba and el-Riqa, and at Kalabsha. A 30-meter stage was identified between Wadi Halfa and Abu Simbel, between Ineiba and Tumas, and at Korosko, Wadi el-Arab, Seiyala, Kalabsha, and Shellal. Gravels were reported from the Sudanese occurrences, from Tumas, and opposite Korosko, with "Chellean" implements *in situ* at the last two localities. A 24-meter stage was described from Dihmit, with a platform on the west bank and wadi gravels on the east bank. Finally, a 15-meter stage was based on wadi gravels with Acheulian artifacts just north of Wadi Halfa, and on platforms at Tushka West, Kasr Ibrim, Korosko, Seiyala, el-Allaqi, and Dihmit.

With the exceptions already noted, most or all of these terraces were identified on the basis of stream-cut platforms rather than of river gravels. In as far as the writers have been able to identify them in the field and on the air photos, it is certain that many of these platforms are pediment surfaces that do not necessarily record definite Pleistocene fluvial stages. In a good many cases, the elevations cited by Sandford and Arkell are grossly inaccurate, particularly in the case of their 15- and 30-meter stages. Sandford and Arkell failed to note numerous exposures of significant gravel terraces and clearly did not have the opportunity to examine the lower courses of any of the major wadis. The impression is obtained that field transects of the Pleistocene "terraces" between the Sudanese border and Shellal were confined to Abu Simbel, Kasr Ibrim, Tumas, el-Riqa, and Dihmit. Most of the other observations are so incomplete, vague, or inaccurate that they must be attributed to casual observation from the riverbanks. In offering this criticism, we are well aware of the field difficulties faced by earlier workers at a time when topographic maps and air photos were unavailable, and, considering the great distances covered by Sandford and Arkell's admirable pioneer survey, the degree and accuracy of detail was necessarily uneven.

The late Pleistocene silts, first studied by Sandford and Arkell in Upper Egypt, were traced through the length of Lower Nubia (Sandford and Arkell, 1933: 37–43, 48–49). These deposits were mapped between Shellal and Aswan (based on Ball's 1:20,000 map, *ibid.*: 58) and between Semna and Dibeira (at about 1:380,000, *ibid.*: 39). Semidetailed sections were described from Wadi Halfa, Dibeira West, Tushka West, and Khor Abu Uruq, near Korosko. Although these sections can be amplified and partly reinterpreted today, they retain their value. As at Kom Ombo, Sandford and Arkell attributed all of the silt aggradation to a single phase of Nile alluviation. But their terminal "degradation" stages, with shingle ridges marked by different Late Paleolithic industries, came within a hair's breadth of recognizing the true complexity of the silts.

After an interruption of thirty years, the Aswan High Dam Project promoted new field activity in Egyptian Nubia. In the 1950's, deep borings were made above the old dam, leading to the recognition of a buried entrenched valley of the Nile to below modern sea level (Georg Knetsch, personal communication in 1958; Voute, 1963; Chumakov, 1965). Early in 1962, the Desert Institute of Egypt carried out a motorized reconnaissance of the Lower Nubian Plain. As well as giving a useful geomorphic overview, the resulting paper by Shata (1962) interprets the great lowland plains west of the Nile by pedimentation processes, conditioned by structural and lithological factors. A brief reconnaissance of the Dakka area by the Geological Survey of Egypt followed in 1962-63. A far-reaching and rather radical structural interpretation of the Nubian Nile Valley was proposed, including Pleistocene block faulting with vertical displacements of several hundred meters and transcurrent faults with horizontal displacements of 2 kilometers or more (Said and Issawy, 1965). These notions will be further discussed below. A reconnaissance of the Pleistocene deposits in the Wadi Halfa area by Fairbridge (1963) late in 1961 was followed by a much-needed, semidetailed study of Sudanese Lower Nubia by Jean de Heinzelin and Roland Paepe during three winter seasons beginning in 1962 (de Heinzelin and Paepe, 1965; de Heinzelin, 1967).

The Yale Prehistoric Nubia Expedition 1962–63

Our Nubian field work, as members of the Yale expedition, began October 8–13, 1962, travelling upon the Shellal–Wadi Halfa steamer. Three days were spent in the Wadi Halfa area visiting local Pleistocene exposures and obtaining an impression of the regional geomorphology. Preparatory to the major field study, we examined the 1:30,000 air photos of Nubia for pertinent geomorphic features and for localization of Pleistocene deposits. The field survey, employing a leased boat, began January 12, 1963, and, with a brief interruption, lasted until February 23, 1963. Detailed mapping of surficial deposits and geomorphology on the

1:10,000 topographic series was completed for a 3-kilometer-wide, 277-kilometer-long stretch of the Nile Valley as well as for the lowermost 30 kilometers of Wadi el-Allaqi. The air photos, over 40 detailed transects at 1:1250, and constant field checking formed the basis of this phase of the work. Local stratigraphic sequences were established and complemented by subsequent laboratory study of 179 sediment samples collected during the total of six weeks spent in the field. The air photo interpretation was completed after returning to Cairo and was later supplemented by Gemini IV satellite photography (June, 1965) of Nubia.

The Nubian phase of our work preceded the archeological salvage work of the Yale Expedition in that area. Much of the geomorphological field work was nonetheless focused on the Paleolithic archeology of Lower Nubia, and some prehistoric associations were established. Our maps were placed at the disposal of the Yale Prehistoric Nubia Expedition during the winters of 1963–64 and 1964–65, at which times Robert Geigengack was responsible for the earth science work. Results of the later Yale missions—other than brief reports by Reed (1966) and Wendt (1966)—are not yet available.

Bedrock Lithology

The Basement Complex underlies the Nubian Sandstone throughout Nubia, occasionally exposed at the surface or evident below the sedimentary mantle at shallow depths. The pink granites, gneisses, and schists of the Aswan area extend upstream into the vicinity of Dihmit, where they can be followed into the Eastern Desert to elevations of over 500 meters. Another boss of the Basement Complex emerges at the Kalabsha Gorge, an area we mapped in detail. Medium- and coarse-grained pink granites form the dominant intrusive rock, with later sheet-intrusions of fine-grained gray granite and dark-gray microgranite. At least two generations of dykes can be identified on lithological grounds alone. These range from white quartz and pinkish-white aplite to very-dark-gray dolerite. The dominant strike directions, in order of prominence, are N 45°–60° E, N 15° W–N 15° E, N 45°–60° W, and E 10° N–E 10° S.

Further outcrops of the Basement Complex occur in the center of the Lower Nubian Plain at elevations of 180 to 230 meters. Granites are marked on most maps at 28 kilometers due south of Dungul, while at the Chephren Quarries, about 40 kilometers northwest of Abu Simbel, amethysts and carnelian were once mined from dykes in a dark-blue gneiss with quartz veins (Murray, 1939). Finally, the Basement Complex is but thinly veneered by sandstones near Tumas, on the west bank of the Nile, at perhaps 140 meters. This can be inferred from intersecting rectilinear drainage lines, similar to those of the Basement Complex southeast of Aswan and east of Kalabsha.

From this evidence it appears certain that the contact between the basement and the sedimentary cover is highly irregular, varying in a vertical range of over 400 meters within the confines of Egyptian Nubia alone. Although Cenozoic tectonics may have roughened this contact, there can be little doubt that much of the relief predates deposition of the Nubian Sandstone in mid–Upper Cretaceous times.

A period of deep chemical weathering preceded deposition of the Nubian Sandstone. This has been inferred by Attia (1955) and Kolbe (1957) at Aswan and can be verified from contact exposures we studied at the Kalabsha Gorge, about 1 kilometer northeast of Nag el-Shima el-Qibli. A light-gray (10 YR–2.5 Y 7/2) alteration zone of over 1.2 meters depth has developed from pink granite, retaining a granitic structure. The feldspars have been kaolinized, limonitic staining is common, and secondary calcite has been enriched in cracks. Quartz, hornblende, and biotite minerals are badly corroded or altered. This weathering horizon recalls the zone of kaolinite enrichment often found below the solum of certain tropical soils (Kubiena, 1955, 1957). It is overlain by 30 to 40 centimeters of light-reddish-brown (5 YR 6/3–4), sandy silt, an unstratified wash of arcosic quartz. The basal conglomerate of the Nubian Sandstone follows, a light-reddish-brown (5 YR 6/3), silty, coarse-grained sandstone with medium-to-coarse quartz pebbles.

The basic facies of the Nubian Sandstone in Lower Nubia is a massive, well-stratified, white to very-pale-brown (10 YR 8/2–3), medium- or coarse-grained sandstone. As a result of weathering, near-surface rocks range in color from pink (7.5 YR 7/4) and reddish yellow (7.5 YR 8/6) to light brown (7.5 YR 6/4). Surface exposures are discolored to a varying degree with desert "varnish" (see Engel and Sharp, 1958), a rind of ferric and manganese precipitates. Normally, the desert varnish is a few millimeters thick, varying from weak red (7.5–10 R 4–5/2–4) and dusky red (10 R–2.5 YR 3/2–3) through reddish brown (2.5–5 YR 2–5/2–4) to dark gray (5 YR 3–4/1) in color. These rinds may be matt or glossy, dense or porous, and the primary components usually include hematite, limonite, pyrolusite, and recrystallized silica. Solutions of this type may impregnate bedrock to depths of 15 to 30 centimeters or Pleistocene sediments to depths of a meter or more.

The development of thick ferromanganese rinds or impregnations varies regionally and appears to be related to the primary occurrence of ferruginous sandstones within certain watersheds. Bands of dark-gray (7.5 YR 4/0), ferruginous quartzite, averaging 15 to 30 centimeters in thickness, can frequently be observed within the sandstone formations of the Eastern Desert. In addition to the iron ores of Aswan, smaller lodes have been reported from Kalabsha, Muharraqa, Korosko, and Abu Simbel East (Attia, 1950). At Wadi Halfa, lenses of oolitic iron up to 1-meter thick have been traced over horizontal distances of as much as 8 kilometers. They are most commonly found between ferruginous sandstones and quartzites (Kleinsorge and Kreysing, 1960). In other words, the presence of significant primary iron concentrations in the Nubian Sandstone is an established fact. Later solution of iron compounds, with precipitation and concentration in joints and bedding planes, points in the same direction as the ferromanganese rinds or ferricrete conglomerates of Pleistocene age, namely, to a very widespread post-Cretaceous redeposition of iron compounds. Kleinsorge and Zscheked (1959) describe analogous phenomena from the central Sudan, suggesting that iron-rich waters may have emanated from springs created in the wake of the Tertiary vulcanism. Although there is no evidence of such springs in Nubia today, vulcanic springs may have been activated during or after the more significant Pleistocene pluvials.

Whatever the origin of the ferromanganese compounds, they have been distributed by running waters throughout certain watersheds in Nubia—as can be clearly seen from the Gemini IV photos. And, although the Nubian Sandstone upstream of Aswan is fairly homogeneous, considerable lithological differences have resulted from ferromanganese precipitation and concentration. As a result, significant differences can be observed in resistance to erosion. Ferromanganese rinds are thin and are underlain by pale, friable sandstones in many sectors of Nubia. The rinds invariably flake off as the softer rock disintegrates with time. By contrast, thick ferricrete or silicic rinds may mantle the rock in an ironstone crust that is virtually indestructible under natural conditions. Consequently, the Nubian Sandstone cannot be considered lithologically uniform.

Structure

The structural geology of Egyptian Nubia was first outlined on a very small scale by Yallouze and Knetsch (1953) and by Knetsch (1957), who postulated a number of NNW–SSE transverse faults across the Nile Valley between Korosko and Dakka, as well as N–S faults in the Aswan-Kalabsha area. Shata (1962) indicated the presence of a NE–SW swell axis—forming the backbone of the Lower Nubian Plain—together with a variety of disturbances and minor faults in the northern part of this zone. Said and Issawy (1965) postulate a number of NW–SW faults intersecting the Nile Valley, together with large-scale transcurrent faults in the Dakka area and a major horst block in the Korosko bend of the Nile.

On the basis of our 1:10,000 mapping of the Nile Valley and our study of the Gemini IV photography of Nubia, we believe that it is not yet possible to do justice to the rather complex minor structures apparent in the geomorphic landscape. Instead we have attempted to map those major features that can be demonstrated with some certainty (Fig. 5–1).

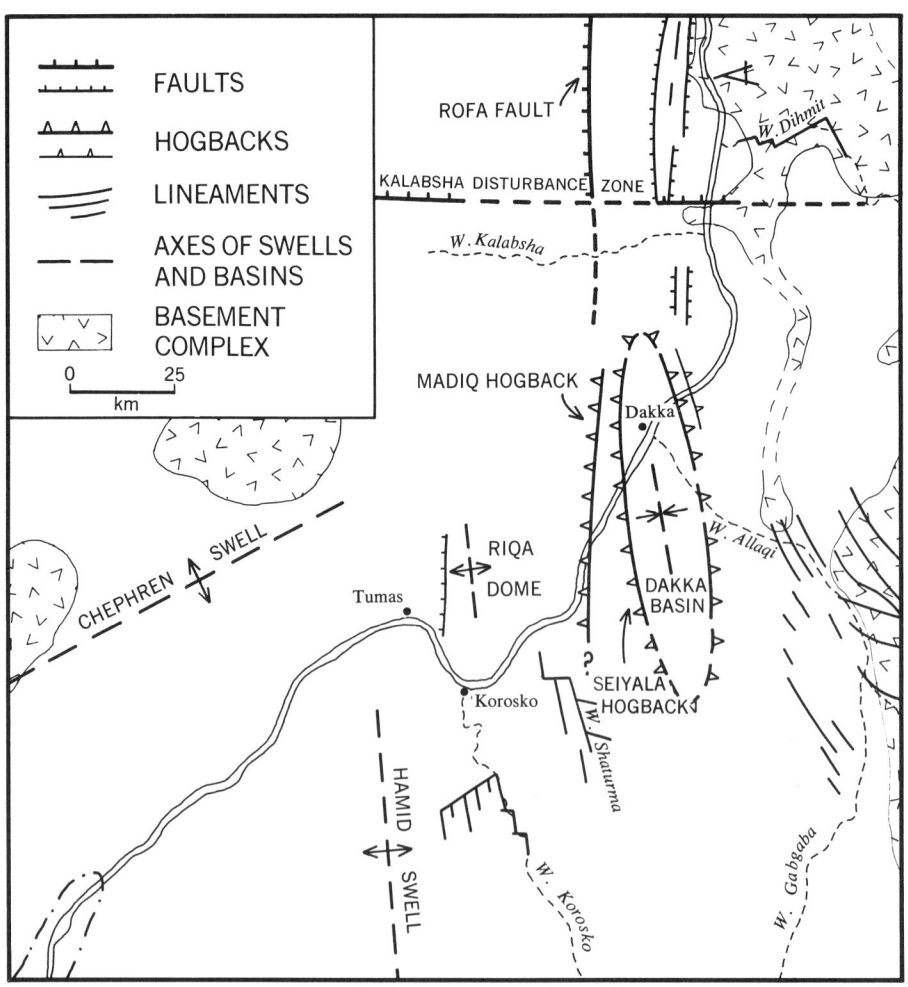

Fig. 5–1. Tentative structural map of Egyptian Nubia. Map: UW Cartographic Lab.

Not even the distribution of the Basement Complex exposures as shown by conventional maps (Attia, 1954; Shata, 1962: Map 1; Said and Issawy, 1965: Fig. 1) can be reconciled with the topographic expression of the Gemini IV photos. Fig. 5-1 attempts to correct this situation in as far as possible.

Throughout the igneous and metamorphic landscape east of the Nile, rectilinear intersecting drainage lines are common. Some of the more striking structural lineaments, representing faults, dykes, shear zones, or folded strata of variable resistance, have been shown on Fig. 5-1. Similar features cannot be detected among the Basement Complex rocks of the Lower Nubian Plain, at least not at the scale of the Gemini IV photography.

The single most striking tectonic feature in the sandstone mantle, whether seen from the ground or the air, is the well-developed, shallow basin structure at Dakka (Figs. 5-1 and 5-2). Average inclination of the strata lies between 2% and 5%, although dips along the conspicuous Seiyala Hogback average 5% to 10% and the extremes are somewhat higher (Fig. 5-3). The western part of this hogback crosses the Nile between Seiyala and Qurta Temple, trending almost due north; the eastern part intersects the Nile at Kushtamna. This hogback is often subdivided into two or three parallel ridges, with an average local relief of 25 to 40 meters. These ridges disrupt the drainage patterns noticeably. The Seiyala Hogback was verified by us in the field, on the 1:30,000 air photos, on the Gemini IV photos, and on the 1:100,000 topographic maps. Its localization, as shown on Fig. 5-1, is irreconcilable with the transcurrent faults shown by Fig. 1 of Said and Issawy (1965).

A second striking feature, also related to the Dakka Basin, is the Madiq Hogback. This delimits the eastern bluffs of the Nile between Sebua Temple and Nag Umm Simbil, from where it can be traced to northwest of Dakka. The great sandstone blocks that tower 200 meters above the Nile have been noticeably tilted in what is obviously a zone of strong and complex disturbance. These presumably are the slickensided and slip-jointed cliff faces referred to by Said and Issawy (1965: 10) as evidence of large-scale Pleistocene block faulting in the Korosko highlands.[1] The Madiq Hogback appears to converge with a fault zone that may be an extension of the Rofa Fault. The regional dip averages 10% to 20% to the east, although gravitational sliding has produced strong local variations in strike.

1. These features are unique in Egyptian Nubia, so that there can be little doubt about their localization. Yet, rather curiously, Said and Issawy's map (1965: Fig. 1) shows no disturbances along this stretch of the valley.

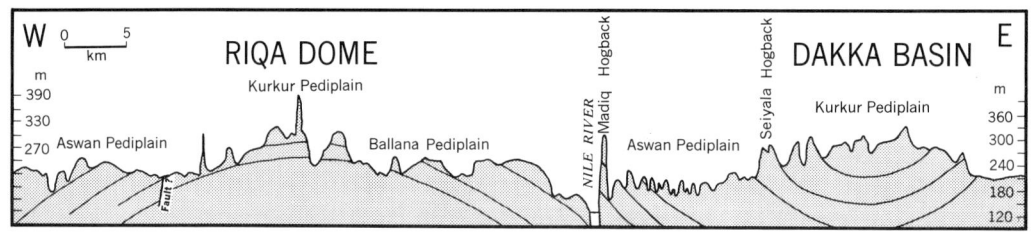

Fig. 5–2. West-East profile through basin and dome structures at approximately 22° 49′ N. Map: UW Cartographic Lab.

A possible complementary structure to the Dakka Basin is the Riqa Dome (Figs. 5–1 and 5–2). The strata are very gently inclined, with an average dip of less than 3%, and, correspondingly, peripheral cuesta-forms are poorly developed. Instead, three Tertiary planation surfaces are superimposed upon the Riqa Dome as they are on the Dakka Basin. These erosional surfaces show no conspicuous tectonic deformation in either area.

Other fundamental structures in Lower Nubia include the Hamid and Chephren Swells (Fig. 5–1), delineated on the basis of the Gemini IV

Fig. 5–3. 134- to 137-meter Nile platforms at Seiyala, veneered with eolian sand or late Pleistocene wash, seen from the crest of the Seiyala Hogback.

photos. These swells are separated from each other by a rather gentle depression, which is followed by the Nile River. The Korosko bend results from deflection of the Nile around the northern end of the Hamid Swell and the southern edge of the Riqa Dome. Further downstream, the course of the river is modified or controlled by north-south structures, as already described from the Kom Ombo area (Chap. 2).

Linear structures are less apparent in the Nubian Sandstone of Egyptian Nubia than they are in the sedimentary cover north of Aswan. Wherever the sandstones are sufficiently thick, drainage is usually dendritic as opposed to the subparallel patterns of the intermediate-order wadis of the Kom Ombo region. Somewhat exceptional are the lineaments of the Wadi Allaqi–Gabgaba area, all of which have been projected into the mantle from the underlying Basement Complex. Also of interest are the trellis drainage patterns of Wadi Shaturma and parts of the Wadi Korosko basin (Fig. 5–1).

A certain amount of small-scale block faulting may be observed along the Nile Valley upstream of Tumas. In particular, many of the lower pediments are bounded by faults with vertical displacements that seldom exceed a few meters. It should be emphasized that in no case were any of the Pleistocene gravels deformed along such fault zones. For reasons of time, it was not possible to systematically measure and map these minor faults in the field. Their direct influence on the morphology is limited, and none of these features can be detected on the Gemini IV photos.[2] These results agree very closely with the observations of de Heinzelin and Paepe (1965) in the northern Sudan, adjacent to our study area. De Heinzelin and Paepe found that vertical displacements of 1 or 2 meters are common, throws of 8 to 10 meters rather exceptional. Three main strike directions were identified: N 25°–48° W, N 63°–73° E, and N 70° W, in that order of importance.

In conclusion, the intensity of tectonic deformation in Egyptian Nubia does not compare with that of the Kom Ombo Plain. Large-scale grabens are absent, and well-defined structural-geomorphic entities, such as the Dakka Basin, are few and shallow. Gentle flexures, with resulting shear and tension-jointing, have been the most common form of deformation. These swell and basin axes have noticeably affected the geomorphic evolution of Nubia, more so than the major fault zones within the sedimentary strata. Along such intermediate and large-scale faults as can be verified, there is no substantive evidence for significant horizontal dis-

---

2. On the basis of considerable work in Khor Adindan, we dispute the existence of a major fault there as postulated by Said and Issawy (1965: Fig. 1).

placements. Vertical throws along fracture lines certainly never exceed a few 10's of meters. Finally, no reasonable claims can be made for significant deformation of any type during middle and late Pleistocene times. Negative evidence to this effect will be cited in subsequent sections dealing with the geomorphic evolution of Nubia. Most of the tectonic deformation in Egyptian Nubia may be attributed to the primary tectogeny of late Oligocene to early Miocene age (Knetsch, 1957), even though minor reactivation may have recurred at intervals throughout the late Tertiary and early Pleistocene.

Geomorphic Units

The salient landforms of Nubia between the old Aswan Dam and the Sudanese border are outlined by Fig. 5–4, based on the Gemini IV photography, the 1:100,000 topographic series, and our detailed mapping of the Nile Valley proper. The geomorphology is dominated by three complex erosional surfaces, dissected in variable degree. To express these two facets of the morphology, Fig. 5–4 shows these erosional surfaces superimposed with the terrain classification as devised by Hammond (1964). In this way, the state of dissection of the different surfaces is readily apparent. The four basic terrain types present are plains (local relief under 100 meters, surface predominantly in gentle slope); tablelands (local relief over 100 meters, surface predominantly smooth in upper part of elevation range); plains studded with hills (local relief over 100 meters, surface predominantly smooth in lower part of elevation range); and, lastly, hills (surface predominantly in intermediate or steep slopes). In addition to the basic map of surfaces and degree of dissection, the inset map of Fig. 5–4 synthesizes the major geomorphic units of Egyptian Nubia.

In order to facilitate identification of erosional surfaces, a study of accordant summits was made of Egyptian Nubia (Table 5–1). Peak elevations were read or estimated from the 1:100,000 topographic series (1960 revised edition) and were grouped in 10-meter intervals for nine arbitrary regions corresponding to the map sheets (listed by name and number). A total sample of some 6,400 elevation points was obtained so that the data given by Table 5–1 is fairly objective.

DABUD HILLS

From Aswan upstream to Girf Husein, the Nile Valley is fringed by irregular, hilly terrain, particularly where the Basement Complex is ex-

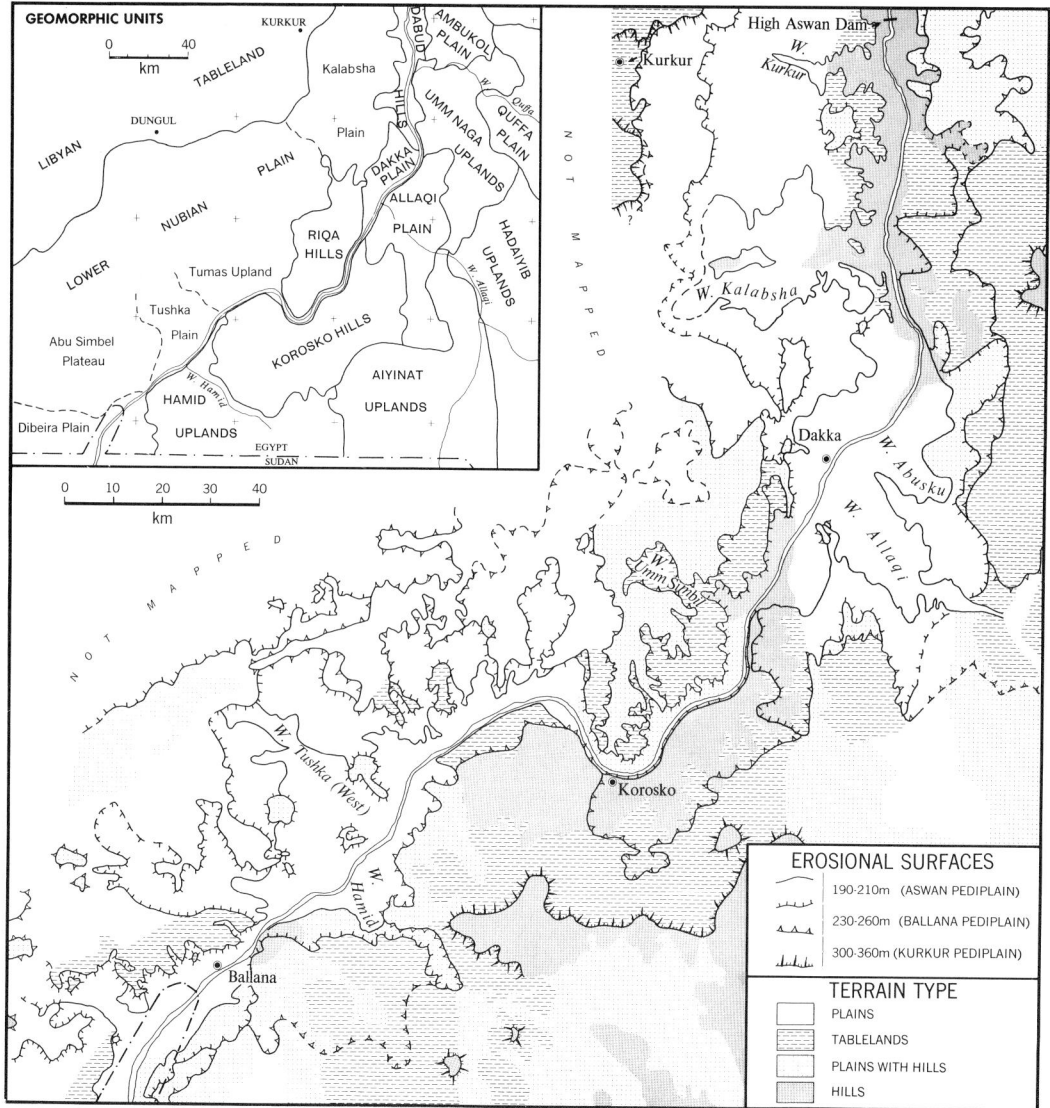

Fig. 5–4. Geomorphology of Egyptian Nubia. Map: UW Cartographic Lab. (This map also available at 1:500,000. See p. 524.)

posed. Local relief in 5- or 10-kilometer squares averages just under 100 meters, but locally attains 230 meters in the Gebel Abu Hor. The intricate wadi network, so well developed along the valley margins, produces a fairly rough terrain type, with 40% to 75% of the surface in generalized slope classes exceeding 8%.

Table 5-1. Accordance of summits in Egyptian Nubia
(Log values of numbers of summits in each altimetric interval.)

Meters	Diberia Sharq (Sheet 5569)	Arminna (Sheet 5669)	Masmas (Sheet 5670)	El-Diwan (Sheet 5770)	Korosko/Hamid (Sheets 5769, 5869)	Seiyala (Sheet 5870)	El-Allaqi (Sheet 5871)	Kalabsha (Sheet 5872)	Wadi Dihmit, etc. (Sheets 5971–73)
171–180	1.20	1.56	1.60	1.20		1.46	1.80	1.93	
181–190	1.62	1.76	1.98	1.28		1.36	1.83	2.28	0.78
191–200	1.56	1.57	1.60	1.34		1.66	1.92	2.38	0.48
201–210	1.79	2.14	2.16	1.61		1.89	1.71	2.15	1.34
211–220	1.34	1.89	2.09	1.56		1.85	1.48	2.01	1.32
221–230	1.30	1.98	1.57	1.74		1.95	1.59	1.59	1.40
231–240	1.71	1.73	1.54	1.93		1.87	1.40	1.20	1.38
241–250	1.58	1.63	1.52	1.85		1.62	1.23	1.28	1.46
251–260	1.46	1.62	1.54	1.86		1.69	1.23	1.43	1.58
261–270	1.38	1.51	1.56	1.72		1.53	1.00	1.20	1.67
271–280	1.20	1.28	1.23	1.75	1.20	1.60	0.90	0.60	1.49
281–290	0.95	1.43	1.43	1.88	1.71	1.53	0.30		1.70
291–300	0.78	1.23	1.26	1.63	1.45	1.32	0.90		1.56
301–310	0.30	1.45	1.30	1.72	1.69	1.57	1.28	0.78	1.69
311–320		1.00	0.85	1.30	1.83	1.36		0.30	1.76
321–330	0.60	1.20	0.85	1.51	1.79	1.26			1.65
331–340		1.15	0.90	1.52	1.80	0.95	0.30	0.30	1.60
341–350		0.60	0.60	1.04	1.85	0.85	0.60		1.52
351–360		0.95		1.15	1.70	0.48	0.30		1.46
361–370		0.90	0.30	0.70	1.93	0.60			1.56
371–380		0.90		1.15	1.88				1.15
381–390					1.52				1.32
391–400		0.30		0.30	1.61				1.08
401–410					1.46				1.28
411–420					1.51				1.00
421–430					1.48				1.08
431–440					1.38				0.60
441–450					1.15				0.48
451–460					1.26				
461–470					0.60				
471–480					0.90				
481–490					1.00				
491–500					0.78				

a 1:100,000 A.M.S. Series p. 677.

River platforms and pediments are locally evident in the sandstone reaches, but, in general, the Dabud Hills form a stretch of broken terrain between the Nile floodplain and higher-lying erosional surfaces such as the Ambukol and Kalabsha Plains. Structural features are everywhere evident, particularly in the form of dykes and ancient faults in the Basement Complex—several of which control the basic morphology of the local Nile Valley. Minor disturbances are also evident in the Nubian Sandstone. These include widespread inclined strata (10% or more) in the vicinity of contacts with the Basement Complex, as well as a number of small, *en échelon* faults, striking north along the edge of the Kalabsha Plain and striking west near the Kalabsha Gorge.

KALABSHA PLAIN

This broad, almost featureless sandstone pediplain extends 30 to 50 kilometers west of the Nile and stretches 125 kilometers from the Gallaba Gravel Plain in the north (Fig. 2-6) to the latitude of Dakka in the south. Rising abruptly from the Dabud Hills along the Nile Valley, the eastern periphery of the Kalabsha Plain has a local relief of about 100 meters and qualifies as a tableland. The greatest part of this bleak, undulating surface at 180 to 210 meters elevation is a plain, however, locally studded with low, irregular hillocks and other irregularities due to minor tectonic roughening or differential erosion. The only major relief feature is provided by the Ras el-Abd hills, a mesaform remnant of a higher erosional surface at 230 to 260 meters. Over most of the Kalabsha Plain, local relief varies from 10 to 30 meters in 5 kilometer squares, and 95% of the slopes are smoother than 1% or 2%. Patterns among the shallow drainage lines are next to incoherent, as the wadis from the limestone scarp to the west deteriorate into broad, flat washes, often terminating in closed depressions that may be of eolian origin. Surface runoff is at all times minimal as a result of both the hyperarid climate and the highly permeable bedrock. In fact, only three wadis cross the Kalabsha Plain to reach the Nile Valley: Wadi Kubaniya to the north, Wadi Kurkur, and Wadi Kalabsha to the south.

The greater part of the Kalabsha Plain represents a well-developed and well-preserved erosional surface at 180 to 210 meters, equivalent to the Aswan Pediplain in the Kom Ombo area. Vestiges of a higher, 230- to 260-meter surface are present at Ras el-Abd and are more extensively developed west of Dakka. Wadi Kalabsha has cut a lower surface into the Aswan Pediplain at about 160 to 180 meters, possibly contemporary with the Gallaba Terrace at Kom Ombo.

Faulting has some significance in accounting for the surface forms of the Kalabsha Plain. A complex zone of microfaults is apparent some 10 kilometers west of, and running parallel to, the Nile (Fig. 5-1). Another fracture, the Rofa Fault, strikes due north about 30 kilometers west of the Nile, forming a subcontinuous scarp with a relief of about 10 meters (Fig. 2-6). Bedrock dips are rather variable in its vicinity. Another significant fault runs west to east across the Libyan Tableland and continues as a complex fracture zone north of the Kalabsha Gorge. This Kalabsha disturbance zone appears to be complemented by a further major fault bounding the limestone escarpment east of Dungul (Said and Issawy, 1965: postscript). A part of this faulting must be younger than the planation responsible for the Aswan Pediplain. This agrees with the evidence from Kom Ombo (Chap. 2) that a certain amount of tectonic activity postdates the Aswan Pediplain.

AMBUKOL PLAIN

Developed in the Basement Complex east of the Nile between Aswan and Khor Dihmit, this is a typical pediplain, studded with low, steep-sided hills. Although 60% to 80% of the surface is gently sloping, local relief in 5 kilometer squares averages 100 to 125 meters. Drainage patterns are rectilinear in some areas, reflecting the complex fault and dyke patterns of the bedrock.

Although vestiges of the Aswan Pediplain are preserved along the Nile margins, the Ambukol Plain is dominated by an erosional surface at 230 to 270 meters. Well-developed accordance of summits at 290, 320, and 370 meters indicates the probable presence of higher erosional levels among the residual hills. And, in the interior parts of the plain, the major wadis move along surfaces that rise gently from 200 to 300 meters elevation with little or no break of gradient.

UMM NAQA UPLANDS

These uplands include the rough tablelands of igneous and sedimentary rocks that extend from Khor Dihmit to Wadi Allaqi along the eastern margins of the Nile. Deeply entrenched wadis follow dendritic patterns between mesaform uplands that describe two well-developed surfaces at 230 to 270 meters and at 300 to 340 meters. Scattered residual hills, with accordant summits at 370, 390, and 410 meters, rise above the tableland proper, culminating in Gebel Umm Naqa (570 m).

Although parts of these uplands have been dissected to a hilly terrain type, bimodal slope distributions are most characteristic: fairly smooth wadi floors and upland surfaces, separated by very steep slopes. Local relief in 5-kilometer squares averages about 150 meters.

QUFFA PLAIN

Although only of peripheral interest to a discussion of Lower Nubia, this plain deserves mention in passing as one of the most extensive geomorphic units of southeastern Egypt. This great, complex pediplain covers an area of over 5000 square kilometers and ranges in elevation from 350 to 600 meters, while the surrounding hills rise 200 to 300 meters above the level of the plain. The Quffa Plain originally developed in relation to the local base level provided by the major through-wadis, Wadi el-Quffa (the major tributary of Khor Dihmit) and Wadi Abu Maru (a tributary of Wadi Allaqi). Judging by the Gemini IV photography, surface drainage over most of this area is extinct today. Veneers of eolian sand and wash suggest disintegration into a multitude of unconnected, shallow basins.

HADAIYIB UPLANDS

Again of only peripheral interest, the area is poorly mapped, but the Gemini IV photos show a rough, irregular terrain developed on the Basement Complex and overwhelmingly controlled by structural features. Geomorphically, these uplands form a part of the Red Sea Hills.

DAKKA AND ALLAQI PLAINS

These plains occupy the Dakka Basin, whose structural aspects have been discussed already (see Fig. 5–1). The northern part of this lowland plain describes a semicircle around a nodal point at Dakka Station, while the southern half is focused on the lower courses of Wadis Allaqi and Abusku. As a result, drainage patterns are of centripetal type, intersected by the Nile. Slopes rise very gently away from the Nile and the major wadis, with average inclinations of 0.5% to 2.0%. Bevelled sandstone surfaces, partly conformable with the gentle dip of the bedrock, are locally veneered with wash or stream gravels. Several groups of conical, mesaform, or serrate hills rise abruptly 100 meters or more above the general level of the plain.

The greater part of the Dakka-Allaqi Plain is graded to a local base level of about 140 to 150 meters. This geomorphic stage can be readily correlated with the rather common ridges of Pleistocene wadi gravels, situated on top of this erosional surface. It appears that Pleistocene denudation was more effective here than in any other part of Nubia. The Aswan Pediplain is well developed on both flanks of Wadi Allaqi at elevations of 190 to 210 meters. Higher erosional surfaces at 225 to 240 meters and 300 to 310 meters are developed in the southern half of the Dakka Basin. The Seiyala Hogback, which runs along the western periphery of the Dakka-Allaqi Plain, delineates one of the most conspicuous geomorphic boundaries in Nubia.

RIQA HILLS

These hills developed in the Riqa Dome through dissection of two conspicuous planation surfaces at elevations of 230 to 260 meters and 300 to 310 meters (Fig. 5–4). Accordance of summits suggests the former presence of higher erosional levels at 325 to 340 meters, 360 meters, and 380 meters. The dominant higher level at 300 to 310 meters has been dissected to a plain with low hills, these mesaform hills corresponding to the 300- to 310-meter surface. The dominant lower level at 230 to 260 meters, on the other hand, rises abruptly above the dissected fringe of the Nile as a tableland, locally studded with tabular residuals of the 300- to 310-meter surface. Despite the intricate character of the resulting terrain, 60% to 75% of the surface is fairly smooth, although local relief in 5-kilometer squares averages 150 to 200 meters along the Nile Valley and 90 to 140 meters further inland.

At least one north-south fault lies along the western margin of the Riqa Hills, and the orientation of the western escarpment of the 300- to 310-meter upland suggests that location was influenced by structural lines striking due north. On the other hand, the development of the 230- to 260-meter surface, penetrating through the Riqa Hills in the form of broad wadi floors or extended pediments, indicates no major deformation. Wadi Umm Simbil is a case in point, with an extensive pediment developed at 240- to 270-meters elevation in the heart of the Riqa Hills. This precludes significant upwarping of the Riqa Dome subsequent to planation of the 230- to 260-meter surface. Equally significant for the tectonic and erosional history of the Nubian Nile is the clear development of the 230- to 260-meter surface on the western, southern, and eastern peripheries of the Riqa Hills, caused by headward erosion radiating back from the modern Nile Valley. This leaves no question about the great antiquity of the Nile in its present course. The Korosko bend of the Nile clearly antedates the 230- to 260-meter erosional surface.

KOROSKO HILLS

These hills constitute the roughest landform unit in Nubia. A belt of intensively dissected hill country extends 10 to 20 kilometers away from the right bank of the Nile. Between 60% and 80% of this surface is in steep slope, and local relief in 5-kilometer squares averages between 120 and 200 meters. Only in the Kasr Ibrim area does a prominent tableland—a part of the Aswan Pediplain at 190 to 203 meters—front directly on the Nile. Elsewhere, two plateau surfaces, with dominant levels of 230 to 260 meters and 310 to 350 meters, account for an extensive tableland within the interior of the Korosko Hills. Local relief here averages 100 to 150 meters. Finally, the highest part of the Korosko Hills is formed by a strongly dissected block of hill country, attaining a local relief of over 250 meters and a maximum elevation of 545 meters.

Three planation levels are conspicuous as extensive tableland surfaces in the Korosko Hills (Fig. 5–4) at 190 to 205 meters, 230 to 260 meters, and 310 to 350 meters. In addition, accordance of summits suggests the presence of a minor surface at 290 meters and of vestiges of high levels at 370 to 380 meters, 460 meters, and 490 meters. Although it is impossible to disentangle the minor erosional stages without detailed field work, the broad, major units in the geomorphic landscape are clearly defined and are quite comparable with those both upstream and downstream. Contrary to the cursory remarks of Said and Issawy (1965: 17), the highlands on either side of the Korosko bend do not interrupt the suite of erosional features developed along the length of the Nubian Nile in Egypt. Rather than attribute these highlands to Pleistocene uplift, we would emphasize their selective preservation from erosion, because of exceptionally resistant rock types. The degree of ferromanganese or ferruginous cementation of bedrock attained in the Korosko Hills is unrivalled elsewhere in Egypt. Even Pleistocene deposits are commonly indurated to a remarkable degree, whereas they are generally unconsolidated in other sectors of the Nile Valley. We are convinced that any regional geomorphic anomalies, other than the structural features shown on Fig. 5–1, can be safely attributed to the lithological distinctiveness of the Korosko Hills.

Structural features of the Korosko highlands are presently understood in a rudimentary way only. Despite a few small zones of rectilinear or trellis-like drainage (Fig. 5–1), patterns are generally of characteristic dendritic type. The Hamid Swell is noticeable in all aspects of the topography, but possible bounding faults remain to be verified.

AIYINAT UPLANDS

These uplands are once again of peripheral interest to Lower Nubia, and their study is impeded by inadequate mapping and incomplete Gemini photography. Possibly this sandstone plateau records some of the highest erosional surfaces in southern Egypt, at elevations between 400 and 500 meters.

TUMAS UPLAND

Like the Kalabsha Plain, this upland forms a part of the major geomorphic entity known as the Lower Nubian Plain. Most of this region is an upland plain, formed by erosional surfaces at 200 to 215 meters (Fig. 5–5), 230 to 260 meters, and 300 to 310 meters. Generalized slopes average less than 5% over about 90% of the surface; the remaining slopes offset horizontal surfaces at different levels and are quite steep. Local relief in 5-kilometer squares averages between 50 and 120 meters, so that some areas of accentuated relief technically qualify as tablelands.

The 200- to 215-meter surface within the Tumas Uplands is equivalent to the 190- to 205-meter surface across the Nile at Kasr Ibrim (Fig. 5–6). In all aspects of its geomorphic development, it resembles the Aswan Pediplain, and we see no reason to doubt the identification of the surfaces at Tumas and Kasr Ibrium with those found at similar elevations downstream of the Korosko bend. Of further interest are extensive pediments at 170 to 180 meters, fringing the periphery of the Tumas Uplands.

Fig. 5–5. Erosional surfaces east of Amada Temple. The sand- and wash-swept, 166- to 169-meter pediment dominates the scene, with dissected remnants of the Aswan Pediplain visible in the background and the immediate foreground.

TUSHKA PLAIN

This plain marks a break in the high erosional surfaces of the Lower Nubian Plain. Two now-defunct wadi systems, debouching at Tushka West and near Nag Farqanda, appear to have effected considerable denudation in early and middle Pleistocene times. As a result, the lowland plain is studded with swarms of residual hills with accordance of summits at 200 to 210 meters and 220 to 235 meters. Extensive pediments at 180 to 190 meters and at 160 to 170 meters dominate the landscape near the Nile Valley, while the Aswan Pediplain is well developed farther west. Except for a few areas of accentuated roughness, local relief in 5-kilometer squares commonly ranges from 50 to 90 meters.

Fig. 5–6. The Korosko Hills seen from Ineiba. The well-developed erosional surface on the opposite side of the river represents the Aswan Pediplain at 200 meters elevation, with spurs of a lower pediment at about 180 meters. Late Pleistocene silts in middle foreground.

ABU SIMBEL PLATEAU

The plateau is rather similar to the Tumas Uplands, except that there are few remnants of the highest surface (at 330 m). Instead, there is an exceptionally distinctive pediplain at about 230 to 255 meters elevation, forming along the Nile Valley a precipitous scarp with a local relief of 100 to 125 meters. The temples of Abu Simbel were cut into this escarpment, the periphery of which preserves remnants of the Aswan Pediplain at 195 to 205 meters (Fig. 5–7).

DIBEIRA PLAIN

The final geomorphic subdivision of the Lower Nubian Plain, this plain includes sections on both sides of the Nile Valley. The two lower erosional surfaces are well represented on the western bank, together with frequent pediment segments at 160 to 170 meters and at 180 to 190 meters. On the eastern bank, dissection has produced a greater number of residual hills, in contrast with the more intact escarpments on the other side of the river. In addition, a 300- to 310-meter surface is widespread on the eastern bank.

Fig. 5–7. The Aswan Pediplain west of Abu Simbel at about 200 meters elevation. Massive accumulations of drifting sand fill the fossilized wadis in the middle.

HAMID UPLANDS

These uplands include an extensive tract of variable high country lying east of the Nile and south of the Korosko Hills. Inadequate maps preclude all but a few generalizations. Except for a rough, hilly zone opposite Abu Simbel, the fringes of the Hamid Uplands have been strongly dissected to an open plain with hills. Less-dissected plateau surfaces are found further inland.

The fairly smooth plains extending back from the Nile at Arminna East have a local relief of less than 50 meters. Pediments at 160 to 170 meters are the most prominent feature, although remnants of a higher level at 180 to 190 meters can also be identified. The Aswan Pediplain is developed at 200 to 210 meters, while extensive older surfaces are found at 220 to 270 meters and at 300 to 340 meters. Accordance of summits suggests that the 220- to 270-meter level has two prominent subdivisions at 230 and 260 meters, while the higher surface exhibits steps at 310 meters and at 330 to 340 meters. Finally, remnants of a higher level at 360 meters are suggested.

The Tertiary Pediplains

The preceding discussion of the small-scale regional geomorphology of Egyptian Nubia has shown the persistence of three broad erosional surfaces along the length of the Nile Valley.

The lowest and best-preserved of these surfaces can be traced continuously from Kom Ombo to Seiyala and again from Amada Temple to south of the Sudanese border (Figs. 5–4 to 5–7). The dominant level rises slowly from 180 to 190 meters at Kom Ombo; 180 to 210 meters southwest of Aswan; 190 to 210 meters in the Dakka Basin; 190 to 215 meters near Tumas and Kasr Ibrim; and 200 to 210 meters between Tushka and the border. Regional variability reflects the gentle rise of this surface away from the Nile Valley. The Nile provided a local base level for erosion, increasing in elevation from less than 180 meters at Kom Ombo to 200 meters or so at the Sudanese border. A drop of about 25 meters along a longitudinal distance of 350 kilometers indicates a gradient of only 0.71 in 10,000, compared with 0.87 today.

Deposits contemporary with the 180- to 210-meter surface have not been identified with any certainty, and they probably do not exist. Instead, all inferences concerning the origin of this surface must be based on the erosional morphology. Through most of Nubia, the 180- to 210-meter surface is studded with butte and mesaform residual masses of variable relief. The surface is commonly projected into dissected higher country through broad wadis that frequently widen into local pediment valleys. Finally, wherever well preserved over larger areas, this broadly horizontal surface suggests a patchwork of merging segments at approximately equal elevations but differing minutely in direction of dip. All in all, these features can best be explained by coalescent pediments, related to the same local base level of the Nile Valley but cutting backward at variable rates and gradients along different wadi systems.

In this restricted sense, we are dealing with a pediplain, a complex planation surface characterized by several minor aspects of arid-zone landform: steep-sided residual masses; rectilinear midslopes; angular breaks of gradient; and broad, extremely shallow drainage lines (Figs. 5-5 to 5-7). This Aswan Pediplain, as it was first designated in Chapter 2, marks a single but protracted stage of planation, presumably associated with a semiarid climate (see Tricart and Cailleux, 1961, v. II: 39-55). The Nile itself flowed at approximately 80 to 90 meters above its present floodplain level in southern Egypt, and wide lowland plains, as much as 50 kilometers in cross section, developed in response to denudation by local runoff. At Abu Simbel and in the Korosko bend, the contemporary valley was sharply constricted, very much as it is today. How far this valley extended upstream in the Sudan, or what its relationships were to the base level of the Mediterranean Sea, will not be known until further investigation, but we are already dealing with a geomorphic entity that is basically compatible with the modern geography of southern Egypt.

The stratigraphic age of the Aswan Pediplain is impossible to determine with any precision. The evidence at Kom Ombo and in Nubia indicates that this surface is considerably older than the earliest Pleistocene gravels. At Kom Ombo (Chap. 2), at least 30 meters of vertical bedrock cutting and appreciable tectonic deformation separate the Aswan Pediplain from the Gallaba stage. In Nubia, a succession of minor lateral pediments and a certain amount of block faulting intervened prior to the oldest known Pleistocene gravels. Furthermore, the Aswan Pediplain at Kom Ombo appears to be older than the Pliocene Gulf deposits, which would suggest a pre-Middle Pliocene age.

One of the most significant corollaries to the existence and late Tertiary age of the Aswan Pediplain is that the Nubian Nile is of very great antiquity. An early Proto-Nile followed the approximate course of the modern river from the Sudanese border northward, past Aswan, and into the present deserts west of Kom Ombo. At Aswan and at the High Dam, the Tertiary age of the river is further documented by the buried fossil valley to 172 meters below sea level, with an infilling of 110 to 160 meters of deposits related to the Middle and Upper Pliocene Gulf.

The intermediate erosional surface of Egyptian Nubia at about 230 to 260 meters elevation (Fig. 5–4) is only imperfectly understood. It appears that two substages at 230 meters and at about 260 meters can be discerned in many areas. The fact that this complex surface can be traced without interruption from the Tumas Uplands through the Korosko bend to the Dakka Basin suggests that we are dealing with a stratigraphic entity. The greater degree of dissection and denudation of the surfaces at 230 to 260 meters precludes confusion with the Aswan Pediplain. Yet it remains impossible to prove that the 230- to 260-meter surfaces of the Abu Simbel–Wadi Hamid area, those of the Tumas-Dakka sector, and the isolated remnants further north are indeed all contemporaneous. The possibility that block faulting or gentle warping have invalidated the altimetric criterion of correlation must be borne in mind. With these reservations, we feel that a complex 230- to 260-meter stage *probably* exists. The evidence, however, is inconclusive. Possibly, a study of further, more detailed remote-sensing data from the latest satellite orbitings will provide more definitive arguments. In the meanwhile, the designation Ballana Pediplain is tentatively assigned to the 230- to 260-meter surfaces as a matter of convenience—bearing in mind that these surfaces may not all pertain to a single stage of geomorphic evolution.

The development of the Ballana Pediplain in the area of the Riqa Dome and the Dakka Basin suggests that major tectonic deformation in these structural units preceded planation of the 230- to 260-meter surface. On these grounds, the Ballana Pediplain may be assumed to postdate the primary orogenic phase of late Oligocene to early Miocene times (Knetsch, 1957). A late Miocene age seems a reasonable possibility for the Ballana Pediplain.

The 300- to 360-meter surfaces pose an almost insuperable problem of correlation and stratigraphic dating. Preservation is largely confined to groups of residual hills or strongly dissected mesaform plateaus, usually isolated from one another (Fig. 5–4). The possible role of faulting or warping is difficult to assess on a regional scale, yet surfaces at these elevations, often suggesting three subdivisions, reoccur along most of the Nubian Nile. Furthermore, distinctive pediments can be identified at the Kurkur Oasis at 360 to 365 meters, 340 meters, and 320 to 325 meters (Chap. 7). For these reasons, the 300- to 360-meter surfaces will be designated as the Kurkur Pediplain, without necessarily attaching any stratigraphic significance to the term. All or most of these surfaces are older than the Ballana Pediplain, and presumably all are of middle to late Tertiary age, but it is improbable that all of these surfaces are contemporary. The Kurkur evidence suggests that the three surfaces between 320 and 365 meters are polygenetic and that considerable intervals of time separated each unit.

L. C. King (1962: 286 ff., Fig. 119), in his sweeping geomorphic hypotheses for the African continent, suggests the presence of one or more late Tertiary denudation surfaces over most of Egypt and the northern Sudan, with Cretaceous and early Tertiary surfaces fragmentarily preserved in the high country of the Red Sea Hills. Conceived on a very broad scale, these generalizations are nonetheless compatible with the detailed local picture presented here for Egyptian Nubia.

The Pleistocene Pediments

The occurrence of pediments in the Nile Valley, at elevations between 150 and 190 meters, provides a link between the Aswan Pediplain and the Pleistocene gravels. All of these pediments form steplike denudation surfaces, studded with residual hills away from the Nile and possibly coalescing with true fluvial platforms near the edge of the modern floodplain. All are considerably younger than the Aswan Pediplain, marking long-term interruptions in the late Cenozoic degradation cycle of the Nile. Development of these pediments is largely restricted to the Dakka Basin and the region between Amada Temple and the Sudan border.

A widespread pediment level can be observed at 65 to 68 meters above modern floodplain. It forms an extensive flat surface east of Adindan at 185 to 190 meters (Fig. 5–11), capped by ironstone and strewn with detrital wash. A similar surface can be traced northwards in the form of a well-developed bench studded with residuals and dissected in turn at its nileward margins. The level of this bench drops to 180 to 182 meters at its downstream terminus at Kasr Ibrim on the east bank (Fig. 5–6), Tumas on the west bank. The same surface can also be detected from accordance of summits between 180 to 190 meters (Table 5–1). This surface bears no contemporary deposits and may best be interpreted as a valley-margin pediment, graded to a former Nile level at about +65 to +68 meters. Minor, zigzagging faults delimit the surface in many areas, so that there can hardly be complete confidence in the stratigraphic value of such a stage. But displacements along such faults are small, seldom attaining as much as 5 to 8 meters where vertical throws can be measured. Conceivably, these minor fractures have exerted more influence on erosional development than they have on fault displacement as such.

A well-developed, pediment-like surface can also be seen at 178 to 181 meters (+57 to +60 m) on both sides of the Nile near Adindan (Figs. 5-10, 5-11) and at a similar relative elevation just north of Abu Simbel (176 m). As in the case of the higher level, ferromanganese solutions have indurated the surface bedrock to ironstone, under a lag or wash of ironstone rubble.

An extensive pediment is developed at Arminna West, terminating at about 168 to 172 meters (+50 to +54 m) at the Nile floodplain and rising slowly landward to knickpoints between 190 to 205 meters (Hansen, 1966: Chap. 2). Although deposits are not preserved on this surface, sinuous bedrock ridges with detrital caps and local relief of 15 to 25 meters wind over a part of it. They suggest selective preservation of ferruginized wadi beds or bed-load deposits in the wake of surface denudation. At Amada Temple there also are traces of a dissected pediment at 166 to 169 meters (+52 to +55 m) (Fig. 5-5). Whether these +50 to +55 meter pediment fragments are all contemporary is debatable. There is, however, a well-developed gravel and platform stage of the Nubian Nile at precisely this level (Allaqi Stage).

From Adindan (Figs. 5-10, 5-11) to Amada Temple, a minor pediment of some stratigraphic value can be traced discontinuously at +44 to +49 meters. Locally, as at Arminna East and Amada Temple, these pediments are unquestionably graded onto fluvial platforms cut by the Nile, sporadically capped with Nile gravels of the +44 to +48 meter Dihmit stage. At Adindan West, on the other hand, a fragment of the same pediment is capped by ironstone and is then overlain disconformably by unconsolidated gravel of the younger, +42 meter Adindan stage (Fig. 5-10). This suggests that these pediments may be the equivalents of the Dihmit stage in southern Egyptian Nubia. The well-developed gravels and platforms of this early Pleistocene stage would be compatible with a protracted period of stability, accompanied by extensive denudation through wadi flow and rill wash.

Lower pediment surfaces are uncommon, and only two examples deserve mention. A pediment adjoins a 151-meter (+34 m) Nile platform at Tushka West, and a similar pediment can be seen west of Ineiba. This suggests a correlation with the middle Pleistocene Dakka stage.

The pediments of the Dakka Basin are separated from those of southern Egyptian Nubia by a river stretch of over 80 kilometers, and Nile gradients during the Pleistocene appear to have been a little steeper than today. A dissected pediment at 160 to 163 meters (+53 to +55 m) along Wadi Allaqi suggests greater antiquity than a well-developed pediment with a dominant level of 153 to 154 meters (+46 to +47 m). Both of these may possibly be equivalent to the substages of the High Terrace of Wadi Allaqi.

Although the post-Aswan surfaces of Egyptian Nubia are not ideally suited for stratigraphic correlation over wide areas, they nevertheless provide an important chronological link between the Aswan Pediplain and the Pleistocene gravels. There is little question that all of these denudational steps must be assigned to the early Pleistocene. Possibly, the 180- to 190-meter pediments were broadly equivalent with the Proto-Nile system that led to development of the Gallaba gravel terrace at Kom Ombo.

The Nature of the Nile and Wadi Gravels

The general erosional tendency of the Pleistocene Nile was interrupted by several periods of aggradation, recorded by terraces of coarse, rounded gravel. These occur within the Nile Valley or along the lowermost courses of the tributary wadis. In many, but not all, cases, such gravels are sporadically found on top of extensive fluvial platforms. The deposits consist primarily of macrocrystalline quartz, with ferricrete sandstone or ironstone locally important. Igneous and metamorphic materials may be abundant or dominant near the embouchures of several Eastern Desert wadis. All of the gravel pertains to the Nubian Sandstone or to the Basement Complex, so that we are dealing with autochthonous gravels, derived from the Lower Nubian Nile and its tributary wadis rather than from the Central Sudan or subsaharan Africa.

Composition and facies of these autochthonous gravels have been remarkably similar through time. In all cases, coarse bed-load deposits are represented almost exclusively, although median pebble calibre varies within defined regional patterns. So, for example, the smaller wadis from the Western Desert carried rather coarse sandstone and ironstone pebbles with only a little quartz, chiefly in the granule class. After entering the Nile, these softer materials were rapidly worn down and dispersed, ultimately contributing to the quartz-sand matrix of the Nile gravels.

Several wadis of the Eastern Desert drain parts of the Basement Complex. In such cases, coarse- to cobble-grade gravels of variable lithology form the wadi deposits. A few kilometers downstream, median pebble diameter has decreased, and less resistant rock types selectively disappear until the remaining gravel becomes exclusively quartz or quartzite. At least a part of the selective removal of nonsilicic rocks can be attributed to chemical weathering. Wherever soil profiles are well-developed, pebble lithology within the (B)-horizon is largely limited to quartz and quartzite. Significantly, if the unweathered C-horizon of the same deposit is exposed, a great complexity of igneous and metamorphic rocks will be present. In such cases, subsequent differential weathering has been the primary factor. Chemical weathering contemporary with alluviation may also be involved. The Pleistocene Nile was a perennial stream, whereas the hydrography of the wadis probably remained episodic or seasonal in type. Nile pebbles may have been immersed in water and liable to chemical attack for a period of as much as twelve months per year. Wadi pebbles, on the other hand, would have been dry for all but a few weeks of the year and, consequently, subject to less thorough or rapid decomposition. As a result, weathering within the stream bed—a major factor in pebble destruction (Tricart and Cailleux, 1960, v. I: 103)—would have been several times more prolonged and intensive within the bed of the Nile as within the bed of a tributary wadi.

The bulk of the gravel which follows the Nile through Nubia can be illustrated from a type case near Arminna Temple, at 165 meters elevation. Granule or fine gravel (2 to 6.4 mm diameter) accounted for 36% by weight of the material coarser than sand grade. Of this, 98% consisted of macrocrystalline (crystals greater than 0.75 mm) quartz, the remainder of ironstone. Medium gravel (6.4 to 20 mm diameter) accounted for 54%, including 82% macrocrystalline quartz, 13% mesocrystalline (crystals between 0.2 and 0.75 mm diameter) quartz, and 5% mesocrystalline, silicified sandstone. Finally, the coarse gravel (20 to 60 mm) accounted for 10% of the total, with 50% macrocrystalline quartz, 34% mesocrystalline quartz, and 16% ironstone. The medium-to-coarse quartz pebbles were primarily derived from pegmatitic veins or dykes intruding the Basement Complex. In other words, local materials are subordinate to pebbles transported into Lower Nubia from the Basement rocks exposed between the Second and Third Cataracts.

Heavy-mineral composition of the sandy gravel matrices corresponds closely to pebble lithology. Heavies are extremely scarce in wadi deposits derived within sandstone catchment areas. In such cases, the heavy-mineral spectrum of the Nubian Sandstone (see Shukri and Ayouty, 1953) is encountered, probably augmented by extraneous minerals introduced through eolian transport. Heavy minerals are abundant near the mouths of Wadis Allaqi and Dihmit, derived from breakdown of igneous and metamorphic pebbles. In the remainder of the Nubian Nile Valley, quartz sands, as well as quartz pebbles, are found almost exclusively.

The silt and clay fraction of these alluvial beds is derived either from decomposition of igneous and metamorphic pebbles or from deep weathering within the various red paleosol horizons. Kaolinite, and a variable admixture of montmorillonite, is the chief clay mineral represented. Silty lenses were not observed in any of the Nubian alluvia, although it should be mentioned that good exposures are few.

The Nile and wadi gravels of Egyptian Nubia were deposited on the beds of broad, high-competence streams. Crude stratification and limited sorting, together with a scarcity of graded or inclined bedding, suggest torrential flow with rather thorough scour of finer bed deposits prior to the fill stage of each flood. Where not fractured by weathering processes, pebbles are quite generally rounded. The moderate degree of pebble flattening suggests that transport involved both rolling and sliding motions. All in all, stream competence was rather high, both in the Nile and in the tributary wadis. Today the wadis are only able to transport sands with dispersed fine to medium pebbles—at least in their lower courses—and the modern Nile transports little or nothing coarser than sand grade along its bed.

Further information on the character of these Pleistocene alluvia can be obtained from the interrelationships between Nile and wadi gravels. The two are clearly interdigited in an inextricable fashion, and gravels in the lower wadi courses are just as clearly graded to the different Nile stages. As at Kom Ombo, the two facies are contemporary and are seasonally in phase. Wadi floods cascaded into the Nile Valley at times when the Nile itself was at bankfull stage. Fanlike zones of deposition were built up in the wadi embouchures, as wadi gravels were shot into the Nile and swept downstream in long, taillike intergradations of Nile and wadi materials. Winter rains presumably fed these floods.

The development of Pleistocene gravels in Egyptian Nubia is geomorphically of limited significance only. Although such alluvial deposits are widespread between Adindan and Korosko, they are usually discontinuous and almost invariably rather shallow (Fig. 5–9). Rarely do the gravel bodies exceed 5 meters in thickness. Downstream of Korosko, gravel terraces are restricted to the Dakka Basin, the Dihmit area, and the First Cataract. The intervening lacunae measure between 30 and 50 kilometers in each case, and nowhere in Nubia are Pleistocene gravels present to the same extent and thickness as around the Kom Ombo Plain.

The fluvial platforms complement this picture to some degree. Nile-cut surfaces can frequently be identified on the airphotos, and they are readily distinguished from other types of erosional surfaces. They appear as convex platforms, swinging into valley embayments, but separated from pediment surfaces by a weak but distinctive knickpoint. The platform surfaces have few, if any, residual masses and are noticeably smoother and flatter than adjacent pediments. It is reasonable to assume that these fluvial platforms and any superimposed gravels are broadly, although not strictly, contemporaneous. Platforms are rather well developed between Abu Simbel and Amada Temple and again in the Dakka Basin. They suggest long periods of dynamic stream equilibrium, during which bedrock planation was accomplished with little or no net aggradation. The gravels may record a final phase of alluviation prior to a renewal of general downcutting within a more restricted channel.

A final insight into the nature of these Pleistocene gravels is provided by the "perched wadis" or "wadi ridges," sinuous lines of gravel or bedrock projecting as much as 30 meters above the general elevation of the land. These ridges converge on the Nile or on a master tributary very much as a meandering, dendritic stream channel would, and they can frequently be traced subcontinuously for several kilometers. Knetsch (1954: part 3, Figs. 6 to 9) first recognized these features during an aerial reconnaissance of Nubia, and provided photographs, a sketch, and a field description from the Dakka Basin.[3] The ridges Knetsch described consisted of a meter or so of well-rounded quartz or ironstone stream gravels resting on a raised bedrock surface, suggesting the bed load of an extinct stream. Knetsch called these gravel ridges, "pseudo-eskers," attributing them to selective preservation of gravel *vis à vis* chemical weathering and deflation.

On the basis of our own field observations, we would distinguish two major types of pseudo-esker: a bedrock and a gravel variety, with a number of intermediate forms. The bedrock type is well developed at Arminna West and is capped by ironstone and veneered with detritus. In all probability, this form of ridge marks a former bedrock wadi floor, once selectively impregnated by ferromanganese solutions. After the stream was abandoned, denudation removed the surrounding, softer sandstone that was not cemented to the same degree. As a result, the wadi bed may be 20 meters higher than the surrounding surface today, whereas originally it had been excavated a little below the surface. The gravel type of

---

3. They apparently are not unique to Egypt but have also been recognized by Knetsch (1954) in the western Cape Province of South Africa.

pseudo-esker is very common and consists of former wadi bed deposits of ferricrete sandstone or ironstone pebbles—the most indurated and resistant detritus originally found in the area. Denudation of surrounding bedrock would leave a wadi gravel of this type relatively intact. Apart from the resistance of the individual pebbles, this grade of material cannot be attacked by deflation, and it remains too permeable for effective erosion by water, at least in the case of unconsolidated deposits. Practically all of the west-bank wadi ridges are unconsolidated, although the few east-bank examples south of Seiyala are ferricreted. In many cases, the gravel rests on distinct bedrock ridges, such as those described by Knetsch (1954).

Some eighteen typical examples of major pseudo-eskers were mapped in Nubia (see Figs. K–1 through K–10); although they are best developed beyond the perimeter of our 1:40,000 maps. These ridges may be graded to the edge of Nile gravel terraces, where the distinction between the two geomorphic entities is often difficult to make. Although pseudo-eskers of the types defined above formed repeatedly during the course of the Pleistocene, they are best developed and most common in relation to the oldest gravel stages, a fact also noted by Knetsch (1954). Understandably, they are also most frequently found on the denuded plains of the Dakka Basin and along the edge of the Western Desert between the Sudan border and el-Riqa.

In conclusion, the autochthonous Pleistocene gravels of Egyptian Nubia provide evidence of repeated moist paleoclimates, accompanied by torrential runoff, upland erosion, greater sediment yield, and higher stream competence. It is questionable whether a genetic distinction should be made between lateral planation of river platforms and the superimposed gravel. Since deposits are generally shallow, it follows that the particular gravels present are younger than the platforms, representing a fairly brief phase of net aggradation preceded by a long period of dynamic equilibrium. But the lateral planation was accomplished by a similar bed load during the course of repeated scour and fill. In this sense, the gravels and associated platforms are broadly contemporary and their interpretation is similar. Rather than periods of aggradation, the autochthonous gravels really record interruptions of the Pleistocene erosional cycle—periods of relative stability with a net balance between scour and fill in the Nile Valley, but with a tendency to floodplain widening and pedimentation along the valley margins.

A more detailed interpretation of Pleistocene fluvial processes has been attempted in Chapter 2.

Stratigraphy of the Nile and Wadi Gravels

Establishing a relative stratigraphy of autochthonous Nile and wadi gravels in Nubia is a difficult matter. Distances are great, and deposits are discontinuous and frequently interrupted by stretches of erosional topography lacking either gravels or platforms. As a consequence, direct correlations are impossible for the length of Nubia and are difficult even in areas with well-developed terraces. On a more regional scale, direct correlations are impeded by the intensive erosion of terrace gravels. Maximum elevations of homogeneous deposits frequently vary by 2 to 5 meters along a 5-kilometer longitudinal reach of the Nile, and local relief on a "level" gravel surface may exceed 5 to 10 meters within a 500-meter square. Altimetric correlation is further complicated by the presence of multiple minor stages. Technical errors of measurement in the field were largely precluded, however, by the use of Abney levels and by the availability of good topographic maps and third-order survey markers.

Indirect correlations are impeded by a relatively uniform lithology, at least on a local scale, and by an almost general lack of good exposures. Heavy-mineral studies are rendered questionable because of the selective weathering of less resistant minerals, so that a high proportion of the heavy minerals are altered. Fossils are absent, and prehistoric artifacts have so far been found *in situ* at only two or three localities. Finally, there is a total lack of materials suitable for isotopic dating.

Faced with these difficulties, we have been hesitant to propose a rigid scheme of terrace correlation. Wherever gravel terraces are well developed, contemporary deposits can be subcontinuously followed in the field or by the aerial photography. On this basis, we have no reservations concerning the internal validity of correlation and differentiation within certain restricted areas: Adindan to Korosko, the Dakka Basin, the Kalabsha-Dihmit area, and, of course, the Kom Ombo Plain. Regional correlation between these local sequences is more tenuous. The relative geomorphic significance of terraces, the degree and character of surface rill erosion, and the depth of weathering proved to be rather useful criteria in linking up these areas. Altimetric correlations were referred to the average modern gradient of the Nile between the Sudan border and Kom Ombo.[4]

---

4. Due to the presence of the First Cataract at Aswan, the Nubian Nile was graded to a temporary base level long before construction of the Aswan Dam. Since higher Nile stages were not affected in this way, Pleistocene stream gradients must be assumed to have been rather simpler, with no local knickpoint of such impor-

The stratigraphic scheme proposed here is expressed by informal stage designations, with relative elevations given with respect to modern Nile floodplain between Adindan and Ballana, or otherwise with respect to modern floodplain elevation at the most southerly exposure of the gravels in question (Fig. 5–9):

Wadi Korosko (WK)	+ 23 to + 25 meters
Dakka (DA)	+ 30 to + 35 meters
Adindan (AN)	+ 40 to + 42 meters
Dihmit (DI)	+ 44 to + 48 meters
Wadi Allaqi (WA)	+ 50 to + 55 meters

Although this sequence is not proven beyond a doubt for Egyptian Nubia as a whole, we have no reservations concerning the intraregional correlation and differentiation between Adindan and Korosko (Fig. 5–9) or the Dakka Basin. Furthermore, the entire scheme seems reasonable and consistent from the border to Aswan. Probable correlations with the Kom Ombo Plain will be discussed, together with the problem of stratigraphic dating, after an outline of the regional evidence.

Pleistocene Terraces Between Adindan and Ballana

The most prominent gravel terrace between the Sudan border and the Abu Simbel gorge belongs to the Adindan stage (+ 40 to + 42 m). It is readily visible on Figs. K–1, K–2 as well as in Figs. 5–9 to 5–14. At the type locality of Adindan West, there are frequent patches of coarse, quartz gravel to a maximum elevation of 163 meters and with a thickness exceeding 7 meters. Two tributary pseudo-eskers, composed of ironstone gravel, adjoin these gravels from the west. On the east bank, gravels of the same stage can be followed subcontinuously from the border to opposite Ballana, with a maximum elevation of 163 meters and a thickness of at least 12 meters in places. Everywhere the gravels rest on an iron-

---

tance. An additional problem is posed by flood-silt accretion within the Aswan Reservoir since 1902. Accelerated silting up of the modern floodplain is apparent from Ineiba northward. In order to reconstruct the approximate floodplain level in 1902 (Fig. 5–9), we have used the mean high-flood levels at Wadi Halfa and Aswan as reported by Willcocks (1899: 44) for the late nineteenth century, complemented by our field observations at Adindan—where floodplain level lay just above maximum reservoir level, even after the last additions to the old dam in 1929–34. To avoid confusion, all subsequent elevations are given in the text both as absolute values and relative to modern floodplain.

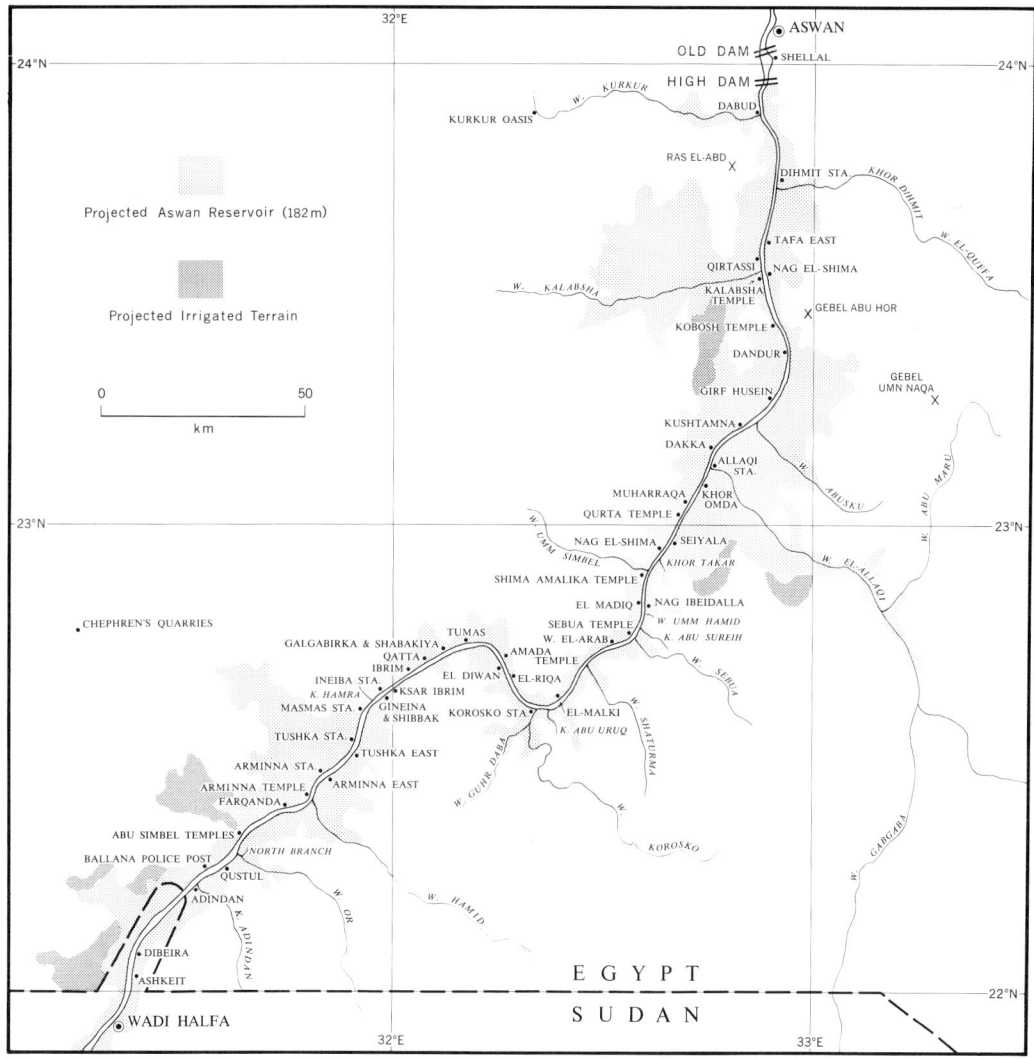

Fig. 5–8. Lower Nubia. Map: UW Cartographic Lab.

stone-capped pediment surface with a dominant elevation of 157 to 168 meters. This older pediment may be related to the Dihmit stage, although it is bounded by zigzagging block faults. Significantly, the + 40 to + 42 meter gravels traverse one such fault at Qustul without any evidence of disturbance.

A deep, red to yellowish-red (2.5–5 YR 5/6) paleosol with a coarse, subangular blocky structure has developed on the gravels of the Adindan

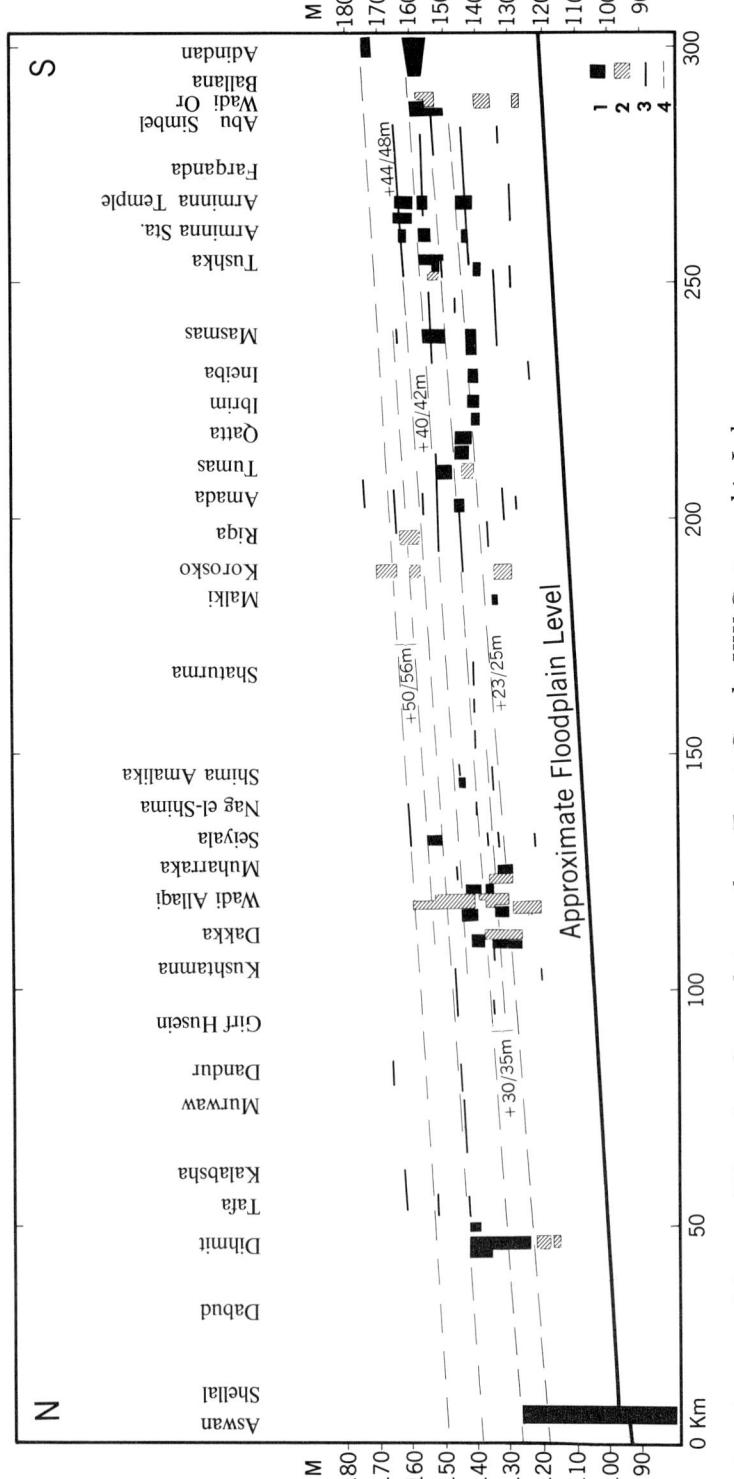

Fig. 5-9. Autochthonous Nile and wadi gravels in southern Egypt. Graph: UW Cartographic Lab.

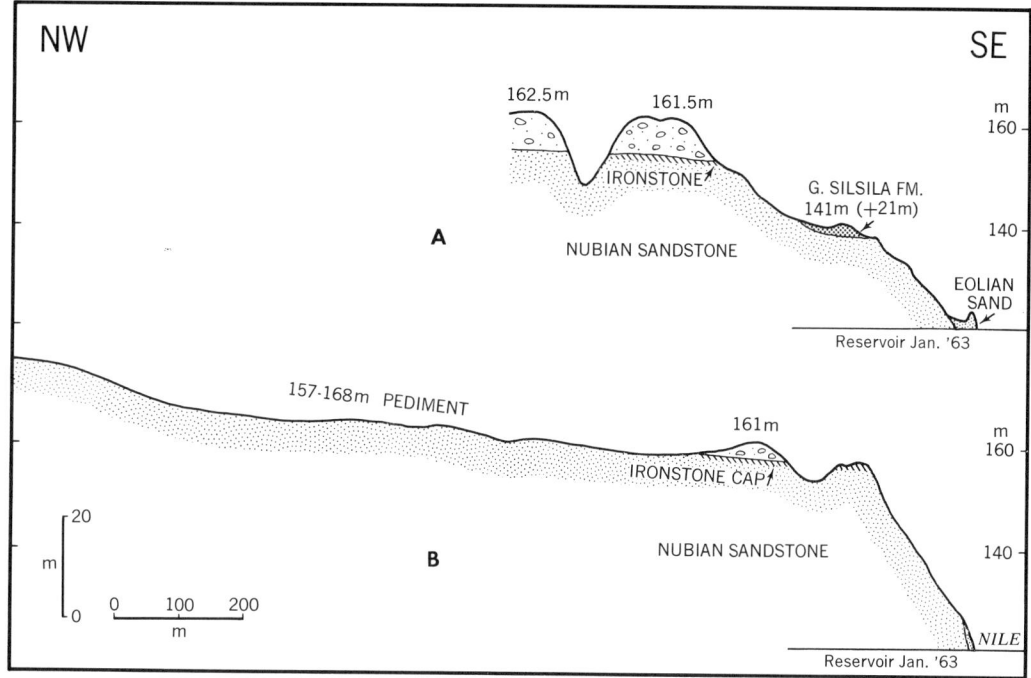

Fig. 5–10. Surficial deposits and geomorphology at Adindan West. A: directly opposite Khor Adindan, B: 750 meters further south. Map: UW Cartographic Lab.

stage. The gravel matrix is a slightly clayey, coarse sand that has presumably been eluviated of fines (No. 158, Table 5–2). The clay minerals present include kaolinite and possibly a trace of hematite. Basically, these soils are similiar in type to the deep, red paleosols developed on the High Terrace complex at Kom Ombo.

The second gravel terrace south of the Abu Simbel gorge is confined to Adindan East, where it rests on a pedimented fault block on either side of Khor Adindan, with an average gravel thickness of 2 to 3 meters (Figs. 5–11, 5–12). Maximum elevation of the pediment is 180 meters. The gravels attain 176 meters (+55 m) and probably form a part of the Wadi Allaqi stage. The constituents of this terrace include coarse quartz and cobble-grade ironstone pebbles. The intensity of former rubefaction can be concluded from the red (2.5 YR 5/6) colluvial soils found washed together in the many gentle draws. These are poorly sorted, silty coarse sands with medium, angular blocky structure. The clay minerals include kaolinite, montmorillonite, and possibly a little hematite. Obviously,

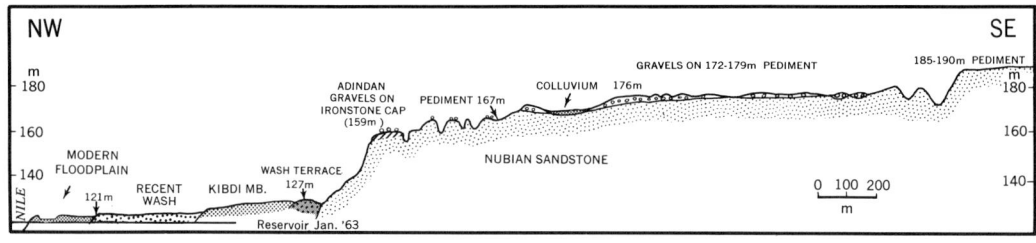

Fig. 5–11. Surficial deposits and geomorphology at Adindan East (north of Khor Adindan). Map: UW Cartographic Lab.

these colluvial soils are polygenetic, and they have probably been eluviated of fines.

Although the paleogeography of the +55 meter terrace is next to impossible to reconstruct, the valley configuration of the +42 meter stage is still fairly evident. Total floodplain width was 3.2 to 3.6 kilometers, compared with 1.6 to 1.8 kilometers today.[5]

Pleistocene Terraces Between Wadi Or and Arminna

Three stages of the Pleistocene Nile are well developed and can be traced with little difficulty from Ballana to Arminna (Figs. K–1 and K–2; 5–9 and 5–14). These include the Adindan as well as the Dihmit and the Wadi Korosko stages.

The Adindan stage is a continuous feature behind Qustul, and a related pseudo-esker, with a possible lower stage, is last seen on the east bank opposite Ballana Station. After a 3.5-kilometer break, these gravels

---

5. Sandford and Arkell (1933) do not specifically mention any Pleistocene gravels between Adindan and Ballana, although Said and Issawy (1965: 12 and Fig. 3) recognize two terraces of autochthonous materials at 32 meters and at 21 meters *above reservoir level* (approximately 152 m and 141 m respectively). Comparing Said and Issawy's Fig. 3 with our own transect (Fig. 5–13), run in the same area at Qustul, it seems that Said and Issawy's 152-meter gravels would be our Adindan gravels—here at 150 to 160 meters elevation. Their 141-meter gravels must refer to late Pleistocene nilotic sands. Said and Issawy describe the so-called 152-meter gravels as consisting of quartzite pebbles set in a matrix of red muds. In point of fact, the gravels consist of quartz with some quartzite (ironstone), and the matrix is a red sand, the color being due to soil development rather than to primary deposition. Finally, the Middle Paleolithic implements occasionally found on this terrace are younger and do not "date" it by any means.

Fig. 5–12. Gravels of the Wadi Allaqi stage (176 elevation) north of Khor Adindan, resting on bedrock (right foreground). The 180-meter pediment forms the skyline to the right.

are again recorded by a ferricrete wadi gravel ridge at 159 meters (+39 m) on the south bank of Wadi Or, by rubefied Nile gravels north of the Wadi Or embouchure (at 160 m, +40 m), and again opposite the Abu Simbel temples (at 161 m, +41 m). A dissected fluvial platform at 159 meters is recorded north of Abu Simbel on the west bank, and impressive 157-meter platforms can be seen on the east bank between Farqanda and Arminna Temple. The bulk of the great rubefied gravel terrace at Arminna Temple pertains to the Adindan stage, with a dominant crest level

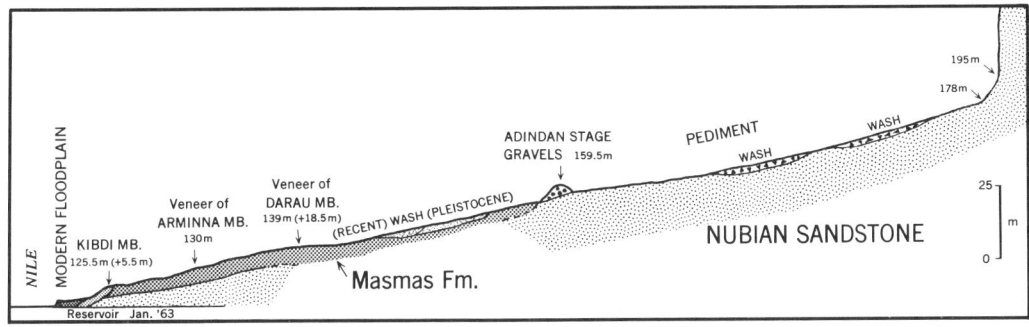

Fig. 5–13. Surficial deposits and geomorphology at Qustul (Kimam Goha). Map: UW Cartographic Lab.

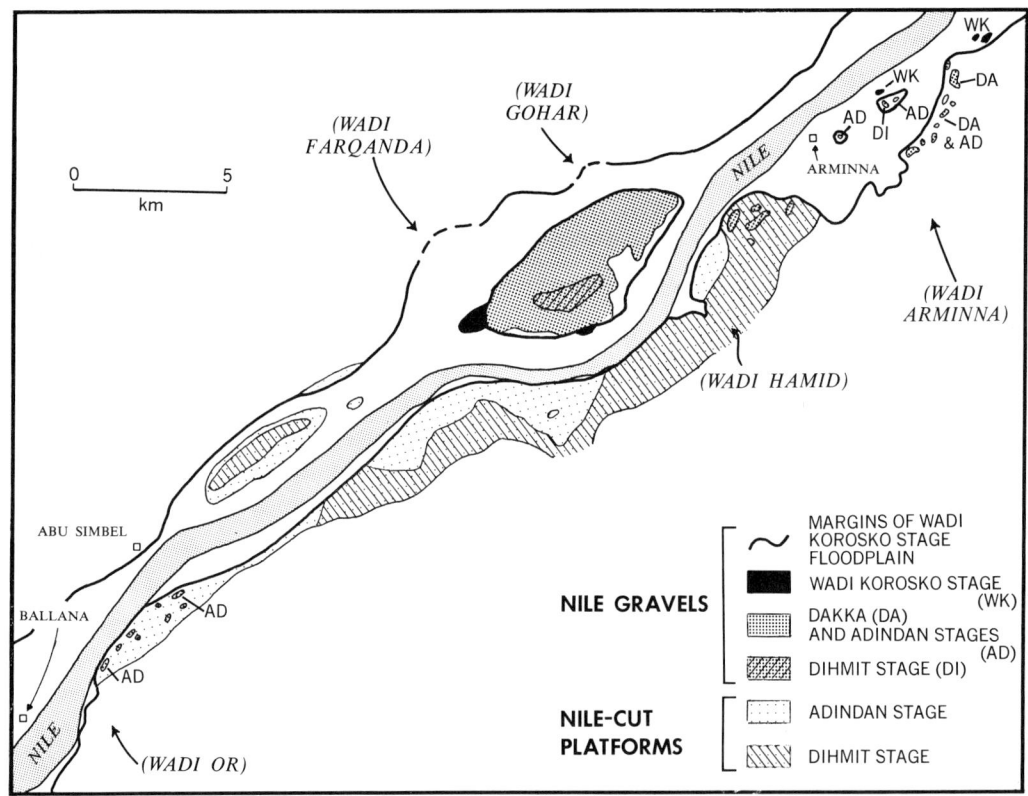

Fig. 5–14. Lower and Middle Pleistocene Nile terraces between Ballana and Arminna. Active wadis indicated. Map: UW Cartographic Lab.

of 156 to 159 meters (Fig. 5–15). Finally, there are patches of Adindan gravels at 155 to 158 meters (+ 40 m) near Arminna East.

In general, the gravels and fluvial platforms of this + 40 meter Adindan stage are sufficiently clear-cut and continuous to warrant reliable correlation throughout this sector of Nubia, so providing a convenient stratigraphic guide horizon. It appears that Acheulian-type artifacts were found *in situ* within these gravels opposite Abu Simbel (Maxine Haldemann-Kleindienst, *personal communication*).

The Dihmit stage is exceptionally well recorded in the form of fluvial platforms at 160 to 167 meters between Abu Simbel and Arminna East (Fig. 5–14), as well as by gravels at Arminna Temple and Arminna East. The deposits in question are at 164 to 166 meters (+ 46 to + 48 m) and consist of quartz and ironstone. The paleosols developed on both major gravel exposures are red (2.5 YR 4/6) (*B*)-horizons with intensive sec-

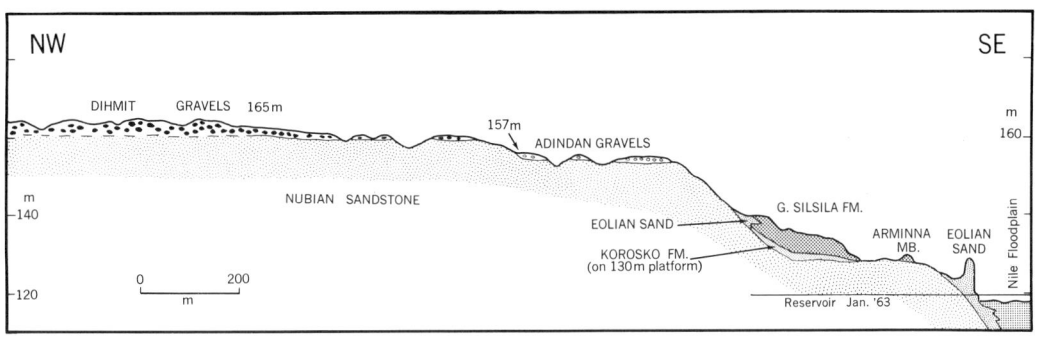

Fig. 5–15. Surficial deposits and geomorphology 1.7 kilometers downstream of Arminna Temple. Map: UW Cartographic Lab.

ondary corrosion evident on quartz-pebble surfaces in general and within fractures in particular. Much of the mesocrystalline quartz is, in fact, somewhat brittle to the touch. Near Arminna Temple (Fig. 5–15), where subsequent eluviation is less prominent, the matrix under 2 millimeters diameter has been altered to a silty clay with 34% clay fraction, primarily kaolinite. Structure is medium, angular blocky. This particular paleosol suggests intensive chemical weathering of Rotlehm type. The Arminna East exposures are eluviated, with sandy silt matrices characterized by single-grain structure. Corrosion of quartz is equally severe, however. Bifacial and flake tools are scattered over the surface of the Dihmit terrace near Arminna Temple, but the body of the gravels is sterile.

The Wadi Korosko stage is readily apparent in the topography of this sector of Nubia (Fig. 5–14), and the various channels of this period provide the geomorphic setting for the late Pleistocene aggradations. Nile-cut platforms are evident on both flanks of the "islands" north of Abu Simbel and west of Arminna Temple. Another platform is cut into the east bank near the mouth of Wadi Hamid. The landward edge of these surfaces is at an elevation of 145 meters (+26 m) near Abu Simbel, 143 meters (+26 m) at Arminna. Gravels are sporadically preserved on these surfaces at Arminna Temple (143 to 147 m), Arminna East (145 m), and Tushka East (141 m) (Fig. 5–14) and may be more extensive underneath the mantles of late Pleistocene silt. All of these deposits are only moderately rubefied. Clay-mineral formation was limited and quartz corrosion absent. The Wadi Korosko–stage floodplain varied in width between 2 and 4 kilometers, suggesting a river of considerable geomorphic impact. Several of the larger east- and west-bank wadis were still rather active at the time, grading their valleys to the Nile base level.

Lower and younger ferricrete wadi gravels can be observed in Wadi Or. A terrace 21 meters above wadi floor culminates at the Nile edge in a 141-meter (+22 m) gravel surface, while a 9-meter wadi terrace is also present. These post–Wadi Korosko deposits may be related to some of the sporadic, but nonetheless distinct, platforms found at low levels elsewhere in Nubia. Their overall significance is uncertain.[6]

Although the Dakka stage is first definitely recorded by Nile gravels and ferricrete pseudo-eskers at Tushka East, several possible correlative features may be cited from more southerly localities. So, for example, a few lower gravel surfaces at 151 to 154 meters flank the Adindan gravels between Adindan West and Qustul. Whether these are simply denuded terraces or are primary deposits of the Dakka stage could not be determined in the field. At the mouth of Wadi Or, a ferricrete gravel pseudo-esker at 156 meters may also be younger than the nearby wadi terrace at 159 meters. Finally, the gravels and platform opposite Abu Simbel have been dissected by an erosional stage related to a Nile level of about 153 meters. It is impossible to decide on the status of these features.

Pleistocene Terraces Between Tushka and Qatta

The Wadi Korosko stage is the most prominent feature between Arminna and Qatta. The auxiliary channel east of Arminna is complemented further downstream by a similar bedrock channel west of Masmas, extending beneath the silt plain behind Ineiba. Related gravels occur east of Tushka at 141 meters (+24 m); between Masmas and Ineiba on the west bank at 143 meters (+26 m); and again between Ibrim and Qatta, at 142 to 143 meters (+26 to +27 m). Wherever preserved, these gravels are strongly denuded, since they were swept over by the late Pleistocene Nile. Despite the variable elevations, related erosional and depositional features can be traced with little difficulty (Figs. K–3 and 5–9). There is little or no evidence of rubefaction.

---

6. Sandford and Arkell (1933: 29) indicate the presence of +45 meter, +30 meter, and +15 meter platforms between Abu Simbel and Tushka but give no details. It is impossible to reconcile these casual observations with the field evidence. Said and Issawy (1965) do not mention any Nile or wadi gravels between Ballana and the Dakka Plain.

The Adindan stage is fairly well developed as far north as Ineiba. Rubefied gravels are preserved on the east bank near Arminna (158 m or + 40 m) and again at Gineina and Shibbak (157 m or + 40 m). Fluvial platforms occur discontinuously on both sides of the Nile Valley between Tushka and Ineiba at relative elevations of + 36 to + 38 meters (Figs. 5–9, 5–16).

Other stages are poorly represented. The Dakka stage appears to be recorded by Nile and wadi gravels east of Tushka at 154 to 155 meters (+ 35 to + 36 m), but this correlation is uncertain. Several low, fluvial platforms occur, including a well-developed surface at about 135 meters (+ 18 m) between Tushka (Fig. 5–16) and Masmas. This may be related to the + 21 meter wadi gravels of Wadi Or.[7]

Pleistocene Terraces Between Tumas and Korosko

The Dakka stage, an insignificant or dubious phase in the far south, assumes a prominent geomorphic role between Tumas and Korosko. Gravels at 147 meters (+ 32 m) are developed on the southwestern side of Tumas, while a wadi gravel ridge enters the Nile Valley just east of the village at 145 meters (+ 31 m), terminating in a broad fan. Midway to Amada Temple there are further exposures at about 147 meters (+ 33

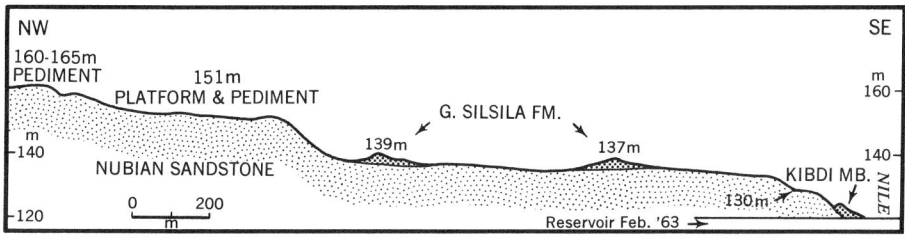

Fig. 5–16. Surficial deposits and geomorphology north of Wadi Tushka (west bank). Map: UW Cartographic Lab.

---

7. Sandford and Arkell (1933: 31) mention this same low platform west of Tushka. Further north, they record a + 30 meter platform, with gravel, from Ineiba to Tumas (*ibid.*: 26). The Wadi Korosko and Dakka stages are both found in this area and may have been grouped as one unit in the earlier survey. Wherever specific localities can be identified—e.g., Plate Va in Sandford and Arkell (1933)—these authors refer to the Dakka stage as the 30-meter terrace.

m), beyond which fluvial platforms at 144 to 146 meters can be subcontinuously followed on alternating sides of the river as far downstream as Korosko (Fig. K–4). Sandford and Arkell (1933: 26 and Plate Va) found early Acheulian bifaces *in situ* in deposits of the Dakka stage between Galgabirka and Shabakiya, west of Tumas. Implements were also collected from such gravel resting on a 145-meter platform opposite Korosko.

The Adindan stage is distinctly recorded between Tumas and Riqa by fluvial platforms on both banks of the river, as well as by extensive and massive gravels at Tumas (153 m, +39 m). Remnants of a 157-meter (+43 m) platform may record the Dihmit stage at Amada Temple (Fig. 5–17). More conspicuous is a dissected 164- to 166-meter (+52 m) platform between Tumas and el-Riqa, verified by a pseudo-esker in 164 meters (+51 m) at el-Riqa. A probable equivalent is the ferricreted 161-meter (+48 m) gravel terrace and related platform at the mouth of Wadi Korosko. These features can probably be correlated with the Allaqi stage. Higher wadi gravels are preserved in Wadi Korosko (Fig. 5–18) and on a 175-meter platform (+61 m) at Tumas. Their significance remains obscure.

The Wadi Korosko stage, as such, is poorly preserved in the Korosko bend, being limited to sporadic platforms at +24 meters, a single patch of Nile gravel at el-Malki (Fig. 5–9), and the type locality situated near a major tributary confluence about 5 kilometers upstream in Wadi Korosko. Here a small exposure of coarse to cobble gravel stands at 14.5 meters above wadi floor, grading into an alluvial cone developed within a steep,

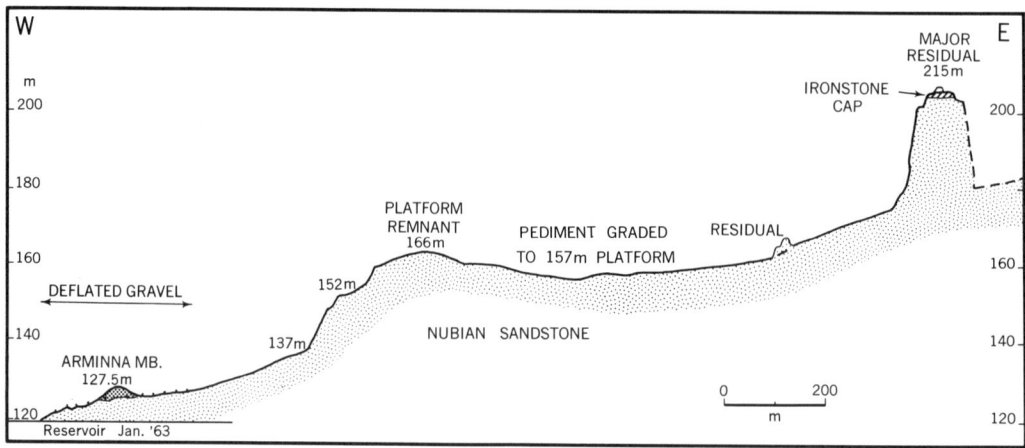

Fig. 5–17. Surficial deposits and geomorphology at Amada Temple. Map: UW Cartographic Lab.

Fig. 5–18. The confluence of Wadi Korosko and Wadi Guhr el-Daba, with Nile Valley in center background. Ferricrete gravels of uncertain correlation in immediate foreground (180 m elevation).

V-shaped minor tributary (Fig. 5–19). Lithology consists of ironstone and derived blocks of older, ferricrete conglomerates. Thickness exceeds 3.5 meters. The cone itself is foreset bedded (30% to 48%) at its base, grading up into gentle topsets (3% to 5%). The lower beds are intensely rubefied and partly consolidated, but incompletely ferricreted. Color ranges from light red (2.5 YR 6/6) to weak red (10 R 5/4). The uppermost meter or so is completely indurated by a ferruginous cement. Surface color is dark reddish brown (5 YR 3/2) to very dark gray (10 YR 3/1), but inside the abundance of soft ferric precipitates in the pore spaces lends a reddish-yellow (5 YR 6/6) color. The dark cement, incomplete as it is, is as resistant as the quartz sand grains themselves, so that the deposit qualifies as a quartzite or an ironstone.

The significance of the Wadi Korosko type section is manifold. Some degree of primary rubefaction presumably preceded impregnation of the alluvial deposits from without by iron solutions, mainly hematite. The deposits themselves imply torrential overland flow, feeding gravel and boulders into the wadi from minor tributaries. Projecting the gradient of

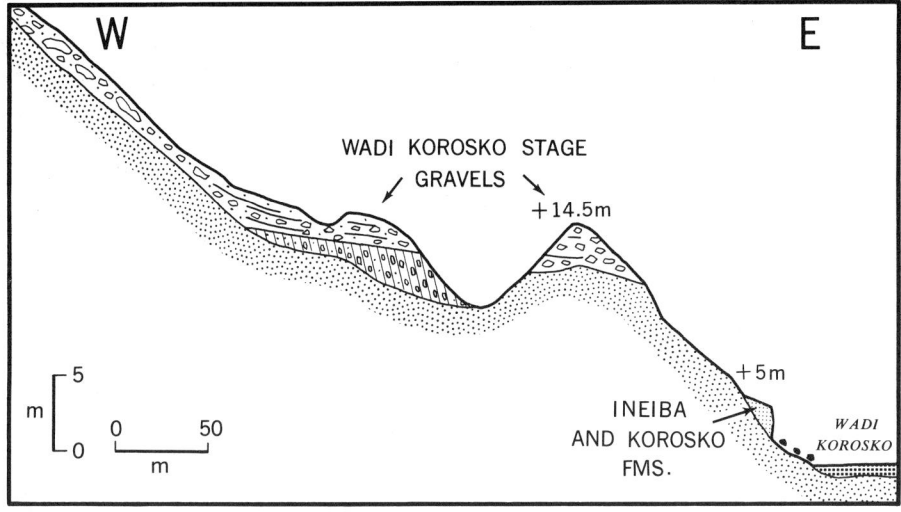

Fig. 5–19. The + 15 meter wadi terrace of the Wadi Korosko stage near the confluence of Wadis Korosko and Guhr el-Daba. Map: UW Cartographic Lab.

the resulting terrace to the Nile, base level would be 135 meters (+ 22 m).[8]

Pleistocene Terraces Between Korosko and Dakka

Preservation of platforms or gravels is extremely poor (Fig. 5–9) in the constricted valley of the Nile between Korosko and the Dakka Basin. Steep cliffs tower above the east bank of the river, while inclined pediments come right down to the west bank. The Wadi Korosko stage is recorded by a platform with gravels at el-Malki, opposite the mouth of Wadi Sinqari, at 136 meters (+ 24 m), and similar deposits are not seen again until Wadi Allaqi. Before attaining further prominence downstream, the Dakka stage appears to be recorded by a set of discontinuous but distinct platforms near the mouths of Wadi Shaturma (142 m, + 31 m) and Wadi Sebua (142 m, + 32 m), opposite Khor Ibeidalla (142 m, + 32 m) and Khor Takar (141 m, + 33 m), and at Seiyala (137 m, + 29 m). A lower substage of the Dakka unit appears in the form of platforms

---

8. Sandford and Arkell (1933: 19–20, 26) recognize + 46 meter and + 30 meter platforms or gravels between Tumas and Korosko. Unquestionably, the lower level can locally be equated with the Dakka stage. The higher platform may refer to either the Dihmit or the Adindan stage, or to both.

at 136 meters (+27 m) opposite Shima Amalika Temple and at 134 meters (+26 m) at Seiyala (Fig. 5–3). A probable correlative of the Adindan stage is found in the form of a bench and gravel cap at the former locality (146 m, +37 m). Finally, several conspicuous platforms at 160 to 163 meters (+53 m) on both banks of the Nile, south of Seiyala, may pertain to the Wadi Allaqi stage. Of all these fragments, only the Dakka-stage platforms can be considered with some confidence to be a stratigraphic link between Korosko and the Dakka Basin (see Figs. K–4 to K–8).

As the name of the type locality suggests, the Dakka stage is the best developed terrace of the Dakka Basin. Nile and intergrading wadi gravel ridges appear north of Seiyala at 137 meters (+29 m); Nile gravels are found at Muharraqa West at 138 meters (+31 m) and south of the Allaqi embouchure at 135 meters (+28 m), with a massive wadi gravel ridge near Khor el-Omda at 140 meters (+33 m). The major exposure begins north of Dakka Temple, extending subcontinuously on a related platform for over 4 kilometers, with an average width of 400 to 500 meters. Attaining 135 meters (+29 m) elevation, this exposure has an average thickness of 5 meters and a total thickness of over 10 meters. The deposits of the type area consist of coarse to cobble gravels with a matrix of unconsolidated, light-brown (7.5–10 YR 6/4), coarse sand. Gravel lithology includes quartz and a variety of igneous and metamorphic rocks, indicating derivation from Wadi Allaqi. Dark-colored intrusives are dominant. Individual pebbles are well rounded and flattish in shape. Said and Issawy (1965: 14) report rolled "Early Middle Paleolithic" artifacts *in situ* in these gravels. There is no evidence of rubefaction, although thin paleosols may well have been stripped off by the waters of the late Pleistocene Nile. On the east bank, there is a gravel pseudo-esker graded to 138 meters (+32 m), as well as a 136-meter (+30 m) platform cut across tilted bedrock strata. The Middle Terrace of Wadi Allaqi, at about 15 meters above wadi floor, is graded to the same base level, and the lithology is identical with that of the Dakka gravels at the type site.

The Adindan stage appears to be recorded by extensive Nile gravels and wadi gravel ridges at 143 to 146 meters (+36 to +39 m) along the west bank from Muharraqa to Dakka.

Wadi Allaqi exhibits four well-developed gravel terraces. The High Terraces are developed as two distinct levels in an apparently homogeneous body of gravel, averaging 20 to 30 meters in thickness. The lower level is graded to a Nile base level of about 153 meters (+46 m), whereas the higher substage, preserved further upstream only, suggests a

Nile base level of 160 to 165 meters (+53 to +58 m). The dominant level of High Terrace I is 35 meters above wadi floor, that of High Terrace II about 25 meters. Gravel pseudo-eskers of minor tributaries are recorded in at least three instances (Fig. K–8). The High Terrace complex of Wadi Allaqi has been selected as the type site of the stage of the same name. The 160- to 163-meter platforms south of Seiyala and the dissected 160- to 163-meter pediment plains of the Dakka Basin presumably are contemporary with High Terrace I. The gravel patch at 155 meters (+47 m) at Seiyala and a well-developed pediment plain with a base level of 153 to 154 meters are almost certainly equivalent to High Terrace II. The High Terrace gravels are all intensively rubefied and are dissected by rill erosion.

The 15-meter Middle Terrace of Wadi Allaqi, contemporary with the Dakka stage, is recorded on both sides of the modern wadi floor and indicates a wadi floodplain 1.8 kilometers wide, compared with a 350- to 850-meter-wide channel today. A minor tributary is preserved as a gravel pseudo-esker. The Low Terrace has a dominant level of 10 meters, grading to a Nile level at 130 meters (+23 m). Preserved on both wadi banks, it indicates a fairly constant wadi channel width of 1.7 kilometers. Successive denudational ridges can be seen closer to the Nile, suggesting that base level fell during the last stages of aggradation. This may explain the total absence of contemporary Nile terraces north of el-Malki. The lithology of these gravels is analogous to that of the Dakka stage, although greenstones and macrocrystalline igneous rocks are more conspicuous. The Adindan stage was not recognized in Wadi Allaqi.[9]

9. Said and Issawy (1965: 14–16) describe gravel terraces at Dakka at 130 meters, 146 meters, 151 meters, and 162 meters elevation, coinciding closely with our own observations. In Wadi Allaqi, they describe wadi terraces at relative elevations of 5 meters, 10 meters, 21 meters, and 32 meters. On the basis of their lithological characterizations, and despite the different elevations cited, their 5-meter terrace must be the Low Terrace, their 10-meter, our Middle Terrace. We cannot follow their correlations between Wadi Allaqi and Dakka, which equate the Low Terrace with the Dakka stage and the Middle Terrace with the Adindan stage. Such correlations can neither be reconciled altimetrically nor lithologically. Said and Issawy emphasize the predominance of quartzite, flint, and chalcedony in the High Terrace gravels and suggest that Wadi Allaqi first began draining the Basement Complex as late as the Middle Paleolithic. This is unreasonable, since fine-grained igneous and metamorphic rocks of considerable variety are common within the unaltered body of the High Terrace gravel. Weathering has destroyed them in favor of resistant silicic rocks within the deep, intensively weathered (B)-horizons, however. None of the gravel variations of the Allaqi terraces can be interpreted through hydrographic changes in that drainage basin. De Heinzelin's (1967) hy-

Pleistocene Terraces Between Kushtamna and Aswan

Between Dakka and the Kalabsha Gorge, the Nile passes through a narrow valley flanked by sandstone bluffs or hills. The Dihmit stage is well developed and can be followed with few interruptions (Fig. 5–9). It is marked by platforms on both banks of the Nile north of Kushtamna (147 m, +42 m), on the east bank at Dandur (146 m, +43 m), and from Kobosh Temple northwards for 7 kilometers (144 m, +43 m). After a small platform at Tafa East (143 m, +42 m), the gravel exposures of the type site first appear east of Qirtassi and reappear at intervals to Dihmit Station (Figs. K–7, K–9, and K–10).

The Dihmit gravels were deposited in an overflow channel about 1.2 kilometers east of the Nile, following a subcontinuous tectono-erosional valley. Consisting of rounded, coarse, igneous gravel and cobble-grade quartz pebbles (Fig. 5–20), the material must have been derived from the Basement Complex via the local east-bank wadis, but the gravel was ultimately redeposited on the Nile floodplain, with a maximum elevation of 143 meters (+43 m). The matrix of these crudely stratified beds is a semicemented, very-pale-brown (10 YR 7/4), coarse quartz sand. Total thickness exceeds 25 meters, although dissection has taken a heavy toll.

A deep, reddish paleosol was once developed in the Dihmit gravels. Where no (B)-horizon is preserved, fine-grained igneous pebbles may show intensive red (2.5–5 YR 4/6) discoloration and thick weathering rinds, very similar to the C_1-horizon evident on the Fatira beds at Kom Ombo (Chap. 2). On better preserved surfaces, the gravel matrix has been altered to a yellowish-red (5 YR 4/6), clayey silt. Original depth of solum cannot be estimated.

The Dihmit stage is absent in the Dakka Basin, probably as a result of significant denudation during the middle Pleistocene. North of Dihmit Station, platforms are entirely absent in the resistant rocks of the Basement Complex.

Other stages are poorly represented between Dakka and Aswan (Fig. 5–9). The Wadi Korosko stage seems to be recorded by a wadi terrace in Khor Dihmit, graded to a Nile level of 122 meters (+22 m). The Dakka

---

pothesis, that the Upper Nubian Nile followed Wadi Allaqi prior to the late Pleistocene, also finds no supporting field evidence.

Sandford and Arkell (1933) make no mention of any Pleistocene gravels in the Dakka Basin.

Fig. 5–20. Gravels of the Dihmit stage south of Khor Dihmit at 140 meters elevation.

stage is last seen in the form of a 136-meter (+31 m) platform between Kushtamna and Girf Husein, and the next presumed correlative is the Shellal-Aswan gravel channel east of the First Cataract, at 126 meters (+33 m above floodplain at Aswan). This area has been adequately mapped by Sandford and Arkell (1933: 57–59).

Of some interest in northern Nubia is the 159- to 163-meter (+62 m) platform developed in the vicinity of the Kalabsha Gorge and continuing across the sandstone and the Basement Complex without interruption. This boss of resistant rock must have formed a cataract at the time, and the preserved topography on top of the bluffs is identical to that of the modern cataracts at Aswan. The waters appear to have surged through irregular bedrock channels studded with rocky knobs and islets. The present Kalabsha Gorge narrows the flood-stage valley from 1.0–1.2 kilometers to about 225 meters. By comparison, the early Pleistocene Nile flowed across a channel varying between 2.8 and 3.0 kilometers in width. During later times, the channel thalwegs were remodelled as spillways for the Wadi Allaqi stage, represented by a 153-meter (+52 m) platform at Tafa East. At an even later date, the Dihmit stage also modelled a minor overflow channel east of the gorge. The age of the modern gorge is uncertain. At least 20 meters of gravel are found under the overdeepened modern channel (Ball, 1903), but more recent data on the depth of the entrenched, buried channel have not been available to the writers. The unrestricted development of the 159- to 163-meter channel suggests that the modern gorge did not exist at that time or, more probably, that it was closed with older fill of unknown facies.

Several other geomorphic features can probably be related to the +62 meter stage at Kalabsha. The wadi of the same name is cut into the Aswan Pediplain and shows a widespread denudation surface graded to a base level of about 160 meters. Similarly, there is an extensive platform on the west bank at Dandur at about 166 meters (+63 m). Possibly this local stage represents the Gallaba terrace (Chap. 2).[10]

---

10. Sandford and Arkell (1933: 20–21, Pl. IVa) recognize a +45 meter platform at and south of the Kalabsha Gorge, a +30 meter platform south of Kalabsha (*ibid.*: 26), a +22 meter platform at Dihmit West (*ibid.*: 29), and a +15 meter platform in the Dihmit area (*ibid.*: 21). Allowing for differences of measurement, we can only reconcile the +45 meter platform (Dihmit stage) with our field observations. The same authors describe a +26 meter wadi terrace in Khor Dihmit (*ibid.*: 29–30), presumably referring to the Wadi Korosko stage. Said and Issawy (1965: 16) recognized no gravels between Kushtamna and Aswan but noted the +60 meter surface atop the Kalabsha Gorge. Although they suggest a waterfall here at

Pleistocene Dissection of the Nubian Nile

The erosional history of the Nubian Nile can only be inferred indirectly from the record of denudation surfaces and gravel terraces. The vertical extent of the downcutting which interrupted periods of stability or aggradation is largely a matter of conjecture. So, for example, the Nubian Nile may have entrenched its bed to below sea level along its whole length, as it did at Aswan in late Tertiary times. If this were so, all the Pleistocene degradation intervals would have involved simple dissection of older fill. On the other hand, if the Nubian Nile south of Kalabsha does not have a buried, entrenched channel, significant bedrock incision may have occurred periodically through most of the Pleistocene. These two opposed hypotheses would imply radical differences in rates of erosion as well as complex changes of hydrological regime or base level. In either case, any period of downcutting between two intervals of stability or aggradation would seldom be a unidirectional affair. And, long-term net degradation would probably exceed the vertical difference between any two levels of aggradation.

Available subsurface evidence does not yet permit a decision in favor of one or the other hypothesis. Those borings from Dihmit, Dakka, and Abu Simbel that have been published or alluded to by Egyptian authors (Attia, 1954: 9–10; Shata, 1962; Said and Issawy, 1965: 6) were either shallow or poorly chosen along the Nile Valley margins. Consequently, we have no proof of whether a buried entrenched channel exists south of Kalabsha. The fact that such a channel has been tapped to depths of 172 meters *below* sea level at Aswan (Chap. 2) raises certain geomorphic problems for Nubia. Either a very narrow graben of block-fault origin terminates abruptly somewhere between Aswan and Dakka, or the Nubian Nile was at least partially graded to this temporary base level in late Tertiary times. There is no evidence at Aswan for a block fault of this kind, and Chumakov (1965) believes the buried, entrenched channel of the Nile must extend upstream past Wadi Halfa. On the other hand, the major wadi tributaries of the Nubian Nile do not have entrenched channels. Bedrock floors accompany Wadis Kurkur, Dihmit-Quffa, Kalabsha, Abusku, Allaqi, Korosko, Hamid, and Or to within a few hundred meters of the Nile floodplain, if we ignore flooding by the Reservoir. Apart from minor rock-cut channels near the Nile, the lowermost kilometer of each of

---

the time, this is precluded by well-developed, 161- to 162-meter platforms cut into sandstone bedrock further downstream. Gradients were maintained, but the uneven nature of the channel almost certainly did generate cataracts or rapids.

these wadis is graded to a Nile base level not more than 10 meters below the flood level of the nineteenth century. It is surprising indeed that, only 5 kilometers from the banks of the Nile, as great a wadi as the Allaqi has a bedrock floor 13 meters above local floodplain level (107 m). North of Aswan this is not so, and none of the major wadis—e.g., Wadi Abu Subeira or Wadi Kubaniya—expose bedrock floors with 5 or 10 kilometers of the Nile. In the Luxor area, Pliocene deposits have filled the deeply entrenched bedrock channels of at least three major wadis to distances of up to 60 kilometers from the Nile (see Sandford, 1934: folding map; Butzer, 1959a). There is little question that the bedrock configuration of the major Upper Egyptian wadis has in fact changed little since the Pliocene.

At the present state of our knowledge, there does seem to be a different tectonic and physiographic setting for the Nile Valley north and south of the First Cataract. Evolution of the Nubian wadi systems has been decidedly slower. This may reflect tectonic accident, relief-energy, distance from ultimate base level, regional climate, or a combination of several of these factors. Whatever the explanation, it seems likely that land surfaces below the 200-meter contour are almost exclusively of Pleistocene age. In this sense, the landscape of Nubia appears to be somewhat younger than that of Upper Egypt.

Allowing for at least some bedrock cutting in Egyptian Nubia during Pleistocene times,[11] can any stratigraphic datum be set for the last significant phase of wadi bedrock incision? Both Wadis Korosko and Or have a cemented gravel unit of late Pleistocene age that mantles the bedrock floor of the wadis to within short distances of the Nile. This Wadi Floor Conglomerate, to be discussed in the following chapter, underlies the Korosko Formation. Dissection during the last 50 millennia has failed to remove this conglomerate except for narrow, often discontinuous segments of the wadi floor. This implies that bedrock cutting has been insignificant since the mid-Upper Pleistocene. On these grounds, it seems likely that most of the bedrock sculpture visible in the lowland landscapes of Nubia dates from the early and middle Pleistocene.

Dissection within the locus of the modern Nile Valley reaches well back into the Tertiary, however. The location of the Ballana and Aswan Pediplains proves that a major river already followed the approximate

---

11. But certainly not 40 meters of bedrock denudation, as suggested by Reed (1966), for a period of 35,000 to 100,000 years duration, terminating 25,000 years ago. Apart from the dubious reasoning by which these time estimates are derived, the geomorphic and stratigraphic evidence definitely does not allow for such rapid or significant *bedrock* cutting anywhere in Nubia during the later Pleistocene.

course of the Nile during early or middle Miocene times, i.e., 20,000,000 to 25,000,000 years ago. As much as 150 meters of vertical cutting separate at least some of the surfaces designated as the Kurkur and Aswan Pediplains. This work was completed in Tertiary times, possibly by the late Miocene. This erosional tendency of the Nile has continued, with protracted interruptions, into the recent geological past.

Red Paleosols

The status of red paleosols in Egyptian Nubia corresponds closely to that of similar soils at Kom Ombo (Chap. 2). Relict and occasional buried soil profiles can be readily exposed on most Pleistocene deposits by a little digging, although good natural sections are rare. Most paleosols have been truncated through denudation, and fine-grained (B)-horizons are commonly absent. Deflation during hyperarid intervals without vegetation has obviously been significant, preconditioned by total soil desiccation and by structural deterioration. As at Kom Ombo, the red paleosols in Nubia commonly exhibit one or more generations of surface wash, in particular a coarse lag horizon that rests on a fine-grained powdery horizon rich in carbonates and evaporites. These fine horizons are 10 to 20 centimeters thick and represent a wash derived both from sandy lag of the former topsoil and from wind-borne dust of later date. The pebble pavement, on the other hand, is mainly a result of extensive sheetflooding.

The paleosols of the Allaqi, Dihmit, and Adindan gravels are remarkably deep or intensively rubefied, despite the advanced degree of denudation and rill cutting. Wherever accessible, the (B)-horizons show red coloration (2.5 YR 4–6/6) to at least 2 to 3 meters in depth. Matrix textures vary from coarse sand to silty clay, depending on horizon depth and on the degree of subsequent, mechanical eluviation of fines. Fine-grained igneous and metamorphic pebbles are selectively decomposed in the upper few meters, and even quartz pebbles exhibit considerable surface corrosion and enlargement of microfractures. Clay minerals are dominantly kaolinite, commonly with a little montmorillonite (Appendix D). As at Kom Ombo, it is difficult to assign these red paleosols to a particular modern soil type, but, in the instance of the Dihmit gravels near Arminna Temple (No. 154, Table 5–2), the various analyses and macroscopic profile observations suggest a deep Rotlehm-type soil developed in a sandy gravel.

Paleosols developed on the Dakka and Wadi Korosko deposits are rather shallow, and (B)- and (B)C-profile depth combined probably did

Table 5-2. Paleosols and soil sediments in Egyptian Nubia
(Compare X-ray diffractograms in Appendix D.)

Sample number	Horizon	Color	Texture (matrix)	Sorting	CaCO₃ (%)	pH
		Arminna Temple, Dihmit stage gravels				
154	(B)	(2.5 YR 4/6)	Medium-sandy clay	Poor	7.1	7.10
		Adindan East colluvial hollow on Allaqi stage gravels				
172	(B)E	(2.5 YR 5/6)	Silty coarse sand	Poor	0.0	7.60
		Abu Simbel, pediment above Hathor Temple				
161	(B)E	(5 YR 7/4-6, 5 YR 5/6)	Medium-sandy silt	Poor	0.8	7.55
162	(B)C	(5 YR 6/6)	Silty medium sand	Moderate	1.1	8.00
163	C₁	(5 YR 7/2, 10 YR 5-6/1)	Silty medium sand	Moderate	1.5	7.85
		Qustul, Adindan stage gravels				
158	(B)	(5 YR 5/6)	(Clayey) coarse sand	Poor	0.8	7.65

not exceed a meter or so originally. In areas with intensive ferruginization, such as Wadis Or and Korosko, ferric oxides appear to have been primarily illuviated after percolation of surface waters, and, in such cases, rubefaction has penetrated several meters. Where rubefaction has resulted entirely from soil development *in situ*, however, colors are less reddish (5 YR hues on the Munsell scale) than in the deeper, older paleosols. Texture and chemical properties are similar, but pebble decomposition is limited, and evidence of secondary eluviation within the (B)-horizon is absent. Finally, red paleosols have also formed on deposits of the Ineiba Formation, as discussed in Chapter 6.

Distinct, truncated (B)C-horizons developed in bedrock can frequently be observed beneath late Pleistocene colluvial wash. A profile of this type, found on top of a Dakka-stage platform at Seiyala, will be described in Chapter 6. Similar truncated profiles with a veneer of soil sediment can also be found on older bedrock surfaces. A profile from the Aswan Pediplain above the Hathor Temple at Abu Simbel illustrates the usual pattern well. The soil is developed in sandstone bedrock, with a veneer of soil sediment under a more recent wash, mingling with a lag of moderately to well-ferruginized sandstone detritus:

a) 0 to 10 cm. (B)E. Stratified, pink (7.5 YR 7/4), sandy silt; coarse, subangular blocky structure; moderately calcareous.

b) 0 to 10 cm. (B)E. Stratified, reddish-yellow to pink, medium-sandy silt (No. 161, Table 5-2); coarse, angular blocky structure. Contains local pockets of silty sand as well as dispersed pebble-grade wash. Such soil sediments extend 40 cm into joint cracks, where red (2.5 YR 6/6), semicemented soil traces can also be found. These sediments are older than Middle Paleolithic surface sites collected here by the Guichards (1965).

  c) *20 cm.* (B)C. Deeply altered, reddish-yellow sandstone (No. 162, Table 5–2), maintaining original structure but highly friable.

  d) *Over 10 cm.* C_1. Weathered, reddish-gray, semicemented sandstone (No. 163, Table 5–2).

In addition to relict and buried soils, colluvial soil sediments are well developed. Most of these appear to be of late Pleistocene age, although some are probably older. They range in facies from redeposited (B)-horizon material found in small, shallow draws to sandy pediment wash incorporating any available soil products. A typical sample from Adindan East, representing a redeposited (B)-horizon, was analyzed in the laboratory (No. 172, Table 5–2). It is similar to the red matrix of the source gravels, although material coarser than 5 millimeters is absent.

Interpretation of the red Nubian paleosols is analogous to that of the Kom Ombo paleosols. Effective chemical hydration and hydrolysis at considerable depth, as well as thorough oxidation and formation of anhydrous ferric oxide, implies a fairly moist rainy season with a mat of vegetation. A semiarid climate is suggested. Apart from their stratigraphic and paleoclimatic interest, the red paleosols ultimately aided denudation by preparing a highly friable bedrock surface, more readily sculptured by water or wind.

Early and Middle Pleistocene Deposits of the Sudanese Nile Valley

The early and middle Pleistocene record of the Sudanese Nile Valley is poorly preserved and understood. South of the Egyptian border, contemporary deposits are either scarce or absent in the former headwater zone of the Proto-Nile (see de Heinzelin and Paepe, 1965; de Heinzelin, 1967; Robinson and Hewes, 1967). A brief reconnaissance by Sandford (1949) suggests that the same is true for the Howar Basin (Fig. 1–1), between the Third Cataract and the mouth of the Atbara. Comparable beds are first reported from the Central Sudan, but even here they appear to be scarce and sporadic. An interesting section has been described by A. J. Arkell (1949a) from Khor Abu Anga, a kilometer downstream of the Blue Nile and White Nile confluence at Khartum. The following beds are found in a small west-bank wadi, their surface graded to a Nile floodplain about 4 or 5 meters higher than that of today:

a) 150 cm. Coarse, ferricreted conglomerate resting on weathered Nubian Sandstone.

b) 90 cm. Coarse gravel with Acheulian hand axes *in situ*.

c) 120 cm. Medium gravel with clayey matrix and calcareous concretions, some 58 Acheulian bifaces, and a faceted-platform core found *in situ*.

d) Lenses of homogeneous, calcareous white silt, recalling the Korosko Formation of southern Egypt. Artifacts found *in situ* and at the surface suggest an evolved Acheulian industry with Sangoan affinities, related to the Nubian Middle Paleolithic (Guichard and Guichard, 1965).

e) Reddish-brown, sandy clay wash with gastropod shells (*Limicolaria flammata*); presumably to be correlated with the Khartum Mesolithic, traces of which occur in the wadi.

The potential significance of the Khor Abu Anga sequence is considerable, but field reexamination is obviously necessary. The gravels suggest greater wadi activity and a slightly moister climate, while the silts almost certainly are late Pleistocene nilotic deposits.

A similar sequence of deposits seems indicated at Khashm el-Girba–Sarsareib on the middle Atbara River (Arkell, 1949*a*). Here older gravels, representing several substages, attain +10 meters above floodplain and are fossilized by a cap of up to 2 meters of travertine. The subsequent deposits include (*a*) coarse gravels forming +4 to +5 meter and +13 to +14 meter terraces, (*b*) high silts to +25 meters, and (*c*) low silts to a few meters above floodplain, with an Early Dynastic surface site. Clearly a potential stratigraphic link between the Egyptian and the Ethiopian Nile does exist along the Atbara River but remains next to unstudied.

Perhaps the most fundamental sequence of Pleistocene deposits in the Republic of Sudan is preserved in the subsurface of the Paleo-Sudd Basin (Fig. 1–1), masked by fossil eolian sands in the north and by fluvial or lacustrine silts and clays to the east and south. The following succession of Pleistocene units can be reconstructed from the existing literature.

UMM RUWABA SERIES

A massive accumulation of extensive fluviolacustrine beds, known as the Umm Ruwaba Series, has been tapped to a depth of at least 280 meters near Umm Ruwaba and Kosti and over 150 meters in the Muglad area (Andrew and Karkanis, 1945; Andrew, 1954; Kleinsorge and Zscheked, 1959) (Fig. 3–1). These unconsolidated deposits rest within a basin, the floor of which is at least 200 meters below the igneous-metamorphic threshold in the region north of Khartum. The facies varies, becoming increasingly fine toward the former basin centers, where fine, well-sorted sands or clays underlie the recent organic clays of the Sudd Swamps. In the Muglad area, materials are poorly sorted, and facies variations between silty clay, sandy silt, coarse sand, and gravel or cobble strata are rapid. Pebbles are mainly subangular, suggesting short transport distances by steeply graded streams or sheetfloods. In addition to these detrital components, bands of oolitic iron and calcareous concretions have been reported. Abundant fresh feldspars and biotite, near Umm Ruwaba, suggest that intensive physical denudation rather than deep chemical weathering was regionally characteristic. Occuring in an area that is semiarid today, the Umm Ruwaba series suggests an extended period of similar or slightly moister climate. Stratigraphic subdivisions have not yet been recognized, and dating is unsatisfactory. A molar of the Lower Villafranchian pig *Omochoerus* (Arambourg, 1948: 344) appears to have come from the top of the series near Kosti. This suggests a late Pliocene or early Pleistocene age or both.

BLACK CLAYS

Locally overlying the Umm Ruwaba series are polygenetic black clays of different ages (Kleinsorge and Zscheked, 1959). In the Muglad area, many of these deposits are apparently rather early, suggesting uniform sedimentation over wide areas with limited relief. Ferruginous weathering on higher surfaces may have been contemporary in part. In the foothills of the Nuba Mountains, similar clays still form in ponded wadi floors (*fulas*) today, whereas the analogous clays of the Gezira Plain (Tothill, 1946; Berry, 1961a) are late Pleistocene to middle Holocene. There is reason to infer that the Pleistocene Black Clays—excluding the local *fula* deposits—may record one or more pluvial periods.

The original differences between these diverse deposits have been very largely obscured by modern soil development. Seasonally impeded drainage and swelling alternate with desiccation, forming networks of dehydration cracks that extend to about 1.5 meters in depth. Original stratification is destroyed, and montmorillonitic clay minerals are produced in the *A*-horizon, with calcareous concretions forming near or below its base. These are mature vertisols, the products of contemporary weathering in the seasonally wet, savanna lowlands.

FRESHWATER LIMESTONES

Scattered lacustrine chalks, tufas, and clays are known from several localities, and a 10- to 12-meter exposure was studied at Sodiri by Huckriede and Venzlaff (1962). The Sodiri deposits consist of semiconsolidated tufas with derived masses of travertine, as well as with windworn quartz sand. Seventeen different freshwater mollusca were identified, none of them extinct, but with one Palearctic form, *Vertigo* cf. *antivertigo*. This fauna requires extensive perennial waters, e.g., a stream or shallow ponds with vegetated banks and a fringe of galeria woodland. Drainage appears to have been integrated with the Bahr el-Arab. Perennial waters are completely absent today, so that the Sodiri beds prove the existence of a Pleistocene pluvial.

KORDOFAN SANDS

The most conspicuous surficial deposits are the Kordofan Sands, an undulating sand sheet with local dunal topography ("tied" lee dunes, transverse dunes, and whalebacks) (Edmonds, 1942; Andrew, 1954; Kleinsorge and Zscheked, 1959; Huckriede and Venzlaff, 1962). Maximum sand depths are on the order of 40 to 60 meters, although average values are considerably less. The materials are windworn quartz grains derived from disintegration of the local Nubian Sandstone. The upper few meters are intensively rubefied, and the formation is now essentially fossil, immobile, and fixed by vegetation. Stream dissection is effective only in the more humid, southern areas.

The Kordofan Sands are in major part younger than either the Black Clays of Muglad or the Sodiri tufas. No evidence of intradunal weathering profiles is available, but it is unlikely that accumulation was confined to one period. Found in areas with 200 to 700 millimeters annual precipitation today, the Kordofan Sands clearly record one or more intervals of distinctly more arid climate, with a 5° to 6° latitudinal shift of climatic zones. No stratigraphic dating is available for this last major eolian phase, although most of these deposits appear to be older than the late Pleistocene silts of the Gezira Plain, laid down by the White Nile and the Blue Nile. The deep, red soil formed on the Kordofan Sands suggests a later, moist interval of considerable duration. This rubefaction also remains undated but probably does not reflect contemporary pedogenesis.

In retrospect, the rather sketchy picture of early to middle Pleistocene deposits in the Sudan suggests that correlations between the different basins of the Nile system will prove to be difficult.

It requires little emphasis that the Paleo-Sudd Basin is filled with deposits rather unlike those of the region further east and north, and only the influx of Ethiopian silts in late Pleistocene times provides a geomorphic link between the White Nile and the Blue Nile. On purely theoretical grounds, Ball (1939: 74–84) argued that the first connection of the Paleo-Sudd Basin ("Lake Sudd") with the Saharan Nile was in late Pleistocene times. The existing geological evidence does not, in effect, contradict this hypothesis, although considerably more field work must be done before any conclusions can be reached. Above all, many deep borings will be necessary from the valley of the White Nile between Khartum and the mouth of the Sobat.

In the case of the Blue Nile and Atbara Basins, the evidence is too fragmentary to allow any conclusions. What little evidence there is seems more comparable with the Pleistocene record of southern Egypt than with that of the Paleo-Sudd, but only future field investigations will tell. The late Pleistocene silts of the central Sudan, discussed briefly in Chapter 6, should provide a promising guide horizon for any earlier deposits.

Contemporary Geomorphic Processes in Nubia

CLIMATOLOGICAL DATA

It requires little emphasis that Nubia experiences a hyperarid climate, situated as it is on the axis of the Sahara Desert. In fact, Lower Nubia lies within the area of overlap between sporadic winter and summer rainfall. The mean annual precipitation (1902–59) for Wadi Halfa is 5 millimeters (British Meteorological Office, 1958; S. P. Jackson, 1961), with some rain recorded in all months except June and December. Maximum amounts have been registered in February, April, May, July, and November, and the maximum fall recorded in 24 hours is 7.5 millimeters. At Aswan, the 1926–59 average is 3 millimeters. No rainfall has yet been recorded at Aswan during the months July through September. There are no rainfall records for the Nile Valley between Aswan and Wadi Halfa, although a cloudburst is remembered at el-Diwan in 1923 (Dunbar, 1941: 79), while heavy rains occurred at Abu Simbel in July, 1962, and September, 1965.

This and other lines of evidence suggest that sporadic summer rains of rather localized distribution occur in the southern half of Egyptian Nubia every few years, although some villages have experienced no rain in living memory. Light winter rains are not infrequent at Aswan and may extend over fairly large areas during the passage of cold fronts. Such rains do penetrate southwards to the border, but only on rare occasions. More intensive winter cloudbursts are also experienced occasionally in Upper Egypt, some of them localized, others rather widespread—so, for example, the notorious rains of January 25, 1901 (Ball, 1902: 33; Hume, 1925: 83). Consequently, it is probable that both summer and winter rainfall have played a role in the geomorphic evolution of Nubia. But whether both types of rainfall were periodically strengthened, and whether in equal degree, is a moot question. Certainly the geomorphic record tells us nothing about this, at least not until late Pleistocene times, and it is questionable whether we will ever know the answers to these questions for the earlier part of the Pleistocene.

Temperature records for several upper Egyptian stations, including Aswan, have been referred to already. Wadi Halfa (122 m elevation), with a 45-year record (1902–47), is probably the most typical station representing Nubian conditions. The mean January temperature is 60.5° F (15.8° C), with a mean daily range of 29° F (16° C) and a mean daily minimum of 46° F (7.5° C). The absolute minimum occurred in December (28° F, $-2°$ C). In July, the mean is 90° F (32° C); the mean daily range, 32° F (17.5° C); and the mean daily maximum, 106° F (41° C). Frosts, although very rare today, do occur and, presumably, had geomorphic significance at some times during the Pleistocene, but there is no physical evidence to this effect. Nile water temperatures (at 1.5 m depth) have been recorded in the Aswan Reservoir at 7 A.M. once weekly from 1922 to 1942 (Hamed, 1950: 149). The lowest monthly mean is 63° F (17.1° C) in February, the highest monthly mean, 82° F (27.9° C) in August. The lowest values on record are 59° F (14.8° C) for January and February. This seems to suggest that, even with an appreciable drop of temperatures during a Pleistocene glacial, the Nubian Nile was probably never frozen over. But, chemical weathering would have been optimal at all times of the year wherever moisture was available.

MODERN WADI ACTIVITY

Overland flow of water has never been described for Nubia, and, consequently, any assessment of current fluvial processes must be based on indirect evidence. Several generalizations may be made. Generally, evidence of wadi-wall undermining and slope retreat is significantly less apparent in Nubia than around the peripheries of the Kom Ombo Plain. Nowhere did we observe active salt-weathering in the deserts south of Dakka. Instead, patination here is generally better developed on rockfree faces than is the case in the sandstone country north of Aswan. Wadi activity appears to rework the bed load of most of the east-bank wadis every few years or decades, transporting sands and possibly undermining Pleistocene fill terraces. On the western side of the valley, most of the wadis are choked with drifting sands, and, even where eolian activity is not apparent, there is little or no evidence of recent wadi flow, even in the largest channels. This contrast of east and west banks is everywhere apparent in Nubia, in both the gross aspects and the microaspects of the geomorphology. It reflects the different degree of aridity in the Libyan and the Eastern Deserts, as well as the greater relief and overall gradients of the country east of the Nile.

Unlike the wadis draining onto the Kom Ombo Plain, the Nubian deserts have very little vegetation except for rare thorny shrubs such as *Zilla spinosa*. Acacias are almost entirely absent in the lower wadi courses. This dearth of desert vegetation probably reflects the extreme aridity of the climate and of the soil environment, as well as the impact of man during the course of millennia.

An intensive study of the geomorphic record of recent fluvial activity at Arminna West (Hansen, 1966: 51 ff.) serves to illustrate the localized character and the insignificance of overland flow in the western deserts today. The alluvial wash mantling the pediments of Arminna West is frequently masked by sand sheets or extended tied dunes, and these eolian sands can frequently be observed resting disconformably on top of fossilized rills. The antiquity of at least a part of these fluvial deposits is shown by an *in situ* Middle Paleolithic flaking site situated on top of a reddish colluvial wash with *Zootecus* shells (No. 260, Table 6–4). At another site, a surface proliferation of *Corbicula* shells, deflated from late Pleistocene silts, was observed in the center of a shallow, bedrock-incised runnel. The fact that these shells had not been disturbed or redistributed by fluvial agencies suggests that no effective water flow has followed this minor drainage line for many millennia. Although a part of the pediment rills may be of post-Pleistocene age, it seems safe to say that most of the alluvial wash and the superimposed rill patterns at Arminna West are fossil or relict, and that, except for the major drainage channels, they have seldom been reactivated during the last few millennia—and possibly only occasionally since late Pleistocene times.[12] Instead, eolian sands have replaced the alluvial wash as contemporary, functional surficial deposits.

RECENT EOLIAN ACTIVITY

Eolian deposits in the Eastern Desert are, with some exceptions, limited to small tied dunes of nebka and rebdou type. West of the Nile, however, there are extensive, rippled sand sheets and well-developed tied dunes, many of them several kilometers in length. In fact, wind streamlining of the eolian mantle over much of the Lower Nubian Plain gives a decided linear pattern to the landscape at the macroscale of the Gemini IV photography.

Although deflation during the course of the Pleistocene has excavated numerous shallow bedrock basins up to several square kilometers in area, recent phenomena are restricted to superficial remodelling of existing surfaces. Deflation is apparent in the Eastern Desert through the lag surfaces commonly developed on Pleistocene beds, but sandblast is seldom evident, and typical yardangs were not observed. On the Lower Nubian Plain, eolian corrasion is abundantly evident, and faceting, fluting, and grooving of bedrock pediment surfaces may locally produce a honeycombed microtopography with a local relief of 50 centimeters (see Hansen, 1966: 41 ff.). Sandblast may also affect the hillslopes—although on a restricted scale and almost always near ground level—and deflation has led to an even greater attrition of late Pleistocene silts than on the east bank of the Nile.

SLOPE RETREAT

The basic mechanisms of cliff weathering and slope retreat in Nubia are similar to those of the Kom Ombo area (Chap. 2 and Knetsch, 1960) and have been discussed in some detail for Arminna West by Hansen (1966: 33 ff., 26 ff., 142 ff.). Although salt hydration and subflorescence are apparently minimal today away from the Nile and its seepage waters, these

---

12. This should not imply that rainfall or overland flow are entirely absent. A dead, thick-stemmed, woody plant, found near the Middle Paleolithic site mentioned, gave a radiocarbon age of 1952 A.D. (Y–1527) (Stuiver, et al., 1968). This suggests that a fair amount of surface water was periodically available in this shallow depression within the last few years. The associated runoff has left no visible geomorphic aftereffects, however.

agencies may have been rather more significant during periods of slightly moister climate.

Judging by the degree of patination generally observed on rock-free faces and talus accumulations, backwearing of slopes by stream undercutting and by release of pressure through unloading or by geochemical exfoliation seems to be minimal today. The frequency of late prehistoric or early historical rock drawings in Nubia (see Dunbar, 1941) provides some measure for the age of the deep desert patina, which at all sites predates the petroglyphs initially, even though patination has been active since. On these grounds, it seems safe to assume that moderately to deeply patinated rock surfaces have been directly exposed to weathering processes since at least late prehistoric times. Since almost all bedrock faces are moderately to intensively patinated—with the exception of occasional sand-scoured surfaces—little or no slope retreat can have taken place during at least the last 5 millennia. By contrast, recent sandblast has faceted many formerly patinated free faces or pediment rock outcrops at Arminna West (Hansen, 1966: 41 ff., 55–56), and patinae have not formed afresh on such bare spots. This implies that fluvial processes have become ineffective and have been succeeded by eolian forces as the dominant geomorphic agency today.

Stratigraphic and Paleoclimatic Conclusions

In overview, the gravel terraces and fluvial platforms of Egyptian Nubia have been recorded in considerable detail. Basic stratigraphic relationships between Korosko and the border are clear, while the links downstream to Wadi Allaqi, Dakka, Kalabsha, Dihmit, and Aswan are reasonable, although impossible to prove conclusively. This probable scheme of internal stratigraphy is summarized by Table 5–3.

Although the different deposits and related erosional features have been mapped in detail (Figs. K–1 through K–10), we have refrained from defining formations. Even at Kom Ombo, where deposits and platforms can be traced with confidence within a restricted area, formal definitions do not appear warranted unless considerably more sedimentological work is done. These reservations are reinforced in the case of Egyptian Nubia by the difficulty of establishing long-distance correlations north of Korosko. Consequently, we have adhered to informal stage designations.

Several aspects of Table 5–3 require explanation. The two periods of rubefaction indicated (units *13* and *18*) are a little arbitrary. There may well have been other phases of soil development, and their temporal relationships to the deposits which served as parent material can only be approximated in a schematic way. Desert varnish probably has formed at all times during the Pleistocene. Large-scale impregnation by ferruginous solutions, as in the Korosko Hills and Hamid Uplands, appears, however, to have been more restricted in time. Obviously we need to know a great

Table 5-3. Geomorphic evolution of Egyptian Nubia during the late Tertiary and early to middle Pleistocene

18.	Chemical weathering, with development of a red paleosol. Ferruginization prominent locally, involving impregnation by ferromanganese solutions.	MIDDLE PLEISTOCENE
17.	Alluviation by wadis and Nubian Nile, accompanied by cutting of auxiliary bedrock channels. Wadi Korosko stage (+ 23 to + 25 m), with possible lower substages. Pluvial erosion of uplands; accelerated activity of desert wadis.	
16.	Dissection of fill or bedrock (vertical differential at least 7 to 12 m).	
15.	Alluviation corresponding to Dakka stage (+ 30 to + 35 m), with a possible lower substage at + 27 m in the type area. Significant fluvial activity of larger wadis and Nubian Nile.	
14.	Dissection of fill or bedrock (vertical differential at least 10 to 12 m).	
13.	Chemical weathering, with development of a deep, red paleosol. Moist rainy season with limited geomorphic activity. Local ferruginization.	
12.	Alluviation corresponding to Adindan stage (+ 40 to + 42 m). Significant fluvial activity of Nubian Nile. Not recorded north of Dakka.	LOWER PLEISTOCENE
11.	Dissection of fill or bedrock (vertical differential at least 4 to 8 m).	
10.	Alluviation corresponding to Dihmit stage (+ 44 to + 48 m). Significant fluvial activity of larger wadis and Nubian Nile. Major pedimentation along margins of Nile Valley.	
9.	Dissection of fill or bedrock (vertical differential at least 6 to 10 m).	
8.	Alluviation corresponding to Wadi Allaqi stage (+ 50 to + 55 m), probably marking two substages. Significant fluvial activity of larger wadis and Nubian Nile. Major pedimentation along margins of Nile Valley.	
7.	Long and complex period of dissection of bedrock, interrupted by denudation of 180 to 190 m pediment between Adindan and Tumas, as well as by planation of the + 62 m platforms and related erosional surfaces near Kalabsha. Minor tectonic activity at times.	PLIO/ PLEISTOCENE
6.	Lateral planation, with development of Aswan Pediplain at 180 to 210 m. Base level of Proto-Nile 80 to 90 m above modern floodplain.	MIO/ PLIOCENE(?)
5.	Dissection of bedrock (vertical differential perhaps 40 to 50 m).	

(*Table 5-3*, continued)

4.	Lateral planation, with development of Ballana Pediplain at 230 to 260 m, probably with two or more substages.	MIDDLE TO UPPER MIOCENE
3.	Dissection of bedrock (vertical differential perhaps 70 to 100 m).	
2.	Lateral planation, leading to development of several polygenetic surfaces between 300 and 360 m elevation, designated as Kurkur Pediplain.	
1.	Primary tectogenic phase with gentle flexures and moderate faulting. Creation of several shallow basins, domes, and swell axes.	OLIGO/MIOCENE(?)

deal more about the process involved before we can hope to unravel the complex stratigraphic aspects of massive ferruginization. The character of the erosional intervals in Egyptian Nubia is more difficult to assess than at Kom Ombo. Estimates of vertical downcutting are minimum estimates only.

The early and middle Pleistocene record of Egyptian Nubia carries no clue concerning the subsaharan drainage basin of the modern Nile. The deposits can all be attributed to an autochthonous—rather than an exotic—river and its tributaries. This does not necessarily preclude a Blue Nile or a White Nile component to the discharge nor, for that matter, does it rule out a late summer flood of exotic origin, competing with a winter flood of local origin. The complete preponderance of bed-load sediments reflects the major role played by the wadi tributaries of Egypt and, at the same time, emphasizes the vigor of the regional discharge that transported gravel and sands into the Nile. Scour and fill at such times would have eradicated any flood silts left by a sluggish, summer-flood regime. As a result, it is uncertain whether or not the Proto-Nile Basin remained isolated from the Blue Nile and Atbara Basins during the early and middle Pleistocene, and the geological record in the central and northern Sudan does not yet serve to fill this hiatus.

A second, unresolved problem concerning the evolution of the present Nile system is the linkage of the Howar and Proto-Nile Basins. Basic to this particular problem is the apparent absence of early and middle Pleistocene deposits between the Second and Fifth Cataracts. Logically, the rifting of the Red Sea, with updoming and upfaulting of the Red Sea Hills, should already have directed the Howar drainage northwards in late Tertiary times, but direct evidence remains unavailable. De Heinzelin and Paepe (1965) believe that the Proto-Nile terminated at the Second Cataract until late Pleistocene times, when a connection was first established across the Batn el-Hagar, with drainage from Upper Nubia. These deductions appear to be based on the absence of lithological indicators at Wadi Halfa rather than on the geomorphology of the Batn el-Hagar itself. Our own examination of a Gemini IV photograph of part of this rocky stretch of the Nile suggests the presence of planation surfaces adjacent to the modern valley and cut into the Basement Complex. These impressions have been greatly strengthened by the work of the Colorado Expedition (Robinson and Hewes, 1967), but only detailed field study and publication can provide a conclusive answer.

Another problem of the Proto-Nile Basin concerns the relative relationships of winter and summer rains in southern Egypt during the course of the Pleistocene. As indicated already, both genetic types of rainfall occur over much of Lower Nubia today, at long intervals of time. How the two regimes would have responded during the Pleistocene pluvials is difficult to predict solely on theoretical grounds. And the early to middle Pleistocene record unfortunately sheds no light on the matter.

Interregional comparison between Kom Ombo and Egyptian Nubia indicates a broad parallelism of geomorphic processes and events, prior to deposition of the Korosko Formation. Three criteria may be applied to compare the Nubian gravels with those of Kom Ombo: geomorphic development; relative elevation; and, where red paleosols are evident, depth of solum. On these grounds, the Wadi Korosko stage compares favorably with the Low Terrace ($+22$ m); the Dakka stage, with the Middle Terrace ($+34$ m) (see Table 2–2). Soils are quite analogous. Again, the Wadi Allaqi, Dihmit, and possibly the Adindan stages seem quite compatible as counterparts of the High Terrace complex. Earlier still, the Aswan Pediplain is common to both areas and provides a useful stratigraphic base. The Pleistocene subdivisions suggested for Egyptian Nubia in Table 5–3 are based on the tentative validity of these correlations with the Kom Ombo Plain.

In conclusion, the geomorphic record of the Nubian Nile Valley, prior to the late Pleistocene, compares closely with that of the Kom Ombo Plain. There is evidence for at least five pluvial periods, during which the desert wadis were active, torrential streams of seasonal or episodic type. Paleoclimatic interpretation of these events can be made by using the same arguments as were presented for the Kom Ombo area. But duration and chronology of these moister and drier intervals of the Nubian early and middle Pleistocene will remain vague and inconclusive.

6

Late Pleistocene and Holocene Sediments of Egyptian Nubia

Introduction

The five late Pleistocene to Holocene formations defined for the Kom Ombo Plain are all present in Nubia, with similar or identical facies. In contrast to Kom Ombo, natural exposures in Nubia are common and stratigraphic relationships are more readily established, particularly along the many wadis that have repeatedly dissected fills of local or nilotic origin. As a consequence, it is not surprising that additional subdivisions can be recognized for several of the formations, while, at the same time, a number of local deposits present in Nubia are nowhere exposed on the Kom Ombo Plain.

The deposits in Nubia will be discussed in the same way as the late Pleistocene beds at Kom Ombo. The lithostratigraphic units will be described in turn, and the major local sequences will subsequently be discussed or outlined in detail, including the type sections for any new subunits. The stratigraphic conclusions follow, together with paleoclimatic interpretations, in as far as they differ from or supplement those already presented for the Kom Ombo Plain. Speaking more generally, it should be pointed out that this chapter presumes the groundwork laid in Chapter 3.

The Korosko Formation

The Korosko Formation of Egyptian Nubia is quite comparable with that of the Kom Ombo Plain in terms of facies and distribution. The highest exposure was found at Ballana at 155 meters (+34 m with respect to modern Nile floodplain), where it rests disconformably on an alluvial wash of local origin and under silts of the Masmas Formation (Fig. 6–1). Sands or marls of the same formation are further preserved in Khor Adindan (148.5 m, + 27 m); in Wadi Or (142 m, + 22 m); within the bedrock channel north of Abu Simbel (141.5 m, + 22 m); in small, localized pockets south of Arminna Temple (*ca.* 131 m, + 12.5 m); in a minor wadi near Tushka West (146 m, + 29 m); at the mouth of a small wadi between Masmas and Ineiba (126 m, + 10 m); and sweeping under the silt plain at Ineiba, where the formation is mantled by later wash and silts. Similar sands or marls are again preserved within the embouchures of eleven different east-bank wadis between Korosko and el-Madiq, attaining a maximum elevation of 137 meters (+ 28 m) in Khor Abu Sureih. Traces are also found at Seiyala and under the surficial wash of the Dakka silt plain—after which there is a 130-kilometer gap until the Kom Ombo Plain is reached further downstream. The Korosko Formation in Wadi Or extends down to or below floodplain level, but vertical thickness in southern Nubia exceeds 34 meters.

The characteristic facies of the Korosko Formation in Nubia (Figs. 6–2 to 6–4) varies in color from white to light gray or light brownish gray (10

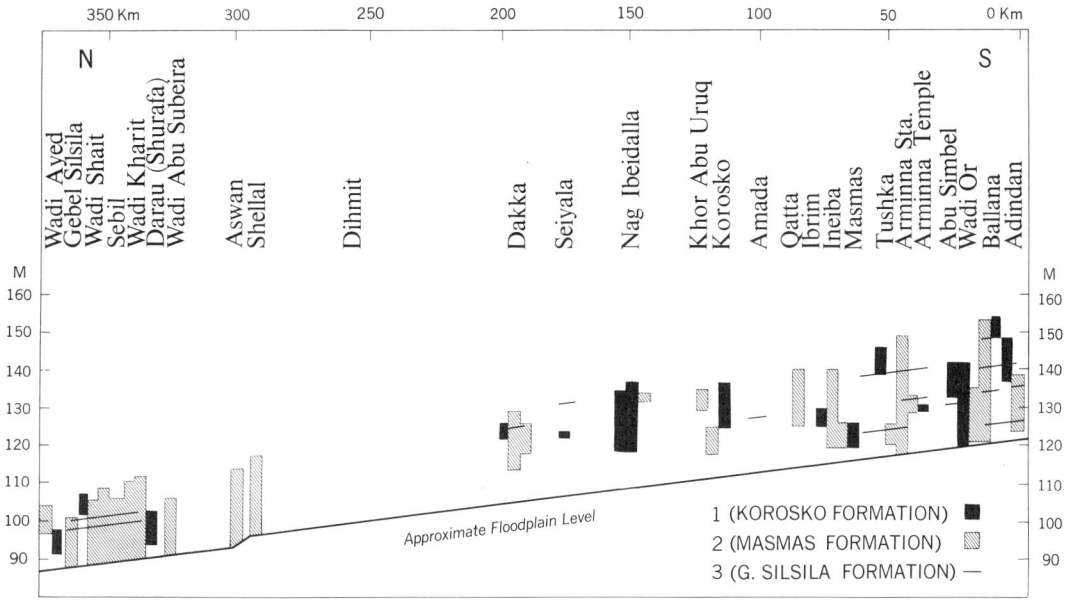

Fig. 6–1. Late Pleistocene nilotic deposits in southern Egypt. Whereas *1* and *2* indicate the vertical range of sediments recorded, *3* shows only the maximum elevation of the different substages. Graph: UW Cartographic Lab.

Fig. 6–2. Marls (top) and marly silts (lower) of the Korosko Formation at Km 0.21 in Wadi Or.

YR–2.5 Y 6–8/2) and ranges from semicemented marl to semiconsolidated, coarse sand. Clay minerals are primarily montmorillonite, with some kaolinite present (Appendix D). Carbonates vary from almost 0% to 45%, and pH values normally lie between 7.6 and 7.9. Depending on the amount and type of local derivatives, sorting may be poor or moderate. Evaporite and concretions are less common than at Kom Ombo, but yellow to reddish-yellow limonitic staining and occasional pyrolusite suggest temporary waterlogging in most deposits. As at Kom Ombo, the usual structure is coarse, angular blocky, with inconspicuous to moderate stratification. Rootlet impressions and root casts are fairly common, but mollusca were not observed in any of the Nubian exposures.

There are a number of subfacies in the Korosko Formation of Nubia. At Ballana, 66% of the noncarbonate fraction of one sample (No. 166, Table 6–2) lies in the medium-sand size between 0.06 and 0.2 millimeters, suggesting possible derivation from eolian sands (but compare Appendix C). Near Arminna Temple, also on the west bank, there are lenses of subrounded gravel near the edge of the former river channel (Table 6–1, No. 2). These are derived from local detritus, washed in from the riverbank. In the wadi embouchures of the Eastern Desert, sands and marls of the Korosko Formation grade upstream into sandy wadi alluvium. Where suitable exposures are preserved, the nilotic and wadi deposits interfinger, suggesting that alluviation was not in phase seasonally. At several localities, however, bedding suggests wadi sedimentation into standing waters. In some wadis, massive gravel beds (Table 6–1, No. 3) are intercalated with the nilotic sands and marls, primarily in the lower half of the sedimentary sequence.

Synthesizing the local evidence, to be discussed in detail below, it appears that there are two members to the Korosko Formation in Wadi

Fig. 6–3. Marly sands of the Korosko Formation (light terrace banks to left) overlain by sands of the Ineiba Formation (fine-grained, gray surface deposits) and detrital colluvium of the Shaturma Formation. Older wash mantles the pediment near the horizon. Km 2.13 in Wadi Or.

Fig. 6–4. Homogeneous sands (foreground and upper half) of the Korosko Formation, intercalated with cobble gravels in lower Wadi Umm el-Hamid (see Fig. 6–22).

Table 6-1. Morphometric gravel analyses

Sample number	Formation[a]	Sample size	ρ (%)	CV of ρ	Detrital component	E/L	E/l	L (cm)
5	Ineiba	75	13.0	87.1	38.6	33.3	44.0	4.53
4	Gebel Silsila	40	—	—	—	58.6	72.3	1.92
3	Korosko	75	26.0	74.1	16.0	30.2	41.3	4.79
2	Korosko	75	30.3	69.9	14.6	35.3	49.3	4.11
1	Older Wash	50	12.4	68.1	38.0	40.5	53.3	2.46

[a] Sample 1 from Seiyala, 2 from Arminna Temple, 3 and 5 from Wadi Or, 4 from Tushka West; all are ferruginized sandstone except number 4, which is quartz.

Or. A basal marly sandy to clayey silt (Fig. 6–2), attaining 131 meters (+11 m) and extending to below floodplain, contains frequent sandy pockets and lenses of subrounded, coarse gravel. In addition to this evidence of periodic wadi sedimentation, irregular bedding suggests dissection and refilling near the center of the wadi course, at least in part out of phase with nilotic deposition. This lower silty member was dissected to bedrock prior to accumulation of the sandy marls of the younger member, characteristic elsewhere in Nubia and at Kom Ombo. A radiocarbon determination from a semicemented, laminated marl from the lower half of this younger deposit gave a date of 25,250 B.C. ± 1000 (I–2061) (Appendix E). The subsequent sequence of nilotic and wadi deposits is generally one of homogeneous sandy marls at the base, followed by coarse-to-cobble gravels (Fig. 6–4) and, ultimately, by marly sands (Fig. 6–3) intercalated with gravel lenses or coarse sands.

Stream competence in Egypt appears to have declined during the sedimentation of the younger member. The occurrence of gravel in narrow lenses or pockets, or the local concentration of coarse sands in laminae or pockets, point to the fact that carbonate precipitation was generally not synchronous with wadi alluviation. In some cases, standing waters were found in the lower wadi courses at the time of wadi activation, but more commonly the Nile floodplain was dry, so that wadi spates cut into the marls just before they spread out their load. The nilotic sands themselves are fairly well sorted and are concentrated in the fine- to medium-sand grade (0.02–0.2 mm), with little coarser material. In other words, interpretation of the Korosko Formation in Nubia must duplicate that already given for the Kom Ombo Plain, and the relationships between nilotic and wadi activity are similar in both areas.

No genuine archeological sites were found within the Korosko Formation, although corroded, waterworn but unpatinated ironstone flakes of Middle Paleolithic type were found *in situ* at two localities in Wadi Or. Archeological sites from the time range represented (from before 50,000 B.C. to a little after 25,000 B.C.?) may be present in the concession of the Combined Prehistoric Expedition between Wadi Halfa and Ballana, but so far no comparable sediments have been recognized south of the Egyptian-Sudanese border (see de Heinzelin and Paepe, 1965; de Heinzelin, 1967).

The Masmas Formation

In Egyptian Nubia, the Masmas Formation is best preserved south of Dakka, and similar deposits have not been preserved above the old reservoir level along the 100-kilometer stretch between Dakka and Shellal. The highest exposure is found at Ballana (154 m, +33 m), where vertical thickness exceeds 31 meters. At Tushka West, on the other hand, these deposits extend to below floodplain level, just as they do on the Kom Ombo Plain. Consequently, a major period of Nile and wadi downcutting can be inferred for the time interval separating the Korosko and Masmas Formations. An isolated body of wadi alluvium in Wadi Or suggests that this erosional phase may have been interrupted by temporary alluviation in the local watercourses.

Exposures of the Masmas Formation (Fig 6–1) are found in Khor Adindan (138 m, +17.5 m), at Qustul (152 m, +33 m), in Wadi Or (*ca*. 135 m, +16 m), near Arminna Temple (133 m, +14.5 m), within the bedrock channel at Arminna East (*ca*. 140 m, +22 m), at Tushka West (125.5 m, +8 m), on the Ineiba silt plain (*ca*. 140 m, +24 m), and west of Ibrim and Qatta, where these beds attain approximately 140 meters under younger wash. Similar deposits occur at the mouth of Wadi Korosko (125 m, +13 m), in Wadi Abu-Uruq (132 m, +21 m), in Khor Abu Sureih (133.5 m, +24.5 m), and, under a veneer of wash, on the silt plains near el-Allaqi and Dakka (to *ca*. 126 m and 129 m respectively, or +11 m and +14 m). Further exposures probably exist in some of the east-bank wadis between Sinqari and el-Madiq. In most of these areas, the Masmas Formation either lies disconformably on the Korosko Formation or under the Gebel Silsila Formation.

The facies of the Masmas are a little more variable than at Kom Ombo, while still falling within the range of variability defined in Chapter 3. Characteristic in Nubia is a brown to grayish brown (10 YR 5/2–3), silty sand, sandy silt, or clayey silt, with as much as 17% calcium carbonate, and a pH of 7.3 to 7.6. Unconsolidated, the structure is coarse, angular blocky, with inconspicuous to moderate stratification. Cracking, with intrusion of coarser wash (Fig. 6–5), is common enough in finer beds, but slickensides are less common than at Kom Ombo, reflecting lower clay content in general. With few exceptions, sorting is poor. Evaporites, concretions, limonitic and pyrolusite markings, and, above all, clay minerals correspond very closely to those properties as represented at Kom Ombo (Appendix D). Although small rootlet zones are common enough, mollusca were seldom found *in situ*, one of the possible exceptions being a few *Corbicula fluminalis* shells found in a wash derived from the Masmas Formation.

Interpretation of this facies in Nubia is identical to that at Kom Ombo (see Appendix B). In Nubia also, there is little or no evidence of contemporary wadi activity, and interdigited wadi beds are absent. In one case, a marly variant was found in Khor Adindan, but elsewhere the Masmas Formation records typical flood silts or sand splays of a floodplain environment.

Fig. 6–5. Vertisol phenomena developed in dark silts of the Masmas Formation at Km 1.2 in Khor Adindan. Reddish-yellow, coarse sands fill the major vertical crack networks.

Radiocarbon dates and artifacts *in situ* are not available from the exposures we studied in Egyptian Nubia. Extensive excavations by the Combined Prehistoric Expedition at Ballana have recovered archeological sites from the Masmas (= Khor Musa) Formation, however, and, in the northern Sudan, occupation floors with Khormusan industry were found within the silts. One such site provided a radiocarbon date of 20,750 B.C. ± 280 (Wendorf, Shiner, and Marks, 1965).

The Gebel Silsila Formation and Its Subdivisions

The Gebel Silsila Formation in Nubia includes three distinct members, the two younger of which are not present on the Kom Ombo Plain. These units may be designated as follows and may be defined in turn:

    I) Darau Member, the exclusive unit of the Gebel Silsila Formation present at Kom Ombo and widely distributed in Nubia;

    II) Arminna Member, localized south of Dakka;

    III) Kibdi Member, localized south of Masmas.

The Darau Member, as such, has already been defined from the Kom Ombo Plain on the basis of sections at Gebel Silsila, Darau, and elsewhere. In Egyptian Nubia, the equivalent deposits are best developed between Ineiba and the Sudanese border (Fig. 6–1), and the highest exposures are found at Arminna Temple at 140.5 meters (+ 22.5 m). Patches of Darau deposits occur at Adindan West (141 m, + 20 m), at Qustul (140 m, + 19.5 m), and at Ballana (140 m, + 19.5 m); and extensive spreads occur at Arminna East (138 m, + 20.5 m), at Tushka West (139 m, + 21.5 m) (Fig. 6–6), and at Ineiba (*ca.* 130 m, + 15 m). A gravel lag of presumed similar age is found at Amada Temple to 128 meters or higher (+ 16 m). Sands of the Darau Member are last found to 125 or 126 meters just south of Dakka and again at Kushtamna, and no further typical deposits were recognized as far downstream as the Kom Ombo Plain. Although the vertical thickness of the Darau Member at Arminna Temple exceeds 10 meters, total depth can be estimated at over 18 meters. The period of Nile dissection separating the Masmas and Gebel Silsila Formations cut down to less than 5 meters above modern floodplain level.

The typical facies of the Darau Member is a pale-brown (10 YR 6/3), medium-sandy silt or medium-to-coarse sand, frequently grading into a medium-grade quartz gravel (Table 6–1, No. 4) with odd pebbles of chert, agate, white chalcedony, carnelian, jasper, aplitic feldspar, and

Fig. 6–6. Sandy gravels of the Darau Member (middle background), resting on bedrock in Wadi Tushka West. Korosko Hills on skyline.

granite. Carbonates average about 1.5%, except where concretions are found, and pH values range from 7.6 to 8.3. Loose to unconsolidated, the structure varies from single grain to coarse, angular blocky, depending on the texture. Stratification is generally conspicuous, and sorting is moderately good in coarser deposits. Although textural classes are variable, vertical and horizontal changes of facies are not as pronounced as they are on the Kom Ombo Plain, and inclined bedding is a little less common. For the most part, the gravels and coarse sands represent bed-load deposits, whereas the sandy silts pertain to levee beds or adjacent alluvial flats.

Calcareous concretions and some salt concentrations occur but are infrequent. Organic vestiges are equally rare, although root drip or rootlet zones are found at some localities. *Corbicula fluminalis* shells were found at Qustul, Tushka West, and Masmas; *Corbicula, Unio willcocksi,* and *Cleopatra bulimoides,* at Adindan West. Rather unusual are the *Bulinus truncatus* and planorbid shells found in a panlike lagoonal deposit at Ballana. Waterworn Late Paleolithic artifacts and rolled Middle Paleolithic artifacts are abundant in some of these beds or in the lag found on their surface. A number of occupation sites have been found in the equivalent Sahaba Formation of the northern Sudan (Wendorf, Shiner, and Marks, 1965; Wheat and Irwin, 1965). They have been discussed in Chapter 4. De Heinzelin (1967) dates the Sahaba Formation between two bounding radiocarbon determinations of 14,500 B.C. and 10,300 B.C., leaving no question about correlation with Kom Ombo.

The Arminna Member is best exposed at the type site, south of Arminna Temple (see below), where it attains a maximum elevation of 133.5 meters (+ 15 m), resting disconformably against Darau deposits as well as against the Korosko and Masmas Formations (Fig. 6–19). A basal conglomerate at this site indicates a previous period of downcutting, the dimension of which is difficult to assess. On the basis of other exposures, however, floodplain level must have been lower than + 5 meters. As a result of deflation, maximum elevation of the Arminna Member at other sites varies, e.g., at Khor Adindan, it is at least 131 meters (+ 10 m); at Ballana, 134 meters (+ 13 m); at Qustul, 131 meters (+ 10.5 m); and north of Abu Simbel, 131 meters (+ 11.5 m). A probable equivalent of the Arminna unit, with fish and mammalian remains, is recorded at 132 m maximum elevation (+ 17 m) at Catfish Cave, east of Ibrim (Wendt, 1966). Further downstream, comparable deposits are last recognized at Amada Temple, where they attain 127.5 meters (+ 14.5 m).

Lithologically, the Arminna and Darau Members are almost identical (see also Appendices B, C, and D). The typical Arminna facies is a pale-brown (10 YR 6/3) sediment, varying between a clayey silt and a gravelly coarse sand. Structure ranges accordingly from single grain to coarse, angular blocky, and stratification is good, although variable. A minor, but fairly consistent, difference is the higher carbonate content of unconsolidated beds, averaging about 4%. Correspondingly, small, incipient concretions of calcium carbonate or calcareous salt (up to 5 mm in diameter) are very abundant in most deposits, forming a distinctive lag. Other aspects are quite comparable to the Darau Member.

Mollusca are uncommon in all beds but, where present, include the Nile oyster, *Unio willcocksi*, *Corbicula fluminalis*, and *Cleopatra bulimoides*. Waterworn and fresh Late Paleolithic artifacts are common among the lag of some exposures, particularly in the case of bed-load deposits. The flood silts tend to be sterile. Living sites were found in the equivalent, lower part of the Arkin Formation of the northern Sudan (Wendorf, Shiner, and Marks, 1965; Wheat and Irwin, 1965). A good radiocarbon date obtained from Nile oyster attached to the basal conglomerate at the type site is 9140 B.C. ± 200 (Y-1526, Stuiver *et al.*, 1968). Nowhere do the deposits exceed 5 to 10 meters in thickness, and, since the facies suggests rapid sedimentation, the time interval represented may not exceed a thousand years.

The Kibdi Member was identified at the type site of Nag Kibdi—at the mouth of Khor Adindan (Fig. 6–7, 6–8), where it attains 127 meters (+5.7 m)—interdigited with one wadi alluvium and resting disconformably against a second. At Ballana, similar deposits attain 127 meters (+6.2 m); at Qustul, 125.5 meters (+5.9 m); at Arminna East, 125 meters (+ 7.5 m); and at Tushka West, 123.5 meters (+7 m). The Kibdi Member has not so far been identified north of Masmas Station.

The typical facies is indistinguishable from that of the Darau Member (see also Appendices B, C, and D), and there is no point in duplication. Deposits are usually limited to low ridges just beyond modern floodplain edge, varying from simple flood silts to levee beds and channel-margin deposits. The latter are commonly littered with Epi-Paleolithic artifacts and occasional bone fragments, commonly waterworn.

Downcutting to below modern floodplain level separates the Arminna and Kibdi Members. In the absence of archeological remains or suitable material for radiocarbon dating at the type site, however, the question may be raised whether all of the deposits attaining $+6$ to $+7$ meters in similar geomorphic situations are in fact contemporary, and, if so, what their exact age is. In the northern Sudan, there seem to be multiple stages in about the same relative elevation range, belonging to the upper units of de Heinzelin's (1967) Arkin Formation and dating younger than 7440 B.C. and older than 3650 B.C. (see Wendorf, Shiner, and Marks, 1965;

Fig. 6–7. Interdigited wadi beds (Shaturma Formation, at right) and nilotic silts (Kibdi Member, at left) near Km 0.91 in Khor Adindan.

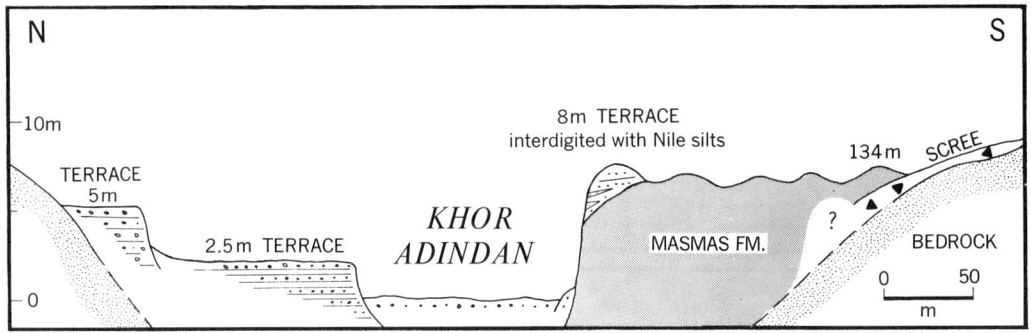

Fig. 6–8. Cross section of Khor Adindan at Km 1.12. Map: UW Cartographic Lab.

Robinson, 1966; Robinson and Hewes, 1967). De Heinzelin (1967) also recognizes a younger, Qadrus Formation of Neolithic age, with an elevation of 126 meters. It is significant that at Ashkeit a fireplace with early A-Group ware was later covered by 120 centimeters of flood silts to an elevation of *ca.* 125 meters. Subsequently, late A-Group burials were interred within this silt (Säve-Söderbergh, 1964). In other words, deposition at about this level was underway as late as 3000 B.C., which seems to mark the terminal date for the Kibdi Member. A corroborating date of 3135 B.C. (new half-life value) has been obtained from a habitation site interstratified within nilotic silts in the Batn el-Hagar (Robinson and Hewes, 1967).

Although the Kibdi unit is entirely of post-Pleistocene age, our own work was insufficient to permit more precise generalizations. Intensive geological and archeological work, supplemented by a large number of radiocarbon determinations, would be necessary to resolve the details of these last subunits of the Gebel Silsila Formation. Possibly the final reports of the other expeditions working in Nubia may contribute detailed, dated sections for these terminal alluvial deposits.

A final problem concerning the younger members of the Gebel Silsila Formation arises in connection with evidence for a brief, very high Nile stage *ca.* 11,500 years ago. At Dibeira East (Site 34), Wendorf, Shiner, and Marks (1965: xix) recognize a channel deposit at about + 30 meters, resting on top of the equivalent of the Masmas Formation. These Nile gravels contain some artifacts of Qadan and Sebilian type, and two radiocarbon runs from shell gave dates of 9460 B.C. and 9250 B.C. We have a similar situation at Ballana (Fig. 6–10), where a broad erosional bench, with a terrace edge at 149 meters (+ 28 m), has been cut into the Masmas Formation. A veneer of sandy silt with some root casts and

Fig. 6–9. Interbedded nilotic silts of the Arminna Member (vertical structure) and wadi alluvium of the Ineiba Formation (horizontally stratified), resting on the Masmas Formation at Km. 1.12 in Khor Adindan.

Fig. 6–10. Stratigraphic section of late Pleistocene deposits at Ballana (Nag el-Nuqta). Map: UW Cartographic Lab.

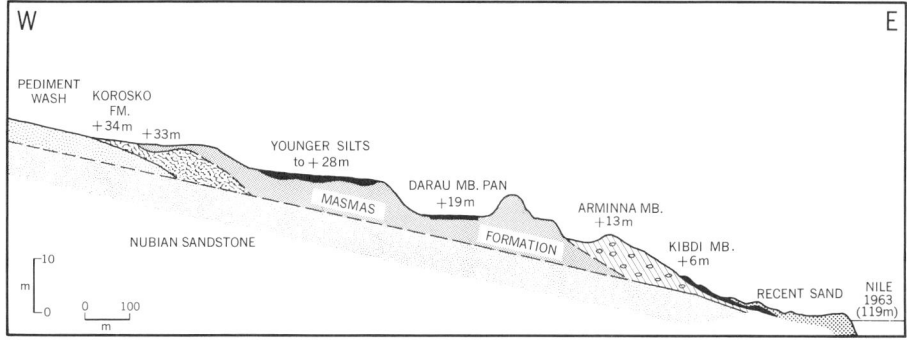

abundant waterworn Late Paleolithic artifacts rests disconformably on older silts over much of this surface. Similar phenomena were recognized at Seiyala, where a lag of medium-grade gravel (quartz and exotic pebbles) with Late Paleolithic implements rests on top of a colluvial wash to 135 meters (+ 28 m), and, west of Wadi Halfa, the Colorado Expedition found a 10-centimeter-thick nilotic silt overlying site 6–G–31 at about + 30 meters (Henry Irwin, personal communication). This particular site has a late Paleolithic industry and a radiocarbon date of about 10,000 B.C. On the basis of accumulating evidence from different parts of Nubia and the Kom Ombo Plain (Chap. 3), there is a reasonable possibility that a brief period of excessively high floods, channel-bank erosion, and limited deposition to *ca.* + 28 meters, marked the end of the Darau or the beginning of the Arminna substage. De Heinzelin (1967), who originally postulated very late tectonic displacement to explain local features of this type, now holds a similar view (*in litt.*, March 9, 1967).

All in all, the Gebel Silsila Formation represents a rather complex period of Nile aggradation beginning about 15,000 B.C. at Kom Ombo and terminating about 3000 B.C. in Nubia. Deposits younger than 10,000 B.C. appear to be absent north of the Korosko bend. The oversteepened gradient of the classical, Darau unit beds—between a relative elevation of + 22 meters in southern Nubia and + 13 meters at Kom Ombo—is probably significant in this regard: if parallelism of gradients was maintained in late prehistoric times, the deposits of the Arminna Member would hardly have risen above modern floodplain level at Kom Ombo and those of the Kibdi Member would have been suballuvial at about the latitude of Seiyala. This oversteepened gradient is in keeping with the evidence from Middle Egypt (Sandford, 1934: 87–94) and the Nile Delta (Butzer, unpublished), where deposits remarkably similar to those of the Gebel Silsila Formation fill out a buried channel of the Nile to considerable depths below modern floodplain. Although glacial-eustatic changes of Mediterranean sea level may have controlled the aggradation-degradation cycles of northern Egypt, climatic stimuli over the upper Nile Basin appears to have determined the vicissitudes of the Nubian Nile. Throughout this time span, climatic conditions in Ethiopia appear to have oscillated repeatedly—protracted periods of heavy monsoonal rains and high, swift floods interrupted by drier phases with low flood levels, leading to Nile downcutting in southern Egypt.

As a matter of convenience, the Darau, Arminna, and Kibdi Members will be informally referred to in time-stratigraphic terms as "substages" in the subsequent discussion.

The Ineiba and Shaturma Formations

A profusion of Late Pleistocene and Holocene local deposits can be recognized in Egyptian Nubia, ranging from wadi fill to slope or pediment wash. Some of these deposits are sufficiently distinctive to serve as local stratigraphic markers, but, in most cases, the deposits are difficult to identify unless underlying and overlying strata are available. In Wadi Or, for example, at least five major wadi alluvia can be recognized, of which the oldest is considerably younger than the middle Pleistocene. This complexity and exceptional development of wadi deposits makes Wadi Or the best stratigraphic norm for Egyptian Nubia.

The Ineiba Formation at Kom Ombo was defined as a fine wadi alluvium resting upon the Masmas Formation and overlain by detrital wash of the Shaturma Formation. A reddish paleosol is commonly developed on top of this alluvium, predating the sheetfloods that were responsible for deposition of the Shaturma beds. Applying this definition to Nubia, we find that two identical wadi alluvia qualify as separate members of the Ineiba Formation, analogous to the Malki and Sinqari Members at Kom Ombo. The two subdivisions can be defined, however, with reference to Wadi Or (Figs. 6–14, 6–15).

The Malki Member of the Ineiba Formation in Wadi Or is characteristically represented by 20 to 40 centimeters of subangular-to-subrounded, basal gravels, overlain by up to 150 centimeters of stratified, coarse sands. The deposits change progressively upstream from mixed wadi-deltaic beds—with strong evidence of groundwater oxidation—in the lower stream course, to a typical reddish-yellow fill with calcareous concretions and root drip at higher elevations. The waterlogging can be attributed to seepage from Nile floods attaining about 140 meters (20 m above Nile floodplain). Although more direct intercalations are lacking, these deposits can be correlated with the Darau substage.[1]

The Sinqari Member of the Ineiba Formation was deposited following a period of channel cutting of unknown extent or duration. A basal gravel, up to 70 centimeters thick, may underlie about 50 to 200 centimeters of reddish-yellow (5 YR 6–7/4–6), coarse sands with local proliferations of *Zootecus* shells. Below an elevation of 133 meters (+ 14 m), the lower third or half of these deposits is limonite-stained, as evidence of seasonal floodwater seepage from the Nile. Consequently, we suggest correlation with the Arminna Member. Ironstone flakes, scrapers, flake blades, and cores of Middle Paleolithic type are found *in situ*. They are all well

---

1. This view is substantiated by the radiocarbon determination of 15,400 B.C. ± 300 (I–2179) from a clayey marl in the lower member of the Ineiba Formation at New Sebua (Chap. 3). Coarse flakes of Middle Paleolithic type are frequently found *in situ* at Wadi Or but are derived.

patinated and are clearly derived. A radiocarbon determination of 6940 B.C. ± 160 (Y-1377) (Stuiver *et al.,* 1968) was obtained for *Zootecus* shells collected from the upper half of this younger fill. In view of this date, it appears that the earliest period of sedimentation was contemporary with the Arminna substage, but that alluviation continued into the seventh millennium.

After deposition had ceased, possibly following an interval of downcutting, decalcification and superficial rubefaction of the upper 30 centimeters or so of these beds record a period of soil development. This reddish-yellow paleosol, formally designated as the Omda Soil, was accompanied by kaolinite formation (Appendix D) and suggests a period of biochemical weathering with some vegetation.

Both units of the Ineiba Formation are frequently current-bedded, and subangular pebbles (Table 6-1, No. 5) are more common than in contemporary wadi wash. Together with the root drip in the lower member and the *Zootecus* proliferations in the upper, this suggests more frequent wadi activity and slightly greater stream competence than today. Broadly speaking, therefore, the Darau and Arminna substages of the Gebel Silsila Formation corresponded to periods of slightly moister climate in Egypt.

The Shaturma Formation in Nubia can best be identified with reference to the Omda Soil Zone and the Ineiba Formation. As at Kom Ombo, the typical facies (Figs. 6-3, 6-7, 6-16, 6-17) is a subangular, coarse colluvial gravel grading into a shallow wadi fill of light-brown to reddish-yellow (7.5 YR 6-7/4-6), gravelly coarse sand. The interpretation of these deposits is similar to that of the same formation at Kom Ombo, but they are not as well developed as they are further north, being confined to the smaller, well-incised wadis of the Eastern Desert. Radiocarbon determinations and archeology are lacking.

Other Deposits of Local Origin

In addition to the classical wadi formations already defined for the Kom Ombo Plain, there are a number of older wash deposits in Nubia.

Conglomeratic beds can frequently be seen at the base of the Upper Pleistocene sequence in Nubia. These are coarse-to-cobble-grade gravels of ferruginized sandstone or ironstone, subangular to rounded in shape, with a white to pink (5-7.5 YR 8/2, 7/4) matrix of calcreted or ferruginized coarse sand or silty coarse sand. Conglomerates of this type line the floors of many intermediate-sized wadis of the Eastern Desert, and they postdate a last period of significant bedrock incision. Locally, such coarse, semicemented fill may attain a thickness of 6 meters, suggesting considerable stream competence and transport distance, with major pluvial activity. Sandford and Arkell (1933: 42) have already drawn attention to a similar deposit in Wadi Abu Uruq. We are inclined to believe that these conglomerates record the wettest Upper Pleistocene paleoclimate in southern Egypt. Unfortunately, the stratigraphy of these coarse wadi fills is not unequivocal. In Wadis Or (Fig. 6–12), Korosko (Fig. 6–21), and Abu Sureih, a wadi-floor conglomerate rests disconformably under the Korosko Formation, while in Wadi Umm el-Hamid (Figs. 6–4, 6–22) and in Khor Adindan, similar deposits are intercalated with the Korosko Formation. There appear to be two such conglomeratic units: an older one with a ferruginous matrix of pink coarse sand and a younger deposit with a calcrete matrix of white silty sand. The latter would be equivalent to the gravelly subfacies of the Korosko Formation at Kom Ombo. The former necessarily postdates the middle Pleistocene Wadi Korosko stage but appears to precede the oldest deposits of the Korosko Formation. Artifacts have so far not been found in any of the conglomeratic fills.

Since the stratigraphy of these conglomerates has not been entirely resolved, it seems preferable not to apply formal definitions. Instead, where relationships are clear, the lower conglomerate will be informally designated as the "Wadi Floor Conglomerate," while the upper unit is identified as the gravel subfacies of the Korosko Formation.

Other sandy wadi deposits, predating the Ineiba Formation and postdating the conglomerates, are uncommon and are difficult to identify. One such unit has been defined in Wadi Or (see details below, Fig. 6–14) but has not been recognized elsewhere. It is represented by at least 160 centimeters of reddish-yellow (5–7.5 YR 7–8/6), well-stratified, sorted, coarse sand. A reddish paleosol, with a 40-centimeter (B)-horizon, developed in these sands prior to deposition of the upper member of the Korosko Formation. These older wadi fills are comparable in facies to the Ineiba Formation, and their interpretation is probably analogous.

Finally, surface wash deposits are well developed on pediments and on other erosional surfaces. These light-reddish-brown to reddish-yellow (5 YR 6/4–6) deposits consist of stratified, silty sand or coarse sand of mixed alluvial-colluvial origin. Thickness generally varies between 15 centimeters and 2 meters. Root drip or rootlet zones are occasionally found, and the land snail *Zootecus insularis* is fairly common. In several instances, this reddish wash was observed over truncated (B)C-horizons developed in bedrock (Fig. 6–18), so that much of the finer material is a true soil sediment. Very probably there are several generations of such surface wash.

At Ballana a semicemented pediment wash (30 cm) of sandy, fine gravel rests on bedrock under a coarse gravel fan (2 m) and a younger, unconsolidated wash (2 m) of coarse sand. This, in turn, is overlain disconformably by the Korosko Formation (upper member). Although there is no way to prove the age of these pediment mantles, it can be speculated that the consolidated granule gravel corresponds to the Wadi Floor Conglomerate, while the unconsolidated beds may be equivalent to the older wadi fills underlying or intercalated with the Korosko Formation. At Seiyala there are surface wash deposits (Table 6–1, No. 1) older than the Gebel Silsila Formation but younger than an important red paleosol, and at Arminna West similar *Zootecus* wash accumulated in shallow depressions prior to human occupation at a Middle Paleolithic flaking site. Like the conglomeratic fills, these pediment deposits do not lend themselves well to formal stratigraphic definition, but they are rather extensive in some parts of Nubia and have consequently been mapped as a separate unit (Figs. K–1 through K–7).

Stratigraphy of Late Pleistocene and Holocene Deposits in Khor Adindan

Khor Adindan is a fairly typical, shallow wadi, extending about 40 kilometers from a poorly defined headwater zone in the Sudan to reach the east bank of the Nile at Nag Kibdi (Adindan), some 4 kilometers north of the administrative border between Egypt and the Sudan. During its lowermost course, the wadi is moderately incised into a series of bedrock pediments, narrowing down to a little less than 200 meters before it broadens again to 500 meters as it debouches onto the Nile floodplain. Late Pleistocene nilotic deposits have flooded back into the lower 3 kilometers or so of this wadi, leaving a record of several high floodplain stages in the form of alluvial terraces of Nile or wadi origin.

Although the bedrock floor in the thalweg of Khor Adindan is mantled by an undisclosed depth of coarse, sandy alluvium, part of the present wadi floor coincides with a bevelled bedrock surface. At Km 2.41 (measured from the Nile bank), 2 to 2.5 meters above wadi floor, there is a dissected bedrock bench, badly potholed and fluted so that the relief exceeds 2 meters. The stream-carved potholes are partly lined by conglomerates of coarse, ferruginized sandstone gravel. The subrounded pebbles were responsible, in part, for churning out the potholes in which they are found. The gravel matrix consists of highly calcareous, semicemented, silty coarse sand, ranging from whitish to pink and reddish yellow (5 YR 8/2, 7/4, 6/6) in color. These calcreted beds are separated from the bedrock proper by a gray (10 YR 5/1), banded, calcite rind about 2 millimeters in thickness. Fragments of the conglomerate extend from present wadi floor to 2 meters above it and must be considered a basal wadi gravel to nearby exposures of the Korosko Formation. Considerable fluvial activity is implied, before the rising Nile floodplain converted the lower wadi into a backwater. The floor of Khor Adindan had already been dissected to about or below modern level before deposition of this conglomerate. The + 2.5 meter bedrock bench itself is well defined and follows the wadi course for about 500 meters. Because it is fossilized by the calcrete fill, it appears to antedate the late Pleistocene beds.

The Korosko Formation proper is preserved as a terrace spur along the north bank of the wadi from Km 2.7 to Km 3.0, maintaining a relative elevation of about + 10 meters and attaining a maximum elevation of 148.5 meters (27 meters above floodplain). The characteristic and fairly homogeneous deposit is a very-pale-brown, semiconsolidated, marly coarse sand (Table 6–2, No. 179), rich in colloidal silica and with dispersed or concentrated gravel. One gravel horizon is 30 to 40 centimeters thick, with coarse pebbles of subrounded to rounded ironstone. Root

Table 6–2. Sediment characteristics of deposits in Khor Adindan

Sample number	Formation/Member and location	Color	Texture[a]	Sorting	$CaCO_3$ (%)	pH
		Khor Adindan				
182	Kibdi (Km 0.91)	(2.5 Y 5.5/2)	Silty medium sand	Moderate	0.9	7.60
181	Shaturma (Km 0.94)	(5 YR 7/8)	Coarse sand	Moderate	0.0	7.95
177	Arminna (Km 1.65)	(10 YR 5.5/3)	Clayey silt	Poor	10.5	7.85
186	Mixed (Km 1.12)	(10 YR 5/3)	Silty coarse sand	Poor	5.2	7.70
185	Ineiba (Km 1.12)	(7.5 YR 8/6)	Coarse sand	Moderate	0.0	8.20
184	Masmas (Km 1.12)	(10 YR 4.5/3)	Silty coarse sand	Poor	0.6	7.30
180–I	Masmas (Km 1.13)	(5 Y 6.5/2)	(Marly sandy silt)	—	16.5	7.90
180–II	Masmas (Km 1.13)	(5 Y 6/2)	Medium-sandy clay	Poor	6.7	7.35
179	Korosko (Km 2.83)	(10 YR 8/3–4)	Marly coarse sand	Poor	17.9	8.35
		Ballana				
170	Masmas	(10 YR 4–5/3)	Silty coarse sand	Poor	3.9	7.20
166	Korosko	(10 YR 7/3)	Marly medium sand	Moderate	15.9	7.90
		Qustul				
159	Darau	(7.5 YR 6/4)	Coarse sand	Moderate	1.6	7.60
157	Masmas	(7.5 YR 4/2)	Silty coarse sand	Poor	3.4	7.55

[a] Texture in parentheses not determined by quantitative methods.

casts and root drip are common in the finer beds. These subaqueous deposits of mixed nilotic-wadi origin suggest considerable wadi activity. Intensive yellow limonitic staining emphasizes the seasonal submergence, at which time carbonates and fine suspended sediments intruded into the wadi with the Nile floodwaters. The remaining materials are local.

Following a period of wadi cutting of unknown dimensions, flood silts of the Masmas Formation were accreted in a large tributary embayment of Khor Adindan east of Nag Kibdi, and scattered vestiges of related deposits may be found upstream to Km 1.81 and to an elevation of 138 meters ($+17.5$ meters). Three subfacies can be recognized. Near the thalweg of the wadi, there is a brown, silty coarse sand (Table 6–2, No. 184) with occasional quartz granules or sandstone detritus up to 2 centimeters in diameter. Away from the thalweg, the major silt body consists of a light-olive-gray, sandy or silty clay (Table 6–2, No. 180–II). Locally, there may be sandy marls (Table 6–2, No. 180–I) of mixed nilotic-wadi origin resting on the silt facies, to a maximum thickness of almost 1 meter. These last beds suggest an interval of slightly accelerated wadi activity following a lengthy period of rather limited wadi flow.

Slickensides, prismatic structure, polygonal crack networks, calcareous concretions (1 to 2 centimeters in diameter), and pyrolusite flecks are common in all exposures of the Masmas Formation, particularly in the more fine-grained deposits. The dehydration cracks occur as a major network at intervals of 50 to 100 centimeters, with minor networks at 15-centimeters intervals (Fig. 6–5). These fissures are 5 to 10 centimeters and 1 to 2 centimeters wide respectively and were observed to depths of at least 130 centimeters. Such networks are filled with coarse wadi sands and are best developed at the base of a shallow wadi channel cut into the surface and filled with deposits of the Ineiba Formation. We believe that these vertisol phenomena are, in general, epigenetic and that their major development was due to periodic wetting—either by local wadi flow or by Nile inundation—following a period of major downcutting.

The Darau substage is not represented in Khor Adindan. Some 50 centimeters of grayish-brown, medium-sandy to clayey silt (Table 6–2, No. 177) with abundant calcareous concretions are found at Km 1.65, resting on a $+9.3$ meter wadi terrace which covers local exposures of Masmas silts, but this highly micaceous deposit—attaining some 136 meters (15 m above floodplain)—is lithologically more similar to the Arminna Member, particularly in view of the high carbonate content. The Darau Member is clearly present on the west bank of the Nile (Fig.

5–1), however, where nilotic beds rest on a rock platform at 137.5 meters, exposing the following sequence from base to top:

a) 130 cm. Very-pale-brown (10 YR 7/3–4), stratified, coarse quartz sand with single-grain structure. A borer-like microblade in agate, of Late Paleolithic type, was found *in situ* in these sands.

b) 10 cm. Pale-brown (10 YR 6/3), stratified, medium-sandy silt with rootlet zones. Coarse, angular blocky to platy structure.

c) 160 cm. As *a* attaining maximum elevation of 141 m (+ 20.5 m).

d) Veneer of nilotic gravel, primarily quartz and ironstone, with *Unio willcocksi*, *Corbicula fluminalis*, and *Cleopatra bulimoides*, up to 141 meters elevation.

The interdigitation of the Arminna deposits and the Malki Member is well exposed at Km 1.12 (Fig. 6–8, 6–9). The following section is exposed, from base to top:

a) Over 320 cm. Silty coarse sand of Masmas Formation (Table 6–2, No. 184). Inconspicuously stratified, with coarse, angular blocky to prismatic structure and slickensides. The coarse sands, rare fine detritus, and occasional pebble lenses indicate very limited wadi activity, comparable to that of today, during the course of flood-silt accretion. The epigenetic crack networks contain reddish-yellow (5 YR 6/6), coarse sand and light-brown (10 YR 6/4), silty sand of wadi origin. These sandy fillings are stratified parallel to the crack faces, and a flint flake was found *in situ* in one, an ironstone microflake in another. Subsequent wadi dissection to below 129 m indicates a local Nile floodplain level below about 126 m.

b) 260 cm. Alternating wadi sands, nilotic silts, and mixed beds, constituting a terrace 7.5 to 8 m above wadi floor. The sequence from base to top is as follows:

  i) 80 to 85 cm. Reddish-yellow, well-stratified or current-bedded, coarse wadi sand (Table 6–2, No. 185), with an impression of a *Colocynthis* fruit and a waterworn, ironstone flake blade *in situ*. Medium, subangular blocky structure. Interrupted by a wedge of brown, silty sand derived from nilotic beds.

  ii) 15 cm. Brown (19 YR 5/3), stratified, clayey silt with lenses of sandy silt, rootlet zones, and occasional limonite or pyrolusite staining. Coarse, angular blocky structure. Dipping 1.5% towards the Nile, the upper end of this nilotic bed attains 131 m (+ 9.5 m), where it is truncated by erosion.

  iii) 35 to 40 cm. Brown, stratified, silty coarse sand with some coarse gravel (Table 6–2, No. 186). Coarse, subangular blocky structure. Traces of root drip in sandy lenses. Wadi deposit derived in part from erosion of nilotic silt (bed *ii*).

  iv) 10 cm. Nilotic silt as bed *ii*, attaining 130.5 m elevation.

v) 100 cm. Wadi sand as bed *i*, with lenses of coarse, subangular to subrounded ironstone gravel and rare root drip, as well as with laminae of pale-brown (10 YR 6/3), sandy silt, derived from nilotic beds.

The alternating nilotic and wadi deposits of unit *b* provide documentation for contemporary alluviation, seasonally out of phase: the wadi frequently reworked older nilotic silts, but the sediments were laid down by regular fluvial agencies on the wadi floor, rather than subaqueously in standing Nile floodwaters. In other words, wadi activity was confined to low Nile stage, i.e., the winter half-year. The lithology and the elevation range of at least 131 meters (+9.5 m), but probably not much above 134 meters (+12.5 m), indicate that the nilotic silts pertain to the Arminna substage. The reddish-yellow wadi sands form part of a fragmentary +8 meter wadi terrace, with Late Paleolithic artifacts. Wadi activity, as suggested by the ratio of wadi sands to Nile silts, was somewhat greater than today. The wadi deposits form part of the upper Ineiba Formation.

After a period of wadi cutting to below 125 meters (+3.5 m), renewed aggradation led to the development of a +5 meter wadi terrace, consisting of light-brown (7.5 YR 6/4), coarse sands with occasional, coarse ironstone pebbles. This terrace is only preserved at Km 1.1 to 1.2 on the north bank and at Km 2.2 to 2.4 on the south bank. It suggests alluviation in response to a floodplain level of 130 meters (+8.5 m), and concentrations of fresh flint artifacts of Late Paleolithic type (lacking blades or microliths) were found on the surface. This terrace probably represents a late phase of the Malki.

A renewed period of wadi incision to below 121 meters, i.e., below modern floodplain level, preceded aggradation of the 2.6-meter wadi terrace. This is a well-developed, fan-like deposit, averaging 2.7 meters in thickness and exposed from Km 0.94 to 1.35. The reddish-yellow, well-stratified to current-bedded, coarse sands are moderately well sorted (Table 6–2, No. 181) and contain occasional, coarse, subangular to subrounded ironstone pebbles. The beds dip 3% towards the Nile. A single, diminutive agate flake found *in situ* was probably derived from an older site.

At the mouth of Khor Adindan, the +2.6 meter wadi deposits are interdigited with nilotic silts (Fig. 6–7). The latter take the form of a light-brownish-gray, micaceous, medium-sandy silt to silty sand (Table 6–2, No. 182), with levee backsets dipping 4% to 5% near the contact with the wadi deposits, but attaining a maximum inclination of 26%. The highest elevation of these flood silts is 127 meters (+5.5 m), and this is the type site of the Kibdi Member. Between Km 0.72 and 0.92, the beds exhibit a thickness of over 4 meters. The contact with the 2.6-meter wadi terrace covers a longitudinal distance of up to 30 meters and is rather significant. The immediate zone of interdigitation is about 5 meters in width. The terminal wadi beds are mainly horizontal or have a slight reverse inclination of 0.5% (away from the Nile), while occasional wadi strata dip as much as 8% nilewards, suggesting local deposition into hollows at the floodplain edge. Redeposited, brown, silty sands are commonly found interfingering with the wadi sands. These lenses may be up to 5 centimeters thick and 50 centimeters long. They include small, intact laminae of flood silts that were eroded by the wadi while in the dry state. Several laminae of wadi sands, about 2 millimeters thick and up to 60 centimeters long, occur within the nilotic beds. Interpretation of these exposures is fairly obvious: the two facies are contemporary but out of phase. Summer alluviation of Nile silts took place at a time when the wadi was inactive, and wadi aggradation on the floodplain periphery was accomplished at low Nile stage. The Nile muds at the wadi mouth were dry and were subject to erosion and reworking when the winter wadi floods surged out onto the sun-baked floodplain.

The implication is that local winter rains were more frequent than today, allowing the wadi to aggrade simultaneously with the floodplain. The contemporary wadi bed contains practically no gravel, suggesting a similar competence at that time. Towards the close of the aggradation phase, stream activity was declining, however. In the uppermost Kibdi silts, there are pockets of up to 100 centimeters of crudely stratified calcareous concretions (up to 2 cm in diameter) with rare quartz pebbles, embedded in a pale-brown (10 YR 6/3-4), sandy silt. Reversed bedding of up to 1.5% suggests these redeposited materials were embanked against the floodplain edge by occasional wadi spates. But wadi alluviation was no longer keeping up with silt accretion, so that depressions were created in the wadi embouchure, behind the levee. Shortly thereafter, Nile and wadi aggradation terminated entirely. Presumably, these Nile silts are contemporary with the A-Group occupation at Ashkeit, where similar silts were deposited to about 125 meters elevation at that time. The wadi deposits appear to be equivalent to the Shaturma Formation, although no firm lithological arguments can be provided.

Later activity in Khor Adindan was confined to wadi erosion and redeposition of fill. A low, +50 centimeter terrace from Km 1.56 to 1.75 is identical in composition to the modern wadi bed and may be related to a younger substage of the Shaturma Formation.

In summary, the following late Pleistocene and Holocene sequence can be recognized in Khor Adindan and its vicinity:

a) Bedrock potholing followed by deposition of wadi conglomerate, including silts and carbonates of nilotic origin in matrix.
b) Accretion of Korosko Formation—subaqueous, marly coarse sands with sands and gravel strata of wadi origin, as well as dissolved and suspended sediments of nilotic origin, to 148.5 meters elevation. Accelerated wadi activity persists.
c) Wadi dissection.
d) Deposition of Masmas flood silts in lower wadi to at least 138 meters. Wadi activity no greater than at present.
e) Wadi dissection followed by vertisol formation.
f) Deposition of Nile sands of Darau unit to 14 meters (on west bank).
g) Wadi dissection.
h) Deposition of Nile flood silts of Arminna unit to 134 meters (initially to 136 m?), interdigited with 8-meter wadi terrace composed of sands of upper Ineiba Formation.
i) Wadi dissection (to below 125 m).
j) Deposition of 5-meter wadi terrace to floodplain base level of 130 meters (terminal stage of Ineiba Formation?).
k) Wadi dissection (to below 121 m).
l) Deposition of Nile flood silts of Kibdi unit to 127 meters, interdigited with 2.6-meter wadi terrace, probably part of Shaturma Formation.
m) Wadi dissection (to below 121 m).
n) Deposition of 0.5-meter wadi terrace.
o) Wadi dissection and redeposition of fill.

Late Pleistocene Deposits at Ballana and Qustul

Late Pleistocene deposits are well developed southwest of Ballana, behind a fringe of dune-veneered floodplain, embanked on the edge of a series of pediments developed below the escarpment of the Ballana Pediplain (Fig. K–1). Broad but shallow fans from the wadis that dissect the escarpment rest upon such pediments at elevations of 150 to 200 meters, sloping nileward at gradients of 0.75% to 1.5%. Three generations of such pediment wash were recognized.

a) A granule conglomerate, with a cemented matrix of reddish-yellow (5 YR 7/6), silty coarse sand and an average thickness of 30 centimeters forms the basal bed.

b) Resting on the eroded edge of this deposit is an average of 2 meters of coarse, subrounded-to-rounded sandstone or ironstone gravel, in an unconsolidated matrix of similar type.

c) A reddish-yellow (5 YR 6/6), coarse sand with dispersed pebbles and local proliferations of *Zootecus insularis* completes this sequence and is far more ubiquitous than the fan-like spreads *a* and *b*. A rolled Middle Acheulian biface was found on the surface of wash *c*.

Younger than these pediment deposits are a suite of nilotic sediments (Fig. 6–10), described here on the basis of a transect run 300 meters north of the former Ballana Police Post (Nag el-Nuqta). Resting against the finer, reddish wash are unconsolidated, very-pale-brown, marly medium-grained sands (Table 6–2, No. 166), moderately sorted but inconspicuously stratified. There is considerable, massive root drip, and lenses or pockets of silt are occasionally present. The clay minerals in this deposit have the X-ray diffractogram typical of the Korosko Formation, leaving little doubt that these are fluvial sands of nilotic origin. But the sand was probably derived from reworked local dunes or wash and laid down in the bed of the Nile. Maximum elevation is 155 meters (+34 m).

Embanked against the Korosko Formation are massive beds of fairly homogeneous flood silts of the Masmas Formation, attaining an elevation of 154 meters (+33 m). Sediments vary from light brown to light yellowish or grayish brown, with texture ranging from clayey to sandy silt (Table 6–2, No. 170). A fine network of dehydration cracks with secondary sand fillings transcends these beds. Although we found no artifacts *in situ,* subsequent excavations have uncovered archeological materials. A superimposed bench with a "shoreline" at 149 meters (+38 m) is littered with unworn Late Paleolithic flint artifacts in a veneer of late silts (terminal Darau substage?) with root casts.

The Darau unit is associated with a small lagoon, demarcated by the 140-meter contour and indented into the Masmas beds. Within this lagoon (which is about 200 meters wide), there is a small silt pan, located 825 meters from the modern Nile and measuring 15 by 27 meters. These slightly warped, basin-like beds at 140 meters (+19 m) elevation represent a small desiccation pan, the edges of which have since been deflated. Semiconsolidated, pale-brown (10 YR 6/3–4), sandy marls with planorbids and *Bulinus* snails suggest a seasonal pond with access to the Nile. The material often has a spongy structure of organic origin, with abundant 1- to 2-millimeter rootlets. A Late Paleolithic flint industry, with diminutive flakes and blades, occurs on, in, and around this silt pan. Near the margins of the lagoon, at 740 meters from the Nile bank and 141 meters elevation, there is a surface site of ironstone flakes (3 to 5 cm in length) with a single core in diminutive Levalloisian technique. This site appears to be related to a Nile riverbank at about that elevation.

The Arminna substage is clearly recorded by a deflated gravel ridge at 134 meters (+13 m) elevation, separated from the Masmas silts by an irregular swale about 75 meters wide. Beneath a lag of quartz pebbles, the constituent sediment is a pale-brown (10 YR 6/3), medium-sandy silt with calcareous concretions and dispersed medium-grade pebbles. Rolled or waterworn flint artifacts of Late Paleolithic type abound in the lag. Finally, the Kibdi Member can be recognized by a similar, but very much less distinctive, ridge at 126 meters (+5 m), littered with mammalian bone fragments and Epi-Paleolithic flints. Active dune sands rest on the edge of these last beds.

Although the excavations of the Combined Prehistoric Expedition in this area are bound to add detail to the above sequence, the basic stratigraphic pattern is sufficiently clear. It is corroborated by deposits on the opposite bank of the Nile at Qustul (Fig. 5–12).[2] Best developed here are the Masmas flood silts, reaching up to 152.5 meters (+33 m), with a thickness of at least 31.5 meters. A good vertisol is developed under pediment wash at the highest exposures (Table 6–2, No. 157), 1.5 kilometers from the Nile. Several detailed exposures were recorded in the nobles' graves of the X-Group cemetery at Kimam Goha, about 200 to 225 meters from the Nile bank. All of these beds were sterile and usually topset, inclined nileward at gradients of 5% to 8%, with a slight downstream component. Individual beds usually vary from 2 to 50 centimeters in thickness. The depositional environment seems to have fluctuated between the embanked levee and the edge of the alluvial flats. About 750 meters from the Nile, there is a low, irregular ridge of light-brown, coarse sand (Table 6–2, No. 159) with dispersed medium quartz pebbles, superimposed on the Masmas beds at 138 to 140 meters (+19.5 m). These younger beds are 40 to 60 centimeters thick, and several sites with slightly waterworn Late Paleolithic artifacts, mammalian and fish bones,

---

2. Sandford and Arkell (1933: 40–42) give a brief but useful description of the major late Quaternary deposits in southern Egyptian Nubia. By way of interest, it may be remarked that their alleged "silt yardangs" at Qustul (1933: 40, Note 7) were entirely artificial, being part of the X-Group cemetery of Kimam Goha. Said and Issawy (1965: 12–14) allude to nilotic clays, silts, and sands, with some implements on the surface or *in situ*, at Qustul, Ballana, Wadi Tushka, Masmas, and Ineiba. These are reported to maintain fairly constant levels of 16 meters, 10 meters, 7 meters, and 1.6 meters above reservoir level (approximately 119 meters in 1962–63). Apart from the fact that it is unreasonable to assume a horizontal

and *Corbicula* concentrations litter the surface. They suggest the Darau substage.

The Arminna unit is recorded at Qustul by surficial accumulations of powdery, pale-brown (10 YR 6/3), medium-sandy silt with small concretions, localized in embayments within the Masmas deposits. These beds attain 131 meters (+10.5 m). The Kibdi unit is represented by a partly erosional, partly depositional ridge along the edge of the Masmas beds at 125.5 meters (+5 m), marked by surface sites with abundant Epi-Paleolithic implements in chert and agate.

In review, the following sequence can be outlined for the Ballana-Qustul area:

a) Accumulation of a granule conglomerate on the pediment surfaces.
b) Erosion.
c) Deposition of gravel fans on local pediments.
d) Erosion.
e) Aggradation of reddish pediment wash with *Zootecus*.
f) Deposition of Korosko Formation, incorporating local sands in a well-sorted fluvial deposit, to 155 meters elevation.
g) Nile downcutting.
h) Accretion of Masmas flood silts to 154 meters across a broad floodplain 4 kilometers in width, accumulating to a thickness of over 32 meters.
i) Nile downcutting, possibly to below modern floodplain level (120–21 m).
j) Deposition of riverbank and other floodplain deposits by Nile (Darau Member) to 140 meters. Late Paleolithic habitation along streambanks.
k) Nile downcutting to below 125 meters.
l) Deposition of nilotic beds (Arminna Member) to 134 meters (initially to 149 m?). Late Paleolithic habitation in riverine zone.
m) Nile downcutting to below modern floodplain level.
n) Deposition of nilotic beds (Kibdi Member) to 126 meters. Epi-Paleolithic (Neolithic?) habitation along river.
o) Nile downcutting and establishment of modern floodplain.

---

alluvial surface over an 80-kilometer longitudinal stretch, we find it difficult to reconcile some of the data given with the field evidence. Many of the low-level deposits are eroded remnants of older formations which are correlated on altimetric grounds, regardless of the local stratigraphy.

Fig. 6–11. General view of Wadi Or (North Branch), looking towards Nile.

Stratigraphy of Late Pleistocene and Holocene Deposits in Wadi Or

Wadi Or, one of the larger Nubian tributaries of the east bank, drains a large basin east of Wadi Halfa and reaches the Nile opposite Ballana. 1.2 kilometers from the valley edge, the wadi bifurcates, with a minor, northern branch terminating after a total length of only 4.1 kilometers. This North Branch (Fig. 6–11) preserves a remarkably complex record of late Pleistocene and Holocene fill. A complete longitudinal section was surveyed, 17 detailed cross sections were made, and 25 sediment samples were subjected to laboratory analysis. This fundamental sequence deserves a fairly detailed discussion, although only a few of the sections can be described or illustrated (see Figs. 6–11 to 6–18 and K–1).

The oldest, late Pleistocene deposit of Wadi Or is the Wadi Floor Conglomerate, which rests directly on bedrock. This conglomerate was carried in the major wadi course and is not found in North Branch except near the confluence. Thickness in the major wadi channel, above the confluence, averages a little less than 1 meter. Just below the confluence, it is well over 3 meters thick, forming a +3.1 meter wadi terrace (Fig. 6–12A). Further downstream, these gravels floor the wadi with the exception of two runnels cut down to bedrock. Relative level drops from +2.7 meters at Km 1.28 to +2.2 meters at Km 0.63, +2.0 meters at Km

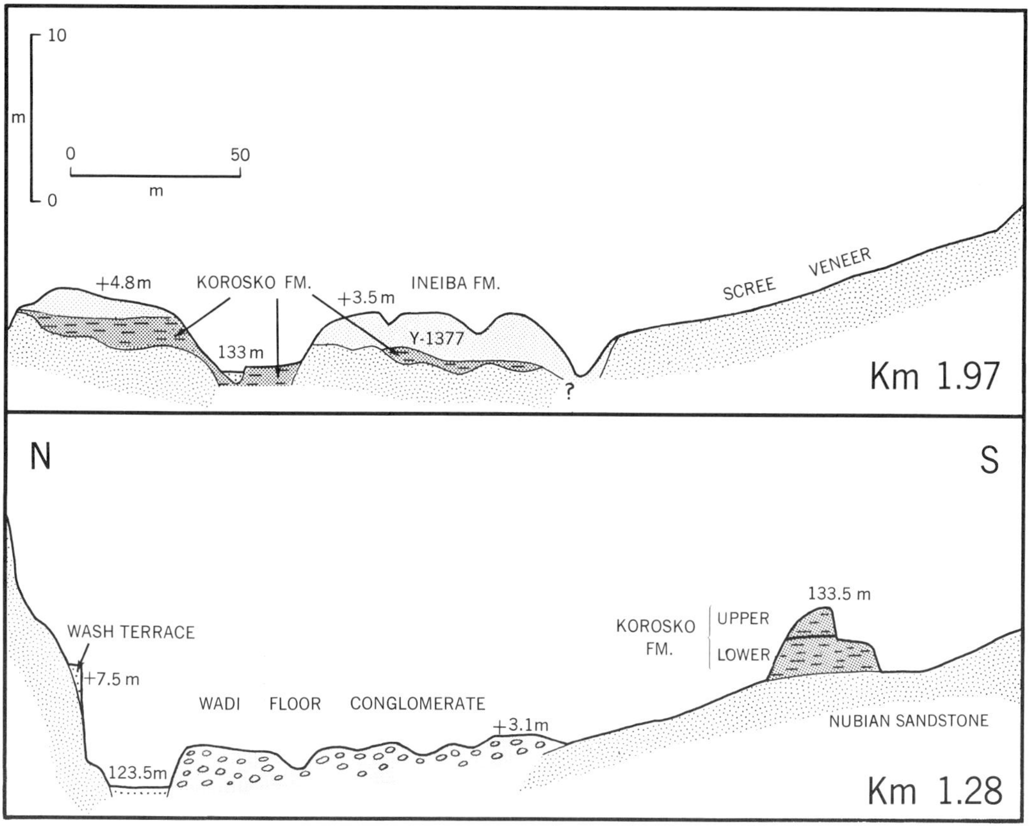

Fig. 6–12. Cross sections of Wadi Or at Km 1.28 and Km. 1.97. Map: UW Cartographic Lab.
Fig. 6–13. Cross section of Wadi Or at Km. 3.92. Map: UW Cartographic Lab.

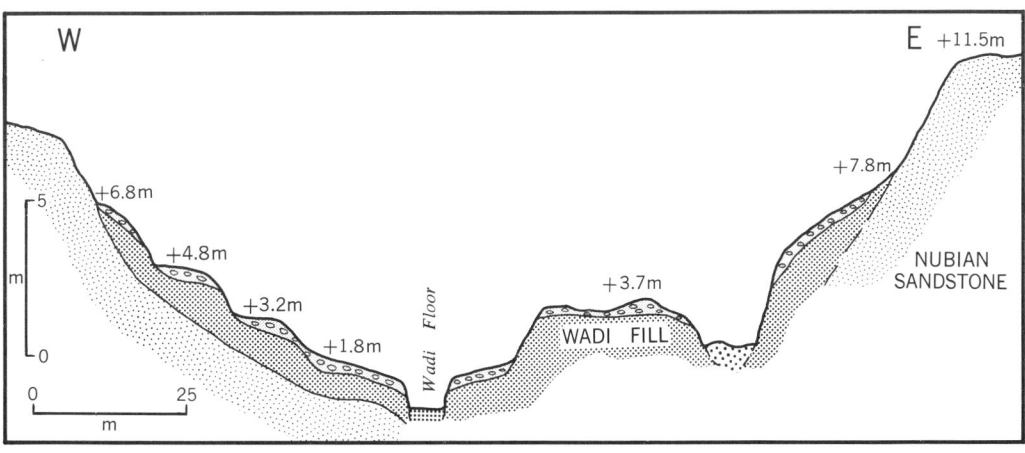

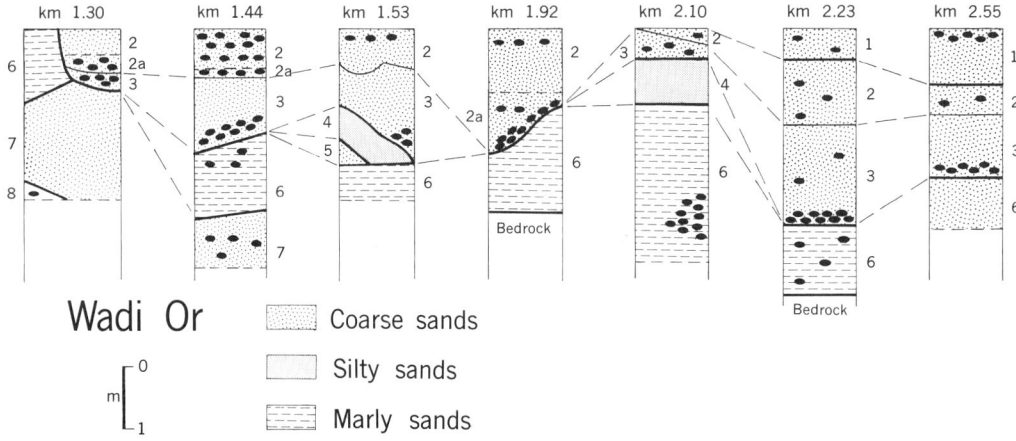

Fig. 6–14. Stratigraphic sequence in Wadi Or (North Branch). *1:* Shaturma Formation; *2, 2a:* Ineiba Formation (Sinqari Member); *3:* Ineiba Formation (Malki Member); *4:* Masmas Formation; *5:* Wadi fill; *6:* Korosko Formation; *7:* Wadi fill; *8:* Wadi Floor Conglomerate. Graph: UW Cartographic Lab.

Fig. 6–15. Section at Km 1.53, Wadi Or (North Branch). See p. 302 for description. Map: UW Cartographic Lab.

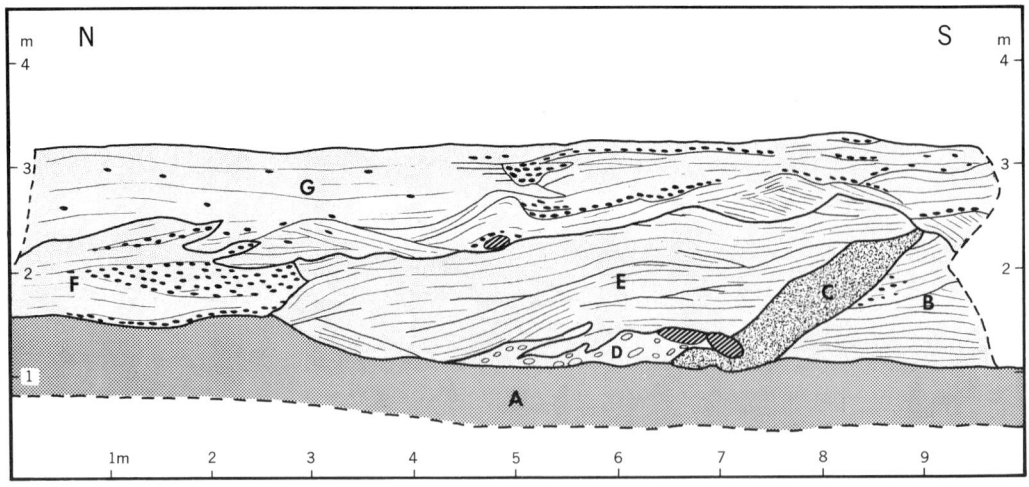

0.31, and +1.3 meters at Km 0.21. In other words, the gradient is oversteepened, suggesting that the local base level provided by the Nile was at least a meter or two lower than present floodplain (120 m). The gravel itself is coarse to cobble grade, consisting of rounded, ferruginized sandstone or ironstone. The matrix is a semicemented, brown (7.5 YR

Fig. 6–16. Section of alluvial deposits at Km 2.55 in Wadi Or (see p. 304). The top of bed *a* terminates 30 centimeters below pick-head, the contact of beds *b* and *c* is distinctly outlined by the zone of calcification, while the contact between *c* and *d* is located at the level of the entrenching tool.

5/2), coarse sand, with reddish-yellow (7.5 YR 6–7/6), limonitic staining. The cement is ferruginous.

The bedrock channel under the Wadi Floor Conglomerate is at 120 meters elevation near Km 0.6, dipping below floodplain level further downstream. Consequently, bedrock incision of Wadi Or must predate the late Pleistocene sequence, and subsequent dissection has not sufficed

Fig. 6–17. Near Km. 3.15 of Wadi Or, facing west. Deep, late Pleistocene wadi fill, forming a 7- to 8-meter terrace, is exposed behind the person. Coarse, dark colluvium of the Shaturma Formation grades upslope into scree and talus.

to remove the late Pleistocene fill. A further implication is that the last major period of Nile bedrock cutting is older than the Upper Pleistocene. The bedrock floor of Wadi Or had been strongly potholed by stream-bed erosion before deposition of the Wadi Floor Conglomerate. Such erosive force speaks for a period of rather considerable wadi flow, just as a comparison of the coarse to cobble gravels with a modern bed load of coarse sand (Table 6–3, No. 209) suggests far greater stream competence. Probably the conglomerate itself is a terminal aggradation related to the period of swiftly moving, gravel-transporting wadi flow. The implication is that a pluvial climate prevailed during a time when the Nile floodplain was at least as low as today, prior to accumulation of the Korosko Formation. At a later date, potholes were cut into the Wadi Floor Conglomerate on a smaller and more localized scale. The stratigraphic position of this erosive phase could not be determined.

The Korosko Formation in the embouchure of Wadi Or begins with a basal unit of white sandy marls and light-brown, sandy to clayey silt

Fig. 6–18. Fossil reddish paleosol near Km 2.1. The entrenching tool extends from the base of the $(B)E$- into the C_t-horizon. The truncated (B)-horizon averages 10 centimeters in thickness, the $(B)C$-horizon, about 25 to 30 centimeters. The wash represented by the $(B)E$-horizon is of late Pleistocene age.

(Fig. 6–2) (Table 6–3, No. 202) to 131 meters ($+11$ m), with a local thickness of over 12 meters. Near the present wadi bed, there is a considerable amount of subrounded, coarse sandstone gravel admixed with these deposits, and the irregular bedding suggests repeated dissection and refilling. Root drip or root casts are frequent in the marly subfacies. Except for the higher silt and clay content in some exposures, these beds are lithologically analogous to the typical Korosko Formation. Since opportunity to study these older Korosko beds was limited while in the

Table 6–3. Sediment characteristics of deposits in Wadi Or

Sample number	Formation/Member and location	Color	Texture	Sorting	CaCO₃ (%)	pH
209	Modern wash (Km 1.53)	(7.5–10 YR 7/4)	Coarse sand	Moderate	0.6	8.45
199	Shaturma (Km 2.55)	(5 YR 7/6)	Coarse sand	Moderate	1.2	8.40
200	Ineiba (Upper) (Km 2.55)	(5–7.5 YR 6/4)	Coarse sand	Moderate	0.0	7.80
207	Ineiba (Upper) (Km 1.53)	(5 YR 7/6)	Coarse sand	Moderate	0.0	8.10
192	Ineiba (Upper) (Km 1.53)	(5 YR 6/8)	Gravelly coarse sand	Moderate	0.0	7.15
198	Ineiba (Lower) (Km 2.55)	(5 YR 6/6)	Coarse sand	Moderate	4.2	8.00
195	Ineiba (Lower) (Km 2.23)	(7.5 YR 7/4–6)	Silty coarse sand	Poor	3.5	7.70
194	Ineiba (Lower) (Km 2.23)	(7.5 YR 7/6–8)	Gravelly coarse sand	Poor	14.2	7.60
189	Ineiba (Lower) (Km 1.53)	(10 YR 6–7/4–6)	Silty coarse sand	Poor	4.8	7.30
191	Ineiba (Lower) (Km 1.53)	(7.5 YR 7/8, 6/6)	Gravelly coarse sand	Poor	8.6	6.90
208	Masmas (Km 2.10)	(10 YR 6/2–3)	Silty coarse sand	Poor	5.8	8.10
196	Masmas (Km 1.53)	(10 YR 5/2)	Silty sand	Poor	15.2	7.35
193	(Wadi wash) (Km 1.53)	(5–7.5 YR 8/4)	Coarse sand	Moderate	0.0	7.60
201	Korosko (Km 2.55)	(5 YR 8/4)	Coarse sand	Moderate	9.9	7.80
197	Korosko (Km 2.13)	(2.5 Y 8/2)	Silty coarse sand	Moderate	9.9	8.10
188	Korosko (Km 1.53)	(10 YR/2.5 Y 8/2)	Marl	Moderate	40.4	7.80
203	Korosko (Km 1.28)	(10 YR 8/2–3)	Marl	Poor	43.9	7.95
206–I	(Wadi wash) (Km 1.28)	(5 YR 7/6)	Coarse sand	Moderate	1.1	7.90
206–II	(Wadi wash) (Km 1.28)	(7.5 YR 8/6)	Coarse sand	Good	0.0	7.90
204	(Wadi wash) (Km 1.28)	(5 YR 8/6)	Coarse sand	Moderate	0.0	7.90
202	Korosko (Lower Member) (Km 1.28)	(7.5 YR 6/4)	Clayey silt	Poor	12.6	7.50

field, we prefer to use the designation "lower member" in an informal sense only. The available evidence indicates that dissection down to the bedrock floor removed most of these earlier silts and marls prior to renewed alluviation.

Between Km 1.25 and Km 1.45 (Fig. 6–14), an older, semiconsolidated wadi wash lies under the classical Korosko beds, with at least 160 centimeters of reddish-yellow, coarse sand exposed. Dispersed gravel and current bedding suggest high-velocity flow, while root drip implies some vegetation. After deposition, a reddish paleosol was able to develop, with a 40-centimeter zone of superficial rubefaction (5 YR 7/6 compared with 5 YR 8/6 in the C-horizon), a limited amount of calcification, and formation of incipient concretions. Although the significance of the deposits and the paleosol cannot be assessed from such fragmentary evidence, the fill may provide a potential link with a 7-meter alluvial terrace in the headwaters of North Branch as well as with older pediment wash at Ballana, Abu Simbel East, and elsewhere.

The upper or typical member of the Korosko Formation is represented in the lower wadi (to Km 1.6) by a consolidated white marl (Table 6–3, Nos. 203, 188), grading upstream into white marly sands (Fig. 6–3) (Table 6–3, No. 197) and ultimately into reddish wadi fill (Table 6–3, No. 201). The typical marls are preserved to an elevation of 134 meters

(+14 m) and vary from white to pale yellow (2.5–5 Y 8/4) in color. Moderately well stratified, they contain fine bands of silty coarse sand or clayey silt, as well as pockets of sand as much as 30 centimeters in thickness. Beds dip nileward at about 5%. Limonitic staining is quite common, frequently concentrated in sandy lenses or in rootlet hollows. Vertical and horizontal root drip or root casts are well developed in all exposures, suggesting a fair amount of vegetation. In general, these laminated marls suggest semilacustrine environments with sedimentation of nilotic silts and carbonates during Nile flood stage and with periodic inruption of wadi sands in the wake of winter rains.

The marly sands appear to be higher in the stratigraphic sequence, attaining a maximum elevation of about 139 meters (+19 m). At Km 1.44, some 60 centimeters of pale-yellow (5 Y 7–8/3), marly sands—interbedded with fine quartz gravel, coarse ironstone pebbles, and derived pieces of marl—rest semiconformably on the laminated marls. As in other exposures of the marly sands, this deposit contained several coarse flakes of Middle Paleolithic type, one of which was quite fresh. The marly sands upstream also vary from white to pale yellow, with intensive limonitic discoloration of some beds. Dendrites of pyrolusite as well as microconcretions provide further evidence of seasonal waterlogging. Although the basic deposits are fairly homogeneous, lenses or massive pockets of coarse, subrounded gravel (Table 6–1, No. 3) are common. Waterworn but unpatinated artifacts are usually associated with such detrital beds, emphasizing the torrential character of occasional wadi spates.

Between Km 2.4 and Km 2.5, the whitish marly sands grade over into a pink wadi fill (see section for Km 2.55, pp. 384–85), with a remarkably high carbonate content but with little stratification or structure. This suggests deposition into standing waters or into a transitional environment and provides a clue to interpretation of the mixed fluvial-subaqueous facies of the marly sands in general. Two suggestions can be made: (a) the clastic sediments were almost entirely a result of winter wadi accretion, with impregnation of these deposits by carbonates and some clays during the annual Nile inundation; (b) the wadi sands (or gravels under more violent runoff conditions) were deposited into shallow standing waters in the wadi embouchure during either summer or early winter. Although fine rootlet impressions and occasional root casts are present, mollusca are conspicuously absent. Whichever interpretation applies, maximum flood elevation recorded in Wadi Or is at about 142 meters (+22 m).

The ensuing period of wadi dissection cut down to bedrock in many parts of the wadi, suggesting a local Nile base level no higher than modern floodplain. Interruption by one or more intervals of wadi deposition is suggested by 125 centimeters of unconsolidated, stratified sands at Km 1.53 (Table 6–3, No. 193) (Fig. 6–15).

The Masmas Formation is recorded by only two exposures, at Km 1.53 and at Km 2.10 (Table 6–3, Nos. 196 and 208). These are pale- to grayish-brown silty sands with flood-silt structure but including an admixture of wadi sands. There is some evidence of oxidation staining; stratification is poor and is interrupted by networks of fine cracks, filled with younger sands. Although these dehydration fissures may record vertisol phenomena, they may also reflect only the fine grain-size of the deposit. Artifacts, fossils, and root drip are absent. Maximum elevation is only about 135 meters (+15 m).

The fragmentary nature of the Masmas deposits is well illustrated by a more or less complete stratigraphic section at Km 1.53 (Fig. 6–15). From base to top:

a) Over 110 cm. White marl (Table 6–3, No. 188) with inclusions of light-olive-gray (5 YR 6/2), sandy marl and bands of oxidized, reddish-yellow (7.5 YR 6–8/6), coarse sand. Dense fine rootlet network and some large, horizontal root casts. Radiocarbon date 25,250 B.C. ± 950 (I–2061). Korosko Formation.

b) 125 cm. Pink, stratified coarse sand (Table 6–3, No. 193), dipping 15% to north, with sandstone detritus in upper part. Very coarse, subangular blocky structure.

c) 130 cm. Wedge of grayish-brown, inhomogeneous silty sand (Table 6–3, No. 196), inclined at 65% to north. Medium, angular, blocky structure with crack network containing light-yellowish-brown (10 YR 6/4), silty coarse sand. Masmas Formation.

d) 40 cm. Reddish-yellow, stratified, gravelly coarse sand (Table 6–3, No. 191). Coarse, subangular blocky structure. Disconformable with *c* but conformable with *e*. Part of Malki Member, Ineiba Formation.

e) 150 cm. Current-bedded, brownish-yellow, silty coarse sand (Table 6–3, No. 189), with abundant limonitic staining. Coarse, subangular blocky structure. Part of Malki Member, Ineiba Formation.

f) 100 cm. Reddish-yellow, stratified, gravelly coarse sand (Table 6–3, No. 192). Coarse, angular blocky structure. The pebbles are subangular ironstone of coarse grade. Disconformable with *e* but conformable with *g*. Part of Sinqari Member, Ineiba Formation.

g) 120 cm. Reddish-yellow, current-bedded, coarse sand (Table 6–3, No. 207) with gravel pockets or lenses. Coarse, angular blocky structure. *Zootecus insularis* and a crude waterworn and patinated ironstone flake of Middle Paleolithic type. Part of Sinqari Member, Ineiba Formation. Topmost 20 centimeters slightly rubefied.

Following a period of wadi cutting that removed all but a few traces of the Masmas Formation, wadi alluviation was initiated with deposition of the Malki Member of the Ineiba Formation. The typical basal deposits include 20 to 40 centimeters of subangular-to-subrounded gravels with a sandy matrix (Table 6–3, Nos. 191 and 194). These are semiconformably overlain by 120 to 150 centimeters of silty sands or coarse sands (Table 6–3, Nos. 189, 195, and 198). At Km 1.44, the beds are almost deltaic in character, inclined at 8% to 17% nilewards at an elevation of 125 meters (+5 m). Up to absolute elevations of 133 meters, this unit is consistently charged with limonite, giving a multicolored appearance averaging a strong brown (7.5–10 YR 5–6/6). At higher elevations, the oxidation staining decreases, colors change to reddish yellow (5–7.5 YR 7/6), and calcareous concretions about 1.5 centimeters in diameter are common. Oxidation phenomena disappear between Km 2.3 and Km 2.4 at an elevation of about 140 meters (+20 m). Colors then remain reddish yellow (5 YR 6–7/6).

The limonitic discoloration marks a classical oxidation horizon pertaining to a seasonally waterlogged floodplain soil (a Rambla in the sense of Kubiena, 1953: 120). Considering the increasing calcification and decreasing oxidation with elevation, there can be no doubt that wadi alluviation was contemporary with a high Nile stage. The local deltaic beds at 125 meters corroborate this interpretation. Nile floods appear to have attained 140 meters, whereas mean low-water level was somewhere between 125 meters and 133 meters. The suggested flood amplitude of 7 to 15 meters compares with a nineteenth century level of 7.5 to 9 meters at Aswan. The resulting seepage and seasonal fluctuations of the groundwater table in relation to the Darau stage explain these phenomena satisfactorily.

Although the lower Ineiba unit forms a +3.5 meter fill in the middle course of North Branch today, there is reason to believe that much of it has been eroded in the lower wadi. A terrace fragment of semiconsolidated, reddish-yellow, gravelly coarse sand at +7.5 meters elevation at Km 1.28 may pertain to the original wadi infilling of this stage. Although molluscs are absent and other organic evidence is scarce, corroded, partly waterworn, ironstone flakes of Middle Paleolithic type are numerous and are frequently found *in situ*. These are badly oxidized when found in the P-horizon, but patination is entirely absent, even where no groundwater oxidation is evident. More or less freshly struck artifacts were obviously embedded in the fill, and their abundance suggests that Middle Paleolithic occupation of the wadi may have preceded alluviation only briefly.

In the central and upper wadi, a zone of calcareous concretions and root drip is apparent at the top of this fill, indicating secondary calcification *per ascensum* during or after the terminal stages of sedimentation. A type section for the two members of the Ineiba Formation, and for related soil zones, is available at Km 2.55 (Figs. 6–14, 6–16). From base to top, four disconformable units can be identified:

a) Over 90 cm. Pink, semiconsolidated, calcareous coarse sands (Table 6–3, No. 201), homogeneous except for dispersed quartz granules in uppermost 40 centimeters. Inconspicuously stratified, with coarse, angular blocky structure. Wadi fill equivalent to Korosko Formation.

b) 120 cm. Reddish-yellow, semiconsolidated, coarse sands (Table 6–3, No. 198) with a 10- to 20-centimeter band of subangular, coarse ironstone and medium quartz gravel at 10 centimeters from base. Stratified, with coarse, angular blocky structure. Several waterworn, corroded artifacts of Middle Paleolithic type are found within this gravelly stratum. The top 40 to 45 centimeters of this bed show secondary calcification, which increases towards the top and concentrates around vertical root drip and hollow rootlet zones. Malki Member, Ineiba Formation. The soil zone is of Calcorthid type (G. D. Smith *et al.*, 1960: 157–58) and suggests arid conditions but, nonetheless, more rainfall than at present. The formal designation Adda Soil Zone is proposed.[3]

c) 45 cm. Light-reddish-brown, semiconsolidated, coarse sands (Table 6–3, No. 200) with dispersed coarse, subangular ironstone pebbles (Table 6–1, No. 5). Stratified with coarse, subangular blocky structure. *Zootecus insularis* shells are abundant, and one crude, patinated ironstone flake was found *in situ*. The uppermost 30 centimeters are noticeably rubefied (5 YR instead of 7.5 YR), and the X-ray diffractogram shows kaolinite, although there is no distinctive textural or chemical change apart from probable decalcification. Sinqari Member, Ineiba Formation. The Omda Soil Zone, as it is designated,[4] is an incipient red paleosol of the type already described from other sections in southern Egypt.

d) 100 cm. Reddish-yellow, current-bedded, coarse sand (Table 6–3, No. 199), grading upwards into a subangular, coarse ironstone gravel. Structure is coarse, angular blocky. Forms a + 3.2 meter terrace grading laterally into coarse, detrital slope wash. Shaturma Formation.

Following a period of erosion (floodplain level below 123 m) and development of the Adda Soil—exact sequence unknown—the Upper Member of the Ineiba Formation was inaugurated by 20 to 70 centimeters of reddish-yellow, gravelly coarse sand (Table 6–3, No. 192). Limonitic staining is evident in the lower part of this basal gravel, to a maximum elevation of 133 meters (+ 13 m). The inference is modest seasonal

---

3. After Gebel Adda, the ruined Medieval town south of the mouth of Wadi Or.
4. After Medinet Omda, another Medieval ruin lying north of the mouth of Wadi Or, but incorrectly substituted for Gebel Adda on some maps.

waterlogging during the early stages of sedimentation, presumably contemporary with the Arminna substage. Although turbulent local alluviation was dominant, presumably in winter, Nile floodwaters must have seeped into the wadi fill during the summer. Equivalent nilotic deposits are recorded in the main branch of Wadi Or to a maximum elevation of 133 meters (see Wendt, 1966: Fig. 13).

Overlying the basal gravels of the upper Ineiba unit semiconformably are 50 to 200 centimeters of reddish-yellow, coarse sands (Table 6–3, Nos. 200 and 207), forming the bulk of the upper Ineiba beds. In the lower wadi, these beds are last seen as a $+2.5$ meter terrace near Km 1.2, at an absolute elevation of 123 to 125 meters. Further upstream, the relative elevation increases rapidly to $+3$ or $+5$ meters. The total lack of pseudo-gley phenomena and the oversteepened gradient suggests that Nile floodplain level was certainly not above 125 meters. This post-Arminna stratigraphic position is corroborated by the radiocarbon date of 6940 B.C. $\pm$ 160 (Y–1377) (Stuiver et al., 1967) from Zootecus shells collected from the top meter or so of this unit at Km 1.92 (Fig. 6–12B). It is probable that the sandy fill of the top of the Sinqari Member is of early Holocene age. Derived Middle Paleolithic artifacts, all well patinated, were found in situ at Km 1.30, Km 1.53, and Km 2.55. A single, patinated flint flake of Late Paleolithic type was found in or on the surface of this bed at Km 1.44.

All of the deposits of the Ineiba Formation, but particularly the basal gravel units, suggest slightly, but distinctly, greater wadi flow, turbulence, and competence than the contemporary wadi. The reddish Omda Soil, described above, further indicates a period of biochemical weathering.

The final alluvial unit of the North Branch succeeds another interval of wadi downcutting. It is the Shaturma Formation, grading from 50 to 100 centimeters of coarse sands and sandy gravels (Table 6–3, No. 199) in the wadis to coarse detritus on the pediment slopes (Fig. 6–17). As a wadi terrace, this unit is only recognizable upstream of Km 2.1, and the relative level increases a little upstream, from $+3.2$ meters to $+3.5$ meters. This, and the sheetflood detritus, suggest a climatically induced aggradation, a period of occasional heavy rains. The beds are completely sterile.

The sequence of late Pleistocene deposits outlined in the preceding pages applies only to the lower half or so of North Branch. At Km 2.69 it is no longer possible to differentiate between the reddish-yellow fills that predate the Omda Soil. Instead, these deposits form a simple $+7$ meter terrace coalescent with the wash on a broad, adjacent pediment surface.

A similar, uniform fill can be followed to the uppermost wadi, where it is last present as a dissected +7 to +8 meter terrace (Figs. 6–13, 6–17). These deposits are sterile and fairly homogeneous, interrupted by occasional detrital bands. Beds are inclined at 3% or 4%, diagonally towards the wadi center. Truncated reddish $(B)C$-profiles, developed in the sandstone bedrock (Fig. 6–18), can repeatedly be seen underneath this wash. This suggests that the +7 meter fill owes its origin to stripping of soils and weathered sandstone, such as may have developed on the uplands and pediment surfaces prior to the late Pleistocene.

Several denudational terraces, with a little detrital wash, are cut into the 7-meter fill of the upper wadi (between Km 3.1 and Km 4.0). Of these, a +3.2 to +3.7 meter level can be traced downstream and correlated with the Shaturma Formation. A low, +0.5 meter level may be of fairly recent age. A higher level at +4.5 to +5.0 meters is generally present in the upper wadi but cannot be followed downstream. It may be equivalent to the upper unit of the Ineiba Formation. Consequently, the +7 to +8 meter fill of the wadi headwaters may either pertain to the lower unit of the Ineiba Formation or to the alluviation simultaneous with the Korosko Formation. We favor the latter view. In any event, most of the later wadi fill was simply derived from this older accumulation of reddish-yellow, coarse sands that had been washed together in the upper catchment zone of the wadi or dispersed over the adjacent pediment surfaces.

In retrospect, the Wadi Or (North Branch) sequence can be summarized as follows:

a) Major wadi activity terminating with aggradation of up to 5 meters of Wadi Floor Conglomerate. Nile floodplain below modern level of 120 m.
b) Erosion?
c) Deposition of lower member of Korosko Formation, including marls and the oldest flood silts which can be attributed to a summer-flood regime. Maximum elevation 131 meters.
d) Wadi dissection.
e) Accumulation of wadi wash; details unknown.
f) Biochemical weathering and minor rubefaction with development of a 40-centimeter (B)-horizon of rudimentary kind and secondary calcification.
g) Wadi dissection?
h) Deposition of the "classical" Korosko Formation, beginning with a fine-grained marly unit (up to 134 m) and terminating with coarse,

marly sands to about 142 meters. The latter unit, at least, was contemporary with general alluviation of wadi sands upstream and mixed fluvial-subaqueous sedimentation downstream. First record of geologically stratified Middle Paleolithic artifacts—unpatinated and partly fresh.

i) Wadi dissection to below modern floodplain level, interrupted by one or more intervals of wadi alluviation. Details unknown.
j) Deposition of Masmas flood silts to at least 135 meters elevation. Wadi activity no greater than at present.
k) Wadi dissection.
l) Aggradation of Malki Member, Ineiba Formation, possibly forming a 7.5-meter terrace downstream. This terrace is graded to a floodplain level of 140 meters (Darau substage).
m) Secondary calcification with development of a desert soil of Calcorthid type (Adda Soil Zone), followed or preceded by wadi dissection.
n) Aggradation of Sinqari Member, Ineiba Formation, forming a terrace of +2.5 meters downstream, +3 to +5 meters upstream. Basal gravels aggraded in response to a Nile level of 133 meters (Arminna stage), with floodplain subsequently lower than 125 meters. Alluviation terminated during seventh millennium B.C. *Zootecus* proliferations.
o) Biochemical weathering: minor rubefaction and formation of reddish paleosol, with an incipient (B)-horizon of 30 centimeters depth, decalcification, and formation of kaolinite (Omda Soil Zone).
p) Wadi dissection.
q) Aggradation of Shaturma Formation, detrital colluvial wash laterally conformable with 3.5-meter wadi terrace. Significant sheetflood activity. Base level unknown.
r) Wadi dissection.
s) Cutting of 0.5-meter denudational terrace upstream.
t) Wadi dissection and redeposition of fill.

Late Pleistocene Deposits between Abu Simbel and Amada

North of Abu Simbel, late Pleistocene deposits are first found in the southern part of the middle Pleistocene bedrock channel (Fig. 5–1). Best developed is a pale-brown (10 YR 6/3), unconsolidated, medium-sandy silt to 131 meters (+11.5 meters), representing the Arminna substage. Further away from the Nile, there are massive, very-pale-brown (10 YR

8/3), semiconsolidated, coarse sands with root drip, to an elevation of 141.5 meters (+22 meters). Embanked against a terrace of wadi fill, these sands very probably belong to the Korosko Formation.

A little south of the ruined mud-brick structure labelled Arminna Temple on the topographic maps, there are several small but significant exposures of late Pleistocene silts (Figs. K–1 through K–4). The Masmas Formation is represented by typical dark flood silts to about 133 meters (+15.5 meters); the Darau Member by brown, silty coarse sand to 140.5 meters (+22.5 meters) (Table 6–4, No. 156), intergrading with light-yellowish-brown, medium-grade sand of eolian origin (Table 6–4, No. 153) along the former riverbank. Several patches of sandy silts belonging to the Arminna substage are found on top of an old rock platform at about 130 meters. A good type section for the Arminna Member of the Gebel Silsila Formation is provided by a deep pit cut into a sequence of late Pleistocene deposits, near the western margin of the 130-meter platform, 500 meters south of Arminna Temple (Fig. 6–19). From base to top, the following beds are exposed.

a) Over 45 cm. Pale-brown, stratified, consolidated, sandy silt (Table 6–4, No. 152) rich in colloidal silica, with some limonitic staining. Coarse, angular blocky structure. Base probably resting on bedrock at 120 centimeters, mixed with angular detritus. Dispersed subrounded pebbles (Table 6–1, No. 2) of rather flattish, ferruginized sandstone in upper part. Korosko Formation. Disconformity.

b) 10 to 15 cm. Coarse quartz and ironstone gravel with salt-cemented matrix of light-yellowish-brown, silty coarse sand (Table 6–4, No. 164). Abundant mollusca, particularly in upper half: *Cleopatra bulimoides, Corbicula fluminalis* ssp., *Unio willcocksi,* and *Etheria elliptica*. Patinated, slightly waterworn, Late Paleolithic artifacts made from exotic pebbles, together with waterworn, ironstone flakes of Middle Paleolithic type.

c) 5 to 20 cm. Eluviated cobble conglomerate of sandstone and ironstone, coated with salts. *Cleopatra* shells and artifacts are concentrated in the contact zone of *b* and *c*, while oyster shells are attached to the individual rocks. Radiocarbon determination from such an attached oyster shell: 9140 B.C. ± 200 (Y–1526) (Stuiver *et al.,* 1967). Disconformity.

d) 260 cm. Pale-brown, well-stratified or current-bedded, alternating beds of semiconsolidated clayey silt, sandy silt, or silty sand. Mica, hornblende, and pyroxenes abundant. Including:

i) 60 to 75 cm. Horizontally-bedded, medium-sandy silt with nodular concretions 2 to 5 centimeters in diameter. Coarse, angular blocky to platy structure.

ii) 60 cm. Backset, thin-bedded (5% to 35% dips) sandy silt and clayey silt (Table 6–4, No. 252), with incipient 2 to 3 mm concretions.

SEDIMENTS OF EGYPTIAN NUBIA · 309

Table 6–4. Sediment characteristics of deposits between Arminna Temple and Wadi Umm el-Hamid

Sample number	Formation/Member	Color	Texture	Sorting	CaCO$_3$ (%)	pH
			Arminna Temple			
254	Arminna	(10 YR 6/3)	Sandy silt	Poor	3.9	8.20
253	Arminna	(10 YR 6/3)	Medium-sandy silt	Moderate	7.4	8.00
252	Arminna	(10 YR 6/3)	Clayey silt	Moderate	3.4	7.80
164	Arminna	(10 YR 6-7/4)	Silty coarse sand (matrix)	Moderate	2.6	7.05
153	Darau (eolian)	(10 YR 6/4)	Medium sand	Good	1.3	8.30
156	Darau	(10 YR 5/3)	Silty coarse sand	Moderate	1.8	8.20
152	Korosko	(10 YR 6/3)	Medium-sandy silt	Poor	8.0	7.60
			Arminna West			
259	(Older wash)	(5 YR 6/6)	Silty medium sand	Moderate	0.0	7.80
260	(Older wash)	(7.5 YR 6/4)	Clayey coarse sand	Moderate	3.9	7.60
			Tushka West			
250	Darau	(10 YR 6/3)	Coarse sand	Moderate	0.9	7.80
249	Darau	(10 YR 6-7/3)	Coarse sandy gravel	Moderate	0.7	7.90
248	Masmas	(10 YR 5-6/2-3)	Clayey silt	Moderate	2.7	7.30
			Amada Temple			
148	Arminna	(10 YR 6/3)	Medium-sandy silt	Poor	3.2	7.55
			Wadi Umm el-Hamid			
222	Korosko	(5 Y 8/2)	Silty medium sand	Poor	1.1	7.55

Coarse, angular blocky to platy structure. Abundant root hollows—frequently filled with gypsum—and subfossilized catkins.

iii) 60 cm. As *i* (Table 6–4, No. 253), with limonite and pyrolusite stains, root hollows, and catkins.

iv) 15 to 20 cm. Gently inclined (5.5% nileward dip), silty medium sand. Very coarse granular structure. With root casts, root hollows, and some catkins.

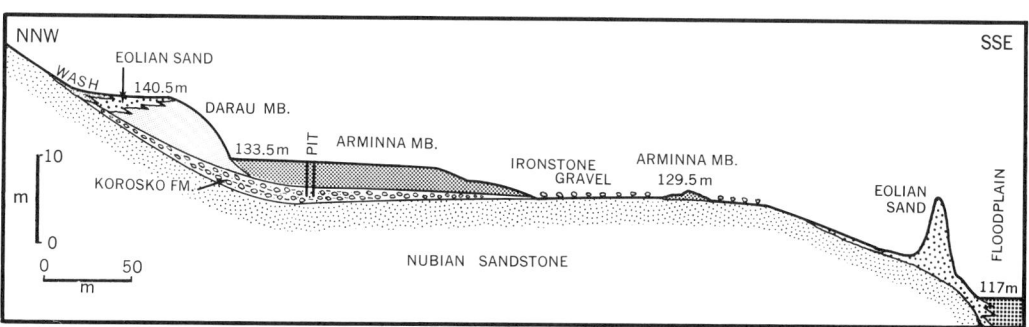

Fig. 6–19. Late Pleistocene deposits 500 meters south of Arminna Temple. The pit exposes the section described on pp. 308–10. Map: UW Cartographic Lab.

v) 60 cm. Horizontally-bedded, clayey silt (Table 6–4, No. 254) with calcareous salt concretions (3 mm) and limonite and pyrolusite staining. Coarse, angular blocky to platy structure. Some rootlet hollows. Maximum elevation 133.5 (+15 m).

Whereas unit *d* marks a conformable sequence of flood silts and levee-type beds, *b* and *c* represent a transgressive gravel deposited in the Nile bed and derived from local detritus. The radiocarbon date approximates the beginning of the Arminna substage, and beds *b* through *d* record a typical change of facies. Beds of the Darau substage are intercalated between *a* and *b* further upslope. The catkins so clearly preserved in the clayey beds are tassel-shaped flowers belonging to trees of the order *Amentaceae*, which includes alder, birch, casuarina, poplar, walnut, and willow. Pending firm generic identification, they can probably be assigned to the only indigenous Egyptian genus of this group, the willow or *Salix* sp.[5]

Further north, at Arminna West, the late Pleistocene silts have been largely deflated, and concentrations of *Corbicula fluminalis vara* at 144 to 148 meters (+27 to +31 m) may have originally been associated with pockets of brown, sandy silt found locally preserved in the general area. Late Paleolithic artifacts, possibly deflated from beds of the Arminna substage, are concentrated near the 133-meter contour. Preservation is excellent on the eastern bank of the Nile, where Masmas flood silts choke a bedrock channel of the Wadi Korosko stage. These silts attain about 140 meters, and average channel width is almost 700 meters. Silts of the Darau substage barely reach up to the saddle point of this channel (139 m), while the Kibdi substage is represented by pale-brown, sandy silts around the village of Arminna, up to 125 meters.

The Darau substage is particularly well developed at Tushka West, masking an older bedrock platform at about 135 meters (+17.5 m) (Fig. 5–1). Two distinct ridges of bed-load material are evident, one about 0.5 kilometers west of the floodplain edge in 138 meters and a second at 1.1 kilometers distance in 139 meters (+21.5 m). The first of these consists of sterile, gravelly coarse sand, the second of gravelly coarse sand with considerable evaporite and abundant Late Paleolithic flakes and blades. An informative section, already illustrated by Sandford and Arkell (1933: 41, Fig. 8), was studied in Wadi Tushka West, which emerges on the floodplain near the steamer station. From base to top, several disconformable beds are exposed.

5. The modern indigenous Egyptian willow is *Salix subserrata* Willdenow [= *safsaf* Forskal] (Täckholm, 1956: 452).

a) Over 5 m, base not seen. Pale- to grayish-brown, stratified, clayey silt (Table 6–4, No. 248), with coarse, angular blocky to platy structure. Dehydration cracks break up the beds into a geometric network, with fissures up to 1 cm wide and filled with sandy silt and small concretions. Charged with black limonite and some pyrolusite. Sterile except for rare rootlet zones. Masmas Formation. The eroded surface attains 125.5 meters (+ 8 m).

b) 35 to 45 cm. Stratified, medium quartz gravel (Table 6–1, No. 4), with occasional pebbles of cherty flint and some metamorphic rocks. The matrix is a coarse sand. Rich in waterworn diminutive flakes and cores—both in ironstone and in flint—of Late Paleolithic type. Blades, burins, and geometric forms are absent.

c) 350 cm. Pale-brown, stratified, coarse sands with dispersed quartz pebbles, and rich in mica, hornblende, and pyroxenes. Loose, single-grain structure. *Corbicula* shells, bone fragments, and partly waterworn flints of Late Paleolithic type.

Beds *b* and *c* belong to the Darau substage and can be followed 750 meters upstream, where they converge into a ridge of sandy gravel approximately 2.5 meters thick, at 139 meters (+ 21.5 m) elevation, resting on bedrock (Fig. 6–6). The Arminna substage is poorly represented in this area, but the Kibdi substage is recorded by low ridges of pale-brown, medium-sandy silt at 123.5 meters (+ 6 m). About 2 kilometers northwest of the steamer station, beds of the Korosko Formation are also preserved to an elevation of about 146 meters (+ 28 m) as a 7-meter wadi terrace in a minor drainage channel. These deposits range upwards from a basal, light-gray, marly sand with rootlet zones to pale-brown, silty sands with *Corbicula fluminalis vara* and sandstone detritus. Waterworn ironstone artifacts of Middle Paleolithic type litter the surface.

An outstanding sequence of late Pleistocene deposits, oriented as longitudinal ridges, can be seen south of the west-bank wadi emerging at Masmas Station. We had no opportunity to study this sector in the field, but the Combined Prehistoric Expedition subsequently carried out excavations there. Another detailed section, here designated as Khor Hamra, is exposed in the west-bank wadi 4.5 kilometers southwest of Ineiba Station, opposite Nag Shurbagi. The Cairo University excavated an X-Group cemetery here, providing deep sections for a composite sequence (Fig. 6–20). From base to top:

a) Over 90 cm. Pink (7.5 YR 6/4), semiconsolidated, coarse sands with dispersed medium quartz gravel. Well stratified, with coarse, subangular blocky structure. Channel facies of Korosko Formation.

b) 350 cm. Light-brownish-gray (2.5 Y 6/2), consolidated, clayey silt or clayey marl, interspersed with veinlike network of pale-brown (10 YR 6/3–4), sandy silt. Banded stratification dips 1% nileward on east, reversing its dip 1%

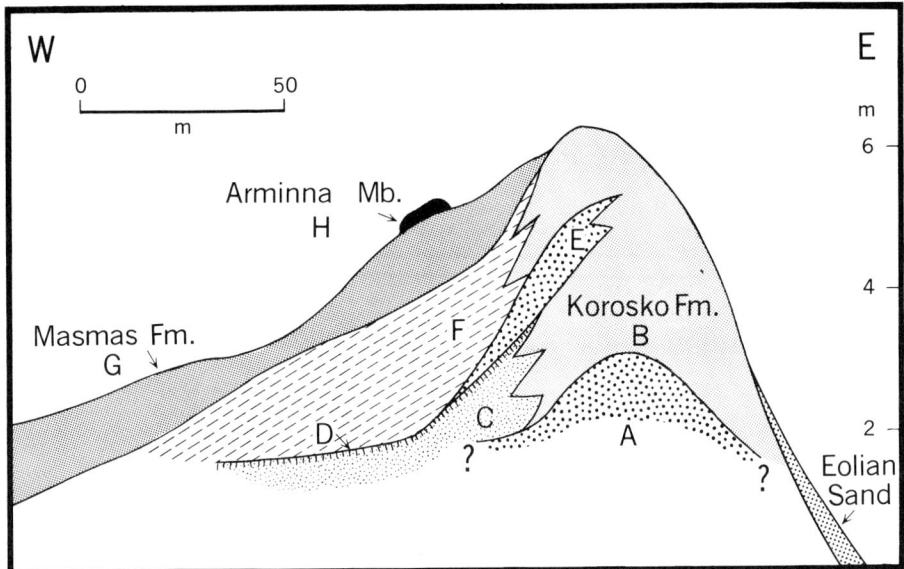

Fig. 6–20. Khor Hamra section, mixed nilotic and wadi facies intersected at the mouth of a wadi 4.5 kilometers southwest of Ineiba Station, opposite Nag Shurbagi. Subfacies of Korosko Formation are described on pp. 312–13. Vertical exaggeration 16 times. Map: UW Cartographic Lab.

to 3% towards the wadi. Coarse, angular blocky structure. Flood silt facies of Korosko Formation, to 126 meters (+ 11 m). Further upwadi, these beds are laterally interdigited with units c to f, in part recording contemporary wadi deposits.

c) Over 90 cm. Brown to very-pale-brown (10 YR 5/3, 7/3–4), semiconsolidated, sandy silt to silty medium sand with subrounded, coarse sandstone gravel. A mixed wadi-nilotic facies.

d) 15 to 20 cm. Semicemented duricrust developed in c, rich in secondary, cryptocrystalline calcite, colloidal silica, and pyrolusite dendrites and with occasional limonitic staining. This suggests a period of low Nile floods and little or no wadi activity, with occasional spilling over of floodwaters charged with carbonates.[6]

e) 50 cm. Lense of coarse, nilotic sand as unit a.

f) 200 cm. Alternating 5- to 15-centimeter-thick beds of:

6. Long-term inundation would almost certainly have led to resolution of the crust, with formation of concretions much further down within unit c by capillary ascent during the low-water stage.

i) Very-pale-brown (10 YR 7/3), unconsolidated, coarse sands with some sandstone detritus. Well stratified, with coarse, subangular blocky structure. Considerable root drip. *Corbicula* fragments indicate wadi wash incorporating older nilotic materials.

ii) Pale-brown (10 YR 6/3), unconsolidated clayey silt. Stratified, with fine, angular blocky structure. Flood silts of Korosko Formation. Beds *b* and *f* terminate the complex beds of the Korosko Formation and are disconformably overlain by

g) 150 cm. Grayish-brown (10 YR–2.5Y 5/2), semiconsolidated, clayey silt. Inconspicuously stratified, with coarse, blocky to prismatic structure and well-developed slickensides. Ped faces are stained with pyrolusite and are marked by dendrites of evaporite. Crack fillings consist of silty coarse sand. Masmas Formation with an epigenetic vertisol. The surface attains 125.5 meters (+ 10.5 m), is veneered by coarse wadi wash—including waterworn artifacts of different types—and is disconformably overlain by

h) 30 cm. Very-pale-brown (10 YR 7/4), unconsolidated, gravelly coarse sand, including microconcretions, rounded quartz gravel, angular sandstone pebbles, and slightly waterworn Late Paleolithic blades. Attains 124.5 meters (+ 9.5 m). Arminna Member.

The Khor Hamra section gives a unique picture of varying nilotic facies within the Korosko Formation, also illustrating the interbedding of wadi and nilotic deposits. Bed *f* shows rather clearly that the wadi wash includes redeposited nilotic materials, while the nilotic flood silts are not contaminated with local wash. In other words, the wadi and nilotic facies are seasonally out of phase. The accelerated wadi activity contemporary with the Korosko Formation is conspicuously absent from the Masmas unit but is once more evident in the deposits of the Arminna substage.

The irrigated silt plain to the west and southwest of Ineiba was originally cut into bedrock during the middle Pleistocene. A complex filling of very-pale-brown, marly sands (Korosko Formation) and brown, clayey silt (Masmas Formation) is veneered by pale-brown, sandy silt (Darau and Arminna substages). Because deep sections or well profiles were lacking, it was not possible to work out a detailed stratigraphy here. Of interest is a reddish-yellow (7.5 YR 8/6), unconsolidated wadi wash of coarse sand, exposed in a shallow wadi 1 kilometer southwest of Ineiba Station. These sterile deposits are current-bedded, forming a fanlike terrace at 124 meters (+9 m), intercalated between the Masmas and Arminna deposits. This is a good example of the Malki Member of the Ineiba Formation.

Shallow mantles of nilotic flood silts are irrigated on the pediment surfaces west of Ibrim and Qatta,[7] where they are veneered by wadi wash. Practically all of these silts belong to the Masmas Formation.

About 1 kilometer east of Amada Temple there is a low platform veneered by a lag of medium-grade quartz and coarse ironstone gravel (Fig. 5–1). This gravel contains frequent, waterworn Late Paleolithic artifacts and extends to the 128 meter contour (+ 15 m) or higher. Possibly it represents the Darau substage. Perched on top of this platform are several patches of typical Arminna flood silts to 127.5 meters (+ 14.5 m). These consist of pale-brown (10 YR 6/3), unconsolidated, sandy silt (Table 6–4, No. 148) locally interbedded with clayey silt. A variety of concretions and Late Paleolithic artifacts are concentrated among the surface lag.

In overview, the late Pleistocene deposits between the Abu Simbel gorge and the Korosko bend do not provide detailed stratigraphic sequences such as at Adindan or Ballana and in Wadi Or. A number of the sections exposed contribute valuable information on facies development within the different formations, however, and, at the same time, the stratigraphic units as such can be readily reconciled with the framework originally proposed.

Stratigraphy of Late Pleistocene and Holocene Deposits between Korosko and el-Madiq

Late Pleistocene sediments are very poorly represented on the western bank of the Nile between Korosko and el-Madiq, in part as a result of deflation, in part due to masking by an almost ubiquitous mantle of eolian sand. Complex sequences of nilotic and wadi deposits are well developed in the wadi embouchures of the east bank, however. Stratigraphic sequences were worked out for Wadis Korosko, Abu Sureih, and Umm el-Hamid, but good exposures are also present in a number of other local wadis.[8]

---

7. A cave site (Catfish Cave), located within a wadi in the Qatta East area, was excavated by W. E. Wendt in 1964–65 (Wendt, 1966). Here, at least 2.5 meters of nilotic silt predate a hearth—radiocarbon dated at *ca.* 7000 B.P.—which was subsequently buried by wadi deposits. The absolute elevation of the silts is 132 m (+ 17 m).

8. From south to north: Khor Tarrat, Wadi el-Sinqari, Wadi el-Dakhlaniya, Khor el-Aqaba, Wadi Shaturma, Wadi el-Sebua, Khor Ibeidalla, and Wadi Umm Gibara (see Figs. K–4 through K–6).

In Wadi Korosko,[9] the typical basal conglomerate of the late Pleistocene is preserved along the wadi floor from Km 3 (measured from the Nile bank) to well upstream of the Wadi Guhr el-Daba confluence (Fig. 6–21B). This Wadi Floor Conglomerate is semicemented with a matrix of pink (5 YR 7/4), coarse sand, embedding coarse-to-cobble-grade gravel of subrounded ironstone. Original thickness may have been as much as 5 meters, and these conglomerates are distinctly younger than the middle Pleistocene gravels of Wadi Korosko.

The Korosko Formation rests disconformably on the conglomerates, a very-pale-brown (10 YR 7–8/4), semiconsolidated, marly coarse sand rising to at least 137.5 meters elevation (24.5 m above floodplain). The deposits are last seen downstream near Km 2.3, where they form a terracelike feature at 130.5 meters (+17.5 m), consisting of white (10 YR 8/2), well-stratified and consolidated marl (Fig. 6–21A). At the same locality, there is a lower terrace at 124 meters (+11 m). The material is a light-gray (10 YR 7/2) to pink (7.5 YR 7/4), stratified, consolidated medium-sandy marl. Possibly this is a semilacustrine facies of the Masmas Formation, which is developed in more classical form at Korosko Station. There we find alternating stratified beds of (*a*) pink (7.5 YR 7/4), unconsolidated, marly medium sand, (*b*) brown (10 YR 5/3), unconsolidated, medium-sandy silt, and (*c*) pink (5 YR 7/3), semiconsolidated marl. Maximum local elevation is over 124.5 meters (+11.5 m).

Following a period of erosion, over 2.5 meters of coarse, subrounded, ironstone gravel were laid down with a consolidated matrix of light-reddish-brown (5 YR 6/4), coarse sand. Resting on top of this are up to 5 meters of light-reddish-brown, semiconsolidated, silty coarse sand with dispersed gravel. Derived conglomeratic fragments of the preceding wash are found in these reddish sands, suggesting an intervening period of calcification. Proliferations of *Zootecus* and a dozen small flakes, many

---

9. Sandford and Arkell (1933: 42–43) allude to late Pleistocene deposits in Wadi Korosko and describe a sequence similar to our own, with a basal conglomerate, from Khor Abu Uruq. Further north, they provide no information except a vague statement that at Khattabab, on the west bank—a locality not found on the 1:100,000 map series—"bones of *Hippopotamus* were added to the usual molluscan fauna and implements" (1933: 42, Note 10). Said and Issawy (1965: 14) indicate the presence of "7 m and 1.6 m clay terraces" in Wadi Korosko and at el-Madiq (Nag el-Sabkha) with "many small and highly polished Upper Paleolithic implements" at the first locality and "small, highly polished Mesolithic (?) implements" at the latter. We know of no such implement-bearing deposits in Wadi Korosko, and, as indicated already, altimetric correlations without regard to sediments or stratigraphy are debatable at best.

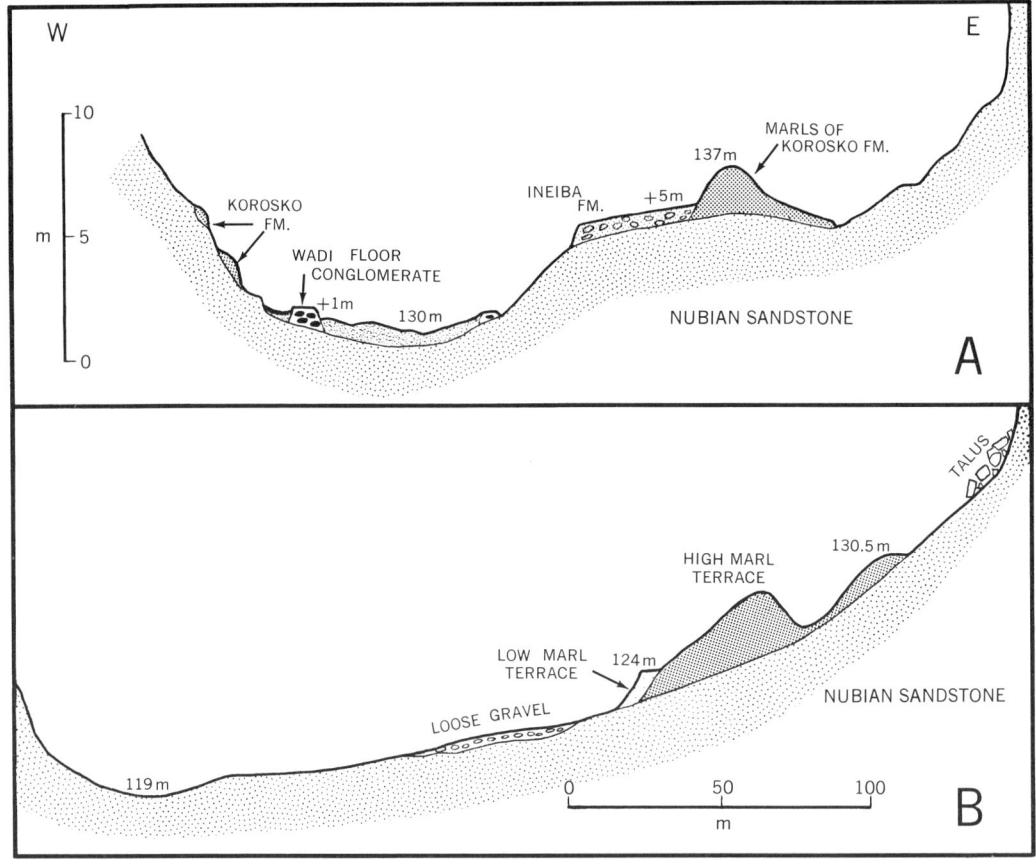

Fig. 6–21. Late Pleistocene deposits in Wadi Korosko. A: at Km 2.4; B: just above Wadi Guhr el-Daba confluence (Km 5.4). Map: UW Cartographic Lab.

with faceted platforms, were found *in situ* in the upper half of this wash at Km 5.4 (near transect Fig. 6–21B). These Wadi Korosko deposits are typical of the Ineiba Formation (both Malki and Sinqari Members?). The surface appears to be rubefied in many sections.

A rather similar sequence can be observed 35 kilometers downstream in Khor Abu Sureih, near Nag Ibeidalla. Here, a Wadi Floor Conglomerate of identical facies forms a +3 to +4 meter wadi terrace, disconformably overlain by the massive, white (5 Y 8/1, with abundant yellow oxidation staining), semiconsolidated, medium sands of the Korosko Formation. Exhibiting stem hollows (0.5 cm in diameter) with fossilized woody material, these stratified, semilacustrine beds have been eroded to form a 5- to 7-meter wadi terrace. Semicemented ferruginous bands of hematite and limonite, varying from red (10 R 5/6) to reddish yellow (7.5 YR 6/6), support a subaqueous depositional environment. These beds can be followed upstream to a maximum elevation of about 137 meters (+28 m), where they grade into light-brown (7.5–10 YR 6/4), coarse sands with some atypical deltaic bedding, including inclined pebble bands deposited into standing waters. In addition to limonitic staining, these coarser beds include calcareous salt concretions (up to 1 cm in diameter). Gravel lenses appear to be intercalated with the lower beds of the Korosko Formation at several localities.

The Masmas Formation may be represented in Khor Abu Sureih by a section of 5-meter wadi terrace cut into the older sands. Some 50 centimeters of white (10 YR 8/2), semicemented, medium-sandy marl, identical in lithology to the lower marly terrace in Wadi Korosko, attain a maximum elevation of 133.5 meters (+24.5 m). But, although we are inclined to relate these beds to the Masmas Formation, this correlation is uncertain in view of the abnormal facies, the lack of direct field correlation, and the known complexity of the Korosko Formation.

The Ineiba Formation (Sinqari Member) in Khor Abu Sureih constitutes a 2-meter terrace upstream, a 6-meter terrace downstream. The material is a light-reddish-brown (5–7.5 YR 6/4), semiconsolidated, coarse sand with dispersed, coarse, subrounded ironstone gravel. *Zootecus* is locally prominent, but artifacts were not found. The upper 30 to 50 centimeters are rubefied, with a color here of 5 YR in the (*B*)-horizon, compared with 7.5 YR in the *C*-horizon of the Omda Soil. Finally, the Shaturma Formation is recorded by a 3-meter terrace upstream, generally limited to 50 centimeters of unconsolidated, coarse, subangular ironstone gravel, with a matrix of light-brown (7.5 YR 6/4), coarse sand. Except for a single shell of *Zootecus*, these deposits are sterile. Laterally, they grade into coarse colluvial wash.

In Wadi Umm el-Hamid, a basal conglomerate forms a 2.5-meter wadi terrace upstream, but the matrix is a pink (5 YR 7/4), coarse sandy silt, with a calcrete cement, some salt, and pyrolusite staining. Further downstream, this unit is intercalated with the Korosko Formation, a relationship illustrated by the following conformable section at Km 0.8, from base to top, resting on bedrock (Fig. 6–4, 6–22).

a) 50 cm. White (10 YR 8/2), semiconsolidated, coarse sand with abundant reddish-yellow (7.5 YR 8/6), limonitic staining and diffuse salts. Subaqueous facies.

b) 80 cm. Coarse to cobble, subrounded ironstone gravel with matrix as *a*, feathering out downstream, thickening to over 200 cm upstream, where it grades into a typical wadi gravel. Inclined nileward at a dip of 1% or 2%.

c) 400 cm. Very-pale-brown (10 YR 7/3–4), semiconsolidated, medium sand with lenses of coarse gravel in lower part. Limonitic staining widespread. Some salt dendrites and small rootlet zones. Upstream, intergrades with a light-reddish-brown (5 YR 6/4), silty coarse sand with gravel bands, originally forming a 6- to 7-meter wadi terrace. Maximum elevation of the lacustrine facies is 135 meters (+ 26 m), and the base is found at or below modern floodplain level. Downstream, the dominant facies is a white (5 Y 8/2), homogeneous, silty medium sand with salty and limonitic microconcretions and with fossilized plant stems (1 to 3 mm in diameter). This subaqueous deposit has the macro-aspect of a marl, and the original carbonates may well have been leached out by iron-rich wadi waters.

These deposits of the Korosko Formation were dissected prior to deposition of the Ineiba Formation (Sinqari Member), represented upstream in Wadi Umm el-Hamid by a 2.5-meter terrace with facies identical to that in Khor Abu Sureih. The same applies to the Shaturma Formation, which forms a 2-meter wadi terrace.

In retrospect, the late Pleistocene and Holocene sequence of Wadis Korosko, Abu Sureih, and Umm el-Hamid can be briefly summarized as follows:

a) Major wadi activity, terminating with aggradation of up to 6 meters of Wadi Floor Conglomerate.

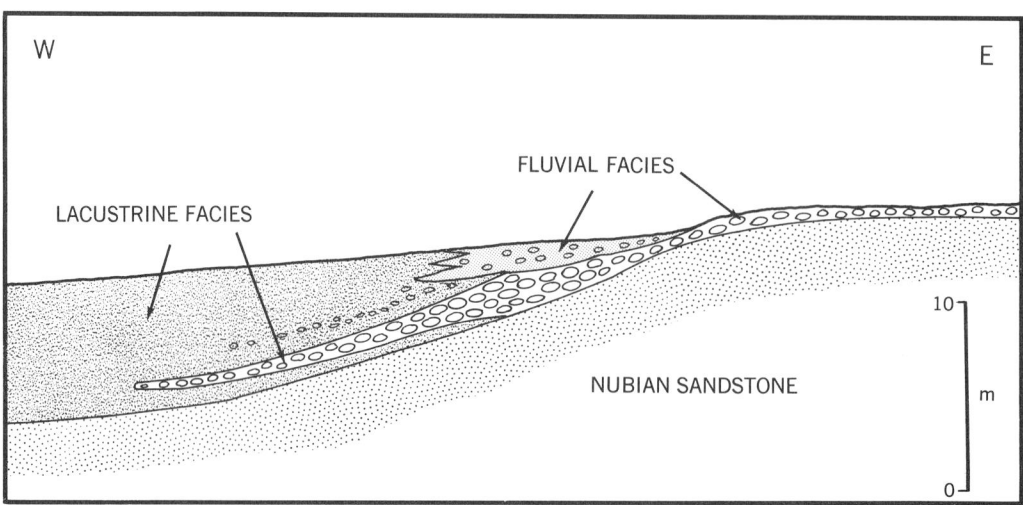

Fig. 6–22. Intergrading subaqueous and fluvial facies in lower Wadi Umm el-Hamid. Not to scale. Vertical exaggeration very approximately 6 times. Map: UW Cartographic Lab.

b) Wadi erosion.
c) Deposition of Korosko Formation (maximum elevation known: 28 m above Nile floodplain). Except for a basal bed of homogeneous nilotic beds, the lower half of this unit is intercalated with coarse conglomerates of subrounded ironstone gravel, grading upstream into true wadi terraces. Accelerated wadi activity, reaching maximum during first half of depositional phase, then waning.
d) Wadi dissection (to below + 10 m).
e) Deposition of sandy marls inside wadis (maximum elevation known: + 25 m), possibly related to Masmas Formation. In any case, typical Masmas flood silts are known to + 11.5 meters.
f) Wadi dissection (perhaps to below modern floodplain).
g) Aggradation of wadi gravels, probably Malki Member of Ineiba Formation.
h) Period of calcification (Adda Soil Zone?).
i) Aggradation of Sinqari Member, Ineiba Formation, forming 2.5- to 6-meter wadi terraces with *Zootecus* and, locally, Late Paleolithic artifacts.
j) Biochemical weathering and minor rubefaction (Omda Soil Zone).
k) Wadi dissection.
l) Aggradation of Shaturma Formation, forming 2- to 5-meter wadi terraces, laterally conformable with coarse sheetwash. Significant fluvial activity. Sterile.
m) Wadi dissection and redeposition of fill.

In general, this sequence is remarkably similar to that of Wadi Or and Khor Adindan, despite certain obvious lacunae. It consequently provides an important stratigraphic link to the Kom Ombo Plain, over 200 kilometers further downstream. Unfortunately, the relationships of the fill terraces to the former Nile floodplain levels are difficult or impossible to establish in these wadis.

Late Pleistocene and Holocene Deposits between Seiyala and Aswan

The late Pleistocene record of northern Egyptian Nubia in 1962–63 was essentially confined to the area of Seiyala and Dakka. A large number of minor, but nonetheless valuable, exposures did once exist below the 1934–1964 dam level, and several of these could still be seen along the

Nile Valley margins between el-Madiq and Seiyala, and again between Dihmit and Shellal, in October, 1962. Adams (1864) reported fossiliferous silts along the valley at Qirtassi, Kalabsha Temple, Dabud, and Shellal, outcrops that were already lost after the third raising of the dam. Sandford and Arkell (1933: 43), who must have seen considerably more along this stretch than was visible in 1962–63, give practically no information whatever on the late Pleistocene beds of northern Nubia.

The deposits at Seiyala include a veneer of medium-grade quartz gravels, with abundant but waterworn diminutive agate flakes of Late Paleolithic type, resting on an older bedrock platform at about 130 to 135 meters (23 m to 28 m above floodplain). These nilotic gravels rest on the edge of a colluvial wash with *Zootecus insularis*, exposed by a Pan-Grave cemetery (*ca.* 1700–1500 B.C.) that was excavated by the Austrian Expedition. From base to top:

a) Over 15 cm. Weathered bedrock, a pink (7.5 YR 7/4), consolidated, medium-grained sandstone, forming a truncated (*B*)*C*-horizon. Similar, truncated paleosols with light-reddish-brown (5 YR 6/4) zones of alteration can be seen in other local exposures under colluvial wash. Disconformity.

b) 60 to 80 cm. Light-brown (5–7.5 YR 6/4), semiconsolidated coarse sand with medium-grade, subangular pebbles of ferruginized sandstone (Table 6–1, No. 1). Abundant *Zootecus* shells. Crudely stratified, with medium, subangular blocky structure, this colluvial wash contains soil sediments and mantles the local 137-meter bedrock platform. Disconformity.

c) 10 to 15 cm. Pink (5 YR 7/4), unconsolidated, coarse sand with dispersed, subangular pebbles. Current-bedded, with very coarse granular structure. This alluvial wash is distinctly younger than *b*, and both beds are older than the nilotic gravels. Disconformity.

d) 5 to 15 cm. Pink (7.5 YR 7/4), coarse sand with dispersed pebbles. Loose, single-grain structure. Mixed colluvial-eolian deposit localized in shallow depressions.

This wash sequence indicates that, following a period of moderate biochemical weathering and rubefaction, torrential rains washed away the soil mantle, redepositing it in lower-lying areas. Since weathering products were removed faster than they could accumulate *in situ*, soil development ceased entirely. The colluvial wash (Figs. K–1 through K–7) represents the material gradually stripped off the nearby hillslopes. Fairly moist conditions during the rainy season are suggested by the snail fauna, but aridity seems to have set in at the end of the depositional period (*b*). The discontinuous patches of alluvial wash (*c*) were probably laid down rapidly by a few spectacular sheetfloods. Unlike the earlier colluvial wash, bed *c* did not accumulate slowly in the presence of a vegetative mat, and snails are conspicuously absent. This brief period of heavy rains may have been contemporary with the Malki Member of the Ineiba Formation, while the colluvial wash may well antedate the Korosko Formation as it does at Ballana. Finally, the red paleosol almost certainly records the last period of general rubefaction noted in Chapters 2 and 5 for the post–Wadi Korosko stage.

Through the courtesy and collaboration of Karl Kromer and Manfred Bietak of the Austrian Expedition, it was possible to establish a succession of Holocene deposits in the A-Group shelters situated below a number of bedrock overhangs or in fissures or small caves on the south side of Wadi Nashryia. The overlying surface at about 130 meters forms part of a 134-meter fluvial platform and is dissected to a depth of about 10 meters. A late Middle Pleistocene stage of the Nile at *ca.* 122 to 124 meters ($+15$ to $+17$ m) favored development of the abris, probably as a result of salt weathering through capillary ascent from the waterline or from ground moisture, combined with undercutting by wadi floods. Previous joint-development of the sandstone bluffs by pressure release and exfoliation had, however, preconditioned the bedrock for differential weathering. The following composite stratigraphy can be determined for the various shelters (abris) and chambers (räume) as distinguished by Bietak and Engelmayer (1963):

a) 50 cm. Multicolored, dominantly light-yellowish-brown (10 YR 6/4), semiconsolidated, marly coarse sand, locally with subrounded sandstone pebbles and angular detritus. Crudely stratified, with coarse, angular blocky structure. Pyrolusite and limonite staining indicate waterlogging. Korosko Formation. (Best represented in Abri 3 and Raum II).

b) 2 cm. White (7.5 YR 8/0), sandstone detritus (averaging 1.8 cm in major axis), derived from the shelter roof, consolidated into a superficially weathered breccia (very pale brown, 10 YR 7/6), limonitic stains. Although the rock ceiling is patinated today, the sandstone was white at the time it was detached. A-Group hearths are cut into this breccia and have locally discolored it (light reddish brown, 2.5 YR 6/4) through heat. Probably late Pleistocene as suggested by groundwater phenomena. (Best represented is Abri 3).

c) 25 cm. Pink (7.5 YR 7/4), unconsolidated, medium quartz sand, with eolian topset-bedding. Coarse, subangular blocky structure. A-Group hearths are located on top of this sand and the topmost 2 cm may be semiconsolidated, moderately leached of carbonates, and darkened by carbonized organic matter. Mid-Holocene. (Best represented in Abri 2).

d) 1 cm. A-Group occupation floor (dating somewhere between *ca.* 3200–2650 B.C.), with inhomogeneous combinations of:

  i) fire-reddened, light-brown to brown (7.5 YR 5–6/2–3), unconsolidated, coarse sand; and
  ii) carbonized organic matter and ash, grayish in color.

A scattering of 1- to 2-centimeter-long sandstone detritus, either charred (very dark gray, 10 YR 2–3/1) or fired (weak red, 10 R 4/4; yellowish red, 5 YR 4/6), include coarse sandstones not present in the immediate local bedrock. They were probably introduced by man. (Best represented in Abri 3.)

e) 20 to 50 cm. Pink to reddish-yellow (7.5 YR 7/4, 8/6), unconsolidated, homogeneous coarse sand with eolian topset-bedding. In Abri 3, this sand, by exception, contains quartz granules and stratified detritus of an intrusive, coarse-grained sandstone (averaging 2.4 cm in length). (Best represented in Abris 1, 2, 3, and 4). Disconformity.

f) 2 cm. Late Roman-Byzantine occupation floor (fourth century A.D.), a dark-gray (10 YR 4/1), unconsolidated mixture of organic matter (carbonized fiber, charcoal, fine ash) and coarse quartz sand. (Present in Abri 1). Disconformity.

g) 0 to 50 cm. Loose, eolian sand was originally found on top of bed *f*, according to data supplied by Manfred Bietak. It was completely removed during the 1962 excavation.

By way of interpretation, the breccia of bed *b* is not sufficiently flattened to conform to frost-shattered rubble (*éboulis secs*),[10] so that a thermoclastic origin is rather unlikely. Instead, based in part upon the evidence of pseudo-gley conditions during or after sedimentation, salt weathering by capillary ascent of water seems to be the responsible agent. This would suggest contemporaneity with a high, late Pleistocene stage of the Nile.

The other point of interpretative interest is the presence of typical topset eolian beds in most of these shelters. Accelerated eolian activity must have begun somewhat before the time of the A-Group occupation and resumed for a somewhat longer interval thereafter. It had completely ceased by the time of the fourth century A.D. occupation, and fully analogous conditions have not occurred since. Presumably, the eolian deposits found along the Nile banks are broadly contemporary. These features recall the evidence for eolian deflation of a Nagada-II cemetery at Hierakonpolis and pre-Hellenistic eolian deposition in western Middle Egypt (Butzer 1959*b*: 69 ff.). The localized post-Byzantine drift sand does not necessarily imply conditions different from those of today.

Since the earliest cave drawings of native fauna at Seiyala are older than the A-Group occupation (Bietak and Engelmayer, 1963), the presence of elephant, giraffe, ostrich, antelope, and gazelle may indicate a period of more abundant vegetation before 3200 B.C., possibly, but not

---

10. Ten measured specimens gave an average thickness/major axis ratio (E/L) of 52%, a thickness/minor axis ratio (E/l) of 65%. Both of these ratios are well above those obtained for ferruginized sandstone pebbles from other Nubian samples (Table 6–1, Nos. 1–3, 5).

necessarily, antedating the first eolian beds. Sand alone is, of course, not necessarily indicative of aridity before and after the A-Group occupation, but marked topset stratification in the shelters would have been unusual if we assumed that there was appreciable vegetation in the adjacent wadi bottom and along the eastern Nile bank. Unfortunately, the Seiyala sequence remains unique in Nubia, since other published excavations of late prehistoric cave shelters and occupation sites have paid no attention to nonarcheological stratigraphy and sedimentation.

North of Seiyala, late Pleistocene nilotic deposits are well developed on the irrigated silt plains in the Wadi Allaqi embouchure and again west of Dakka. Unfortunately, study is complicated by intensive cultivation, a veneer of wash, and a lack of suitable exposures. The Masmas Formation—with occasional outcrops of the Korosko Formation along the edge of the cultivated land—is the major sedimentary unit, although there are patches of nilotic sands with medium-grade chert and chalcedony pebbles opposite the mouth of Wadi Allaqi and just north of Kushtamna West, to elevations of 126 meters (+20 m) at both localities. These suggest the Darau Member. Adams (1864) found no mollusca on the plain behind Dakka—then open desert—but abundant *Corbicula* and *Bulinus truncatus* were collected from the Allaqi silt plain.[11]

The only other possible late Pleistocene deposit that we were able to recognize between Dakka and Shellal is a low wadi terrace at +3 to +4 meters in Khor Dihmit, graded to a Nile floodplain of about 112 meters (+13 m). We could not examine it in the field at the time it was exposed in October, 1962. At Shellal itself, there are extensive deposits of the Masmas Formation, mapped and studied by Ball (1907: 58–59, Pl. 2) and Sandford and Arkell (1933: 43, 57–59). These attain a maximum elevation of about 117 meters.

Late Pleistocene and Holocene Deposits of the Sudanese Nile Valley

In Sudanese Lower Nubia, de Heinzelin and Paepe (1965) and de Heinzelin (1967) have outlined a succession of late Pleistocene deposits that can be fairly readily correlated with our own sequence to the north

---

11. Said and Issawy (1965: 15–16) recognize 10-meter, 7-meter, and 1.6-meter clay terraces in the Dakka area, as measured above reservoir level and supposedly correlated with the terraces at identical absolute levels in southern Egyptian Nubia. Our comments in footnote 2 above apply here as well.

of the border. The following terminological correlations can only be proposed:

Units of de Heinzelin and Paepe	Units of Butzer and Hansen
Qadrus Formation (stabilization and soil)	(?) Kibdi Member (+6 to +7 m)
Arkin Formation (+13 m)	(?) Arminna Member (+15 m)
Birbet Recession	(Downcutting)
Sahaba Formation (+23 m)	Darau Member (+22.5 m)
Ballana Formation (Eolian)	(Downcutting)
Khor Musa (Dibeira) Formation (+36 m)	Masmas Formation (+33 m)
Ikhtyaryia Formation (Eolian)	(Downcutting)
Proto-Nile Group (no data)	(?) Korosko Formation (+34 m)

Problematical relationships are indicated by a question mark. Of considerable interest are the eolian formations found in the Wadi Halfa area, contemporary with the intervals of downcutting. Eolian sands are widespread along the western margins of the Nile Valley today and have periodically advanced onto the floodplain in historical times (de Heinzelin, 1964). During the periods of downcutting (and lower Nile floods?), similar dunes appear to have encroached upon the abandoned floodplain. Publication of detailed profiles and sedimentological data from such eolian deposits will be welcome.

Similar deposits appear to continue upstream above the Second Cataract (Robinson and Hewes, 1967), and field studies are or have recently been underway in parts of Upper Nubia. Older work is essentially limited to the observations of Sandford (1949), who describes the following sequence from the Berber-Atbara stretch of the Nile: (*a*) Coarse sterile gravels, with Hudi Chert, attaining 9 meters above floodplain; (*b*) medium gravels, with Levalloisian flakes, to +5 meters or so; and (*c*) silts with bands of quartz and chert gravels, with abundant Late Paleolithic implements *in situ* and on the surface, and attaining +4.5 meters. The upper two units are semiconformable, while the upper silts can be followed discontinuously downstream to the region of Merowe, where they attain relative elevations of +17 meters. A lower, +6 meter stage is present here as well. The facies description leaves no doubt that these are beds of the Gebel Silsila Formation.

The most impressive late Pleistocene pluvial of the Sudan are the clays of the Gezira Plain (Tothill, 1946), between the Blue and the White Nile. Their stratigraphy can be tentatively interpreted from sections studied at Singa and Abu Hugar by A. J. Arkell (1949a). Approximately 6 to 9 meters of unstratified clay form the surface of this old Blue Nile floodplain, lying at 7 to 9 meters or more above modern flood level. Archeologically sterile, these clays contain shells as well as secondary features (color changes, calcareous concretions, and cracking systems, see Finck, 1961) reflecting formation of margalitic soils or vertisols during and after sedimentation. At the base of these dense clays, there are up to 4 meters of stratified sands and silts with calcareous root drip, resting, in turn, on travertine or water-laid calcareous concretions, derived from erosion of primary travertines or secondary soil concretions. At Abu Hugar, these calcareous sediments attain 2.5 meters thickness, with unworn core tools, flakes, and utilized cores of Levalloisian facies (Lacaille, 1951) *in situ*. At Singa, a megacephalic "Bushmanoid" skull (Wells, 1951) was found in a limestone block from the same stratigraphic unit. Older, sterile clays underlie the calcareous beds at both localities.

The fauna from Singa and Abu Hugar was studied by Bate (1951), and it includes several extinct genera or species, namely, a buffalo (*Homoioceras singae*), a porcupine (*Hystrix astasobae*), an unidentified giraffoid, and an antilopine. Other riverine forms include the Nile crocodile and hippopotamus, while open country forms are represented by several genera, including *Equus, Oryx, Gazella,* and *Antilope*. Finally, there is also a large, unidentified rhinoceros. All in all, this Pleistocene fauna suggests perennial waters with a galeria woodland, bordering on savannas or dry grasslands. Ecologically, it corresponds closely to the faunas of Qau and the Kom Ombo Plain, even though ethiopian species are largely absent in the Egyptian collections.

The implications of the massive clays of the Gezira Plain have been studied by Tothill (1946). From the abundant distribution of the gastropod *Cleopatra bulimoides* to about 40 km beyond the modern Blue Nile banks, Tothill (1946) concludes that these areas were seasonally inundated for about five months, during the first third of the alluviation stage. At the same time, the presence of two amphibian species, *Pila (Ampularia) wernei* and *Lanistes carinatus,* suggests that the entire Gezira Plain was seasonally inundated for several months throughout the aggradation period. Finally, *Limicolaria flammata,* a land snail, is present in the upper clays, suggesting a 300-kilometer northward shift of the zonal vegetation belts, with a much moister climate in the Khartum area. The *Limicolaria* floodplain stage can be correlated with the mid-Holocene site of Mesolithic Khartum (Arkell, 1953: 7–9). The earlier clays, deposited under more subaqueous conditions, rest on the Singa–Abu Hugar strata and are almost certainly late Pleistocene in age—correlatives of either the Masmas Formation or a part of the Gebel Silsila Formation in Lower Nubia. They suggest a Blue Nile flood lasting somewhat longer than that of today. Blue Nile gypsum is found everywhere in the Gezira clays, implying that Blue Nile waters spilled over the whole plain, as far west and south as Renk, into the White Nile Valley (Tothill, 1946). This fact is relevant for an understanding of the semilacustrine terraces of the White Nile, further discussed below.

The mid-Holocene sites at Khartum and Shaheinab have provided a badly fragmented, but nonetheless highly varied, faunal spectrum for the fourth millennium B.C. (Bate in Arkell, 1949*b*; 18–27; in Arkell, 1953: 11–19). In the "Mesolithic" site, several swamp-loving species deserve mention: the Nile lechwe (*Onotragus* cf. *megaceros*), white-eared cob (*Adenota leucotis*), water mongoose (*Atilex* cf. *paludinosus*), and an extinct reed rat (*Thryonomys arkelli*). The remainder of the typically ethiopian fauna of both sites ranges from carnivores to ungulates, including several antelopes and gazelles, buffalo, hippopotamus, warthog, elephant, giraffe, and two-horned rhinoceros. Once more, the nilotic juxtaposition of riverine and savanna environments is indicated.

Two well-developed alluvial terraces of the lower White Nile have been recognized by Berry (1961*b*). These consist of silty clay and extend for some 600 kilometers along the White Nile upstream of Khartum. Both are practically horizontal, suggesting a lake up to 25 kilometers wide, presumably dammed up by the more powerful Blue Nile flood at Khartum. The lower terrace, at 22 meters above the Khartum datum (363 m), is about 5 meters above modern flood level; the higher terrace is at 9 meters above floodplain. The lower terrace has radiocarbon dates of 9350 B.C. ± 400 and 6420 B.C. ± 350 at depths of − 1.45 to − 1.70 meters and − 1.6 to − 1.9 meters respectively (Williams, 1966). The uppermost sediments may be derived from the Blue Nile, being rich in gypsum and mica. They can probably be related to the identical mid-Holocene level of the Blue Nile at Khartum (Arkell, 1949*b*), indicating that the higher terrace also is of late Pleistocene age. Two interpretations are possible: either the Blue Nile flood was higher and lasted appreciably longer than now, or the White Nile flow was smaller, leading to choking up of the river with sediment. The obvious relationship of the higher terrace to the Gezira Clay Plain supports the first hypothesis. Furthermore, Berry (1961a) has described large, nonfunctional alluvial islands at 2 to 3 meters above modern flood level. He attributes these to an undated, late prehistoric or historical period with a White Nile discharge calculated to have been about ten times greater than at present. Since some 47% of the local White Nile waters are derived from the Sobat, primarily from Ethiopia, a broad parallelism of climatic change can be expected with the Blue Nile and Atbara basins.

In overview, there is a wealth of late Pleistocene deposits in the Sundanese Nile Valley. Although these deposits have barely been studied so far, there can be little question that, in view of the similarity of facies, correlation of sediments between southern Egypt and the central Sudan will prove possible. Current information suggests that the late Pleistocene nilotic formations extend from the Ethiopian border well into Upper Egypt, providing a potential guide horizon over wide areas.

Stratigraphic and Paleoclimatic Conclusions

The stratigraphic conclusions derived from the local sequences can be synthesized in tabular form (Table 6–5). It is immediately apparent that this framework is more detailed than that for the late Pleistocene and Holocene evolution of the Kom Ombo Plain. In part, of course, this reflects the advantages of having already established the basic units on the Kom Ombo Plain and of having subsequently completed this picture with further detail from Nubia.

There are, however, regional differences in facies development which affect the stratigraphic record. So, for example, the Arminna and Kibdi substages of the Gebel Silsila Formation are completely absent at Kom Ombo, where there is next to no information on the physical history of the Nile after about 10,000 B.C. Another drawback at Kom Ombo is the lack of direct associations between wadi and nilotic alluvia younger than 15,000 B.C. Perhaps most significant is the lack of deep sections at Kom Ombo, so that units 20 through 23 of Table 6–5 are nowhere exposed and the oldest late Pleistocene sediments seen at Kom Ombo probably date from the middle of unit 24, i.e., midway in the classical Korosko Formation. This fact also emphasizes that almost all of the late Pleistocene sediments in Upper Egypt are telescoped into the last 20,000 years of the Pleistocene, recording only the second half of the Würm Glacial.

Despite the number of stratigraphic entities for the Early Würm in Nubia, even here the picture is vague and uncertain, except perhaps for the Wadi Floor Conglomerate. The lower member of the Korosko Forma-

Table 6–5. Geomorphic evolution of Egyptian Nubia during the late Pleistocene and Holocene

38.	Wadi dissection and redeposition of fill. Very limited wadi activity.	HISTORICAL
37.	Alluviation of 50 cm or more of wadi wash, possibly equivalent to Member III, Shaturma Formation. At this time or earlier, first accumulation of modern Nile silts.	
36.	Dissection of wadi fill (as much as 5 m or more), with Nile floodplain level no higher than today. Local accumulation of eolian sands (Seiyala).	
35b.	Alluviation of Shaturma Formation (Member I) by wadis, as a result of significant winter sheetflooding (thickness over 5 m). Contemporary with last part or all of Kibdi unit, phase *35a*.	HOLOCENE
35a.	Alluviation of Kibdi Member (Gebel Silsila Formation) by Nile (thickness over 6 m, floodplain elevation + 6 to + 7 m). Slightly more vigorous and possibly higher summer floods. Terminal stage *ca.* 3000 B.C.	
34.	Dissection of Nile (to below modern floodplain, vertical differential at least 15 m) and wadi fill (total cutting since beginning of phase *32b* over 8 m). Local accumulation of eolian sands (Seiyala).	
33.	Biochemical weathering with red paleosol (Omda Soil Zone). Frequent, gentle rains.	
32b.	Alluviation of Upper Member, Ineiba Formation, by wadis (thickness over 8 m). Accelerated wadi activity. Possibly subdivided into two fills, separated by over 8 m of downcutting. Earlier fill contemporary with phase *32a*, later fill terminating *ca.* 6000 B.C.	
32a.	Alluviation of Arminna Member (Gebel Silsila Formation) by Nile (thickness over 12 m, floodplain elevation + 15 m). More vigorous and possibly higher Nile floods *ca.* 9200–8000 B.C.	
31.	Dissection of Nile (to below + 3, vertical differential at least 20 m) and wadi fill (at least 4 m).	LATE WÜRM
30.	Calcification and development of Calcorthid paleosol (Adda Soil Zone). Some rains.	
29b.	Alluviation of Lower Member, Ineiba Formation, by wadis (thickness over 7 m). Accelerated wadi activity. Contemporary with phase *29a*.	
29a.	Alluviation of Darau Member ("classical" Gebel Silsila Formation) by Nile (thickness over 18 m, floodplain elevation + 22.5 m). More vigorous and higher Nile floods (with amplitude of 9 to 15 m), *ca.* 15,000–10,500 B.C., possibly terminated by period of exceptionally high floods (28 m above modern floodplain), *ca.* 10,000–9500 B.C.	
28.	Development of hydromorphic paleosol on floodplain and wadi floors (a vertisol of Mazaquert type).	
27.	Dissection of Nile (to below + 5 m, vertical differential at least 28 m) and wadi fill (amplitude uncertain).	MIDDLE WÜRM
26.	Alluviation of Masmas Formation by Nile (thickness at least 33 m, floodplain elevation + 33 m). Flood regime similar to that of today, with little or no local wadi activity, *ca.* 22,000–16,000 B.C.	
25.	Dissection of Nile (to below modern floodplain, vertical differential at least 34 m) and wadi fill (amplitude uncertain), the latter interrupted by at least one phase of temporary aggradation.	

(*Table 6–5*, continued)

24.	Alluviation of "classical," upper member of Korosko Formation by Nile and wadis (thickness at least 34 m, floodplain elevation + 34 m). Greater Nile velocity with significant wadi activity, attaining a maximum during the second quarter of the deposition period, a minimum during the initial stages of nilotic alluviation. Termination somewhat after 25,000 B.C. Mixed nilotic-wadi deposits grade upstream into alluvial wadi terraces, probably contemporary with extensive mantles of alluvial or colluvial wash.	MIDDLE WÜRM
23.	Dissection of Nile (to about modern floodplain or lower, vertical differential at least 11 m) and wadi fill, interrupted by a period of wadi aggradation followed by an interval of biochemical weathering with formation of a red paleosol.	EARLY WÜRM
22.	Alluviation of lower member of Korosko Formation (thickness at least 11 m, floodplain elevation about + 11 m). Greater Nile velocity and first record of summer flood regime with Ethiopian silt influx.	
21.	Wadi erosion (amplitude uncertain).	
20.	Alluviation of Wadi Floor Conglomerate (local thickness at least 6 m), following period of intensive wadi flow with bedrock cutting. Frequent heavy rains. Nile floodplain level below that of today.	
19.	Dissection of Nile (to below modern floodplain; vertical differential at least 25 m since maximum of Wadi Korosko stage) and wadi fill (vertical differential over 15 m). Lower Nile floods and limited wadi activity.	EARLY UPPER PLEISTOCENE

tion was only seen in Wadi Or, and its existence and character is not established beyond doubt. There can be little question that one or more periods of wadi alluviation marked the Early Würm, and here the Wadi Or record is clear, even if unique. Prior to all of these enigmatic deposits, there are the ferruginized Wadi Floor Conglomerates which, at least in Wadis Or, Korosko, and Abu Sureih, are considerably older than any nilotic marls. The geological record for the Early Würm is unsatisfactory, but, as tentative as it may be, it serves to emphasize the complexity of aggradation and degradation by the Nile and its tributary wadis. If it can be accepted as Early Würm, the lower unit of the Korosko Formation in Wadi Or may indicate the existence of a summer-flood regime 50 millennia or more ago; if we could accept this as a fact on the basis of chance preservation at a single locality, it becomes apparent how misleading the seeming absence of flood silts in the older Pleistocene record can be. For this reason, it is only possible to consider the Korosko Formation as a *terminus ante quem* for the first major influx of summer-flood waters of Ethiopian origin.

The basic stratigraphy of three late Pleistocene nilotic formations, albeit with additional subdivisions, is unquestionable in Nubia. On sedimentological and stratigraphic grounds, these fundamental units are unmistakable. The different nilotic substages are often less easy to distinguish among themselves, however, and field correlations become more difficult. Detailed archeological excavations with full publication of materials should, in the coming years, serve to clarify the minor details of the Gebel Silsila Formation and to relate these precisely to occupation sites and specific cultural complexes.

The variable facies composition of the three basic nilotic formations suggests repeated hydrographic, as well as climatic, changes in the Sudanese and Ethiopian parts of the Nile Basin. The probable morphology and dynamism of the Nile floodplain during these different stages has already been discussed in Chapter 3 and amplified in the preceding sections of Chapter 6. The first appearance of bed-load gravels and elements from the Hudi Chert within beds of the Darau Member (see also Appendix C) is as dramatic as the first appearance of massive Ethiopian flood silts with the Masmas Formation. The periods of Nile downcutting that preceded these two novel facies can almost certainly be attributed to changes upstream of Egyptian Nubia, unrelated to the ultimate base level of the Nile. The nature of such changes must necessarily remain enigmatic until more is known about the late Pleistocene history of the Blue Nile and Atbara basins—in particular, why three different types of flood regime succeeded one another during a time span of about 10 millennia.

Equally puzzling are the rapid ups and downs of the Nile floodplain in Lower Nubia between 15,000 and 3,000 B.C. during the course of the Gebel Silsila stage. Repeated aggradation and downcutting within a vertical range of at least 10 to 15 meters and with a cyclical amplitude of 2 or 3 millennia (de Heinzelin, 1967) are difficult to explain. Base-level stimuli from northern Egypt could not possibly be reflected 1200 kilometers or more upstream in so short a time. This suggests climatic changes in the upper Nile drainage basin as the best possible explanation. The deposits themselves invariably indicate much greater Nile capacity and competence, a rapidly shifting channel carrying coarse sands and gravels, and sandy levees, frequently breached by violent floods.[12] Presumably,

---

12. Fairbridge (1963) suggests the opposite interpretation, that the late Pleistocene silts indicate declining Nile floods, deposited as the river choked in its own

the periods of Nile downcutting came in response to decreasing Nile discharge (shorter flood seasons? lower flood amplitudes?), decreasing sediment yield, or a changing ratio of bed load to suspended load. Possibly all of these factors were combined.

Turning to the local deposits of the Egyptian wadis, it can be said that all of these sediments suggest a moister climate with more frequent rains of some duration or intensity. This can be argued on several premises. For one thing, greater stream competence would be required to transport a bed load which was far coarser than the one moved in the same wadis today. Similarly, the preserved deposits in small or intermediate-sized wadis commonly indicate a wadi bed several times broader than the wadi floor is today. This can best be explained by greater discharge and a greater proportion of bed load. In the case of the Shaturma beds, the alluvial fill can again be traced upslope to colluvial screes and slope detritus, providing evidence for exceptionally significant overland flow and upland denudation. It must be emphasized, however, that every deposit does not suggest equally moist "pluvial" conditions. The bedrock erosion and transport of well-rounded, coarse gravels prior to, and at the time of, deposition of the Wadi Floor Conglomerate marks a long period of moister climate, presumably semiarid in character and quite comparable to the middle Pleistocene pluvials. Conditions have never been as moist again, and subsequent wadi alluvia point to less significant climatic fluctuations, with moister oscillations never passing the arid/semiarid threshold. The gravels interdigited with the Korosko Formation (Table 6–1) are subrounded only, while those of the younger formations tend to be subangular.[13]

---

load. The deposits and their stratification speak eloquently against such an argument, as has been repeatedly emphasized in Chapters 3 and 6. Why should a river "reduced to a mere trickle" carry masses of pebbles from the Hudi Chert exposures of the central Sudan as far downstream as Cairo, a distance of 1800 kilometers or more? Today, the Nile bed exhibits almost no pebbles whatever, because the stream lacks the competence to transport them. Why should a stream that maintained a host of overflow channels on a floodplain several times wider than that of today be considered to have been choking in its own silt due to lower flood discharge than at present? This seems particularly unreasonable when the deposits indicate frequent levee breaching, channel shifts, and crevasses, all of which point to exceptionally violent floods. In short, Fairbridge's interpretation cannot be reconciled with the field facts nor with geomorphological theory.

13. The exceptional degree of flattening of the Nubian gravel samples (Table 6–1) can be attributed to lithology. The ferruginized sandstone or ironstone preserved in gravel form was originally derived from thin ferruginous beds within the sandstone or from deeply patinated surface shells, both of which produce a slablike detritus.

The question can also be raised whether an entire alluvial unit is not the product of a single spate or of, at most, a few exceptional floods. Certainly, not all of the local wadi deposits were laid down at the same rates. The period of bedrock incision associated with the potholing, prior to deposition of the Wadi Floor Conglomerate, probably involved a long period of time. Bedrock cutting of this type requires many periods of protracted, vigorous wadi flow. Deposition of fairly homogeneous, poorly stratified sands or silts usually requires a longer time than that of well-stratified, current-bedded, coarse-grained deposits. So, for example, the radiocarbon and general stratigraphic evidence suggests that the Ineiba Formation was gradually accumulated over an interval of as much as 8 millennia, which implies a rate of between 1 and 2 meters of sediment—along the length of the stream—per 1000 years. But, needless to say, innumerable erosional disconformities exist, so that the resulting deposits are the net result of repeated scour and fill over the centuries. Significantly, no massive graded beds were observed anywhere, and well-defined, thick, uniform lenses—of the type suggesting a single major phase of flood accretion—are absent. It should also be emphasized that catastrophic floods today do not lead to deposition of great masses of coarse detritus. They may breach older deposits or man-raised obstacles, but they seldom leave more than a thin layer of material, invariably of finer grade than adjacent terraced alluvia.[14] These geomorphic arguments that aggradation by desert streams is a slow process are fully borne out by the temporal relationships between nilotic and wadi units, both of them seen within a radiometric framework.

A final problem concerns the relationship between periodic wadi dissection and the changes of Nile floodplain level. Did the local Egyptian climate oscillate repeatedly, usually in phase with the Ethiopian floods, or did Egyptian climate remain more or less constant, with the aggradation-degradation cycles solely a response to fluctuations of the local base level? There is little direct evidence to resolve this point. The type section of the Kibdi Member in Khor Adindan shows, however, that wadi alluviation, after at first keeping pace with nilotic aggradation, ultimately

14. The great spates reported by Hume (1925: 83) and others to have deposited several kilotons of rock at a time are essentially restricted to the northern half of the Eastern Desert, and there is no evidence for such fluvial activity anywhere in Lower Nubia in post-Pleistocene times. When such exceptional spates do occur, deposition remains restricted to a small part of the wadi floor, usually to a few hundred meters of the wadi length. Even such floods would require repetition each year for centuries before an appreciable sedimentary layer could accumulate along the whole wadi.

ceased before the initiation of a renewed period of Nile dissection. In other words, local climatic changes were underway during the last part of the alluvial phase. Although this example is unique, we favor consistent application of the geomorphic argument (see Chap. 2) that dissection accompanied the waning stages of each pluvial oscillation.

The late Pleistocene and Holocene paleosols of Nubia provide few new interpretive vistas beyond those discussed in Chapters 2 and 3. The last red paleosol, formally defined as the Omda Soil, is reasonably well established in Nubia, as it is near Kom Ombo. In Nubia, its absolute age can be fixed with slightly greater accuracy as no older than 6000 B.C. and little younger than 5000 B.C. A similar red paleosol is described from the Batn el-Hagar by Robinson and Hewes (1967). Of some interest is the stratigraphic isolation of a Calcorthid paleosol, dating *ca.* 10,000 B.C., in Wadi Or. It may well be equivalent to the duricrusts or subsurface calcium accumulations occasionally noted on the topmost beds of the Darau Member at Kom Ombo. Carbonate enrichment at the top of a sedimentary unit suggests increasing aridity. Yet comparable soils do not form today in a hyperarid environment such as southern Egypt, another factor supporting our overall interpretation of wadi alluviation and dissection. A final word may be said about vertisols in Nubia. Our own work uncovered little substantive information on these phenomena in Egyptian Nubia, although incipient vertisols of epigenetic and syngenetic type undoubtedly exist in beds of the Masmas Formation. Publication of detailed profiles and analyses by de Heinzelin and Paepe from the excavations in similar sediments near Wadi Halfa should contribute to our understanding of these paleosols.

7

The Kurkur Oasis

Introduction

Kurkur, one of several minor oases dotting the Libyan Desert of southern Egypt, is located at 23° 54′ N, 32° 19′ E, about 55 kilometers west of the Nile (Fig. 2–5). Unlike the great oases of Kharga, Dakhla, Bahariya, Farafra, or Siwa, Kurkur is not a closed depression of eolian origin. Instead, it is a vegetated gorge where Wadi Kurkur descends from the limestone uplands and intersects a modest aquifer. The resulting slow seepage of subterranaean water supports some vegetation and a number of brackish wells.

Historical information on the Kurkur Oasis seems to be entirely lacking until the Franciscan missionary Theodor Krump arrived here on December 15, 1700, with a caravan on its way from Asyut and Esna to Sennar (Gumprecht, 1850). During the later Islamic period, the oasis served as a relatively important watering place for caravans on the Darb el-Bitan, which ran from Farshut or Esna through Kurkur and Dungul to Sheb, where it joined the Darb el-Arbain to the Sudan. Secondary caravan routes from the Nile Valley (Darau, Aswan, Dihmit, and Tafa) also converged on the Darb el-Bitan near Kurkur. In addition to the lucrative Sudanese traffic in slaves, ivory, gum, hides, gold, and ostrich plumes, considerable local traffic from Darau and Aswan was directed to the natron deposits at Sheb, used for soap production in Upper Egypt (Murray, 1939).

The Mahdist Wars in the Sudan and the subsequent establishment of the British hegemony effectively terminated the Libyan Desert caravan trade after about 1880. Today, the only traffic is single nomad families or individuals moving between Kurkur and Dungul, Sheb, and possibly Selima. After the Mahdist raid on Kharga during the summer of 1893, the British maintained an outpost of Ababda at Kurkur until the termination of hostilities in 1898. A number of huts and a trench remain from this one period of known historical habitation (Ball, 1902; Evans-Pritchard,

1935). During the present century, there has been sporadic pasturing of sheep, goats, and camels by nomadic Ababda herders moving between Darau, Gebel el-Barqa, Kurkur, and Dungul. There were no traces in 1963 of recent occupation.

Archeological remains of historical age are limited to potsherds of coarse, utilitarian ware found by David Boloyan in the oasis and to a hellenistic amphora reported by Murray (1939) from the Kurkur foreland. Even though more materials may be present in the eolian sands masking the oasis, these isolated finds tend to emphasize the lack of extended habitation at Kurkur throughout the historical era. By contrast, archeological evidence at the Nuq Maneih clay pan, west of Gebel Barqa (Fig. 2–5), suggests that cultivation was certainly once practiced after sporadic rainfalls in ancient times (Murray, 1939). In fact, Schweinfurth (1901) was informed in 1882 that the Ababda sowed crops on the western foothills of Gebel Barqa after rains.

Early Work on the Kurkur Pleistocene

The first and only early systematic geomorphic and geologic study of the Kurkur Oasis was made by John Ball (1902) in January, 1901. Ball drew a rather rough form map of the area at 1:25,000 and erroneously concluded that the oasis occupied a closed depression with no surface drainage outlet. In the absence of visible tectonic deformation, he ascribed the primary origin of the valley to wind action. He recognized and partially mapped the widespread calcareous tufas with their plant impressions and land shells, however, and interpreted these as freshwater deposits of a Pleistocene period of moister climate. He also recognized age and facies differences in these tufas, distinguishing three well-marked "forms": a dark, indurated tufa found on the upland surfaces, contrasting with two facies of softer, lighter colored material on the actual oasis floor. These brief remarks held the key to the Pleistocene of the oasis. On the basis of Alfred Lucas' analyses of water from the wells, Ball suggested its indirect origin from rainwater on the plateau, rather than from artesian sources. No explanation of the hydrogeology was given. Temperature and barometric pressure were recorded several times daily for four days during Ball's stay and suggested that winter temperatures here are systematically lower than those at Aswan by some 3° to 5° F (2° to 3° C). A slight rainfall was noted on January 25, 1901.

Although William F. Hume passed the Kurkur escarpment on a geological reconnaissance from Kharga to Edfu in 1908 (Hume, 1908), he did not enter the oasis. The next visitor was the geologist Kurt Leuchs (1913a, b), early in January, 1911. Leuchs mapped some of the conglomerates at the foot of the limestone escarpment. Although he rendered a good geomorphological description of the cuesta and its western dip-slope, Leuchs did not enter the oasis via Wadi Kurkur and failed to discover the drainage outlet. He noted that quartz sands of eolian origin were embedded in the tufas and found a rich Paleolithic site resting on top of the spring deposits.

The normal drainage of the oasis was first recognized when Richard Uhden (1930) entered via the camel track of Wadi Kurkur in mid-October, 1927, while on a geomorphologic reconnaissance from Dabud to Dungul Oasis. Uhden elaborated on the fluvial origin of the depression and described steplike erosional surfaces in the limestone plateau at, and just west of, the oasis. Well-rounded limestone gravel as well as detritus were found under and, to a more limited extent, within the tufas. This led Uhden to postulate that considerable erosion had gone on simultaneously with tufa deposition in the valley bottoms. In another paper, Uhden (1929) gave an excellent account of the retreat of the limestone escarpment. Curiously, his rather pertinent observations passed unnoticed by later authors.

In January of 1934, Jean Cuvillier (1935a) was the first geologist to travel by car in the Gebel Gharra and Kurkur area. Although his work was primarily devoted to the local bedrock geology, the sycomore or mulberry fig (*Ficus sycomorus* L.) was identified from leaf impressions he had collected from the oasis tufas. Suprisingly enough, Cuvillier also considered Kurkur a closed depression.

In 1938, the oasis area was surveyed by Khalaf Mursi for the Desert Survey of Egypt. The director of the survey, George W. Murray, visited Kurkur on two occasions, in May and December of that year, and subsequently published a *ca.* 1:87,500 form map, emphasizing the normal drainage of the oasis (Murray, 1939). It is amusing to note his remark that "neither Ball nor Cuvillier would believe me when I told them of the result of our 1938 survey" (Murray, *in litt.*, May 24, 1963).

Interest in Kurkur was recently revived as a result of the High Dam project at Aswan. In the first part of 1962, the Desert Institute of Egypt carried out a geomorphologic study of the sandstone plains west of the Nubian Nile, touching on Kurkur somewhat peripherally (Shata, 1962). Shata identified two distinct tufas, one a plateau tufa, capping the rim of the limestone plateau, the other forming a +15 meter wadi terrace. It was hypothesized that the groundwater supply of the oasis was related to faults which allegedly provided a direct connection between the sandstone aquifer and the limestone strata. In mid-January of 1963, the oasis was briefly visited by Said and Issawy (1965), who also distinguished the plateau and terrace tufas in addition to extensive "lake" deposits. The excavation of the oasis was attributed to the terminal Upper Pleistocene. In 1963–64, these same authors studied the Dungul Oasis in some detail, and Bahay Issawy also accompanied James Hester's brief archeological survey at Kurkur during a part of December, 1963 (Hester and Hobler, 1965).

The Work of the 1963 Yale Expedition

Despite the sound preliminary study by Ball (1902) and the cursory visits of earth scientists since then, information and understanding concerning the geomorphology and the Pleistocene record of Kurkur have remained rather rudimentary. The well-known study by Elinor Gardner (1932b) at the Kharga Oasis, some 200 kilometers to the northwest, seemed to suggest that detailed work at Kurkur might permit another such local paleoclimatic sequence, independent of external complications such as Ethiopian pluvials. And Paleolithic archeology seemed equally promising in view of the work at Kharga and Leuchs' discovery at Kurkur.

Accordingly, from its inception in May, 1961, the Yale Prehistoric Nubia Expedition made plans to study Kurkur, and the trip was undertaken early in March of 1963 (see Reed, 1964). We mapped the geology and geomorphology of an 110-square-kilometer area (Figs. 7–2, 7–8)—much of it in detail (Fig. 7–11)—and recorded sections by detailed Abney traverses totalling 22 kilometers in length. Some seventy sediment and rock samples, six gravel collections, and a dozen potential pollen samples and macrobotanical specimens were retrieved by us. Part of our field work was carried out in close collaboration with Egbert G. Leigh, Jr., (malacologist) and David S. Boloyan (archeologist). Subsequent sedimentological study in Aswan and at the University of Wisconsin compensated for the tight field schedule dictated by restricted supplies. If more time had been available, the work would preferably have been extended to other wadis issuing from the scarp both north and south of the oasis and to the Libyan Desert further west.

The palynological and macrobotanical materials are being studied by Madeleine Van Campo. The results available to date are outlined in Appendix I.

Bedrock Lithology

The bedrock lithology and stratigraphy exposed along the cuesta escarpments and in the oasis area can best be described in a little detail since geological work has so far been restricted to cursory visits. Unless otherwise stated, lithological data are based on regional mapping, sections, and rocks samples studied in 1963. With some modifications, the stratigraphy itself follows Cuvillier (1935a, b). The bedrock geology, mapped at a medium scale, is shown by Figs. 7–1 and 7–2. Lithostratigraphic units, beginning with the base of the sequence, are as follows.

a) *Nubian Sandstone* (over 50 m, resting disconformably at depth on a Precambrian basement). White to very-pale-brown, coarse-grained sandstones of moderate consolidation and with both high porosity and permeability. Exposed 15 kilometers east of Kurkur on the Kalabsha Plain (Fig. 2–1) in facies similar to those elsewhere in Egyptian Nubia (Chap. 5). Upper Cretaceous.

b) *Dakhla Shale* (90 m). Primarily light-gray (10 YR 7/1, 2.5 Y 7–8/2), fissile, clayey siltstones and shales of moderate consolidation, with extensive limonitic staining and commonly charged with evaporites. Intercalations include ferruginous, brown (7.5 YR 5/4) to reddish-brown (2.5 YR 4/4) or weak red (10 R 4/3) siltstone (1 to 3 cm thick) and, more rarely, lenses of white (10 YR) gypsum (several mm thick). Carbonates are almost completely absent, so that the adjective "marly," used by Leuchs (1913a, b) and Cuvillier (1935a), is inappropriate. Although locally sterile, these shales are interbedded with a marly limestone containing the Mesozoic oyster *Exogyra overwegi* at Gebel Gharra (Cuvillier, 1935a). The impermeable Dakhla Shale is widely exposed at the foot of the major escarpment from south of Kurkur to north of Gebel Gharra, forming a distinctive geomorphic unit designated as the Kurkur Foreland (Fig. 2–5).

c) *Kurkurstufe* (4 to 5 m). As correctly defined by Cuvillier (1935a, b), this unit is a duplicated bed of fossiliferous, ochrous limestone found at the top of the Dakhla Shale in the Kurkur-Gharra area. The typical facies is a massive, brownish-yellow (10 YR 6–7/6), indurated limestone (to 2.5 m) with dispersed sand-size aggregates of shale. Limonitic staining is general, and oolitic concretions of limonite and hematite occur. Intercalated with and resting on these highly fossiliferous limestones are shaley and marly beds, including: (i) brownish-yellow (10 YR 6/6), limonitic, thin-bedded, calcareous sandstones and marls (to 1.5 m), and (ii) olive (5 Y

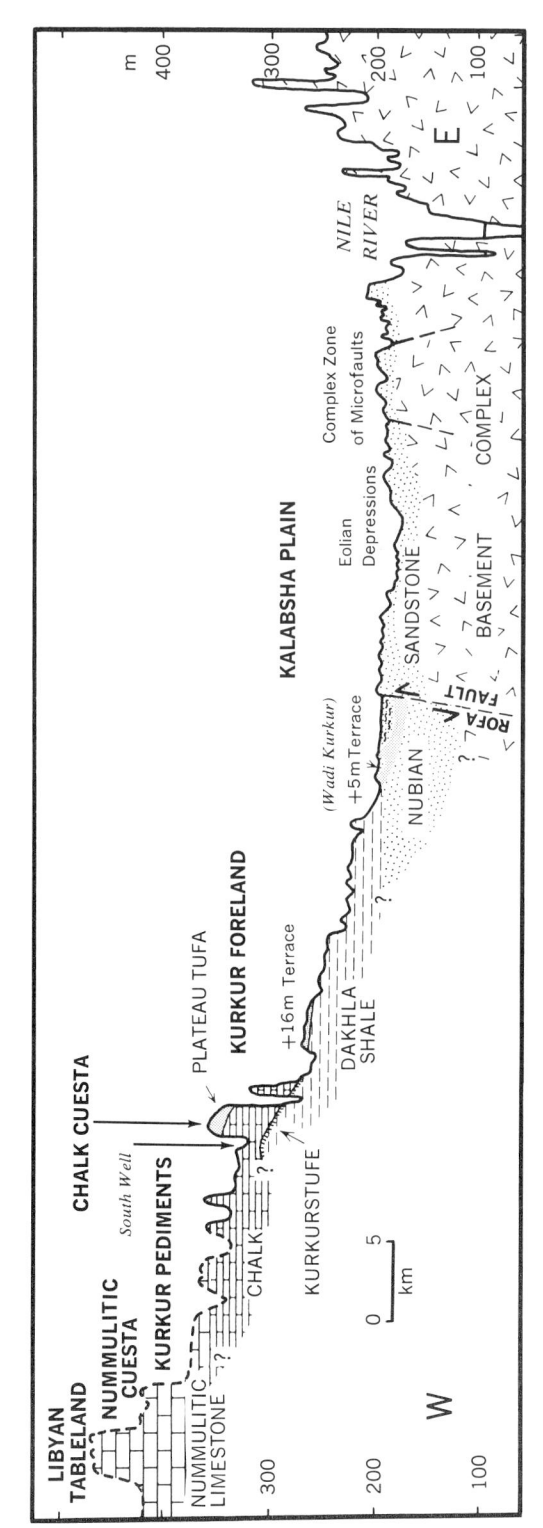

Fig. 7-1. Geological section from Kurkur to the Nile Valley. Topography west of Kurkur is uncertain Map: UW Cartographic Lab.

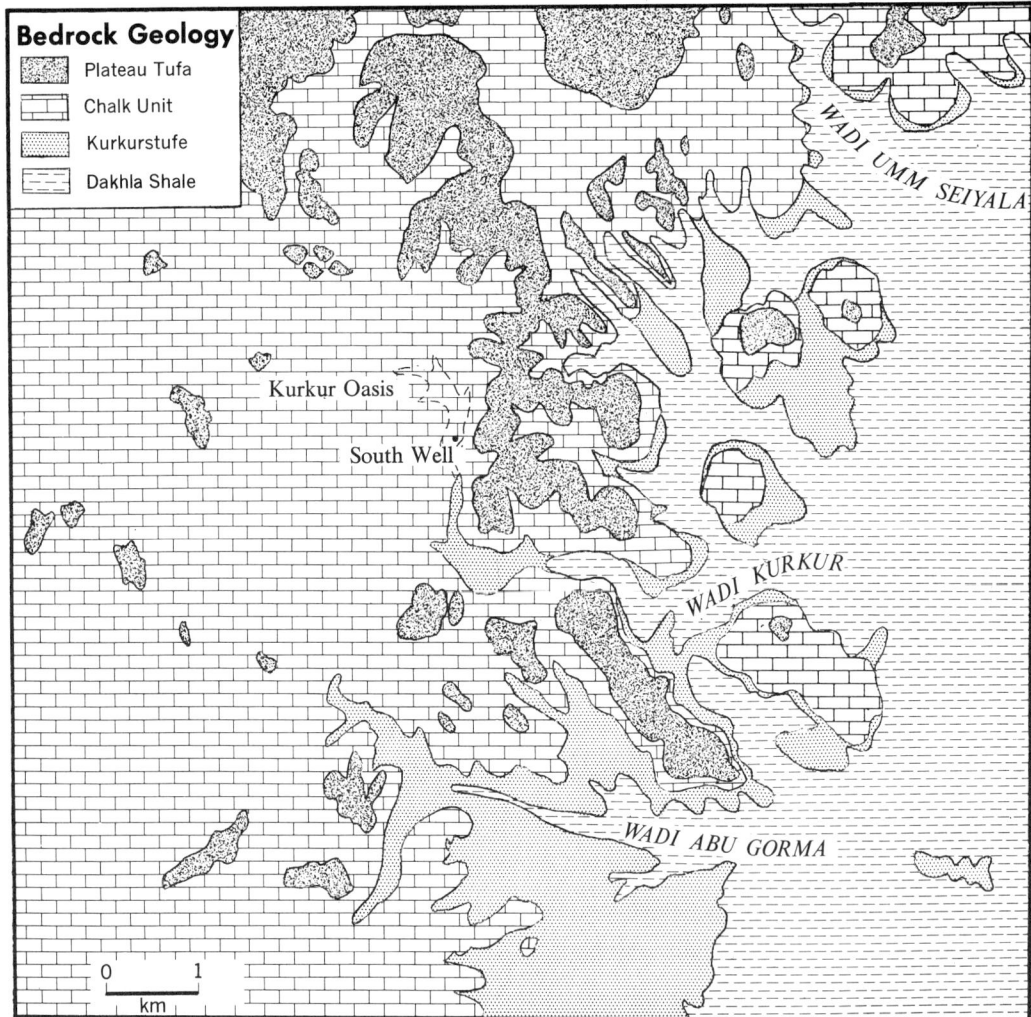

Fig. 7–2. Bedrock geology of the Kurkur area. Map: UW Cartographic Lab.

4–5/3) to very-pale-brown (10 YR 7/3), glauconitic marls and siltstones (60–80 cm). This massive and impermeable unit forms a minor cuesta at the foot of the great escarpment between Kurkur and Gharra and is critical to an understanding of the hydrogeology of the oasis. Cuvillier (1935a) has identified the fauna, with four echinoderms and some thirty-five molluscan genera, as Danian (terminal Cretaceous).

d) *Chalk Unit* (70 m). Mainly a homogeneous, moderately thin-bedded, white (2.5 Y 8/2) to very-pale-brown (10 YR 8/4), chalky limestone, occasionally intercalated with massive, light-brownish-gray (2.5 Y 6/2) limestones (2.5 to 5 m). Some medium quartz grains are present, as well as sodium and magnesium sulfates and halite (see Ball, 1902: 33 ff.). This major aquifer forms the great cuesta at Kurkur itself. Rare *Nautilus* sp. and frequent globigerinas of the genus *Rosalina* suggested a Danian age to Cuvillier (1935a). Resting disconformably on the Kurkurstufe and of a rather different facies than the underlying shaley and limonitic sequence, the local Chalk should probably be attributed to the Paleocene.[1]

e) *Nummulitic Limestone* (local thickness 6 m, attaining 100 m further north and west). This massive, indurated, light-brownish-gray (2.5 Y 6/2) limestone contains *Nummulities biarritzensis* var. *praecursor* (Leuchs, 1913b). It forms the great cuesta at Gebel Gharra and south of Kurkur. Small sections of the Nummulitic are preserved a little north of Kurkur, but the formation as such begins at the second escarpment some 15 kilometers west of the oasis. Lower Eocene.

f) *Plateau Tufa* (40 m). Very-pale-brown (10 YR 7/4), semicemented freshwater limestones, consisting of cryptocrystalline calcite with an admixture of quartz sand, clay, and occasional conglomerates. The dominant travertine facies is moderately impermeable. Probably of Pliocene age, this formation will be considered in further detail below.

Structure

Although broadly horizontal, the Cretaceous to Pleistocene sedimentaries of the Kurkur area exhibit local warping and, in some areas, minor faulting. The average dip is up to 1% WNW, but structural instability of the Dakhla Shale at the foot of the scarp has induced sliding and sagging of the overlying strata. As a result, local bedrock along the cuesta peripheries dips eastwards at 1% to 3%, in contrast to the overall westward dip. Slumping may be locally apparent in the Chalk at the scarp face (Fig. 7–8) and, more commonly, among the irregular shale masses perched along the Kurkur Foreland. Near the South Well, karstic solution has also led to local slumping of the Chalk. Faulting is absent in the Chalk, the Plateau Tufa, and all younger Pleistocene deposits in the oasis area west of the scarp, however.

1. The Lower Tertiary of Egypt is currently being subjected to sweeping stratigraphic revisions on microfaunal grounds. At Luxor, the Esna Shale, which underlies the lowermost Nummulitic Limestone (Lower Libyan or Theban Formation), is now thought to occupy all of the Lower Eocene (Berggren, 1964). Similarly, there are micropaleontological grounds for including the Danian within the Paleocene (Berggren, 1965).

Although local bedrock structure appears to be fairly uniform and uncomplicated, it plays a significant role in the overall pattern of geomorphic units. Apart from the terrain details of the Kalabsha Plain (Fig. 2–5), the Chalk and Nummulitic escarpments are structurally controlled cuestas, reflecting lithological differences in slightly inclined sedimentary strata. The location, and possibly the origin, of the major cuesta escarpment is almost certainly related to the Rofa Fault (see Chaps. 2 and 5). The total vertical displacement at this fault is unknown, but, from the latitude of Kom Ombo to that of Kalabsha, the scarp remains unmistakably aligned with the Rofa Fault, despite subsequent erosion and retreat by some 12 to 35 kilometers. Systematic geophysical exploration will be essential for a clearer understanding of the structural geomorphology here as elsewhere in southern Egypt.

General Geomorphology

An analysis of the terrain of the Kurkur area has already been made (Butzer, 1965), and discussion of the regional landforms may be limited to an outline of the different geomorphic units (Fig. 2–5).

a) The *Kalabsha Plain,* discussed in detail in Chapter 5.
b) The *Kurkur Foreland* is primarily an area of irregular, dissected shales, located 5 to 15 kilometers east of the cuesta rim. The concentrated scarp drainage and the impermeable bedrock have permitted a dendritic drainage pattern to develop. Local relief in 5-kilometer squares is between 25 and 35 meters, increasing abruptly to over 100 meters among the cuesta outliers at the foot of the scarp.
c) The *Limestone Cuestas* are formed by the Nummulitic Limestone, indurated massive Chalk strata, the Plateau Tufa, and the Kurkurstufe (Figs. 7–2 and 7–8). The wadis have cut 50- to 70-meter-deep, steep-sided, "youthful" valleys into bedrock. Local relief of the Chalk escarpment at Kurkur averages 100 to 150 meters, increasing to a maximum of 300 meters at the prominent Nummulitic ridge known as Gebel Gharra. Drainage patterns are dendritic, often with subparallel trends near the cuesta rim—where second- to fourth-order wadis may be perpendicular to the escarpment. Even in this area of maximum dissection, texture is very coarse and the drainage density is only 2.3 (Butzer, 1965).
d) The *Kurkur Pediments* are developed adjacent to the oasis, between the Chalk and the Nummulitic cuestas. Well-developed surfaces, in

major part exhumed from under the younger Plateau Tufas, can be identified at 360 to 365 meters and at 340 meters (Fig. 7–8). A later, incipient pediment level occurs at 320 to 325 meters. The erosional surfaces themselves are almost horizontal, offset from the flat-topped residual masses by steep rectilinear slopes and sharp, angular knickpoints and crest slopes. Local relief is on the order of 50 meters. The watershed lies some 2 to 6 kilometers west of the Chalk Cuesta, and further west the drainage follows the dip slope, terminating in depressions near the foot of the Nummulitic Cuesta.

e) The *Libyan Tableland*, beyond the Nummulitic Cuesta, is a flat, gently undulating limestone hammada at 400 to 500 meters. In the absence of wadis or bedrock rills, extended gentle swales—with no apparent patterns and of complex origin—control any drainage.

The Groundwater Resources

Water is presently available in two shallow waterholes near the place Ball (1902) called the South Well. In the more northerly of the two pools, the water table is at an elevation of 310 meters; in the more southerly pit, 35 meters away, it is at 309 meters. Ball's North Well was originally located some 330 meters further north, at approximately 317 meters (as determined by the Desert Survey in 1938). This particular waterhole was already sanded up in 1938 (Murray, *in litt.*, June 21, 1963). A later pit, dug 1.8 meters in depth near its former locality, is now dry,[2] although Muhammad Said Suleiman, the expedition guide, claimed that water can still be tapped at about 2 meters or so. If this is the case, the standing water table here would be at about 314 meters.

According to Ball (1902), the North Well had the better water. The analysis of water from the South Well (by A. Lucas) indicates a high mineral content (in parts per million): silica, 30; iron and aluminum sesquioxides, 9; calcium carbonate, 245; magnesium, 246; chlorine, 365;

---

2. Because Ball (1902) recorded water at -1 meter, this implies that the groundwater table dropped by at least 1 meter in about 60 years. In 1934, the water level was at about 1.5 meters (Cuvillier, 1935a). It is noteworthy that, in 1700, Krump's caravan—supposedly including 2000 camels and 200 donkeys—spent a day and two nights watering here (Gumprecht, 1850). Although no information is available concerning the rate of water flow at South Well today, it seems incredible that the oasis would now have sufficient water for a caravan even half that size. There are a number of sand hills mixed with dead remains of *Tamarix amplexicaulis* in the upper reaches of Northwest Wadi. Boulos (1966) attributes these phytogenetic hillocks to a past period with greater moisture.

and sulfates, 779. Soluble salts are precipitated on the adjacent bedrock exposures, and salt efflorescences are common on the surficial deposits of the South Well area. The mineral content of the groundwater appears to increase downstream, and the decline of the North Well may be related to the failure of the Ababda to visit Kurkur regularly in recent years.

In order to establish the origin of groundwater at Kurkur, longitudinal sections of the uppermost wadi and its tributaries were prepared, and the vegetation was physiognomically mapped. The plant geographical data were subsequently complemented by the observations and species identifications of Boulos (1966). The profiles (Figs. 7-4 to 7-6) show that sandy alluvium or eolian quartz sands underlie much of the wadi floor to depths of 3 meters or more. The lowermost 700 meters of Northwest Wadi are alluvial, overgrown with acacia trees (*Acacia raddiana*), dum palms (*Hyphaene thebaica*), small succulents (*Zygophyllum coccineum*), low evergreen thorns (camelthorn, *Alhagi maurorum*), acacia brush (*Acacia flava*), and halfa grass (*Desmostachya bipinnata*). The vegetation terminates at a Chalk sill (Fig. 7-4), and, on the bedrock reaches further upstream, plant growth is limited to acacia shrubs, halfa grass, and rare succulents. Identical conditions prevail along the first 400 meters of North Wadi, above the confluence with Northwest Wadi. The heart of the oasis, in Central Wadi, is a 650-meter stretch of eolian quartz and lime sands between Km 0.13 North and Km 0.51 South (Figs. 7-3, 7-5 and 7-7). Dum palms are dominant here, with a scattering of stunted date palms (*Phoenix dactilifera*) and an undergrowth of tall halfa grass. Reeds (*Typha australis, Phragmites communis*) and cattails (*Juncus arabicus*) are prominent in the salty area near South Well, with camelthorn dotting the higher-lying, peripheral dune sands.[3]

There is a small depression at Km 0.48, lying some 50 to 80 centimeters below a threshold of bedrock and sand. A trace of light-gray alluvial soil

3. The spontaneous vegetation of Kurkur has been somewhat degraded by the pasturing of camel, sheep, and goats and by the making of charcoal from acacia branches and fibre obtained from dum leaves (see Evans-Pritchard, 1935; Murray, 1939). Even more devastating is the apparent practice of passing travellers to burn a strip of vegetation—"to reduce water loss" according to the expedition guide. Over sixty dead, charred dum trunks were counted in Central Wadi, the victims of one such senseless conflagration. Our guide added his share during the expedition's stay by destroying the pretty cattails at the southern end of Central Wadi. Boulos (1966) reports a further conflagration a few months prior to his stay at Kurkur during December of 1964. According to his information, the burning practice was intended to destroy old halfa grass and promote new growth, more palatable as animal pasture. The burning of palms would be "accidental" in this context.

rests on an unknown depth of sand at this point. The salt-encrusted Chalk bedrock standing over the southeastern edge of this shallow swale shows sag and slump effects, suggesting a fossile sinkhole, formerly some 50 meters or so in diameter. The downstream part of Central Wadi confirms this impression; water and wash are largely absent at the surface, with water reemerging from a sandy stretch over the impermeable Kurkurstufe at Km 1.4 to Km 1.5 South, an area that supports a growth of reeds, cattails, camelthorn, and some low tamarisk bushes (*Tamarix amplexicaulis*). In view of the peculiar wadi gradient and the nature of the

Fig. 7–3. The South Well area of Kurkur Oasis, looking south. The karstic depression is located left of the crest of the major parabolic dune. Dead and living dum palms amid halfa grass, with camelthorn in foreground and on the dunes.

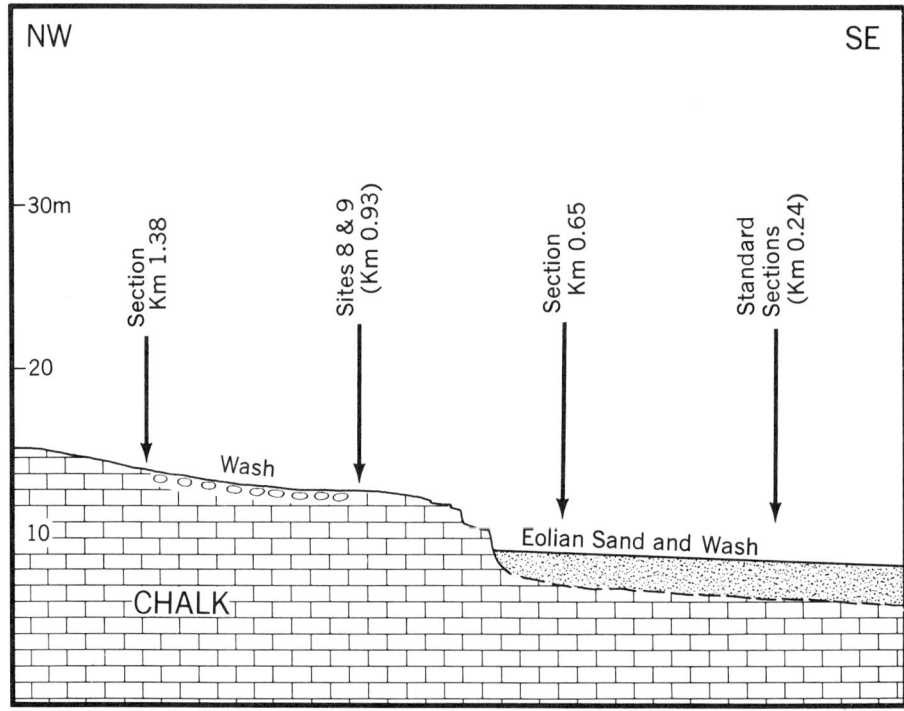

Fig. 7–4. Longitudinal profile of Northwest Wadi. For cross sections at Km 0.65 and Km 1.38 see Fig. 7–18. Map: UW Cartographic Lab.

bedrock geology, there must be a small subterranean passage at the base of the Chalk, fed by the former South Well sinkhole (Fig. 7–5).

The lowermost 200 meters of Tufa Wadi support dum palms, succulents, and camelthorn on eolian sands which mask the contact between the Chalk and the Kurkurstufe (Fig. 7–6). Finally, the sandy alluvium of South Wadi is sufficiently deep that, with the limited supply of seepage over the sill of the Kurkurstufe in Tufa and Central Wadis, little or no water is available for vegetation.

In conclusion, water obviously accumulates in the Chalk, particularly at its base, over the impermeable Kurkurstufe. Consequently, wherever the wadi tributaries dissect the lower strata of the aquifer, some vegetation is found. Near the contact with the Kurkurstufe, locally at 302 to 308 meters, seepage is sufficient to support palm growth wherever moisture can be retained in, and slowly released from, alluvial or eolian sands. The oasis is thus located along a spring line for about 2 kilometers of the

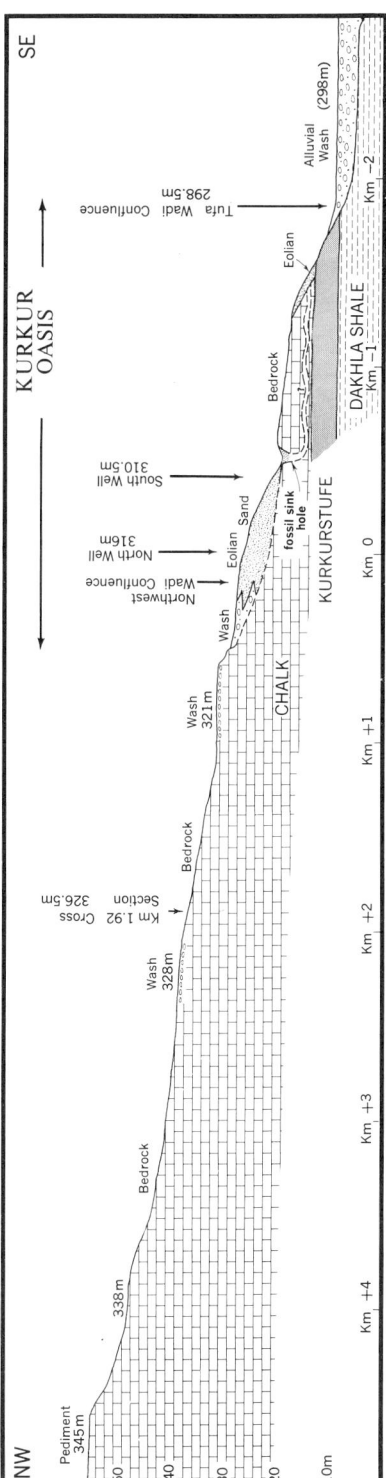

Fig. 7-5. Longitudinal profile of upper Wadi Kurkur (North, Central, and South Wadis). Map: UW Cartographic Lab.

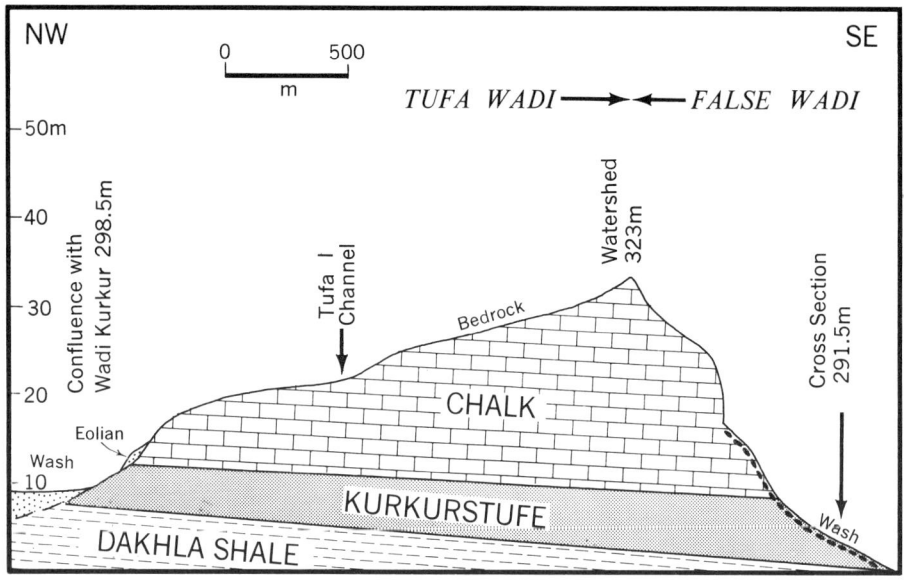

Fig. 7–6. Longitudinal profile of the watershed between Tufa Wadi (Wadi Kurkur) and False Wadi (Wadi Abu Gorma). Map: UW Cartographic Lab.

wadi's length. Vegetation ceases abruptly at the point where the wadi floor cuts through the Kurkurstufe into the "dry" shales below.

Because not all of the scarp wadis tap a seepage line just above the Kurkurstufe, Wadi Kurkur appears to drain a large groundwater basin within the Chalk, one at least as extensive as the modern 25-square-kilometer drainage basin of the uppermost wadi. The WNW dip of the strata beyond the western Kurkur watershed seems to preclude a distant origin for the local groundwater resources which must, theoretically, be attributed to intermittent rainfall over the drainage basin and its environs. A long-term mean annual precipitation of a few millimeters[4]

4. The mean annual precipitation for Dakhla Oasis is 1 millimeter; for Kharga, 0 millimeters; for Aswan, 3 millimeters (Jackson, 1961). Yet, according to the expedition guide (a fellah from Darau), the Ababda only visit Kurkur "if it has rained," as the vegetation will then improve as the water rises. Since the Egyptian peasant is not wont to associate rainfall with the groundwater table, there may be some truth to the idea. Evans-Pritchard (1935) appears to have heard a similar report from his guide, as he notes that pasture is reputedly better in summer than in winter. It seems most logical, however, that an occasional winter rainfall of some intensity or duration will thoroughly wet the sand veneer at the oasis, thus providing extra moisture for local plant growth as the waters rapidly percolate downwards.

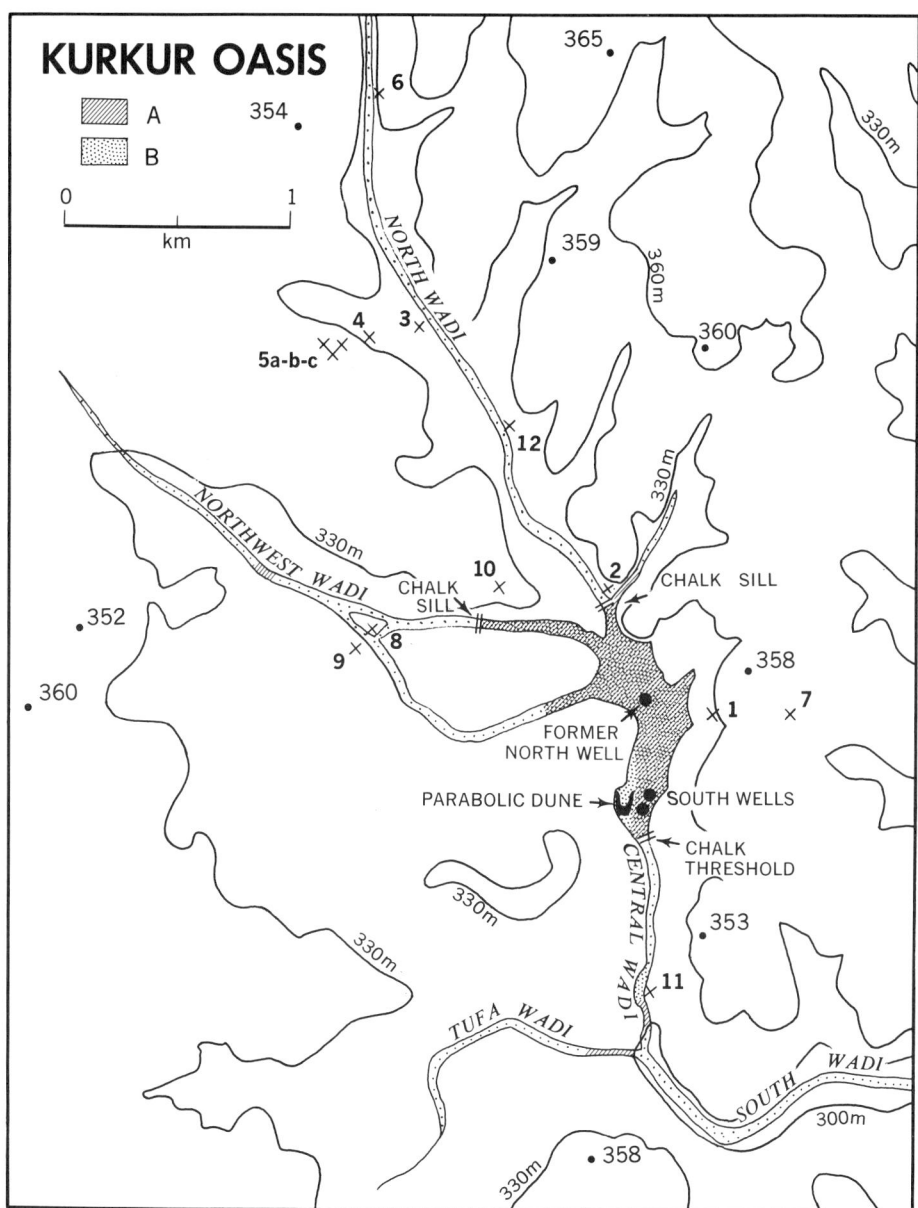

Fig. 7–7. The modern oasis and location of archeological sites. A: Major areas of vegetation, B: Recent eolian deposits. Topography modified after Murray (1939) and 1:100,000 Survey of Egypt map no. 12/72 (AMS 5772). Contour interval 30 meters. The sites are discussed on pp. 390–92. Map: UW Cartographic Lab.

cannot conceivably recharge a groundwater table or supply the seepage water evident at Kurkur, however, and, although the groundwater cannot be related to the artesian waters of the Nubian Sandstone, some obscure hydrogeological factor of tectonic, or possibly karstic, origin must be involved.

Evolution of the Kurkur Pediments

The oldest surface of the Kurkur area is the structural platform of the Libyan Tableland. The general level of the Nummulitic Limestone over much of the hammada between Kurkur and Kharga appears to be 450 ± 25 meters. This suggests that the highest elevations of the Nummulitic Limestone (Gebel el-Barqa, 507 m; Gebel Gharra, 550 m; and the unnamed ridge at 521 m elevation, about 30 Km NW of Gebel Gharra) may be erosional remnants of the original limestone platform. If this were so, the 450-meter surface might be interpreted as an erosional plain cut 75 meters or so below the original structural platform. Because southern Egypt has been emergent since the middle Eocene, the age of the possible 450-meter surface could well be Oligocene. Further speculation based on the information presently available seems futile, however.

The Kurkur pediments are obviously younger than the Libyan Tableland *sensu strictu*. The earlier, higher-lying pediment is best represented by a number of small, mesaform Chalk residuals lying south and west of the oasis (Figs. 7–8, 7–9), merging into a subcontinuous surface farther north. These range in level from 352 to 358 meters. Flat Chalk residuals, capped by Plateau Tufa, range between 355 and 368 meters elevation, indicating better preservation under these later sediments. The residuals are part of an ancient erosional plain at 360 to 365 meters. By inference from the younger, more intact surface, pedimentation was the process responsible. It is no longer possible, however, to reconstruct the paleogeography of Pediment I.

The major pediment (II) is recorded by the broad, level plain extending over at least 75 square kilometers to the west and southwest of the oasis (Figs. 7–8, 7–9) at 330 to 340 meters. Where fossilized by the overlying Plateau Tufa, the level ranges from 339 to 341 meters. The uppermost bedrock stretch of North Wadi is graded onto a level of 337 meters (Fig. 7–5), while the headwaters of Wadi Kurkur and Wadi Umm Seiyala are nothing but a smooth pediment pass at 345 to 350 meters, fingering between the dissected residual masses of Pediment I (Fig. 7–10). Analogous erosional surfaces are preserved in the wadi

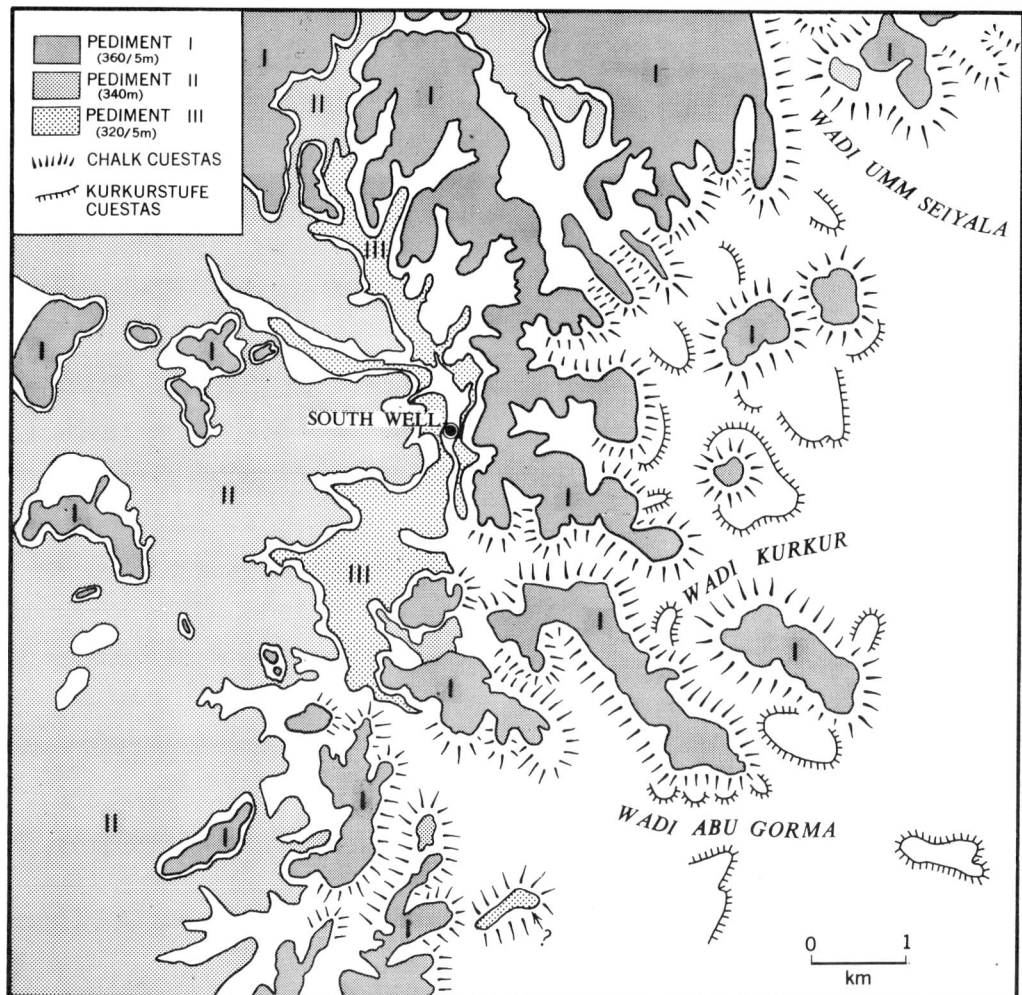

Fig. 7–8. Generalized geomorphology of the Kurkur area. Map: UW Cartographic Lab.

between the oasis and Wadi Umm Seiyala (Fig. 7–8), suggesting that the same base level pertained for the whole area. Although several Pediment I residuals stud the divide between South Wadi and False Wadi, the uppermost Kurkur and Abu Gorma drainage basins were not yet independent.

Despite the fact that Pediment II surfaces are verified overlooking the modern Chalk Cuesta, it is not certain whether the drainage moved east or west at the time of pedimentation. As does the bedrock inclination, Pediment II dips westwards at about 0.5%, towards the foot of the

Fig. 7–9. Pediment I (residuals in left and right backgrounds) and Pediment II (dominant surface), seen from 376 meters elevation in the False Wadi

Nummulitic Cuesta, which lies 15 to 20 kilometers to the west. Whether this regional dip was also present at the time of development is debatable.[5]

The relationship of Pediments I and II to the overlying Plateau Tufa is critical. The latter rests on both surfaces, and various sections show that

---

5. The form lines of the revised topographic sheet "Kurkur" at 1:100,000 (A.M.S. Series P 677, No. 5772) indicate eastward drainage from the Nummulitic Cuesta. A gently undulating plain with very shallow depressions, peripherally studded with hills, appears to lie between Kurkur and this cuesta. Possibly a north-south disturbance zone is situated a few kilometers east of the Nummulitic Cuesta, playing a role similar to that played by the Rofa Fault near the foot of the Chalk Cuesta. This could adequately explain the local existence of shallow depressions from which water-transported materials are ultimately removed by deflation.

headwaters, looking south. Plateau Tufa in left foreground with talus veneer grading downslope into colluvial scree.

the freshwater limestones were deposited on an irregular bedrock surface ranging in elevation from 339 to 368 meters. The two distinct surfaces thus fossilized can be readily observed in the field. Subsequently, both pediments were partly exhumed and were denuded another 10 meters or so. Since the Plateau Tufa is probably of Pliocene age, Pediments I and II could have formed anywhere between the late Oligocene and the early Pliocene. Since the Ballana and Aswan Pediplains are widely developed in Lower Nubia (see Chap. 5), the older pediments at Kurkur may well have been cut during the early Miocene. In fact, the Kurkur pediments can be chosen as the *locus typicus* of the oldest Nubian erosional surface, the Kurkur Pediplain.

The final erosional level, designated as Pediment III, is limited to the headwaters of Wadis Kurkur and Abu Gorma (Fig. 7–6, 7–8). The evidence consists of the lower bedrock floors of North Wadi and North-

Fig. 7–10. Pediment pass between North Wadi (left) and Wadi Umm Seiyala, looking north from Km 4.5. The lower level is part of Pediment II; the upland consists of remnants of Pediment I. The cuesta of the 425-meter surface of the Libyan Plateau is faintly visible on the left horizon.

west Wadi, both graded to 321–22 meters, and the erosional surface and related remnant hillocks at 322 to 325 meters between False Wadi and Tufa Wadi (Fig. 7–6). The very nature of these surfaces indicates that both Wadi Kurkur and Wadi Abu Gorma were already in existence. Pediment III is, in fact, nothing but a broad, flat valley cut about 10 to 20 meters below Pediment II and apparently formed after or during a long period of headward wadi erosion in the Kurkur area.

Nowhere does the Plateau Tufa rest on Pediment III. This suggests, but does not by itself prove, that the pediment is younger. Since the modern drainage lines have no relation to the Plateau Tufa, and since Pediment III grades over fairly smoothly onto the denuded 330-meter surface of Pediment II in Northwest Wadi, it seems likely that Pediment III was denuded simultaneously with, or subsequent to, exhumation of

Pediments I and II, long after deposition of the Plateau Tufa. Pediment III may therefore tentatively be considered as Pleistocene in age.

The Plateau Tufa

The lithofacies of the Plateau Tufa is best illustrated by a composite of exposures from several tufa residuals in the headwaters of False Wadi (see Fig. 7–11 and Figs. 7–14 to 7–16).

a) Lower Unit. Up to 5 meters of pink to reddish-yellow (7.5 YR 7–8/5–6), semicemented, clastic tufa products embedding dispersed or locally concentrated coarse limestone gravel with large quantities of calcareous nodules (averaging 0.5 cm in diameter). The characteristic matrix is a mass of coarse sand to granule-grade, subangular tufa debris, consisting of calcite with a considerable residual content of clay and medium-to-coarse quartz sand. The original crumblike structure has been cemented into a highly porous, semicontinuous network. Lateral and vertical facies variations may include lenses of subrounded tufa debris in a marl matrix, as well as dense, travertine flowstones.[6] The gravel is uniformly rounded but only moderately flattened, suggesting transport by both sliding and rolling (Table 7–1). The sandy matrix, as well as the dominantly clastic character of the tufa, implies torrential water flow. Rare limonitic staining may suggest former groundwater oxidation.

b) Upper Unit. Some 10 meters of very-pale to light-yellowish-brown (7.5–10 YR 6–7/4–6), semicemented-to-cemented, massive travertines, organic tufas, and clastic beds. The subfacies deserve separate description.

   i) Dense, amorphous, freshwater limestone, with veins of secondary crystalline calcite and traces of original flowstone structure.
   ii) Undulating flowstones of cryptocrystalline calcite, with lam-

---

6. *Travertines* are here defined as dense, banded, cryptocrystalline calcite precipitated as horizontal, undulating, or bulbous bands, commonly alternating from dense calcite, crystallized with columnar structure, to porous calcite with little or no macroscopic crystalline structure. *Tufas*, on the other hand, are formed through precipitation of calcite onto growing or decaying plants, leaving an inhomogeneous, spongy, cellular, and often brittle rock. The stems, twigs, grass blades, or leaves are preserved as open casts, by calcite replacements or as a partial cast-filling of lime sand. Soft, porous calcite sands commonly fill the spaces between the vertical or horizontal plant structures.

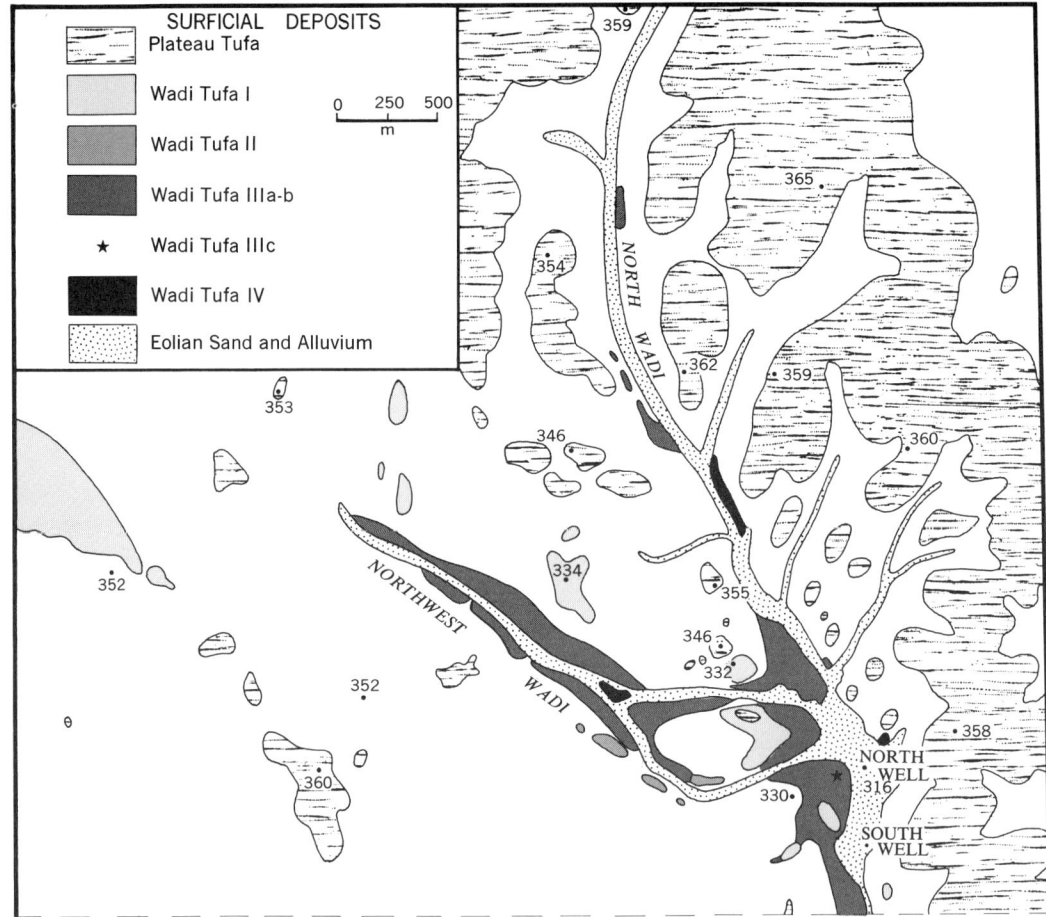

Fig. 7–11A.

inae of alternating, dense or porous, vertically crystallized calcite, varying from several millimeters to 20 centimeters in thickness.

iii) Variable tufas related to former organic structures, in which fine, horizontal grass, stem, branch, or rare leaf impressions have been partially filled with secondary cryptocrystalline calcite. The once porous or cellular matrix of the calcite precipitates and lime sand has been altered by solution (subcontinuous hollows up to several millimeters in diameter), recementation (crust-like surfaces), and alteration *in situ* (reddish-yellow silty products).

iv) Finally, the clastic beds are analogous to the Lower Unit but frequently show topset and foreset bedding and include de-

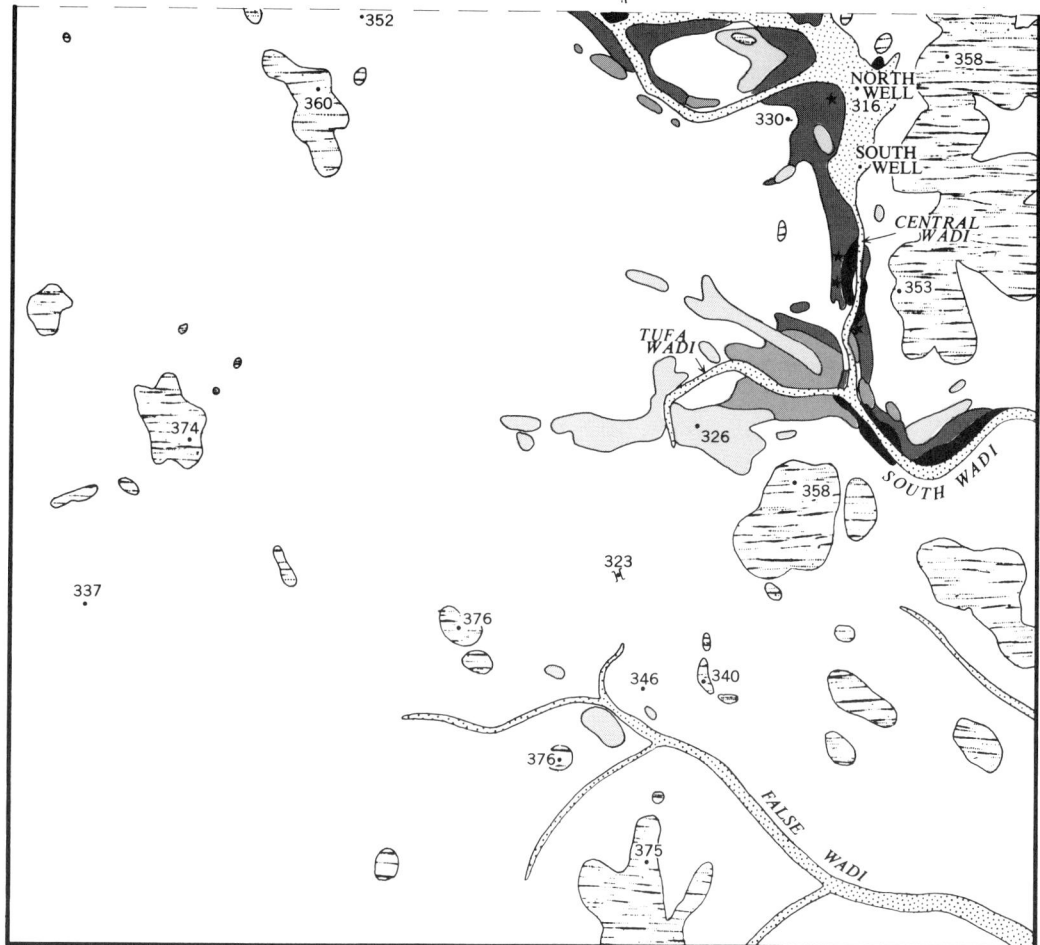

Fig. 7–11B.

Fig. 7–11. Surficial deposits of the Kurkur Oasis. All elevations in meters. Map: UW Cartographic Lab. (This map also available at 1:25,000. See p. 524.)

rived, calcareous stem casts. The nodules are strongly corroded, and the gravel is less well sorted and less homogeneous in terms of rounding characteristics.

Subfacies *i* and *iv* are dominant at the base, while *ii* and *iii* are more characteristic of the higher portions of the Upper Unit. In general, the clay and silt content is rather less than in the Lower Unit, and the quartz sand inclusions are finer grained. Two leaf impressions of *Ficus* sp. were recognized from subfacies *iii* in the Plateau Tufa residual shown in Fig. 7–14.

Table 7-1. Morphometric gravel analyses

Locality	Stratigraphy	Sample size	ρ (%)	CV of ρ	Detrital component	E/L	E/l	L (cm)
South Wadi	Tufa IV	50	35.8	37.9	0	44.4	60.9	3.4
North Wadi	Tufa IV	60	12.9	80.4	42	36.8	49.9	3.7
Foreland	Tufa IIIa	85	63.4	25.1	0	45.8	58.2	4.8
Central Wadi	Tufa IIIa(2)	75	53.5	32.5	0	45.8	58.4	3.9
Central Wadi	Tufa IIIa(1)	40	30.1	53.2	3	38.7	51.4	3.9
False Wadi	Plateau Tufa	75	53.3	29.8	0	53.8	68.7	3.5

Surface weathering of the Plateau Tufa (Figs. 7–12, 7–13) has produced a light to brownish-gray (10 YR 5–6/1–2) patina in the case of the Lower Unit and a pinkish-gray or light-gray to light-yellowish-brown (7.5 YR 6–7/2, 10 YR 6/1, and 10 YR 6/4) patina on the Upper Unit. A fair amount of coarse talus is lodged on the flanks of these residuals, although patination is generally found on most rock faces, suggesting limited movement only.

A unique but significant remnant of an ancient red paleosol is preserved near the site of the *Ficus* impressions. It is a typical, cemented

Fig. 7–12. Plateau Tufa resting on Chalk east of North Well. Contact under overhang. Note vertical "organ-pipe" travertines.

Fig. 7–13. Looking north across South Wadi, with Plateau Tufa resting on Chalk at horizon. Tufa I represented by three dark, irregular, low hills (+ 22 m) in middle to right center; smooth-surfaced Tufa II remnant (+ 16 m), left center; and Tufa III forming distinct continuous terrace (+ 9 and + 6 m substages) along north bank of wadi. A low wedge of Tufa IV gravels (+ 2 m) rests below the Tufa III terrace, right center.

limon rouge (as known from the Mediterranean Pleistocene) (Butzer, 1963a, b), occurring in 5-centimeter bands and cementing corroded 4- to 20-millimeter fragments of Plateau Tufa detritus. The matrix is a reddish-brown to light-red (2.5 YR 5/4, 6/6) soil product of lime sand, quartz sand, and clay, fully cemented into a cryptocrystalline groundmass and later intruded with clear, secondary calcite in veinlets and hollows. This Red Breccia is either a surface deposit or a crack filling and is considerably younger than the Plateau Tufa. It demonstrates the development of a *terra rossa* soil on top of the tufa. Similar material, widely used as a raw material by Paleolithic flint knappers, was found in place on bedrock at Km 4.48 in North Wadi, and Leuchs (1913b) reports numerous outcrops from Gebel el-Barqa.

Some form of soil development preceded the Plateau Tufa. In Central Wadi, near North Well, the Upper Unit rests on a 1-meter-deep zone of former physical and slight chemical alteration in the Chalk. Despite recementation, this appears to be a C_I-horizon, the base of a truncated soil once developed on Pediment II. Occasional hematite-stained quartz sand occurs within the local tufa.

Practically all of the Plateau Tufa exposures pertain to the Upper Unit, suggesting that the Lower Unit is limited to former channels or to topographic hollows. The maximum local thickness for the Upper Unit is 17 meters, measured just east of North Well. Its maximum elevation is 379 meters, south of False Wadi; its apparent minimum elevation, 339 meters, west of Km 1.92 in North Wadi (Fig. 7–24). Consequently, the Upper Unit was originally at least 40 meters thick, and, if the Lower Unit has true stratigraphic significance, total thickness of the Plateau Tufa must have been over 50 meters.

The Plateau Tufa is extensive in the Kurkur region, covering 12 square kilometers in the area mapped (Fig. 7–2). Similar tufas also occur along the limestone cuesta 50 kilometers further south, at and near Dungul (Shata, 1962; Said and Issawy, 1965). In terms of appearance, location, and stratigraphy, the Plateau Tufa here is identical to that described from below the Eocene scarps at Kharga by Miss Gardner (1932b and Caton-Thompson, 1952: 140 and Plate 128). At Bulaq Pass, Plateau Tufa occurs over a 2.5-square-kilometer area at elevations of up to 376 meters and is at least 10 meters thick. At East Rizeikat, the same formation extends over a surface of 8.5 square kilometers to a maximum elevation of 360 meters. Rather insignificant patches are also found at Gebel Umm el-Ghenneiem and at the Matana Pass (0.3 square kilometers at an elevation of 339 meters). At Kurkur, Dungul, and Kharga, the Plateau Tufas are all found in analogous topographic situations—resting on Chalk bedrock, near or at the major cuesta of the Libyan Tableland, and at elevations of about 340 to 360 meters.

The Plateau Tufa at Kurkur must be attributed to a very long period of complex alluviation of stream and spring deposits that extended for at least 15 kilometers as a fairly level surface beyond the Chalk escarpment towards the Nile. As at Kharga and Dungul, the lateral extent of the Plateau Tufa along the edge of the scarp is sharply restricted, suggesting relation to long-extinct drainage lines. During an extended period of dissection and denudation, most of the Plateau Tufa was removed, prior to deposition of the younger tufas in the wadis. As in the case of the

Wadi Tufas, selective erosion of unconsolidated silts may have favored preservation of resistant travertines and calcified tufas. The dominance of spring deposits may therefore be misleading. Although much of the Plateau Tufa was probably deposited by widespread, shallow sheets of lime-charged waters, at least a part of the Plateau Tufas represent clastic bed-load deposits. In fact, the multitude of facies emphasizes that deposition was polygenetic.

Age of the Plateau Tufa and the Local Cuestas

Prehistoric artifacts and fauna are lacking in the Plateau Tufa, and, despite the obvious correlations with Kharga and Dungul, no direct stratigraphic link with the Nile Valley is available. However, Sandford (1934: 23–35) has published systematic observations on massive travertine and marl sequences—often resting on or capped by red breccias—from the Nile Valley margins west of Nag Hammadi and Asyut. These are interpreted as freshwater deposits of major tributaries emptying into the Upper Pliocene Gulf at elevations of less than 180 meters. As is implicit from the work of both Miss Gardner (1932b, 1935) and Sandford (1934: 16–17), the similarity of the Nile Valley travertines to the Plateau Tufa is striking. On the basis of their elevation, about 200 meters higher than the Pliocene Gulf beds, the Plateau Tufas at Kurkur and Kharga can hardly be correlated with the regressive estuarine or alluvial deposits of the later Pliocene. Neither is there a compelling genetic relationship between the two series, unless it be climatic. But, in the absence of contradictory evidence, the Plateau Tufas may tentatively be considered as Pliocene in age. This is in accord with the palynological results (see Appendix I).

The general surface of the Plateau Tufa at Kurkur tends to dip eastward near the scarp at an angle similar to that of the underlying Chalk (Figs. 7–1, 7–2). In other words, the tufas have also been subject to sag and slumping, following encroachment by the local cuesta. Large, dislocated blocks of tufa occur at low levels in the Bulaq Pass (Caton-Thompson, 1952: Plate 128) and near Dungul, 10 kilometers southeast of the scarp (Shata, 1962). This evidence of landslips suggests subsequent dissection of both the tufas and the Chalk bedrock. The implication at each locality is that appreciable scarp retreat, on the order of 10 to 15 kilometers, followed deposition of the Plateau Tufa.

The cuestas as such, however, must predate the Plateau Tufa. The very existence of Pediment I in Miocene times presupposes a Nummulitic cuesta further west, and the fact that the Nile Valley at Kom Ombo and Aswan had been excavated to at least modern sea level, prior to the Middle Pliocene Gulf, shows that by the early Pliocene, a Chalk cuesta must already have existed between Kurkur and Aswan. But this cuesta may have been located midway between its present position and the Rofa Fault—i.e., some 10 to 15 kilometers further east—during the development of the Kurkur pediments and the subsequent period of tufa deposition.

Landslide and Breccia Formations

The overall dissection of the Plateau Tufa at Kurkur included exhumation of the ancient pediments, retreat of the Chalk Cuesta, cutting of deep bedrock valleys near the scarp edge, and, probably, development of Pediment III. The exact temporal and genetic relationships between these several phenomena cannot be reconstructed today, although a reasonable but hypothetical scheme of evolution is suggested in Table 7–2. The cuesta wadis had apparently attained their present dimensions prior to deposition of the earliest Wadi Tufas[7] during the middle Pleistocene, however. This is illustrated by landslide and slump phenomena preserved within False Wadi, to below 290 meters elevation, along the sides and below the floor of the wadi. They have been partly fossilized by beds of Wadi Tufa I (Figs. 7–14, 7–15). Great masses of Chalk strata sag with 5% to 10% dips or have moved 30 meters or more by rotational sliding down the wadi face. In addition to downcutting and cliff undermining by active wadi erosion, lubrication and sliding were involved, suggesting a moister climate. The absence of slumped blocks of Plateau Tufa among the landslide debris implies that the Plateau Tufas had already been largely exhumed. This also indicates that the period of dissection which preceded the landslides postdates Pediment III.

Possibly contemporary with the landslide activity in False Wadi are the karst hollows and caverns that developed between the Chalk and the overlying Plateau Tufa, with heights of as much as 2 meters. Groundwater solution was probably related to spring feeders between the tufa uplands and the wadi floor. Such spring feeders were later filled with breccias and *terra rossa* soil sediments derived from the uplands.

---

7. The term *Wadi Tufa* is used to designate all alluvial deposits younger than the Plateau Tufa. These deposits include both freshwater limestones and gravels, generally following wadis previously excavated into the older tufas or into bedrock. The designations Tufa I to IV serve to indicate stratigraphic units of complex facies.

Table 7–2. Tentative geomorphic evolution of the Kurkur area during the late Tertiary and early Pleistocene

20.	Dissection of Tufa II (vertical differential 15 to 20 m).	MIDDLE PLEISTOCENE
19.	Wadi Tufa II. Aggradation of complex + 16 to + 20 m wadi terrace. Pluvial erosion of uplands followed by accelerated spring activity.	
18.	Limited fluvial activity with accumulation of eolian quartz sands.	
17.	Dissection of Tufa I (vertical differential 25 m) and establishment of modern topography.	
16.	Wadi Tufa I. Aggradation of complex + 25 to + 35 m wadi terrace: clastic basal facies followed by tufas and travertines. Pluvial erosion of uplands (possibly with related Red Breccias on surface and in karst caverns), followed by accelerated spring activity.	
15.	Limited fluvial activity with accumulation of eolian quartz sands.	
14.	Bedrock dissection at scarp edge from Pediment III to foothill level (vertical differential 40 to 80 m); active retreat of Chalk Cuesta. Deformation of local strata through sag and slumping, as a result of undermining by stream and groundwater erosion. Local landsliding. Karst development.	BASAL AND LOWER PLEISTOCENE
13.	Lateral planation with development of Pediment III at 320 to 325 m.	
12.	Bedrock dissection between Pediments II and III (vertical differential 15 to 20 m).	
11.	Dissection of Plateau Tufa, with exhumation of Pediments I and II (vertical differential 40 m).	
10.	Chemical weathering, with development of *terra rossa* paleosol.	
9.	Alluviation of Plateau Tufa, Upper Unit (over 40 m of travertines with organic tufas and clastic beds). Significant spring activity.	PLIOCENE
8.	Erosion.	
7.	Alluviation of Plateau Tufa, Lower Unit (over 5 m of clastic beds with some travertines). Pluvial erosion on uplands.	
6.	Limited fluvial activity, with accumulation of eolian quartz sands.	
5.	Lateral planation, with development of Pediment II at 340 m.	MIOCENE
4.	Bedrock dissection between Pediments I and II (vertical differential 25 m).	
3.	Lateral planation, with development of Pediment I at 360 to 365 m. Development of Nummulitic Cuesta.	
2.	Bedrock dissection between Libyan Tableland and Pediment I (vertical differential between 60 and 110 m).	
1.	Cutting of structural/erosional platform of Libyan Tableland at 425 to to 475 m.	OLIGOCENE(?)

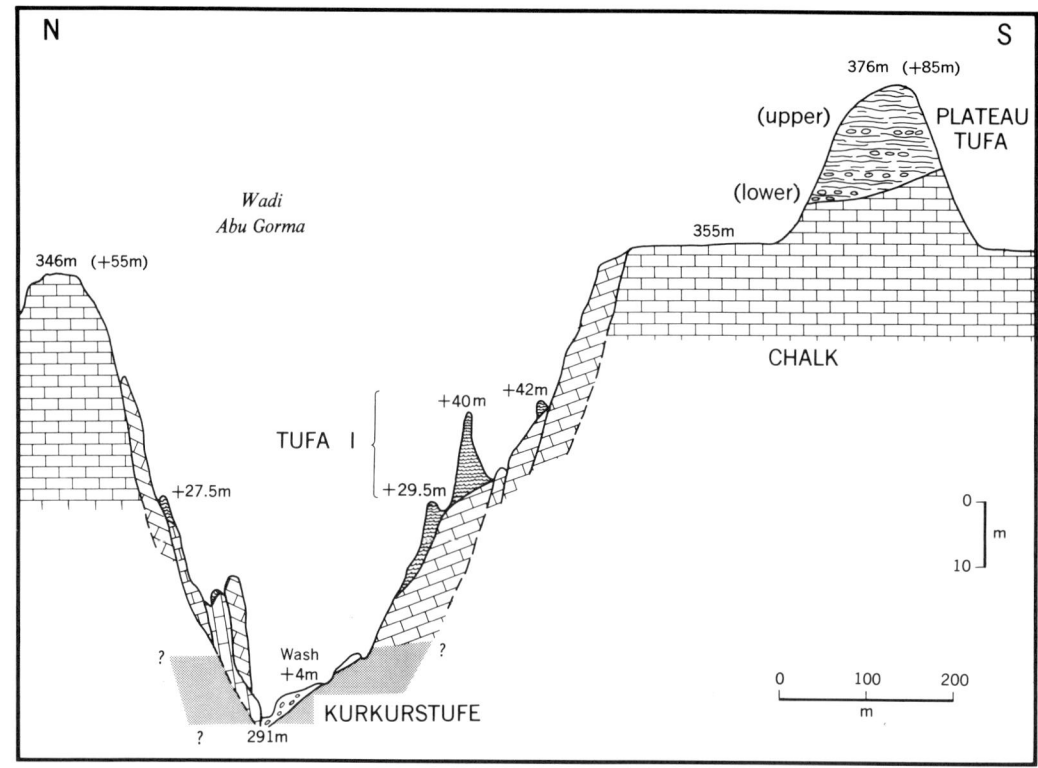

Fig. 7–14. Section of False Wadi. Map: UW Cartographic Lab.

An example of karst cavern breccias is provided at the abri overlooking Central Wadi and designated as Site 1. Here, 5 to 25 centimeters of semicemented, crudely stratified, limestone detritus and subrounded, medium gravel are set in a matrix of pink (5 YR 7–8/3–4, 5–10 YR 7/6), coarse-sandy silt. Structure of the fines is coarse, angular blocky, leaving little doubt about *terra rossa* inclusions. Another interesting exposure is provided by breccias lodged between the Chalk and the Plateau Tufa, within a great talus block lying northeast of South Well. An older, semicemented breccia, up to 1 meter thick, rests on the Chalk. The limestone detritus is embedded in a white marl. A partly consolidated breccia with a matrix of reddish-yellow (7.5 YR 6/6), sandy silt and with coarse, angular blocky structure rests disconformably on the older breccia. This younger wash has a thickness of 30 to 70 centimeters and is identical with the Red Silts found on the upland and further described below.

Fig. 7–15. Chalk slump blocks, in part surmounted by Tufa I (dark), on north side of False Wadi (see Fig. 7–14). Looking east.

The older, cemented karst cavern breccias suggest pluvial erosion on the upland surfaces. They may be contemporary with the earliest Wadi Tufas as well as with the *limons rouges* resting on the Plateau Tufa near False Wadi. Such correlations remain inferential, however.

At the Kharga Oasis, Miss Gardner (1932b; also, Caton-Thompson and Gardner, 1932; Caton-Thompson, 1952: 9) describes a "breccia" formation attaining 15 to 30 meters in thickness, deposited in broad valleys dissecting the Plateau Tufa and the Chalk. These breccias are said to show "no sign of water action," owing their origin to "weathering under arid conditions" during a long period with "little or no rain" (Caton-Thompson and Gardner, 1932: 391). Expressed more professionally by Gardner (1932b: 402), they are attributed to rapid mechanical weathering with extreme diurnal changes of temperature. The large, angular limestone blocks so produced "slid slowly downhill to fill the empty valleys formed in the previous period of erosion." The 1:40,000 map of Refuf Pass in Caton-Thompson (1952: Plate 127) shows that these breccias form two lobes projecting some 300 meters from the scarp face. In the

Bulaq Pass (Caton-Thompson, 1952: Plate 128), they form a great, dissected fan 4.5 kilometers across and 3 kilometers deep, extending down from 300 meters to 100 meters elevation.

Stratigraphically and lithologically, the breccias at Kharga are quite similar to the rather modest landslide phenomena in False Wadi. Contrary to Gardner's interpretation of greater aridity, large-scale landslips and slumping can only be explained by basal cliff sapping through active stream erosion, local karst activity, and bedrock lubrication, all under conditions of abundant moisture.[8]

The Older Wadi Tufas (Tufas I and II)

The earliest Wadi Tufas of the Kurkur Oasis are recorded by two distinct wadi terraces at relative levels of about +25 to +35 meters and +16 to +20 meters. Depending on the modern wadi gradient, these values fluctuate within a wide range (Figs. 7–14 to 7–18).

A type case is provided by the Tufa I fill preserved in False Wadi (Figs. 7–14, 7–15). The highest exposures in the gorge attain 42 meters above present wadi floor, with the relative level dropping rapidly upstream as the modern wadi gradient breaks abruptly. Two facies alternate through the former 30- to 40-meter infilling:

i) organic tufas, consisting of both horizontal and vertical stem replacements in a cellular matrix of lime sand and precipitates; and

ii) travertines, with alternating laminae of dense cryptocrystalline and semiporous, vertical mesocrystalline calcite, averaging several millimeters in thickness. Individual stratifaction of the travertines may be horizontal, undulating, or highly contorted. The latter type is confined to the perimeter of former spring vents.

In either facies, the basic material is a semicemented calcite, but the residual content varies from traces of medium-to-coarse quartz sand in the travertine, to abundant quartz sand and considerable silty clay in the organic tufa. Reflecting both the impurities and a more advanced degree of weathering, the organic tufas are reddish yellow, pink, or light brown (7.5 YR 8/6, 7/4, 6/4), while the travertines are very pale brown (10 YR 7-8/3-4).

8. Wright (1951) comes to similar conclusions in his analysis of Pleistocene landslide phenomena at Ksar Akil, Lebanon. Rognon (1967) even evokes solifluction processes to account for landslide breccias at 2,200 meters elevation in the Hoggar Mountains.

Although both facies of Wadi Tufa I occur at all levels in False Wadi, organic tufas dominate among the basal beds—which are steeply inclined along the slopes—while travertines dominate the final 7 meters or so of horizontal fill. This suggests increased ponding of the waters. The former stream ultimately deteriorated into a richly vegetated, low-gradient channel-complex of ponds, marshy ground, and spring vents with intermittent fluvial activity. Even during the earlier phases of deposition, clastic materials were rare. This emphasizes the nontorrential, spring-fed discharge.

Further upstream in False Wadi, and throughout the upper Wadi Kurkur drainage, Wadi Tufas I and II are almost invariably preserved as sinuous stream channels of travertines and organic tufas, from which the former lateral facies (less consolidated sands and silts?) have long since been removed by fluvial and eolian erosion. Fig. 7–17 shows a cross section of the Tufa I channel as intersected by a later, transverse wadi. It illustrates how one or more episodes of bedrock dissection (some 8 m) preceded deposition first of undulating travertines and then of 1.5 to 6.0 meters of blocks and boulders (derived both from Chalk and Plateau

Fig. 7–16. Block of bulbous travertine and organic tufa resting at base of Tufa I terrace, south flank of False Wadi.

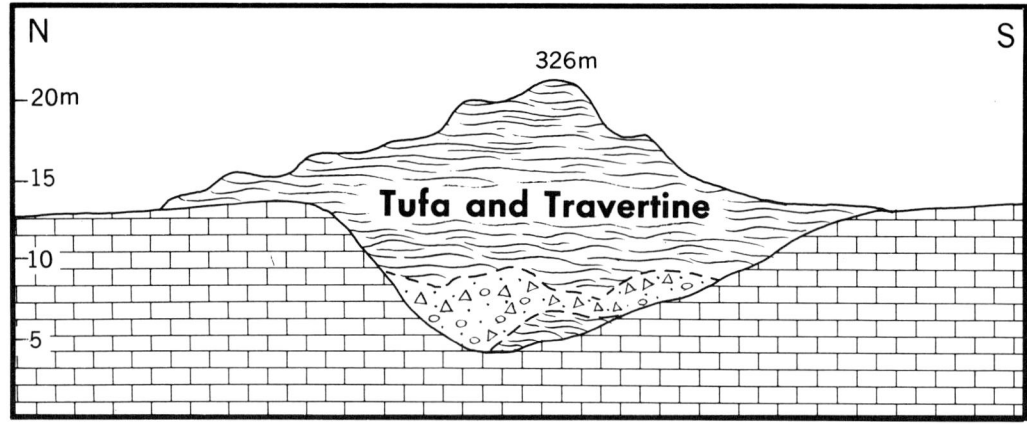

Fig. 7–17. Transverse section through Tufa I channel, now intersected by Tufa Wadi (see Fig. 7–6). Map: UW Cartographic Lab.

Tufa) interbedded with plant tufas and marls containing dispersed, medium-grade pebbles. The basal detritus is conformable with the overlying 10 to 15 meters of freshwater limestones, dominantly of undulating travertine facies. These represent the deposits of slowly moving or ponded waters in a former stream channel. Since even pebble lag is absent adjacent to the channel beds, the peripheral facies must have been both poorly consolidated and fine grained.

The crests of the Tufa I channel-fill are graded to approximately the level of Pediment II, dipping eastward at about 1%. The Tufa II channel, on the other hand, appears to be graded to the 320- to 325-meter surface. The Tufa II beds emerge below the escarpment on the Kurkur Foreland in a series of 16-meter terrace spurs, preserving several meters of rounded and homogeneous, coarse-to-cobble gravel. The gravel is flattened despite the pebbly bed load, implying nontorrential, high-velocity water flow.

As far as can be determined, the Tufa I and Tufa II terraces maintained gradients generally similar to those of modern graded reaches of Wadi Kurkur. As a result, the freshwater beds of the upper wadi can hardly be attributed to impeded drainage. Instead, they can best be interpreted through spring activity and seepage within the stream bed—particularly after the close of the rainy season—subsequent to episodic or seasonal surface runoff and wadi discharge. Thus, the surface of Pediment II may have corresponded to the water table of the Tufa I stage, the surface of Pediment III to that of the Tufa II stage. The quartz

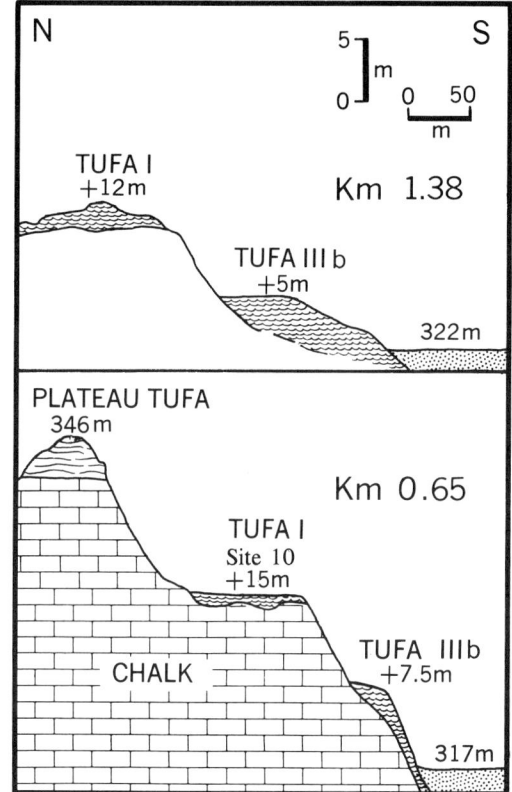

Fig. 7–18. Sections in Northwest Wadi at Km. 1.38 and Km 0.65. Map: UW Cartographic Lab.

sands included in the freshwater limestones, particularly in their basal parts, appear to have been derived from older eolian accumulations in the wadis, identical in morphology and origin to the modern eolian sands of the Kurkur Oasis. A certain proportion of this quartz sand may also have been blown into the tufas during the period of accumulation. The karst cavern breccias and the *limons rouges* must be attributed to the kind of pluvial erosion and associated slope deposition that accompanied basal aggradation of the older Wadi Tufas. At any rate, the older, cemented breccias and *limons rouges* predate Tufas III and IV.

Major dissection (at least 10 to 30 m) of both fill and bedrock, as well as changes in the local drainage, separate the Tufa I and II terraces. Two important tributaries, without modern counterparts, emptied into South Wadi from the west and the northwest during the times of Tufa I (Fig. 7–11), but a fully modern wadi geography had been established before deposition of Tufa II.

The surface of the older Wadi Tufas has a grayish (10 YR 5–6/1–2) patina, but paleosols appear to be absent. Instead, the surface has been selectively weathered, corroded, scoured, and deflated to a jagged *lapiès* with microkarstic appearance. A red paleosol is preserved, however, within the 16-meter gravel terrace on the Kurkur Foreland, e.g., within the 3 meters of gravel resting on Dakhla Shale at the point where the auto track swings south from Wadi Kurkur. The matrix is uniformly weathered to a reddish-yellow (7.5 YR 8/6), sandy to clayey silt, with coarse, subangular blocky structure. Only the gravelly nature of the deposit has preserved the paleosol from erosion. On the basis of this soil, Tufas I and II must be older than the last phase of deep, general rubefaction as recorded on the deposits of the Wadi Korosko stage in Lower Nubia. This suggests that Tufas I and II are no younger than middle Pleistocene.

Although artifacts, fauna, and macrobotanical remains were not found in the older Wadi Tufas of Kurkur, the paleosol and other, general geomorphic criteria suggest that the Tufa II terrace may, in fact, be equivalent to the Wadi-Korosko-stage gravels of the Nile Valley. Extremely probable parallels at Kharga can be cited on the basis of photographs, sections, and maps by Miss Gardner (1932b; Caton-Thompson, 1952). The Upper Sheet Gravels at Kharga are recorded by extensive and complex alluvial fans in the Refuf and the Bulaq Passes. The older of these gravels contains Upper Acheulian implements *in situ;* the younger, an Acheulio-Levalloisian industry. Unfortunately, much of the Khargan stratigraphy appears uncertain at this point, primarily because of stratigraphic identification on the basis of limited archeological typology. But the older of the Upper Sheet Gravels, with the associated Wadi Tufa I, seem to be the exact geomorphic counterpart of Tufa I at Kurkur. The younger Upper Sheet Gravels and Tufa II at Kharga can probably be correlated with Tufa II at Kurkur.

The Intermediate Wadi Tufas (The Tufa III Complex)

Although absent in False Wadi, the semiconformable Tufa IIIa/b complex is well preserved in the upper stretches of Wadi Kurkur and can be followed out onto the foreland as a series of shallow, coalescing alluvial fans extending about 20 kilometers east of the escarpment. As in the case of Tufas I and II, gravel facies dominate below the scarp (Fig. 7–21), while the dual facies of basal detritus and overlying travertines (Figs.

7–19, 7–20) is characteristic of the oasis area.

Tufa IIIa is defined as the basal aggradation of some 5 meters of marls, silts, subrounded or rounded gravel, and lenses of plant tufas. Whereas the plant tufas are almost intercalated at random, the marls and gravels commonly are laterally conformable. Pebbles include Chalk, as well as rolled fragments of older travertines, root casts, and derived nodules. The matrix of the clastic facies is illustrated at Km 0.5 South to Km 0.6 South in Central Wadi (Fig. 7–19): a very-pale-brown (10 YR 7/4), stratified and current-bedded, rather coarse lime sand with a fair admixture of silty clay and coarse quartz sand. The freshwater snail *Melanoides* is common, the land snail *Zootecus*, rare (see Appendix H). The gravel itself is homogeneously rounded near the center of the former stream, while subrounded and with a heterogeneous index of rounding closer to the channel periphery (Table 7–1). In both of these samples, pebble shape is flattish, suggesting a dominance of sliding motions.

The subaqueous marl facies of Tufa IIIa can best be described from the standard section at Km 0.24 of Northwest Wadi (Figs. 7–4, 7–11).

Fig. 7–19. Typical facies of Tufa IIIa in Central Wadi (Km 0.5 South), with well-stratified, marly lime sands underlying current-bedded gravel. James Hester recovered two small bifaces from similar deposits nearby.

Fig. 7–20. Former spring vent from Tufa IIIb terrace southwest of North Well. Surrounding beds have been deflated.

Over 5.5 meters of semiconformable Tufa III beds are exposed here, dipping downstream at inclinations of 2% to 5%:

a) Over 1.7 m (base not seen). White, unconsolidated, homogeneous marl with a trace of salts, pyrite, and a nonsoluble residue of clayey silt (No. 361–I, Table 7–3). The residue consists largely of montmorillonite, with quartz sands accounting for about 10% by total weight. Occasional cobble detritus is derived from Tufa II. Generally well-stratified and laminated, the structure is coarse, subangular blocky to platy. The deposit is marked by frequent stem casts of reeds or grasses, while *Melanoides* is very common. Intercalated with the dominant facies are

  i) White (10 YR–2.5 Y 8/2) plant tufas or travertines occurring as beds up to 20 cm thick and
  ii) thin lenses of gray (10 YR 5.5/1) to light-gray (10 YR 7/2), semi-consolidated marls with sulfides.

A marl sample from the middle of unit *a* yielded a radiocarbon age of "greater than 39,900 years" (I–2063). Even if there has been contamination with dead

Table 7-3. Sedimentological characteristics of deposits from the Kurkur Oasis

Sample number	Location	Stratigraphy	Color	Texture (noncarbonate)	Sorting	CaCO₃ (%)	pH
359	North Wadi	Tufa IV	(2.5 Y 8/4)	Coarse sand	Moderate	21.7	7.7
350	Central Wadi	Tufa IIIc(?)	(10 YR 7/4)	Coarse sand	Moderate	37.7	8.2
365–II	Central Wadi	Tufa IIIc	(2.5 Y 8/2, 10 YR 8/1)	Clay	Good	64.9	7.7
365–I	Central Wadi	Tufa IIIc	(2.5 Y 5.5/2)	Sandy clay	Moderate	59.2	7.9
364	Northwest Wadi	Tufa IIIb	(2.5 Y 8/4)	Clayey silt	Poor	25.0	7.6
358	Central Wadi	Tufa IIIb	(10 YR–2.5 Y 8/2)	Sandy clay	Moderate	82.5	8.2
361–I	Northwest Wadi	Tufa IIIa	(2.5 Y 8/2)	Clayey silt	Poor	75.7	8.2

carbonates, the true age of the sample must exceed the range of radiocarbon.

b) 2.5 m. White (10 YR 8/2), semiconsolidated, cellular plant tufas consisting over 80% of calcite. Irregularly stratified, the structure is dominated by vertical and horizontal cast fillings or impressions. Intercalated bands of white marl may attain 10 cm in thickness. *Melanoides* shells occur in local concentrations.

c) 1.5 m. Very-pale-brown (10 YR 8/3), semicemented travertines, over 88% calcite, occurring as bulbous or irregular, horizontal laminae 3 to 4 mm thick. Massive organic impressions occur locally. Bands of white marl, similar

Fig. 7–21. Conglomerates of +6 meter (Tufa III) terrace on south bank of Wadi Kurkur (19 km ENE of Kurkur Oasis, 2 km west of the Darb el-Bitan). The gravel consists of Kurkurstufe and Chalk limestones.

to facies *a*, may attain 40 cm in thickness, while lenses of pale-yellow, marly clay (No. 364, Table 7–3) with gypsum crystals, limonitic staining, and fragments of the land snail *Pupoides* may attain 30 cm in thickness. These lenses contain over 20% medium and coarse sand and suggest shallow pool deposits between flowstone sheets.

Units *a* and *b* pertain to Tufa IIIa, unit *c* to Tufa IIIb.

In general, Tufa IIIa can be interpreted by means of episodic, torrential stream flow at the height of the rainy season, leading to the removal of detrital materials along the valley margins. At the close of the rainy season, marls were deposited in stagnant wadi pools. As this stage advanced, such ponding in the center of the streams grew more important and was locally prominent, as in Northwest Wadi.

The Tufa IIIb stage refers to 5 meters of travertines (No. 358, Table 7–3), plant tufas, and calcified clays that culminated to form a +8 to +10 meter stream terrace (Figs. 7–13, 7–22, 7–24). The most typical exposures are preserved between North Well and the standard section in the form of resistant mounds of undulating or bulbous flowstones that rise 1.5 or 2 meters above travertine-choked subsurface fissures. These record the fossil vents of springs once rising under pressure from the wadi floor. These miniature "mound springs" (Fig. 7–20) show complex lateral intergrading of horizontal plant tufas and lenses of calcified lime sands or marls, probably representing small pools of spring-fed water. The silty pool deposits may contain freshwater or land snails (Appendix H) and, less commonly, chaotic masses of stem and leaf impressions, the latter tentatively identified as those of *Ficus*.

During the Tufa IIIb stage, there was little or no fluvial activity in the

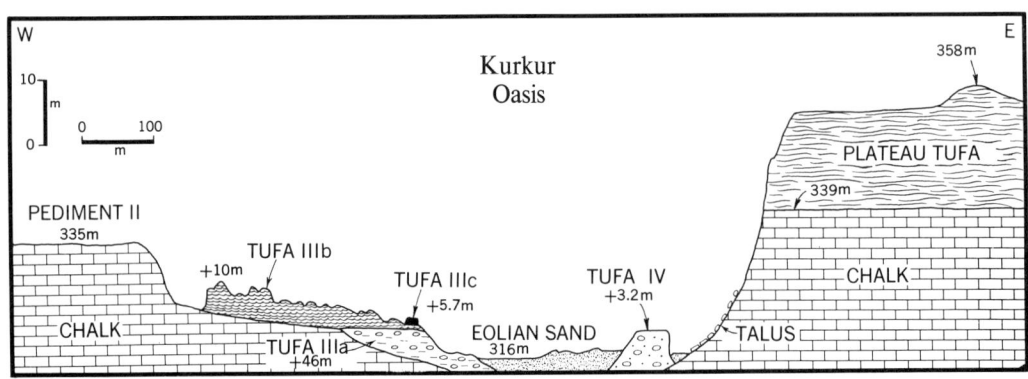

Fig. 7–22. Section of Central Wadi at former North Well. Map: UW Cartographic Lab.

oasis area, and alluviation was confined to the "mound springs" and to nearby ponded parts of the stream channel. A good mat of herbaceous vegetation was probably available to permit percolation of rainwater into the subsoil and to impede surface runoff. As a result, stream discharge was presumably distributed more evenly, and there was little or no erosion. Stream discharge was seasonal or even perennial, fed by spring waters during the dry season. There are no true lacustrine beds in Tufas IIIa or IIIb, and even the marls are stratified, and often laminated, dipping conspicuously downstream. Most of the ponds must have been rather ephemeral and shallow.

In the Kurkur Foreland, the Tufa IIIa/b complex is represented by a broad, fan-like, +6 to +7 meter terrace of coarse, rounded limestone gravel of Chalk and Kurkurstufe origin (Fig. 7–21, Table 7–1), dropping in level to +4 to +5 meters at a distance of 20 kilometers from the scarp. Even here, occasional stream-transported blocks of limestone attest to the former competence of the Wadi.

The radiocarbon date from Tufa IIIa indicates that the formation must be older than the "classical" member of the Korosko Formation in the Nile Valley, possibly corresponding in age to the Wadi Floor Conglomerate. James Hester (personal communication; also Hester and Hobler,

Fig. 7–23. Exposure of Tufa IIIc west of North Well (see Fig. 7–22). Bed *a* (gray) at level of notebook, bed *b* (white) to top.

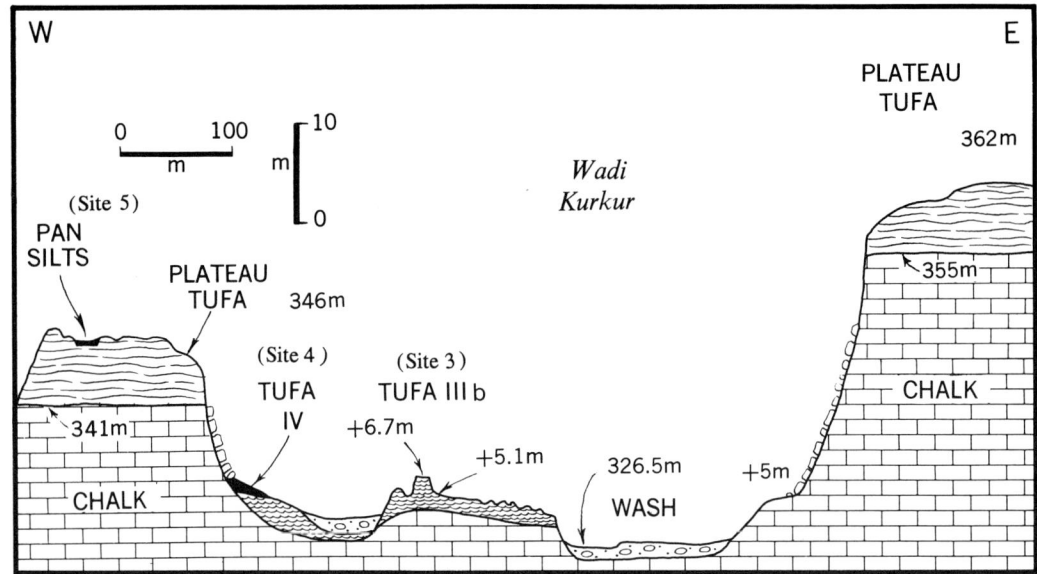

Fig. 7–24. Section of North Wadi at KM 1.92. Map: UW Cartographic Lab.

1965) found two small bifaces of Acheulian type *in situ* in Wadi Tufa IIIa in Central Wadi. Whether or not an Acheulian or Levalloisian industry of Acheulian facies was contemporary with the deposit is, of course, impossible to say. These rare artifacts may well have been derived.

Correlation of Tufa IIIa/b with analogous formations in the Kharga area seems possible, although the stratigraphic separation of gravels and tufas at Kharga (Caton-Thompson, 1952: 4–5) is unfortunate. The Lower Sheet Gravels with Wadi Tufas 3 (Lower Levalloisian) and 4 (Upper Levalloisian)—as illustrated from the Refuf Pass, Localities IV and VII, and from the Matana Pass (Caton-Thompson, 1952: 95–97, 103–05, 139–44)—seem remarkably similar in facies and morphology to the Tufa IIIa/b complex at Kurkur.

A further Tufa IIIc stage is recognized at Kurkur, although the occurrence of what appear to be contemporary sediments is sporadic and confined to the oasis area. Tufa IIIc was separated from the Tufa IIIa/b unit by a phase of downcutting which was interrupted by development of an erosional bench at +5 or +6 meters (Fig. 7–13) and was followed by incision to modern wadi floor. The type site of Tufa IIIc attains +5.7 meters west of the former North Well (Figs. 7–22, 7–23):

a) 50 cm (base not seen). Light-grayish-brown, semiconsolidated, stratified marl (No. 365–I, Table 7–3) with some pyrite. Coarse, platy structure.

b) 100 cm. White, weakly stratified marl (No. 365–II, Table 7–3) interbedded with several minor facies:

> *i*) horizontal concentrations of calcareous nodules (up to 2 cm in diameter), often coinciding with stem impressions;
> *ii*) bands of limonite-stained marls resembling facies *a* and very rich in aquatic snails;
> *iii*) 5- by 15-cm pockets of organic tufas.

A sample of white marl from the middle of bed *b* gave a radiocarbon age of "greater than 39,900 years" (I–2064).

Both beds have a nonsoluble residue of montmorillonitic clay or silty clay (Table 7–3) with some medium-grained quartz sands.

The pyrite, abundant organic matter of moder type, and a trace of hygrophile salts are common to the marly beds and suggest long periods of stagnation, resulting in temporary reducing environments. The six species of aquatic snails, unique in the Kurkur record (see Appendix H), suggest still water with decaying vegetation. A semilacustrine origin seems to be indicated.

A second exposure, possibly equivalent to Tufa IIIc, rests on Tufa IIIa/b beds at Km 1.2 South in Central Wadi, forming a +9.5 meter terrace. The sediment is a very-pale-brown, consolidated, poorly stratified, coarse sandy marl (No. 350, Table 7–3) with fossilized stem rinds and internal casts up to 8 millimeters in diameter. The 62% nonsoluble residue is a moderately sorted, coarse sand, probably derived from older eolian sands. Of a collection of artifacts found on the surface, one crude core appears to have been weathering out of the sediment.

A general interpretation of the Tufa IIIc stage does not appear possible, nor can equivalent features be recognized in the Kharga publications. The radiocarbon determination indicates that Tufa IIIc antedates the late Pleistocene deposits exposed on the Kom Ombo Plain. It may be equivalent to the lower member of the Korosko Formation.

Although the deposits of the Tufa III complex are somewhat eroded as a result of wind and water action, there is little evidence of appreciable chemical weathering. Little or no epigenetic recrystallization has taken place in the tufas, and patination is limited to light-brownish-gray (10 YR 6/2) or light-gray (2.5 Y 7/2) tones. Superficial pink discolorations (7.5 YR 7–8/4) are fairly common and may suggest modest rubefaction, but there has not been any red soil development of significance since the onset of the Tufa III stage.

The Younger Wadi Tufas (Tufa IV)

The youngest Pleistocene deposits of the Tufa IV formation are primarily developed as a low, +2.5 to +3 meter wadi terrace, reappearing at intervals from North Wadi down onto the Kurkur Foreland. The facies resembles that of Tufa IIIa but includes few travertines or plant tufas. Laterally, Tufa IV seems to be conformable with colluvial slope wash. Considerable dissection of the Tufa III beds preceded this last aggradation stage.

Tufa IV is well exposed in North Wadi between Km 1.24 and Km 1.77. The following section of the 3-meter terrace at Km 1.61 illustrates the semiconformable sequence (Figs. 7–25, 7–26):

a) Over 10 cm (base not seen). Crudely stratified, coarse limestone detritus.

b) 10 to 20 cm. White (10 YR 8/1), semiconsolidated, crudely stratified, marly coarse quartz sand.

c) 50 to 70 cm. White (2.5 Y 8/2), sandy marl with coarse, subangular, detrital limestone gravel, moved primarily by sliding motions (Table 7–1). A white, chert denticulate, with striking platform removed, was found *in situ* (identified by L. G. Freeman). It had been utilized but not waterworn.

d) 15 to 20 cm. Pale-yellow, unconsolidated, stratified, moderately well-sorted, marly coarse quartz sand (No. 359, Table 7–3). Dispersed, subrounded, fine-to-medium pebbles confirm a fluvial origin, even though the foreign sands must be attributed to reworked eolian beds. Moderate limonitic staining and some pyrite suggest a temporary reducing environment.

e) 25 to 35 cm. White (10 YR 8/2), semicemented, stratified and laminated calcite crusts (79% $CaCO_3$), including traces of quartz sand, silt, and magnesium sulfate. The crusts consist mainly of "free" and cryptocrystalline calcite accreted around small, unidentifiable organic impressions. Bands of 1-cm-thick, gray (10 YR 6/1) marls with sulfides are intercalated. A sample of the crust, taken from 15 cm below surface, gave a radiocarbon age of 29,850 B.C. ± 1700 (I–2062). Presumably, a proportion of the material consists of dead carbonate and this age may therefore be several millennia too high. This North Wadi exposure indicates a gradual facies change from coarse detritus to fluvial beds and, ultimately, to pond deposits. Scattered surface artifacts recall the local Late Paleolithic assemblages.

Exposures of Tufa IV in southern Central Wadi and in South Wadi consist of gravels with a marly matrix. For example, at Km 2.0 in South Wadi, the Tufa IV matrix consists of a white (5 Y 8/2), semiconsolidated, gypsum marl (92% $CaCO_3$) with considerable limonitic staining. The gravel itself is coarse, subrounded, and rather flattish in shape, implying a dominance of sliding in transport (Table 7–1). Equivalent deposits in the Kurkur Foreland are characterized by coarse sandy silt with dispersed, rounded gravel.

There commonly is a semiconsolidated, white colluvial wash underlying scattered, loose talus along the footslopes of the Kurkur Oasis. The material is a white to very-pale-brown (10 YR 8/2–4), unstratified, homogeneous "chalk," i.e., an accretion of calcite in the silt and fine-sand grades, intermingled with medium quartz sand, salt, and montmorillonitic clay. Seldom more than 10 to 15 centimeters thick, these silts can be attributed to inorganic accumulation of calcite by gentle surface washing and by wind. Plant structures are absent, although land snails, particularly *Zootecus*, are common (Appendix H).

Despite the lack of direct association with the Tufa IV terraces, these White Silts presumably were contemporary. Arguments are indirect. The White Silts would hardly have survived the pluvial erosion contemporary with Tufa IIIa, and, although frequently adjacent to Tufa IIIb, the travertines and the White Silts are never linked. There also are striking differences in degree of consolidation, of weathering, and of subsequent erosion between the White Silts and Tufa IIIb. Near Km. 1.92 (Site 4) in North Wadi, over 15 centimeters of homogenous white wash with *Zootecus*, base not seen, rest under 30 centimeters of reddish-yellow (7.5 YR

Fig. 7–25. Tufa IV (+ 3 m) terrace at Km 1.61 in North Wadi, facing east.

Fig. 7–26. Detail of Tufa IV terrace (see Fig. 7–25) showing surface crust. A white chert denticulate was found *in situ* near the hammer head.

6/5), unconsolidated detrital wash. A later scatter of colluvial scree and boulder talus rests on the sloping (8% to 12%) surface of the reddish wash. On various grounds, the White Silts could be correlated with either Tufa IIIc or Tufa IV. The latter correlation seems most reasonable, since typical plant tufas are absent in each case.

All in all, Tufa IV represents a considerably briefer time interval than Tufa IIIa, and—judging by the significance of salts; the angularity of the gravel; and the absence of aquatic shells, plant tufas, and typical travertines—the associated climate was far less moist. There also was no subsequent "mound spring" activity. Both Tufa IV and the White Silts show modest patination or a pink (7.5 YR 8/4) surface discoloration. If the one radiocarbon determination is accepted as approximately correct, Tufa IV would be contemporary with the "classical" member of the Korosko Formation. In that case, the last pluvial gravels in the Nile Valley and at Kurkur would be equivalent.

Caton-Thompson (1952: 123–27; also Gardner, 1932b) describes a final gravel-tufa accumulation at Kharga, attributed to damming by a landslip. The facies' change from basal clastic to upper freshwater calcites is remarkably analogous to the Tufa IV terrace in North Wadi, however. The industry found *in situ* at Kharga is described as Levalloiso-Khargan, an assemblage into which the isolated North Wadi artifact would readily fit. At the Dungul Oasis, the slightly later Khargan industry is found in surface sites resting on a tufa terrace with a radiocarbon determination of 20,950 B.C. ± 600 (Hester and Hobler, 1965).

Upstream in Northwest Wadi, James Hester (personal communication) appears to have found a younger tufa unit—a feature better represented at Dungul, where an apparent correlative yielded a radiocarbon age of 8300 B.C. ± 260 (Hester and Hobler, 1965).

The Red Silts

Red wash, with a variable addition of detrital material, is common on the upland surfaces at Kurkur, lining the floors of shallow depressions of all sizes. This wash consists of reddish-yellow to pink (5 YR 7/8, 7.5 YR 6/4–6, 5–7.5 YR 7–8/4), medium to coarse-sandy silt, including lime and quartz sands mixed with red soil products. Calcium carbonate varies from 25% to 90%; pH values lie about 7.8. Structure is coarse, subangular blocky. Lime grit and even coarse detritus may be present, but homogeneous beds are more common. Lack of consolidation and crude stratification are characteristic. These Red Silts average 20 to 40 centimeters in thickness but may attain over 70 centimeters in the surface swales

Fig. 7–27. Red Silts accumulated in crevice of Plateau Tufa southeast of former North Well.

on the uplands. They are also found in rock crevices (Fig. 7–27) and, occasionally, in abandoned spring feeders which once drained down into the oasis. The surface wash is littered with patinated detritus, some of it developing as a lag from the Red Silts, some a product of sheet-flood activity and derived directly from higher-lying tufa or limestone outcrops. *Zootecus* snails are often rather abundant.

The Red Silts represent a colluvial wash derived from eroded paleosols of *terra rossa* type, with an admixture of quartz sand ultimately of eolian origin. These silts are genetically identical with the older Red Breccias and can be attributed to washing and gradual accumulation by rare but appreciable rains. At least two generations of unconsolidated Red Silts are evident. On the east bank of North Wadi at Km 1.92 (Fig. 7–24), crack fillings 3 meters or so in depth contain reddish-yellow (5 YR 7/8) wash with three species of land snails. The top 50 centimeters of this filling are, however, distinctly coarser, pink (5 YR 8/4) in color, and contain only *Zootecus*. Multiple reworking of paleosol materials is suggested. The presence of Red Silts over White Silts at North Wadi Km

1.92 (Site 4) shows that the youngest Red Silts are younger than the Tufa IV stage. Late Paleolithic surface sites abound on the Red Silts. Although artifacts were not found within the wash, many implements show a distinctly reddish discoloration on one or more faces, and many of these pieces represent a desert lag. Systematic excavation would be necessary to prove the point, however.

The Red Silts of Kurkur are equivalent to the several generations of pediment wash in Nubia (Chap. 6) and to the Pan Silts at Kharga, as described by Caton-Thompson and Gardner (1932; also, Caton-Thompson, 1952: 4, 128–32) from Sites A through H at the Bulaq Pass and from the plateau at Gebel Yebsa. Pan Silts had already been recognized by Collet (1926) on the limestone uplands between Kharga and the Nile Valley. The Kharga Pan Silts are as much as 3 meters deep, although usually less than 1 meter. Aterian and Khargan assemblages were found on, and as much as 40 centimeters below, the surface of these deposits at Bulaq, while "Microlithic" implements occur on their surface at Gebel Yebsa.

Paleoclimatic Interpretation

CONTEMPORARY PROCESSES

The geomorphic forces at work in the Kurkur area today are rather limited and practically ineffective. Some fluvial redeposition takes place along the wadi floors following rare spates. So, for example, a flood of considerable dimensions carried several dead, charred dum palm trunks about 6 kilometers downstream at some time prior to 1934 (when they were photographed by Cuvillier, 1935a). The wadi bed materials, however, are seldom coarser than granule sand. Eolian activity in Holocene times led to accumulation of several meters of drift sand and there are two well-developed parabolic dunes, partially fixed by camelthorn, near the center of the oasis (Fig. 7–3). The prevailing wind direction would appear to be north by west. The effects of wind erosion are everywhere evident, and deflation must ultimately have been responsible for the hammada surfaces that dominate the region today.

Altogether, such limited eolian and fluvial morphogenesis in a hyperarid climate would be next to inconspicuous in the Pleistocene record. In all probability, however, conditions during a great part of the Pleistocene were very much like those of today, as is suggested by derived quartz sands of eolian origin dominating several beds of both Tufa III and Tufa IV.

GRAVEL ALLUVIATION

In general, the Pleistocene gravels of Kurkur imply greater stream competence and capacity and, indirectly, greater discharge. The morphometric gravel analyses (Table 7–1) indicate that most of the samples qualify as rounded and that only the Tufa IV gravels and mixed lateral components from Tufa IIIa are subrounded or subangular. In all cases other than the Plateau Tufa, the gravels tend to be flat (E/L under 50%, E/l under 65%), suggesting a dominance of sliding (rather than rolling) motions in stream transport. Each of the samples can be classified as a coarse gravel.

Most of the Kurkur gravels are similar in terms of shape and wear to those found in larger streams of subhumid and semiarid environments today. Although absolute generalizations are not possible, a well-defined stream network must have been present at Kurkur during several stages of the Pleistocene. Discharge was certainly torrential in character, judging by the dominantly crude stratification and lack of sorting by size. These observations, in turn, suggest irregular rains and a lack of effective vegetation mat. But, far from being quasi-extinct, the hydrography was probably characterized by periodic or seasonal runoff. Appreciably moister conditions are also suggested by the intercalated tufas and the contemporary slope wash. Rainfall, however, was certainly no higher than during the periods of tufa sedimentation.

TRAVERTINES AND PLANT TUFAS

The freshwater limestones of Kurkur can be attributed to two major agencies.

a) Carbonate precipitation by surface runoff, derived in part from spring discharge upstream and in part directly from local rainfall. Such inorganic precipitation can take place beneath shallow, slowly moving waters or in pools along the wadi floor, either stagnant or overflowing *en échelon*.[9] This process could be responsible for the

---

9. Stream-bed precipitation of cellular tufas and travertines can still be observed in the Mediterranean area today. In the middle course of the Torrente de Canyamel, eastern Mallorca, over 2 meters of soft, fibrous or cellular tufa are still in the process of deposition in a strongly ponded, perennial stream, partly choked with mosses and algae that aid in carbonate deposition. Leaves and twigs of adjacent trees occasionally float on the surface. The waters are derived, via the groundwater table, from Mesozoic limestone uplands with 700 to 900 millimeters precipitation.

cellular plant tufas. Silty or marly beds would be more common at the bottom of deeper pools; white travertines would be more likely to form along the sills between adjacent ponds, where the more rapid release of carbon dioxide in shallow, rippling waters would accelerate precipitation. Lenses of organic tufas may have been deposited in shallow pools at times of partial emergence and colonization by reeds or sedges. Other plant tufas, with horizontal arrangement of stems, presumably represent organic materials washed into the pools after rainstorms. The pyrite content and other evidence for reduction suggest that some of the pools became rather stagnant at times.

Areal spreads of travertine or related crusts are common in the Pleistocene record of North Africa and the Mediterranean Basin, where they are known as *croûtes zonaires* (Durand, 1959).[10] Much of the Plateau Tufa and some of the younger tufas are exceptionally well-developed examples of *croûtes zonaires*. They can be attributed to surface washing and sheetflooding by lime-charged waters, generally moving over a soil mantle protected from erosion by a good rooting network, if not a sod cover.

b) *True spring deposits.* Localized carbonate deposition would result from the seepage and spillover of waters around the spring vents breaking through the alluvium of the valley floor. The bulbous, pseudoscoriaceous beds and a part of the undulating travertine laminae can be attributed to such an origin. Seepage waters might collect in small pools with marl deposition, so forming the many lenticles of calcite and silt intercalated with bulbous spring travertines. Eolian quartz would be added to such waters on occasion, while pockets of bedded, dead leaves would accumulate after periodic storms.

Both classes of deposition imply the presence of flowing or ponded waters in the wadis during all or most of the year. They also require more or less perennial spring activity. A large proportion of the total rainfall appears to have entered the groundwater reservoir after acquiring dilute humic acids from the vegetation mat. Then, charged with dissolved carbonates, these waters emerged later as springs or seepage lines within the valley alluvium or at the valley margins. The annual rainfall must

---

10. They are not necessarily related to modern subsurface travertines caused by groundwater precipitation, as described from semiarid parts of Spain by Rutte (1958).

have been fairly appreciable and the groundwater table higher.

The limited importance of clastic deposits and erosional disconformities suggests that there was little torrential discharge during the travertine-tufa episodes. Moderate but protracted flow can be readily explained by an effective vegetation mat impeding rapid runoff, complemented by equalized spring flow well into or throughout the dry season. There is, then, no need to assume more than a seasonal rainfall.

It is difficult to estimate the rainfall amount. Miss Gardner, in her interpretation of the Kharga tufas, attached considerable importance to the absence of karstic phenomena (Caton-Thompson and Gardner, 1932). Despite the apparent absence of sinkholes and solution caverns at Kharga, karstic features are present at Kurkur. There is, in other words, no discrepancy between the evidence for limestone deposition and that for corrosion and solution. Nevertheless, we agree with Miss Gardner that the geomorphic evidence does not allow for a mean annual rainfall exceeding 350 to 400 millimeters at any time during the more recent Pleistocene.

Miss Gardner's analysis of the gastropod fauna and the flora of the Khargan tufas—three species of *Ficus*, hackberry (*Celtis* cf. *integrifolia*), an unidentified palm, reeds, and a fern (Gardner, 1935; Caton-Thompson, 1952: 14)—is also relevant to the problem of rainfall. She suggests that a minimum annual precipitation of 200 millimeters, with associated spring activity, could explain the biological evidence (Gardner, 1935).[11] This estimate appears to be confirmed by the fossil groundwaters tapped by the present artesian wells at Kharga and studied by Knetsch *et al.* (1963). Three radiocarbon dates of 23,450 B.C., 22,750 B.C., and 17,450 B.C. (H–1267, H–1269, H–1270, Münnich and Vogel, 1963) verified earlier deductions that these waters must be attributed to the last Pleistocene pluvial. C^{14}/C^{13} ratios of these pluvial-age waters suggest a surface accumulation environment moister than the contemporary Mediterranean littoral of Egypt (*ca.* 100—200 mm precipitation). Furthermore, an O^{18}/O^{16} isotopic temperature determination of snails from the mound spring deposits at Kharga suggest a former mean annual temperature of 62° F (16.5° C) (Degens, 1963), which is 11° F (6° C) colder than the modern mean of 73° F (22.5° C) (Jackson, 1961). From the

---

11. It is of more than passing interest that, near Hureidha in the Hadhramaut, a small, valleyside limestone spring has deposited a soft, spongy mass of tufa. The adjacent vegetation includes a few palms, two species of *Ficus*, and a fern (Caton-Thompson and Gardner, 1939).

geomorphologic viewpoint, these Khargan deductions appear acceptable for Kurkur as well.

DISSECTION

Judging by the fairly uniform patination or discoloration of the Tufa III and Tufa IV surfaces, significant dissection has not taken place for a long time. As already discussed for the case of the Nile Valley (Chap. 2), there is good reason to believe that *dissection of fill* was characteristic of the periods of slightly greater, but waning, rainfall immediately following the moist aggradation phases. *Bedrock dissection,* on the other hand, requires far more vigorous erosion and has been ineffective at Kurkur since the establishment of the modern topography, subsequent to the Tufa I stage. As in the case of the Tufa I channel shown in Fig. 7–17, bedrock dissection may have marked the onset of the more important wet phases. The evidence for slumping and landslips—for example, during the period of bedrock dissection prior to Tufa I—also suggests a moister climate.

TERRA ROSSA DEVELOPMENT

Paleosols are poorly preserved in desert environments because deflation almost inevitably removes all fine, nonconsolidated materials. Nevertheless, there is reasonably good evidence that red soils of *terra rossa* type (see Kubiena, 1953: 257 ff.) developed at Kurkur on at least one occasion, even though no undisturbed soil profiles were found. Superficial rubefaction of the Chalk and the Plateau Tufa by hematitic iron is common. More convincing are the soil sediments, designated as the Red Breccias and the Red Silts and recording one or more phases of intensive chemical weathering. *Terra rossas* no longer appear to develop in the Mediterranean limestone areas today but seem to reflect more intensive weathering under a seasonally wet, subtropical or tropical climate (Klinge, 1958; Butzer, 1964*b*: 89–90). The brilliant red hues are due to the presence of anhydrous iron (hematite, goethite), probably derived from the hydrated limonitic form through extreme desiccation during a hot, dry season. Yet the iron compounds are dispersed within colloidal silica, liberated during breakdown of clay minerals by intensive decomposition. Secondary calcification is the rule rather than the exception in *terra rossa* sediments today.

Unfortunately, the lack of fossil soil profiles or of relatively unadulterated soil sediments at Kurkur limits the interpretive possibilities of the red paleosols.

Stratigraphic Conclusions

The sequence of geomorphic evolution outlined for the Kurkur area in Table 7–2 is essentially a relative stratigraphy with few firm time-markers. Pediments I and II can probably be related to some of the very high erosional surfaces in Lower Nubia, grouped as the Kurkur Pediplain. Again, the deep red soil developed on Wadi Tufa II at Kurkur is the last distinctive local paleosol, suggesting a correlation of Tufa II with the Wadi Korosko stage in Nubia and the Low Terrace at Kom Ombo. Theoretically, regional climatic changes in southern Egypt should have been synchronous, at least within the winter-rainfall belt. Because of the contrasting bedrock lithologies, however, facies differences between the alluvia of Kurkur and of the Nile Valley are considerable. So fundamental are they, in fact, that the Kurkur record is geomorphically more analogous to that of the limestone country of the southern Mediterranean Basin than it is to that of the sandstone reaches of the Nile Valley or of southeastern Libya (see Williams and Hall, 1965). Correlation across the Lower Nubian Plain is also impeded by the lack of molluscan or vertebrate faunas in the early and middle Pleistocene deposits of both areas. Accordingly, the Pleistocene stage correlations of Table 7–2 are reasonable but uncertain.

The geomorphic events of the late Pleistocene, as summarized by Table 7–4, are better understood, and correlations with the Nile Valley are facilitated by several radiocarbon dates. Correlations with the Kharga Oasis are satisfactory, and close comparison with the analogous sequence at Dungul Oasis should be possible after publication of the detailed field studies of Hester and others.

A major contribution of the Kurkur sequence is towards an understanding of the paleoclimatic history of the Libyan Desert. The variability of Pleistocene climate (Tables 7–2, 7–4) evident in the heart of this hyperarid region is surprising. Furthermore, it is simpler to assess the comparative magnitude of late Pleistocene pluvial oscillations at Kurkur, since wadi alluviation or dissection are independent of everything but climate and are not modified by the behavior of a local base level such as Nile floodplain. Similarly, the Kurkur evidence contributes to a reinterpretation of the Kharga sequence, long regarded as a standard for the Sahara. Several of the phenomena recorded at Kharga are absent at

Table 7-4. Geomorphic evolution of the Kurkur area
during the late Pleistocene and Holocene

32.	Limited fluvial activity and accumulation of eolian quartz sands.	HOLOCENE
31.	(At Dungul and possibly at Kurkur) Wadi Tufa V. Age 8300 B.C. or younger.	
30.	Red Silt accumulation in surface swales and crevices by occasional, but appreciable, rains.	UPPER PLEISTOCENE
29.	Dissection of Tufa IV (vertical differential 3 m).	
28.	Wadi Tufa IV. Aggradation of + 3 m wadi terrace with gravel, sand, and marl. Moderate spring activity. Age 29,800 B.C. or younger. Deposition of White Silts on slopes.	
27.	Dissection of Tufa IIIc (vertical differential 6 to 9 m).	
26.	Dissection of Tufa IIIc. Aggradation of semilacustrine marl and sand at + 6 to + 9.5 m. Accelerated spring activity, with extensive valley-ponding. Age greater than 38,000 B.C. Deposition of White Silts on slopes(?).	
25.	Limited fluvial activity, with accumulation of eolian quartz sand.	
24.	Dissection of Tufa IIIa/b (vertical differential 6 to 10 m), interrupted by cutting of erosional bench at + 5 meters.	
23.	Wadi Tufa IIIb. Further wadi aggradation of + 8 to + 10 m terrace with travertines and tufas. High water-table, accelerated spring activity, and development of "mound spring" vents on floodplain.	
22.	Wadi Tufa IIIa. Aggradation of wadis to + 5 m with gravel, marl, and organic tufas. Age greater than 38,000 B.C. Pluvial erosion of uplands (with possible Red Silt accumulation in pans).	
21.	Limited fluvial activity, with accumulation of eolian quartz sand.	

Kurkur—in particular, the complicating factor of artesian wells with fossil water from the Nubian Sandstone. Despite the smaller size of the Kurkur study area, the development of individual formations is better and stratigraphic field relationships are clearer.

Prehistoric Occupation of the Kurkur Area

THE ARCHEOLOGY

Although eleven surface sites with several thousand pieces of worked stone were collected at Kurkur by David Boloyan and by us in March, 1963, these extensive collections are still being studied. With the exception of C-Group potsherds found near the modern oasis, all of the archeological materials found by the Yale group are of Middle and Late Paleolithic typology. James Hester, of the Combined Prehistoric Expedition, spent a week at Kurkur during December, 1963, collecting twenty-four sites. The following inventory of sites and cultures is given by Hester and Hobler (1965):

a) Acheulian. One site with two small bifaces found *in situ* in Tufa IIIa, near Km 0.5 (South) in Central Wadi.
b) Kedabian (Middle Paleolithic). Three surface sites, resembling Caton-Thompson's (1952) Lower Levalloisian in typology.
c) Khargan (Late Paleolithic). Four surface sites near Kurkur, corresponding typologically with the original definition of Caton-Thompson (1952).
d) Libyan (Pre-Pottery Neolithic?). One surface site with narrow blade cores, retouched and backed blades (sickle flints?), and numerous grinding stones. Radiocarbon date on charcoal: 5950 B.C. ± 150.
e) C-Group. Twelve surface sites with flake tools, incised "Nubian Ware," and some black-topped red ware. Larger sites are found only near modern water sources. Radiocarbon date: 1690 B.C. ± 180.

THE PHYSICAL SETTING OF THE SITES

The sites collected by the Yale group are found on the surface of all the Pleistocene formations. Their settings are of some interest in relation to the past geography of the area (Fig. 7–7). Consequently, physical descriptions of these sites are briefly given below.

Site 1 (Abris I and II). Located beneath a 1.5-meter-high overhang of Plateau Tufa on a Chalk floor, some 450 meters ESE of former North Well. The archeological inventory included potsherds as well as four flakes and scrapers in chert. These typical abris are relatively well protected, situated on a steep slope overlooking the greater part of the oasis.

Site 2. Situated on Tufa IIIb terrace at +7 meters on east bank of North Wadi (Km 0.6), 100 meters upstream of the rock step interrupting the wadi profile (see Fig. 7–5). A moderately rich lithic assemblage was collected from this small platform in a rectangular area of about 7 by 25 meters.

Site 3. A rather limited collection scattered across Tufa IIIb deposits that form +5 and +7 meter surfaces on the west bank of North Wadi at Km 1.92 (see Fig. 7–24).

Site 4. A moderately interesting concentration of artifacts was found near a minor tributary channel at the base of the gebel (Fig. 7–24), 150 meters WSW of Site 3, resting on top of both the Red and the White Silts exposed here. The artifacts may have been derived from Site 5a on the nearby gebel top.

Site 5 (a, b, c). Located on top of a mesaform hill, capped by Plateau Tufa, situated at 346 meters elevation or 20 meters above North Wadi at Km 1.92 (Fig. 7–24). Very shallow depressions on this small mesa are filled with 40 centimeters or more of light-brown or pink (7.5 YR 6/4, 5 YR 8/4), coarse-sandy silt, unconsolidated, moderately well stratified, and with a very coarse granular structure. Carbonates account for 34%, and the pH is 7.75. These Red Silts contain abundant *Zootecus* shells and may be attributed to reworking of a local soil with eolian components by fluvial activity. The surface is littered with a medium-grade, corroded lag of Plateau Tufa fragments, together with masses of artifacts in brownish chert and Red Breccia (*brocatelli*). Although never found within the Red Silts, a part of the lithic assemblage has a reddish sheen on one or more faces and appears to have been deflated from the wash. The gebel top has a triangular shape and an area of about 1500 square meters. This was divided into three sectors (*a*, northeast; *b*, northwest; and *c*, south) for collecting purposes. Of particular interest at Site 5b was a crude, subcontinuous ring of Plateau Tufa slabs, with an oval plan, approximately 4 meters in major axis. Possibly the slabs were used to weigh down the sides of a tent or a similar shelter.

Site 6. Between Km 3.0 and 3.2 there is a +6.5 meter terrace on the east bank of North Wadi, formed by 2 meters of Tufa IIIb deposit. A collection of Khargan implements was made from this terrace surface.

Site 7 (*Complex*). A number of rich sites on top of the Plateau Tufa east of North Well, an area known as Gebel Kurkur. The various concentrations were collected separately, but were invariably found on the lag-strewn rocky surface, or, more commonly, on shallow accumulations of Red Silts. Steep cliffs offset the upland surface (at about 350 meters elevation) from the oasis below. Gazelles still frequent this upland today and may have been trapped on the rolling, but uneven, terrain. Crude structures, possibly used as animal traps in more recent times, were found near the site area (see Reed, 1964: 12).

Site 8. At Km 0.93, Northwest Wadi bifurcates before entering the oasis proper. Wash has accumulated in midstream at this point, forming a +1.5 meter bar about 100 meters long and 25 meters wide. The material is a coarse-to-cobble gravel of rounded limestone, possibly contemporary with Tufa IV and certainly no older. An interesting, homogeneous lithic collection was made from a part of the surface of this gravel, most of the pieces in mint condition. Occupation clearly postdates accumulation of the gravel.

Site 9. About 100 meters southwest of Site 8, a modest collection of surface implements was made from an irregular +5 to +6 meter Tufa IIIb terrace, near the south bank of Northwest Wadi. The materials were rather scattered.

Site 10. The richest collection comes from a +15 meter terrace on the north bank of Northwest Wadi at Km 0.65 (Fig. 7–18). The terrace itself belongs to Tufa I, but there are moderate accumulations of Red Silts in the hollows and fissures of the tufa or travertine surface. This colluvial wash averages about 20 to 30 centimeters in thickness and consists of a pink (7.5 YR 7/4), unconsolidated, stratified sandy silt with coarse, subangular blocky structure and abundant *Zootecus* shells. The artifacts are found in extremely rich concentrations on top of this wash, among a lag of tufa fragments. Three 10-meter squares were collected by the Yale group, and the remainder of the site was collected by Hester and others in December, 1963 (Hester, *personal communication*). The raw material is predominantly a brownish chert, with a fair proportion of artifacts also made from Red Breccia. The latter are always strongly corroded. A high proportion of the artifacts has a reddish sheen, suggesting derivation from the upper part of the wash. Although lacking many *belles pièces*, the overall typology is that of a Khargan assemblage, probably a workshop rather than a habitation site.

Site 11. Located near Km 1.2 South in Central Wadi, Site 11 is a moderate surface concentration of white chert artifacts resting on a possible exposure of Tufa IIIc (No. 350, Table 7–3), forming a +9.5 meter east-bank terrace. Several artifacts were embedded in the top of this semiconsolidated deposit, and one crude core may have been *in situ* within it.

Site 12. At Km 1.61 in North Wadi, a white chert denticulate was found *in situ* in Tufa IV deposits (see site description above), some 60 centimeters below surface. It is neither patinated nor waterworn. Nearby was a crude piece of identical material, possibly a core or a waste flake. On the surface were two artifacts of patinated, brownish chert—a small waste flake and a flake with a dihedral platform and possible protoburin.[12]

---

12. The artifacts of Site 12 were kindly examined and described by Leslie G. Freeman.

PHYSICAL INTERPRETATION OF THE SITES

Several factors are common to all or many of the sites studied:

a) The majority are located on terrace surfaces adjacent to the modern wadi channels (Sites 2–4, 6, and 8–12). Under more humid conditions, vegetation would have been abundant here, and there was possibly some surface water or groundwater. Vegetable foods and game were certainly available.

b) The remaining sites can be explained by other factors. Site 1 is situated in a defensive location overlooking the oasis, under a rock overhang. Site 5 was possibly a habitation site, admirably suited for overlooking a large sector of the North Wadi valley. Site 7 on the Gebel Kurkur can be considered as potential game country. In other words, given a moister climate, all of the sites are logically situated in relation to the basic geography of the area.

c) Several of the sites, including the richest assemblages, were located on the surface of Red Silts (Sites 4, 5, 7, and 10), with a proportion of the artifacts exhibiting a reddish sheen or patina. It is highly probable that such artifacts were originally embedded in the upper part of the wash, as are the Khargan industries in the Pan Silts at Kharga.

d) The dominant raw material of Sites 1 through 10 is a light-brownish-gray chert, derived from concretions in the Nummulitic Limestone. Most of this material is not locally available near any of the sites and had to be carried in to the oasis by man. The Red Breccia used for many implements at these sites was possibly more common in the vicinity of the oasis, but is far less suitable as a raw material. The white chert employed in the tools of Sites 11 and 12, and forming a minor component at the other sites, appears to come from the local Chalk.

e) With the exception of the white-chert artifacts at Site 12 (and possibly those at Site 11 also), all of the assemblages are younger than Tufas III and IV. Since these assemblages are typologically similar to Caton-Thompson's Khargan, the Late Paleolithic Khargan occupation at Kurkur was contemporary with, or later than, the Red Silts (phase 30, Table 7–4). Although the other collections (11 and 12) are not very diagnostic, they may be Middle Paleolithic, and they would fit into Hester and Hobler's Kedabian. These stratigraphic conclusions are also significant for the geomorphic sequence, since they indirectly support the validity of the Kurkur radiocarbon dates.

In retrospect, Lower or early Middle Paleolithic populations appear to have visited the Kurkur area sporadically, but the scarcity of artifactual remains suggests nothing but ephemeral residence of small, itinerant bands. Even during the pluvial climate contemporary with the first half of the Würm Glacial, there was little or no habitation at Kurkur. Periodic occupation began about 30,000 years ago and reached a peak during the terminal Pleistocene, when semipermanent settlement may have been possible for the makers of a Khargan industry. The very existence of small hunting-gathering groups at Kurkur over extended periods of time speaks for rather better ecological conditions. Since the Red Silts record one of the last periods of greater rainfall, it is probable that the Khargan populations entered the Kurkur Oasis at about this time.

The Khargan groups may have been dispersed over various oases of the Libyan Desert, but they do not appear to have penetrated into the Nile Valley. Their ultimate disappearance from Kurkur may have been a result of declining game and water resources. The absence of Aterian sites at Kurkur—thought to be contemporary with, or slightly later than, the Khargan (Caton-Thompson, 1952: 29–31; Hester and Hobler, 1965)—suggests that the oasis may already have become too unattractive, and the subsequent Epi-Levalloisian cultures of Kharga are entirely absent both at Kurkur and at Dungul.

An evaluation of later, post-Pleistocene attempts to settle the oasis must await publication of the pottery collected by the different expeditions.

8

The Coastal Plain of Mersa Alam

Introduction

The Red Sea littoral of Egypt has attracted travellers from the Nile Valley since late Predynastic times, when rock drawings with Nile motifs were first inscribed along the major wadi routes. An inscription halfway between Edfu and Barramiya, dating from the First Dynasty pharaoh, Djet, suggests an expedition into the Eastern Desert (Edwards, 1964: 19), and the Fifth Dynasty monarch, Sahure (*ca.* 2488–2475 B.C.), sent a fleet down the Red Sea to the land of Punt (W. S. Smith, 1962: 44), somewhere in East Africa. Although a host of graffiti give testimony to pharaonic commercial activity in the Red Sea country, large-scale mining operations near the coast do not seem to have begun until Ptolemaic times.

About 275 B.C., Ptolemy Philadelphus built a new terminus to the Eastern Desert routes at Berenice (see Fig. 1–2), destined to become the entrepôt for marine commerce with Arabia and India. A sizeable settlement with over 1000 houses flourished here until the third century A.D., when nomadic unrest disrupted the southern desert routes, deflecting traffic to more northerly ports such as Quseir. The famous Emerald Mines of Gebel Zabara, to the northwest of Berenice, are first mentioned at the end of the first century B.C. by Strabo and were only abandoned in 1342 A.D. The gold mines at Gebel Sukkari (22 km southwest of Mersa Alam) and at Umm Rus (55 km northwest of Mersa Alam) were also chiefly exploited in Ptolemaic and Roman times. A mining settlement at Umm Rus, with ruins of over 300 houses, was associated with the harbour of Nechesia (Mersa Mubarek), some 7 kilometers away (Murray, 1925, 1967; Meredith, 1958).

Mining came to a standstill well before the Arab invasion, and, in Islamic times, Quseir in the north and Aidhab in the south served the pilgrim traffic across the Red Sea to and from Jidda, with some export of Egyptian wheat from Quseir also. The inauguration of the Suez Canal in

1869 dealt a deathblow to the desert traffic, and only the resumption of mining along the Red Sea littoral during recent decades has brought new activity to the area.

The vicinity of Mersa Alam is almost devoid of archeology and history. The present hamlet owes its existence to a recent phosphate mine that has attracted some fishermen to the vicinity.[1] A single Predynastic grave, cut into the low terrace of a tributary to Wadi Samadai, some 8 kilometers from its mouth, was found and described by G. W. Murray (Murray and Derry, 1923). The inventory suggests Badarian affinities (Resch, 1963b). Also of interest are the ruins of an Arab village at Mersa Nakari, halfway between Ras Samadai and Gebel Dirra. These may belong to Shawna, a harbour mentioned by a Portuguese voyager of 1541 (Murray, 1925). Paleolithic archeology seems to be virtually nonexistant, and we found only a single flake blade of dioritic rock.

Early Work on the Coastal Pleistocene

Scattered observations concerning submerged or raised coral reefs on the Red Sea littoral of Egypt date back to the nineteenth and early twentieth centuries (see Felix, 1904; Blanckenhorn, 1921: 148–49; Hume and Little, 1928).

Of fundamental importance for the coastal sector south of Quseir is the geological survey of H. J. L. Beadnell (1924). Beadnell (1924: 8, 23–26) recognized several sedimentary units of late Cenozoic age: (a) Recent wadi alluvium; (b) recent wadi terrace gravels; (c) recent coral limestones; (d) raised coral limestones intercalated with gravels (attaining over 20 meters elevation).

Due to the difficulties of field differentiation, unit d was mapped together with the *Laganum-Clypeaster* Series; similarly, units a, b, and c were grouped and mapped as one. Additionally, many raised coral beaches are incorrectly shown as "Recent Deposits." The resulting map at 1 : 100,000 has limited applicability for the post-Miocene deposits. Beadnell made no serious attempt to study Pleistocene shorelines, although his observations on the local geology remain pertinent.

The only systematic study of Pleistocene shorelines and wadi terraces along any part of the Red Sea littoral of Egypt was carried out by K. S. Sandford near Suez and between Quseir and Safaga (Fig. 1–2) at various

---

1. The heaps of spider conch (*Lambis truncata* and *L. lambis*) shell commonly found near the shore can be attributed to these fishing folk and to beduin that have recently set up encampments north of Mersa Alam.

times between 1927 and 1933 (Sandford and Arkell, 1939: 60–68). Sandford recognized marine-erosional platforms at 21 to 23 meters ("70 to 76 ft."), 14 to 15 meters ("46 to 50 ft."), and 3 to 4.5 meters ("10 to 15 ft."), as well as a possible high platform at 27 meters ("90 ft."). Abundant fossil coral reefs and gravel bars are attributed to a major shoreline stage at 6 to 7.5 meters ("21 to 25 ft."). Numerous wadi gravels, exposed to maximum levels of over 30 meters above wadi floor, are mentioned, and a 15-meter stage seems to be most prominent. Some Middle Paleolithic artifacts were found within the gravels of a 3- to 4-meter terrace near Quseir, in both Wadi Ambagi and Wadi Hamrawain. A sequence of older wadi terraces at 6 to 9, 15, 21, and 30 meters—among which the 21-meter terrace is most prominent—was further recognized in Wadi Ambagi (*ibid.*: 67).

Sandford noted that the relative levels of the wadi gravels often decreased downstream, near the coastal plain. In Wadi Safaga (*ibid.*: 62), a 34-meter terrace drops off to 7.5 meters near the coast as a supposed result of redeposition to a new base level. Near Gebel Abu Shukaili (*ibid.*: 64), horizontal wadi gravels were laid down over the coastal deposits of the 7.5-meter beach. In Wadis Kuwai and Hamrawain, the 15-meter wadi terrace drops to 7.5 meters at the coast. This is explained by aggradation "onto a sloping submarine plain that was not built up with coral to [a +15 m] sea-level" (*ibid.*: 65). At the mouth of Wadi Hamrawain, however, a 7.5-meter wadi gravel and a coral beach at the same level are thought to be synchronous, even though no direct link exists (*ibid.*: 65). In short, Sandford believes that each wadi stage was, or should have been, graded to corresponding high sea levels. The rather common and disturbing phenomenon of oversteepened wadi gradients is explained by subsequent regrading of wadi deposits or, rather incredibly, by nondeltaic submarine sedimentation, which maintained wadi gradients at depth on the submarine shelf.

Ball (1939: 29–30) reports several raised beaches and coral reefs from the Quseir-Safaga area at elevations of 24, 72, 90, 114, 168, and 238 meters above sea level. Only the three lowest levels are considered to be of Quaternary age. No details are given. Further observations near Quseir were made by J. Büdel (1952), who recognized seven fluvial terraces in the lowermost 10 kilometers of Wadi Ambagi, each with a "core of coral limestone." The lowest terraces, at 1.5 and 4 meters, are thought to be Holocene in age; the intermediate levels, at 6, 10 to 12, 30 to 40, and 60 meters, Pleistocene; the highest, at 120 meters—consisting entirely of coralline limestone—Pliocene. The Pleistocene terraces are believed to indicate eustatically controlled stream gradation. Since these observations conflict with those of Sandford and Arkell (1939: 67) and of Ball (1939: 29), it is unfortunate that no sections are described or illustrated. Büdel (1954a) also mentions the presence of deformed terraces further north, near Safaga.

The status of Pleistocene coastal studies of the Sudanese Red Sea coast is rather different. Systematic field observations by L. Berry and others in 1959, 1960, and 1962 permit on overview of this area, the stratigraphic details of which are reasonably well fixed by several radiocarbon dates and by one Th^{230}/U^{234} determination (Berry, Whiteman, and Bell, 1966):

a) Mukawwar Stage (16 m above high-water mark). Raised reef with radiocarbon age "greater than 37,000 years."

b) Shinab Stage (11–12 m above high water). Raised reef or bench cut into older Mukawwar reef. A Th^{230}/U^{234} determination from *Tridacna* shell in growth position on a bench culminating in 9.5 meters at Mersa Fijja gave an age of 92,000 ± 3000 years (L–945, D. L. Thurber, *in litt.*, August 31, 1966; Berry, Whiteman, and Bell, 1966). The same shell gave a radiocarbon date of "greater than 37,000 years." It appears that Berry, Whiteman, and Bell (1966) feel that the shell dates the reef proper (Mukawwar stage?) rather than the bench.

c) Mohammed Qol Stage (7–8 m above high water). Marine benches apparently cut into older reefs. No isotopic information.

d) Dissection of reefs by the local wadis, cutting down in response to a glacial-age regression to at least − 95 meters.

e) A + 3.5 to + 4 meter bench was apparently cut into the older reefs after the period of dissection (d), although a similar bench may also have formed between phases c and d.

f) A widespread + 2 meter bench and a very minor, localized, + 0.75 meter bench are of fairly recent age.

Berry, Whiteman, and Bell (1966) emphasize that the emerged reef deposits show no signs of faulting or warping and that the constant elevation and sequence of abrasional benches at many points along the Sudanese coast indicate tectonic stability in recent geological times. Unfortunately, considerable unclarity exists concerning whether the post-Mukawwar benches coincided with periods of reef growth, or whether they were entirely abrasional. The matter is not clarified by the radiocarbon dates, since these are automatically referred to the

Mukawwar reef whenever a nonfinite age is obtained. Uncertainty also seems to exist regarding the stratigraphic position of a body of "younger terrace gravels."

Bedrock Lithology

The Red Sea littoral in the Mersa Alam area comprises a belt of broadly horizontal, late Tertiary sedimentaries merging with the modern coral reefs at the coast and abutting on the igneous and metamorphic rocks along the edge of the Red Sea Hills. This littoral zone forms a coastal plain some 4 to 7 kilometers wide, and the underlying sedimentaries are several-hundred-meters thick. The local lithology was studied by H. J. L. Beadnell (1924), and limited stratigraphic revisions based on external areas were made by Sandford and Arkell (1939: 22–26), Said (1962: 116–19), and Sestini (1965).

a) *Basal Calcarenite and Detrital Series* (over 100 m). Coral or algal limestones and calcarenites of littoral facies, alternating with, and ultimately succeeding, a detrital facies of sands and gravels. Middle Miocene.

b) *Evaporite Series* (100 to 200 m). Massive white gypsum and anhydrite with coral impressions, intercalated with dolomites or limestones. Middle or Upper Miocene.

c) *Ostraea-Pecten Series.* Resting on a basal unit of 30 to 60 meters of shales, marls, and sandstones are some 130 meters of sandstones, calcarenites, and marls, intercalated with estuarine conglomerates. Oysters and abundant *Pecten* casts provide a useful stratigraphic index. Either Upper Miocene or mid-Pliocene.

d) *Laganum-Clypeaster Series* (over 100 m). Coral limestone, sandstones, and some conglomerates, with proliferations of sea urchins, primarily *Laganum depressum* and *Clypeaster scutiformis*. Upper Pliocene or Basal Pleistocene.

Except for the Evaporite Series, facies have remained remarkably similar to the present day. The gravels record intermittent uplift or periodic changes of climate in the Red Sea Hills. The existence of the modern horst-and-graben structures has been manifest since the deposition of the Middle Miocene strata, although uplift of the Red Sea Hills appears to have continued well into the Pleistocene. Communications between the Red Sea and the Mediterranean Sea appear to have broken off during the Middle Miocene.

The lithology of the Red Sea Hills adjacent to the coastal plain north of the 25th parallel has been mapped by Amin (1955). Pre-Cambrian metamorphics and, to a lesser degree, granitic intrusions are the basic rock types. Lithologies and structures of this Basement Complex are incredibly intricate.

Structure

The late Tertiary sedimentaries of the coastal plain show a fairly uniform dip of 5% to 12% seawards, the greater part of which can be attributed to the original angle of deposition. Only limited tectonic disturbance is evident in the Tertiary beds, and this is primarily due to sag and compaction (see Beadnell, 1924: 32–33). Although one or more fault lines undoubtedly demarcate the western edge of the Red Sea Graben, related fractures are not apparent in the cover sedimentaries of the Mersa Alam area, a situation which is repeated along the Red Sea littoral of the Sudan (Berry, Whiteman, and Bell, 1966). In fact, several factors have lead Berry, Whiteman, and Bell (1966; see also Sestini, 1965) to conclude that the main Rift Valley fault zone lies offshore, probably along the submarine trough which runs down the center of the Red Sea (see Drake and Girdler, 1964). These factors include (*a*) the absence of fault contacts between the Basement Complex and the cover sedimentaries; (*b*) the submarine topography; and (*c*) gravity and magnetic observations.

Instead of being a series of *en échelon* fault blocks, the structural geomorphology of the coastal plain at Mersa Alam is dominated by several discontinuous and highly dissected cuestas trending parallel to the coast (Fig. 8–1). The Middle Miocene Calcarenite and Detrital Series is only fragmentarily preserved, adjacent to the crystalline rocks of the foothill zone, but the Evaporite Series is prominent as a broad, dreary plateau in some areas and as isolated mesaform hills in others. The *Ostraea-Pecten* Series forms a well-developed, subcontinuous cuesta culminating at about 55 to 59 meters above sea level. Locally, the *Laganum-Clypeaster* Series may also form a final cuesta 400 to 800 meters inland from the modern coast.

General Geomorphology

Most of the geomorphic sculpture of the coastal plain is a result of fluvial processes. The wadis embouching in the foothill zone derive their epi-

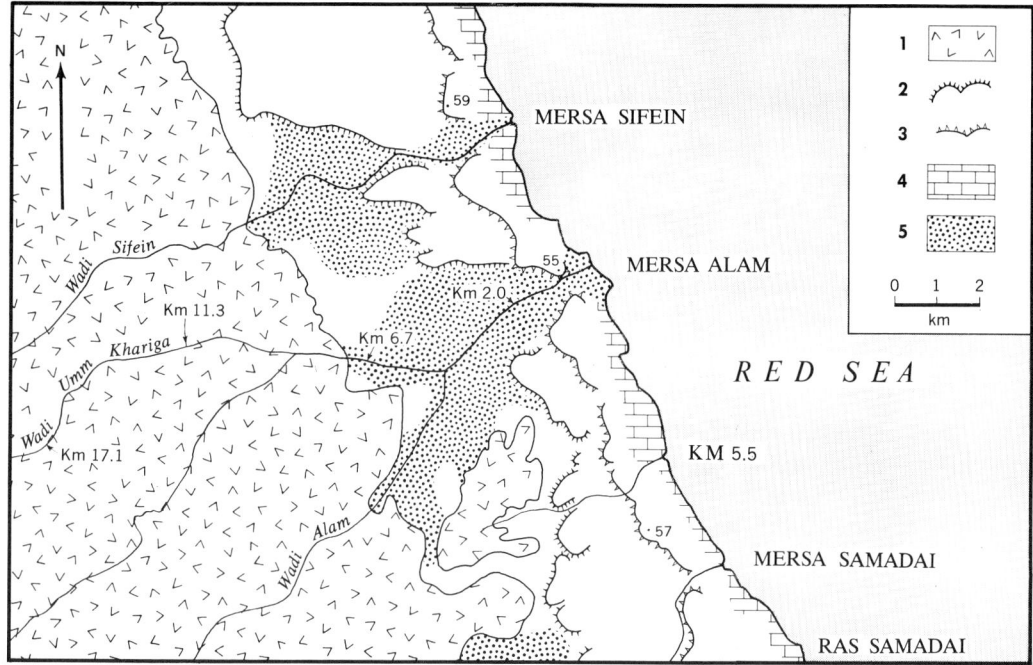

Fig. 8–1. Geomorphology of the Red Sea coastal plain at Mersa Alam. *1:* Basement Complex; *2:* Middle Miocene Cuesta (Evaporite Series); *3:* Mio/Pliocene Cuesta (*Ostraea-Pecten* Series); *4:* Upper Pleistocene coral reefs; *5:* Upper Pleistocene and Holocene alluvial plains. Map: UW Cartographic Lab.

sodic waters from a rugged catchment area at 100 to 900 meters elevation, with steep longitudinal gradients. The energy of these waters is rapidly dissipated on the gently inclined, permeable surface of the coastal plain. As a result, extensive fans of Pleistocene age are common in the foothill zone, narrowing down to alluvial terraces near the coast. Elsewhere, extensive pediment plains have developed, notably around Gebel Dirra. In general, the erosional and depositional plains are proportional in size to the significance of the wadis responsible. Such fluvial surfaces are best developed near the foothills (Fig. 8–1) and are restricted in size or even absent near the coast.

Coastal landforms are confined to the seaward slope of the inclined *Laganum-Clypeaster* deposits. In large part, they pertain to a series of fossil coral reefs up to about 18 meters above present sea level. Higher abrasional surfaces are present but occur only sporadically. They are commonly difficult to distinguish from structural platforms or fluvial benches, and related marine deposits appear to be absent.

The shoreline itself trends about N 25° W, parallel with the axis of the Red Sea Hills. Articulation is very limited, being mainly a matter of gentle promontories and broad, shallow bays. The edge of the modern fringing reef is located several hundred meters offshore (see Appendix F), and is interrupted near the slightly indented mouths of the wadis. Such embouchures are suitable for mooring sailing craft and are designated by the word *mersa* (literally, "anchorage"). In some areas, the wadi mouths form bottlenecked inlets known as *shurum* (singular *sharm*, literally "break-off") (see Rathjens and von Wissmann, 1933). The development of these specialized coastal forms will be further illustrated below in the example of Sharm Sheikh (see also Berry, Whiteman, and Bell, 1966).

In general, the sedimentary strata of the Red Sea littoral form an irregular lowland plain, dominated by innumerable smooth structural, erosional, or depositional surfaces which are gently inclined seawards at slopes of less than 5%—and commonly on the order of 1% to 2%. Small hills with local relief of 25 to 75 meters are not uncommon, particularly in areas of dissected cuestas or their outliers. Level surfaces are invariably offset by steep, rectilinear slopes, such as the wadi walls or the clifflike escarpments of the cuesta landforms. It is a typical desert landscape, despite its proximity to the sea.

Alluvial Terraces of Wadi Alam–Khariga

The asphalt road from Mersa Alam to Edfu provides an excellent opportunity to study a longitudinal sequence of alluvial deposits. It first follows Wadi Alam proper from the coast inland to Km 4.0; then, the major tributary Wadi Umm Khariga to Km 26.0; and, finally, another major tributary, Wadi Abu Quraiya, to the watershed at Km 40.4.

Alluvial deposits begin at Km 0.8 (Fig. 8–2) and are well developed to the gorge at Km 7.5, where the wadi enters a constricted course in the Red Sea Hills. Throughout the coastal plain sector, two striking gravel terraces accompany the broad wadi bed on one or both sides. The Low Terrace (LT) maintains a uniform relative level of +6.5 meters from Km 0.8 to the gorge (Figs. 8–3, 8–4). The Middle Terrace (MT) stands at +10.5 meters at Km 0.8; rises rapidly to +14.5 meters at Km 2.0; then drops gently to +13.2 meters at Km 6.7 (Fig. 8–3) and to +12.5 meters at Km 8.9, the last exposure preserved. Both terraces expose alluvium to below the modern wadi floor. Their material composition is similar, although they are separated by a period of dissection which, at Km 0.8,

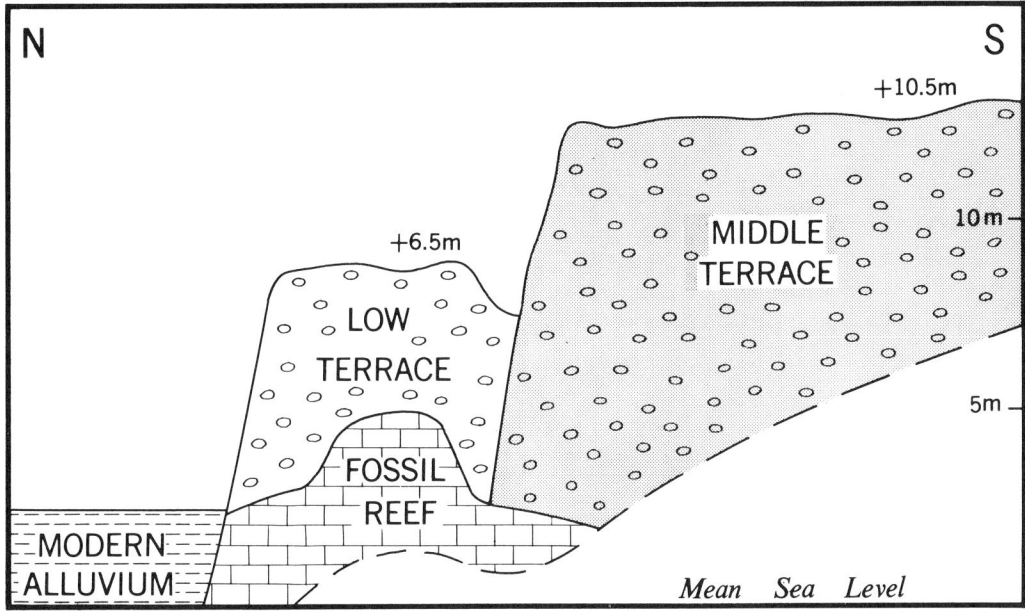

Fig. 8–2. South bank of Wadi Alam at Km 0.8. Vertical exaggeration approximately 4 times. Map: UW Cartographic Lab.

Fig. 8–3. North bank of Wadi Umm Khariga at Km. 6.7. Map: UW Cartographic Lab.

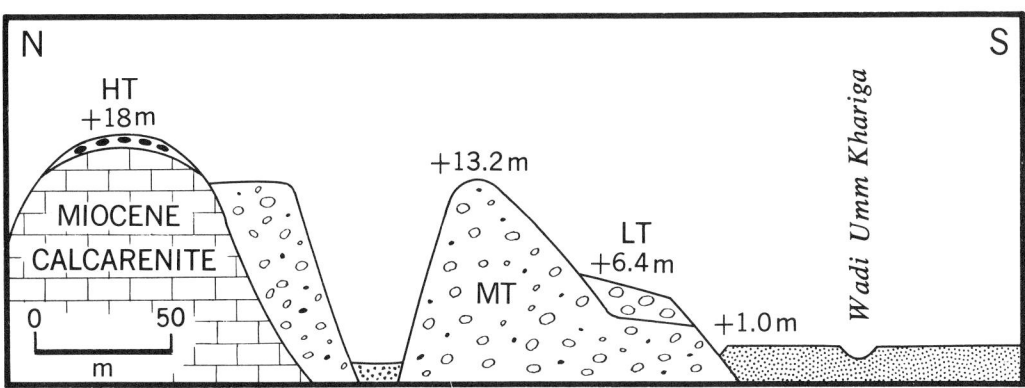

extended to below wadi floor. The gravel is of heterogeneous lithology, coarse to cobble grade, stratified and occasionally current-bedded, rounded, and moved primarily by sliding motions (Table 8–1). The matrix of the Low Terrace is a very-pale-brown (10 YR 7/4), unconsoli-

Fig. 8–4. Wadi Alam and Wadi Umm Khariga confluence (middle background), looking southeast from Middle Terrace. Low Terrace in middle ground; Low and Middle Terraces well developed on south bank.

dated, silty coarse sand, with 15% calcium carbonate. The Middle Terrace has a matrix of reddish-yellow (7.5 YR 8/6), semicemented, silty coarse sand, with 24% carbonates. This supports the general impression that the Middle Terrace is calcified and is possibly more intensively weathered.

The relationships of both terraces to sea level are incompletely recorded. The oversteepened terminus of MT is not confined to the surface

Table 8–1. Morphometric gravel analyses
(Lithology: fine-grained, igneous rocks other than quartz, all moderately metamorphosed.)

Location and stratigraphy	Sample size	ρ (%)	CV of ρ	Detrital component	E/L (%)	E/l (%)	L (cm)
Wadi Alam, Km 2.0							
Low Terrace	100	43.5	47.9	6.0	43.4	61.8	5.26
Km 5.5 south of Mersa Alam							
+ 8.5 meter Beach Gravel	30	67.3	31.2	0.0	34.7	48.2	5.40
+ 4.5 meter Beach Gravel	100	62.0	27.3	0.0	35.0	46.8	3.96

gradient but is evident in the dip of the strata as well. The same phenomenon was studied in detail in Wadi Sifein and is further discussed below. Significant here is the fact that both MT and LT rest disconformably on a fossil coral reef at Km 0.8 (Fig. 8–2). This reef consists of calcarenite and coral, intercalated with medium-grade, well-rounded, estuarine gravels—quite distinctive from those of LT and MT—attaining a maximum of 5 meters above sea level. Dissection to well below modern wadi floor preceded aggradation of both wadi terraces.

As LT progresses upstream into the Red Sea Hills, deposits are only preserved in discontinuous segments—thickening at major confluences, thinning in constricted valley segments. Thus, LT stands at $+4.0$ meters at Km 8.9, $+5.8$ meters at Km 11.3 (Fig. 8–5), $+3.8$ meters at Km 13.2, $+3.2$ meters at Km 17.1, $+3.0$ meters at Km 19.6, and, finally, $+1.9$

Fig. 8–5. Wadi Umm Khariga at Km 11.3 (facing south). Spur of 5.8-meter terrace at right center, with broad, 1-meter floodplain terrace in middle ground (with *Acacia tortilis*).

meters at Km 29.2. LT is absent in and has obviously been eroded from the uppermost 10 kilometers of the wadi. In general, the materials between Km 7.5 and Km 29 are notably less well rounded than those on the coastal plain. They are unsorted and grade laterally into detrital screes. Coarse to cobble grades are characteristic.

The upper wadi, from Km 19 to Km 40, is dominated by subrecent piedmont alluvial fans of coarse, arcosic quartz sands; coarse, subangular-to-subrounded gravel; and related detrital cones. These have been partially dissected by as much as 75 centimeters. In the lower wadi, broad, shallow rills of 50 to 75 centimeters in depth dissect the extensive wadi floor (Fig. 8–5). The latter seems to serve as a floodplain during exceptional spates and is younger than the piedmont alluvial fans upstream. These fans, although still apparently activated on occasion today, seem to record a period of greater fluvial activity, somewhat later than LT.

Both LT and MT postdate a 5-meter Pleistocene coral beach at Km 0.8. Although the local stratigraphy will still be discussed in detail, this suggests a late Pleistocene date for both alluvia. Older gravels (High Terrace or HT) are only fragmentarily preserved on the coastal sector of Wadi Alam and are totally absent further upstream.

The Wadi Alam-Khariga Pleistocene sequence illustrates how poorly recorded wadi terraces are in the gorges of the Red Sea Hills. Even though of considerable paleoclimatic interest, such discontinuous gravels are little suited for longitudinal analysis. Instead, the excellent development of alluvial fans and terraces on the coastal plain itself offers the best prospects for geomorphic investigation.

Alluvial Terraces of Lower Wadi Sifein

The alluvial geomorphology of the Mersa Alam area can also be illustrated by a second case study from Wadi Sifein. The lowermost 2 kilometers of that wadi were surveyed in detail by running two longitudinal transects, one along the wadi floor, the other across the low terrace. A dozen transverse sections were made, and good coastal exposures, relating gravels and coral reefs, are available. The average wadi floor gradient is 0.77%.

Three alluvial complexes occur in lower Wadi Sifein (Figs. 8–6, 8–7): two low terraces (LT I and II), two middle terraces (MT I and II), and a complex high terrace, incompletely recorded in the lower wadi.

The Low Terrace, although present in two persistent levels some 0.5 to 1.5 meters apart, is a single aggradation unit more than 10 meters thick

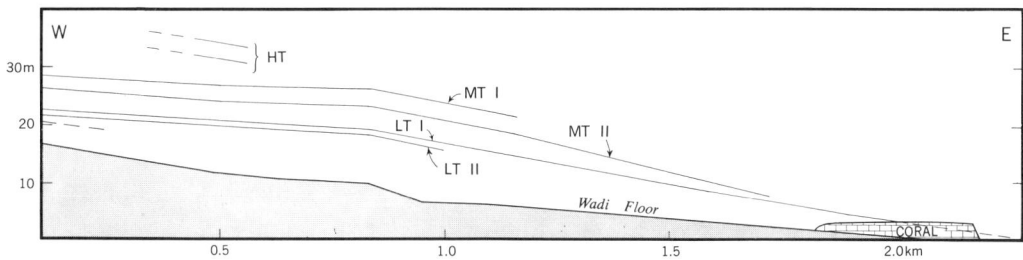

Fig. 8–6. Longitudinal section of lower Wadi Sifein. Map: UW Cartographic Lab.

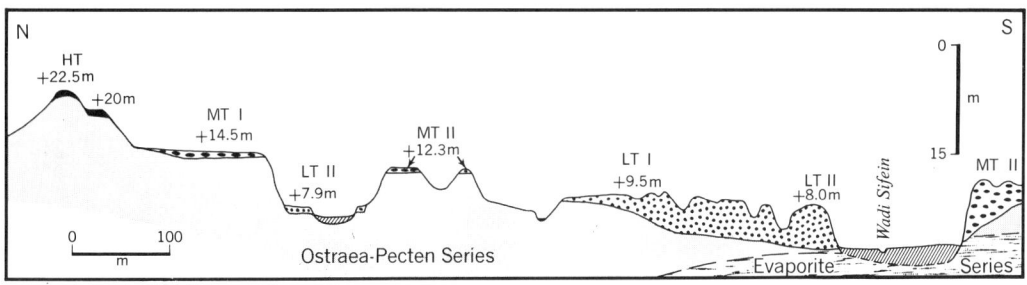

Fig. 8–7. Wadi Sifein terraces at Km 1.68. Map: UW Cartographic Lab.

(Fig. 8–8). It consists of well-stratified, rounded, coarse gravel, with a semiconsolidated matrix of very-pale-brown (10 YR 7–8/3–4), silty coarse sand. Between Km 1.4 and Km 2.1, the gradient of the low terrace is only 0.48%, so that the relative elevation of LT I increases downstream from + 6.5 to + 10.5 meters, that of LT II from + 6.0 to + 9.5 meters (Fig. 8–6). Between Km 1.3 and Km 1.4, there is a sharp break of gradient, and the oversteepened terrace appears to have once dipped to below sea level some 100 meters offshore. The gradient of the lowermost 1.3 kilometers is 1.33%. The alluvium of the Low Terrace postdates shorelines at + 3.7 and + 5.6 m and is separated from them in time by dissection to below modern sea level. Two exposures documenting this fact are described in the next section (Figs. 8–9 and 8–12).

At the time of deposition of the Low Terrace, the active channel of Wadi Sifein was much wider, e.g., 375 meters at Km 1.68 compared with 130 meters today (Fig. 8–7). Today, the wadi bed contains coarse sands and granules with dispersed medium gravel. This suggests both greater stream competence and capacity—and hence a moister climate at the time of alluviation. The inconsistent longitudinal gradients can only be explained by aggradation of extensive, shallow fans on the coastal plain with deposits dipping suddenly near the coast in response to a lower sea level, i.e., a glacial-age regression. The earlier, raised beach deposits are related to interglacial-age transgressions.

The Middle Terrace is analogous to the Low Terrace. Two persistent substages are recorded upstream of Km 1.0, with a vertical difference of 2.5 to 3 meters. Between Km 1.4 and Km 2.1, the longitudinal gradient is identical with that of the Low Terrace (0.48%), and the relative elevations here are $+15$ to $+16.5$ meters for MT I and $+10.5$ to $+13.5$ meters for MT II (Fig. 8–6). Again, the gradient increases markedly below Km 1.3 to 1.7%, until the terrace disappears at Km 0.5. As will be shown further below, the associated gravels rest disconformably on coral reefs and estuarine beds related to former sea levels at $+5.5$ and $+8.0$ meters. Interpretation again involves a broad wadi floodplain dipping to below modern sea level at the coast.

The Middle and Low Terraces are separated by periods of weathering and dissection to below modern wadi floor. Unlike the Low Terrace gravels, the older alluvium is commonly lightly cemented by secondary carbonates, salt, and silica. True soil profiles were not observed.

The significance of local terraces at $+1$ and $+4$ meters found upstream of Km 2.0 is uncertain. They postdate dissection of the Low Terrace and may be fairly recent. High Terraces along Wadi Sifein are recorded by gravel veneers on erosional benches at $+20$ and $+22.5$ meters at Km 1.68 (Fig. 8–7) and at points further upstream (Fig. 8–6). Although the gradient of these High Terraces could not be reconstructed with any degree of confidence, it is considerably greater than that of the Middle and Low Terraces. The stratigraphic position of these High Terraces is unknown, although they probably antedate the local Pleistocene coral reefs.

Pleistocene Littoral Deposits Near Mersa Sifein

The Pleistocene littoral deposits between Mersa Alam and Mersa Sifein are dominated by three parallel reefs, indicating former sea levels at $+3.7$, $+5.6$, and $+9.6$ meters. The lowest reef is limited to both shores of Mersa Sifein, attaining a width of 300 meters or so. It is composed almost entirely of calcarenite. The older, higher-lying reefs are more continuous and, although invariably interrupted by all wadis, can be readily followed along the coast to well north of Mersa Sifein. The 5-meter reef consists primarily of calcarenite, while the 9-meter reef is dominantly composed of coral limestone. Seen in detail, the littoral de-

Fig. 8–8. The Low and Middle Terraces of Wadi Sifein at Km 1.37 (facing north). A remnant of the Mio/Pliocene cuesta (59 m) on the horizon.

posits are intricately interwoven with estuarine beds. Two major sections can be used to illustrate this point.

Immediately south of Mersa Sifein, the following exposure can be seen (Fig. 8–9):

a) Fossil coral reef (Figs. 8–9, 8–10), consisting of cemented, very-pale-brown (10 YR 7/3) calcarenite and white (10 YR 8/2), coral limestone, with some dolomitic cement (see list C.1, Appendix F for the molluscan fauna). The highest deposits attain 2.7 to 3.2 meters above mean sea level near the coast, then dip slowly inland at 0.4%. Surface discoloration is slight (very pale brown, 10 YR 7/4).

b) Estuarine conglomerate (Figs. 8–9, 8–11), resting disconformably on *a*, filling in major cracks on a wave-cut bench developed in *a* at about modern mean sea level. The rounded medium gravel has a heterogeneous lithology (quartz, chlorite schist, basalt, diorite, pink granite, felsite) and a matrix of calcite-dolomite cemented, white (10 YR 8/2–3), coarse sand. Bedded foreset, dipping 22% to 28% seawards, these beds attain 30 centimeters above mean sea level and are submerged at high water. They extend to below modern low-tide level. Fragmentary molluscan remains, mainly waterworn, underscore the estuarine nature of the deposit (see list *D.1*, Appendix F).

c) Laminated calcareous crust (Figs. 8–9, 8–11), usually some 0.8 to 1.5 centimeters of very-pale-brown (10 YR 6–8/3–4), cryptocrystalline calcite and dolomite, suggesting beach rock.

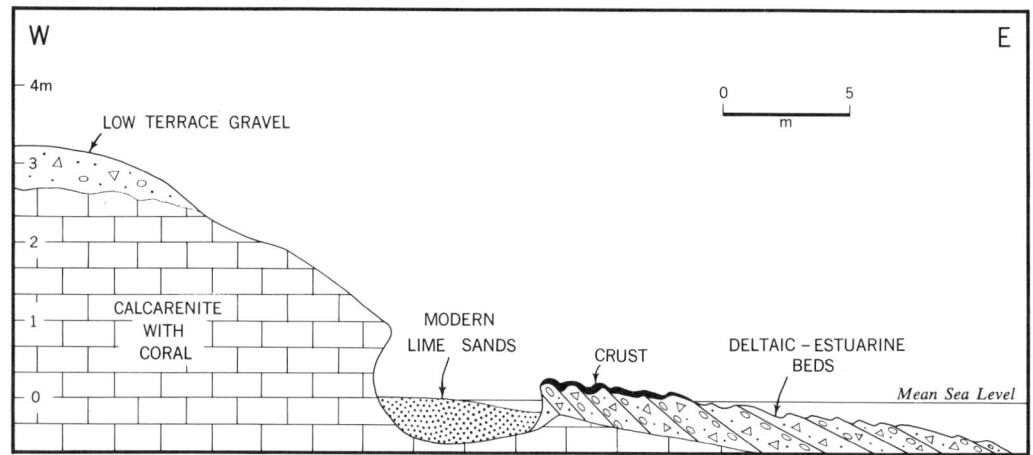

Fig. 8–9. The 3-meter calcarenite reef with littoral and alluvial deposits at Mersa Sifein (South Shore). Map: UW Cartographic Lab.

Fig. 8–10. The 3.2-meter reef at Mersa Sifein (see Fig. 8–9), capped by veneer of low terrace gravel.

d) Wadi gravels, resting disconformably on *a* (Figs. 8–9, 8–10) and forming part of the Low Terrace of Wadi Sifein. Unconsolidated, very-pale-brown (10 YR 7/3), silty coarse sand as a matrix to stratified, topset-bedded, rounded, medium-to-coarse gravel. Although *d* is clearly younger than the estuarine conglomerate, the latter may possibly represent the basal facies of unit *d*. In any case, this estuarine conglomerate is younger than, and apparently not related to, other estuarine beds located further inland and described below.

The sequence of events recorded by this exposure is significant: a nip was cut into a homogeneous 3.2-meter coral reef, prior to deposition of a body of estuarine gravels. Both the nip and the gravels are at about modern mean sea level. At an even later date, the Low Terrace of Wadi

Fig. 8–11. Estuarine conglomerate at Mersa Sifein at low water (see Fig. 8–9). Laminated crust is visible left center.

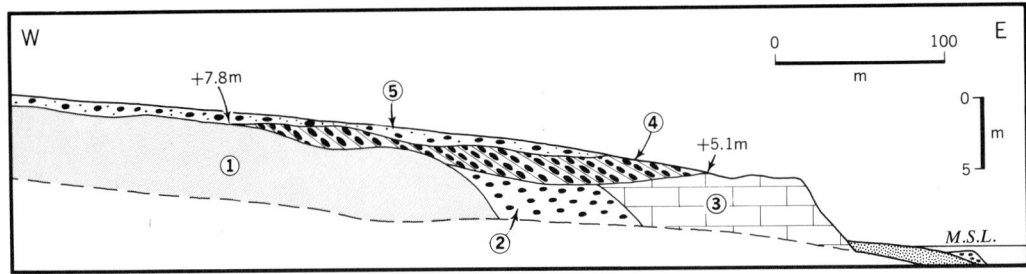

Fig. 8–12. Littoral and alluvial deposits 500 meters south of Mersa Sifein. *1:* Tertiary gypseous sandstones; *2:* beach conglomerate, older than *3; 3:* calcarenite and coral reef; *4:* estuarine gravel; *5:* middle terrace wadi gravel. Map: UW Cartographic Lab.

Sifein was aggraded during a glacio-eustatic regression by gravels sweeping across the 3.2-meter reef.

Another informative record of interrelated littoral and alluvial deposits is recorded in minor wadis 200 to 500 meters south of Wadi Sifein (see Fig. 8–12). The bedrock is formed by a pale-yellow gypseous sandstone of late Tertiary age, dipping seawards at 8% to 12%. The superimposed Pleistocene beds can be described in stratigraphic sequence:

a) Beach conglomerate, consisting of white to very-pale-brown, stratified, semicemented, coarse gravels that dip 2% seawards. Masses of rolled coral and shell debris are intermixed. The Pleistocene sea level suggested is + 5.5 meters, although this deposit may have been truncated by later wave or fluvial erosion.

b) Fossil coral reef, embanked against *a*. Primarily a cemented, white calcarenite, attaining 5.1 meters above mean sea level. The mollusca are described in Appendix F (list A.1).

c) Estuarine gravels resting disconformably on *a* and *b*. The basal facies consist of well-stratified, foreset- (16% to 17%) and topset- (5% to 8%) bedded, well-rounded, medium gravels set in a gypseous matrix of very-pale-brown, coarse sand (faunal list *B.1*, Appendix F). The upper facies includes topset, rounded, coarse gravels with a matrix of consolidated coarse sands, bedded 16% to 17% seaward near the modern coast and 3% to 8% further inland. Rolled coral and unrolled mollusca are abundant, dominated by *Stombus gibberulus* (see list B.2, Appendix F). Attaining at least 7.8 meters above mean sea level, these estuarine gravels suggest a sea-level stage of + 8 meters, *subsequent* to the 5.6-meter sea level. The bar or reef once separating this lagoonal environment from the open sea has not been preserved.

d) Wadi gravels, resting disconformably on *a*, *b*, and *c*, forming part of the Middle Terrace. Semiconsolidated, stratified, subrounded-to-rounded, coarse gravel forming topsets inclined 1.5% seawards.

The significance of these sections 200 to 500 meters south of Mersa

Sifein (Fig. 8–10) is manifold. The oldest beach deposit, predating the 5-meter reef, is represented by a beach gravel including abundant detritus washed in from the reef platform. A little later, a coral reef formed to $+5.1$ meters, indicating a long-term mean sea level at about 50 centimeters higher, i.e., $+5.6$ meters. Younger still is an $+8$ meter stage recorded by active alluviation of bed-load materials into coastal lagoons. This would indicate greater wadi competence than today. The Middle Terrace of Wadi Sifein is younger than either the 5.6- or 8-meter beaches, clearly demonstrating that the typical wadi alluvia have no relationship whatever to any of these littoral deposits. In conjunction with the Mersa Sifein section, it is evident that both the Low and Middle Terraces postdate successive sea-level stages at $+5.5$, $+5.6$, $+8$, and $+3.7$ meters. The exact temporal relationship of the final estuarine conglomerate now found at modern mean sea level to the Low and Middle Terraces cannot be determined.

Pleistocene Littoral Deposits: Mersa Alam–Mersa Samadai

At Mersa Alam, fine wadi alluvium extends to the coast, where it is intermeshed with a number of gravel bars of littoral origin. Fossil reefs are only exposed further south. There are traces of coral platforms at about 3 and 9 meters, but the 5-meter stage is dominant as far as Mersa Samadai. Locally, as at Mersa Sifein, beach gravels are superimposed upon the 5-meter reef, indicating a subsequent shoreline at 8.5 meters, without associated coral formation. At the mouth of almost every wadi, coarse alluvium can be observed resting disconformably on dissected surfaces of the 5-meter reef.

This general pattern can be illustrated by important sections exposed on the left bank of a small inlet at Km 5.5 of the highway south of Mersa Alam (Fig. 8–13). The depositional units may be described as follows:

a) Beach gravels with little or no matrix. The material consists of well-stratified, sorted, well-rounded, and exceptionally well-flattened, coarse gravel (Table 8–1), primarily basalt and diorite. Gravel rounding is homogeneous. These beds attain $+4.5$ meters and dip at 12% to 15% towards the adjacent wadi, where they merge with unit *d* and suggest a former sea level of $+8.5$ meters. Morphologically, the pebbles constitute typical shingle, and the steep bedding suggests an arcuate ridge built into the wadi mouth. A rich fauna of bivalves, almost exclusively *Trachycardium* and *Gafrarium*, is present (see list A.4, Appendix F), indicating a typical beach accumulation. A Th^{230}/U^{234} de-

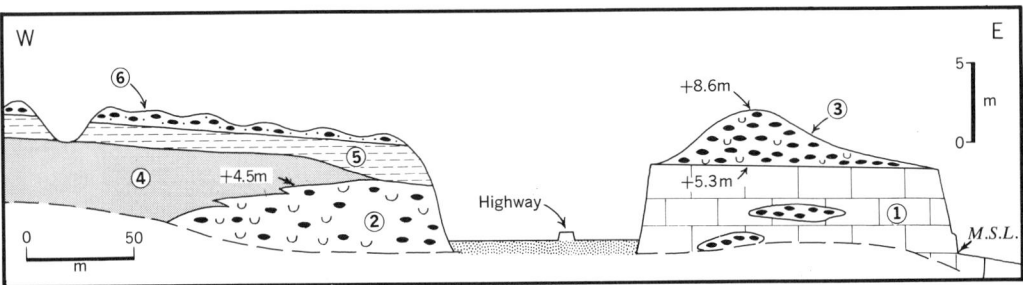

Fig. 8–13. Littoral and alluvial deposits at Km 5.5, south of Mersa Alam. *1:* calcarenite and coral reef; *2:* beach gravel, older than *1; 3:* shingle beach with related lagoonal beds of gypsum (*4*) and silty gypsum (*5*); *6:* wadi gravel. Map: UW Cartographic Lab.

termination was obtained from shell about 1 meter below the top of this formation: 118,000 ± 10,000 years (L 1062–C, D. L. Thurber, *in litt.*, October 12, 1966).

b) Fossil coral reef, mainly cemented calcarenite and coral limestone, attaining 5.3 meters above mean sea level. Molluscan fauna is abundant (see list A.3, Appendix F). A second thorium-uranium determination was obtained from shell within the reef calcarenite, approximately 1.5 meters below the top of the deposit: 80,000 ± 8000 years (L 1062–B, Thurber, *in litt.*) The lateral juxtaposition of beaches *a* and *b* recalls that of Mersa Sifein. The thorium-uranium dates suggest that the beaches are not, in fact, contemporary.

c) Gravel bar, consisting of stratified, sorted, well-rounded, and well-flattened, cobble gravel (Table 8–1) of basalt and diorite. The matrix is a pink (7.5–10 YR 6/4), coarse sand. Fragments of semicemented, light-brown (7.5–10 YR 6/4), silty lime sand and subangular, fine gravel are probably derived from the former reef platform, further offshore. The fauna consists almost entirely of *Trachycardium* and *Gafrarium*. Although this is a typical shingle gravel, it is different from the gravel of unit *a:* sorting is pronounced, with less medium gravel, and rounding is more homogeneous (Table 8–1). The former sea level indicated is about + 8.6 meters.

d) Lagoonal gypsum, a massive series of alternating 1- to 2-centimeter bands of mesocrystalline gypsum and 2- to 4-centimeter bands of silty or sandy gypsum with yellow (10 YR 6/6) oxidation stains. The crystalline gypsum is white (10 YR 8/2), the detrital subfacies grayish (5 Y 5–6/2). Over 6 meters thick, the gypsum attains 8.5 meters above mean sea level further inland (Fig. 8–14). It appears to be laterally and vertically conformable with *a*. Since coarse detritus is entirely absent, the beach gravels *a* and *c* must be derived from longshore drifting or currents (see Table 8–1) rather than from local wadi activity.

e) Detrital gypsum, some 2.5 meters of consolidated evaporites similar to *d* but with small pockets of arcosic coarse sands, powdery gypsum, and secondary halite crystals. Diffuse silts determine the light-yellowish-brown (10 YR 6/4) color. Possibly contemporary with the gravel bar (*c*), this silty evaporite also attains + 8.5 meters.

Fig. 8–14. Wadi gravels overlying lagoonal evaporites at Km 5.5, south of Mersa Alam (see Fig. 8–13). Looking upstream.

f) Wadi gravels, resting disconformably on *d* and *e* and forming a 7-meter wadi terrace (Fig. 8–14). These crudely stratified, unsorted, subrounded and heterogeneous, medium-to-coarse gravels are morphologically quite different from the beach gravels. Lithology is also distinctive, with schistose rocks dominant.

Gravel lithology is of general interest in this case of a small wadi basin limited to two Pre-Cambrian rock types, viz., serpentines and chlorite schists. The beach gravels are dominantly weakly metamorphosed basalt and diorite, with a fair amount of pink microgranite and green schist. All of these rock types are exposed in the Wadi Samadai drainage and are injected into the littoral zone further south at Km 7.5. A northerly current appears to have transported these materials at least 2 kilometers along the coast. Since the pebbles and even the feldspar sands of the +8.5, +5.8, and +8.6 meter beach or reef deposits were not derived from the local wadi, stream competence was limited to silt-and clay-size particles, implying a dry climate. The wadi gravels indicate renewed activity by the local watercourses, subsequent to the high interglacial sea levels.

Some sporadic features suggesting Holocene or even historical sea levels at or above present mean sea level were observed between Mersa Alam and Mersa Samadai. Near Km 2.5 of the highway, an extensive fossil reef of calcarenite and coral limestone is exposed to +4.6 meters (see list A.2, Appendix F for fauna). Embanked against a notch at about present mean sea level are 20 centimeters of subrounded and stratified, medium gravel of marine origin. The matrix is a light-brownish-gray (2.5 Y 6/2), semicemented, coarse sand, and the cement is primarily gypsum. Some 2 to 3 meters wide and inclined at 5%, this subfossil beach deposit attains about 25 centimeters above mean sea level. The fauna is discussed in faunal list E.1, Appendix F.

In a similar situation at Km 2.3, coarse, rounded beach pebbles and coralline rubble were found with a rolled but fresh-looking fauna +1.8 meters above mean sea level. The unconsolidated matrix is a very-pale-brown, coarse sand. Whereas the Km 2.5 deposit suggests a bonafide strandline—possibly equivalent to the younger estuarine beds at Mersa Sifein—the Km 2.3 feature may be only a storm beach.

Although high Pleistocene shorelines other than the +3 to +10 meter reefs and associated deposits were not observed, some high conglomerates between Km 2 and Km 4 of the highway deserve mention. These well-rounded, cobble gravels may attain a thickness of 2 or 3 meters and are fragmentarily preserved on top of ridges formed by the *Ostraea-Pecten* Series, at 20 to 22 meters above mean sea level. They may represent either wadi or beach gravels. A red paleosol postdates the gravels. It is recorded by soil sediments only: a stratified, yellowish-red (5 YR 5/6), decalcified, silty coarse sand with single-grain structure.

Observations South of Ras Samadai

Although systematic observations were not carried out further south than Ras Samadai, transects were run in Wadi Ambaut (Km 14.5), and a reconnaissance mapping was made of the Sharm Sheikh (Sharm Luli) area (Km 62).

Three alluvial terraces are present in lower Wadi Ambaut, all younger than a massive fossil reef of coralline limestone, with an abundant bivalve fauna, at 9.5 meters above mean sea level (Fig. 8–15). The highest, +10 meter terrace lies disconformably on the reef, while the intermediate terrace at +4 meters is embanked against the reef, within the wadi channel that cuts through the coral. In contrast to the 10- and 4-meter gravel terraces, the lowest, +2 meter stage consists of fine-grained materials—coarse sands such as form the wadi bed. The position of the alluvial fill implies dissection of the coral reef to below modern sea level, prior to aggradation of the 4-meter gravels and probably before the 10-meter stage also. The significance of the unconsolidated, 2-meter wash terrace is obscure. It is certainly not of any great antiquity.

Gebel Dirra, at Km 24, rises to 71 meters above mean sea level and, from a distance, resembles a truncated reef. On closer examination, the base consists of horizontal gypsum and anhydrite, the upper parts of gypsum with coral impressions. It is an outlier of the Miocene Evaporite Series (Beadnell, 1924). Similar reef-like geomorphic features as high as 55–60 meters are provided by the *Laganum-Clypeaster* and the *Ostraea-Pecten* Series, both north and south of Mersa Alam, and may have been erroneously identified as Pleistocene reefs by some workers in the Quseir region. Even the typical flat surfaces on these outliers are structural benches rather than wave-cut platforms.

Pleistocene reefs are remarkably well recorded at Sharm Sheikh, an area deserving of more intensive study. A massive reef at +8 to +9 meters forms a 0.5- to 1.0-kilometer-wide coastal platform (Fig. 8–16), intersected by the wadi and the sharm inlet. The coral limestone and calcarenite are rich in mollusca. Since lower reefs occur a little north of Sharm Sheikh, it is fairly probable—on the basis of facies (coral limestone dominant) and altitude—that the 8- to 9-meter reef at Sharm Sheikh is the same as the reefs in Wadi Ambaut and north of Mersa

Fig. 8–15. Fossil coral reef (9.5 m above present sea level) dissected by Wadi Ambaut and overlain by a + 10 meter wadi terrace. At Km 14.5, south of Mersa Alam, facing southeast.

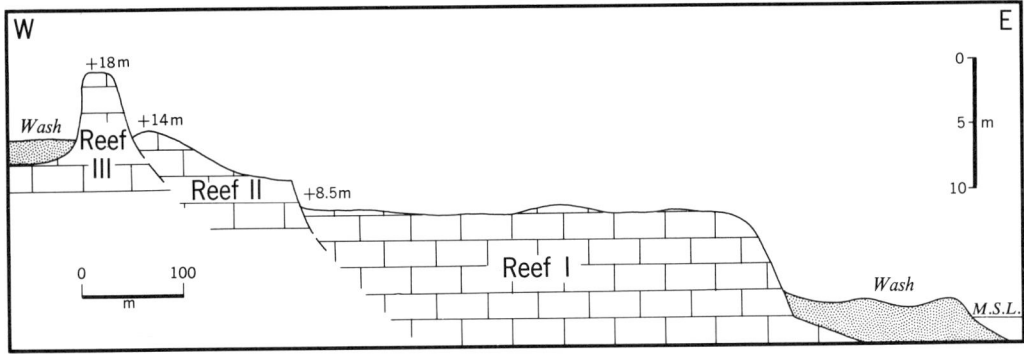

Fig. 8–16. Fossil coral reefs at Sharm Sheikh (South Shore). Map: UW Cartographic Lab.

Alam. Segments of a + 14 meter reef, with few mollusca, are well developed on the south shore, together with two small islands of a + 18 meter reef (Fig. 8–16).

The 8- to 9-meter reef was dissected by Wadi Sharm Sheikh prior to aggradation of a 9-meter gravel terrace. This fill is preserved within the 600- to 900-meter-wide erosional valley and is embanked against the truncated reef. Dissection to below modern sea level evidently preceded alluviation of the low- to middle-terrace complex. The sharm itself is nothing but a drowned wadi mouth. Dissection to below modern sea level was renewed after deposition of the Low Terrace. Only limited quantities of coarse sandy wash have since been deposited by the almost defunct wadi. Consequently, the post-Pleistocene rise of sea level has flooded the lowermost wadi, and the sea still penetrates some 750 meters inland today. Typical shurum are best developed further south along the Sudan littoral. Here too, they are attributed to wadi dissection to well below sea level during the last major glacial-eustatic regression, followed by submergence during post-Pleistocene times (Berry, Whiteman, and Bell, 1966).

Pleistocene Sea-Level Stratigraphy

In reviewing the above material on shorelines, any features which predate major dissection of the local wadis to below modern sea level may be considered older than the last major Pleistocene glacial-eustatic regression. Employing this criterion, Table 8–2 compiles the Pleistocene sea levels identified. Levels cited are all based on Abney transects, with reference to contemporary mean sea level. Since deposits, rather than erosional features, are used as watermark criteria, true sea levels may have been a little higher. Maximum elevations are given, and, in the case of reefs (calcarenite or coral limestone dominant), 50 centimeters have been added to the maximum exposure to give approximate mean sea level. In each case, however, the detailed sections will bear out the sea levels suggested.

Since tectonic deformation was nowhere evident, altitudinal and facies correlation is permissible within the restricted area and limited elevation range under consideration. The burden of proof rests with those who might propose a tectonic hypothesis for either the multiplicity of sea levels or their relative positions. In fact, the major reefs—serving as a datum for the less-frequent estuarine and shingle deposits—can be traced as morphological units from Mersa Samadai to north of Mersa Sifein (Fig. 8–1).

The oldest, 4.5- to 5.5-meter beach gravels at Mersa Sifein and Km 5.5, as well as the levels higher than 10 meters in general, are too sporadic to permit generalization. But definite stages can be recognized at 10 meters (mainly coral limestone), at 5 to 6 meters (calcarenite and coral limestone), at 8.5 meters (a shorter stand without reef development, recorded by shingle and possibly by lagoonal deposits), at 3.5 meters (minor calcarenite reef), and at approximately the high-water mark (sporadic evidence only of notches, estuarine beds, and shingle). Preced-

Table 8–2. Pleistocene levels of the Red Sea in the Mersa Alam area
Dominant facies: L (coral limestone), C (calcarenite), S (gravel bar or shingle ridge), D (estuarine gravels), E (evaporite). Levels in meters with reference to mean sea level.

Mersa Sifein	Wadi Alam	Km 2.5	Km 5.5	Wadi Ambaut	Sharm Sheikh	
0.3 D		(?) 0.3 S				UPPER PLEISTOCENE
3.7 C						
8.0 SD			8.6 SE			
5.6 C	5 C	5.1 C	5.8 C			
9.6 L				10 L	9.5 L	
(?)5.5 S			(?)4.5 SE		14.5 L	MIDDLE PLEISTOCENE
					18.5 L	
			(?)22.0 S			

ing—as they do—the major regression corresponding to the Last or Wurm Glacial (see Donn, Farrand, and Ewing, 1962), all of these stages can probably be attributed to the Last Interglacial or possibly, in part, to an older interglacial. The single thorium-uranium date of 80,000 years from the coral reef at Km 5.5 seems to suggest that the dominant 5- to 6-meter stage is related to the Eem or Last Interglacial. The stratigraphic position of the older, 5.5-meter beach gravels, with a date of (at least) 118,000 years, is reminiscent of the Tyrrhemian IIb in the Mediterranean region (see Butzer and Cuerda, 1962), recently studied by Stearns and Thurber (1965). The 92,000-year date from a 9.5-meter bench or reef platform in the Sudan may possibly suggest that our 10-meter stage is younger than the ancient 5.5-meter beach gravels.

Correlations with the Quseir evidence are quite rewarding. In general, correspondence with Sandford and Arkell's (1939: 67) results is very good. A 5- to 6-meter reef capped by gravel bars at $+7$ or $+8$ meters appears to be characteristic of the Quseir area, a useful stratigraphic indication that relative levels are identical to those at Mersa Alam, some 110 kilometers further south. Sandford and Arkell also tentatively identify a $+3$ meter stage. Slightly higher coral reefs seem to be better preserved at Mersa Alam, while high, wave-cut platforms are better represented at Quseir. The high coral reefs proposed by Büdel (1952) seem incompatible with both our own and Sandford and Arkell's results, unless they are identical with the *Laganum-Clypeaster* Series reefs recorded at elevations of 70 to 170 meters in the Quseir area by Felix (1904).

Comparisons with the sea-level stages of Berry, Whiteman, and Bell (1966) along the Sudanese littoral are not readily apparent. Conversion to mean sea level can be readily made by subtracting 40 to 50 centimeters from these Sudanese elevations. Even then, stratigraphically comparable stages appear to be 1–2 meters higher than at Mersa Alam. Possibly the Shinab and Mohammed Qol stages correlate with our 10-meter and 5- to 6-meter coral reefs. Beyond this, no counterpart is found at Mersa Alam for the abundance of postregressional shorelines indicated by Berry, Whiteman, and Bell (1966). It is possible that the dating of some of these Sudanese low-level benches is equivocal, but it is also possible that the Sudanese littoral was subject to epeirogenic uplift, relative to the Egyptian coasts, in Holocene times. These questions can only be answered after systematic fieldwork south of Sharm Sheikh. A second problem of correlation is posed by tentative interpretation of all Sudanese benches below 12 meters as abrasional platforms rather than as primary reef surfaces (Berry, Whiteman, and Bell, 1966). The multiple stages recorded at Mersa Alam, and described here in detail, are all represented by primary deposits. This is repeatedly borne out by the alternation of organic and detrital sediments, by the intercalation or superposition of estuarine beds, by the differing degree of lithification in the various coral reef units, and to a certain extent also by the faunas. In fact, our own impressions at Mersa Alam have been that depositional features are better recorded than abrasional phenomena.

Relevant to an understanding of the Mersa Alam littoral deposits is the deep-sea core stratigraphy established for the southern half of the Red Sea by the Lamont Geological Observatory, based on fourteen cores removed in 1958 (Herman, 1965). By examining six plankton tows from the modern Red Sea waters, cold and warm faunal zones were identified on the basis of certain indicator species. The following zones were recognized (Herman, 1965), employing the Lamont terminology:

>   Faunal zone "z"   (warm): Postglacial
>   Faunal zone "y"   (cool): Last Glacial
>   Faunal zone "x"   (warm): Last Interglacial
>   Faunal zone "w"   (cool): Penultimate Glacial

A radiocarbon determination from the top of zone "y" gave 16,000 years. Extrapolating the duration of the "Last Interglacial" from a comparison of the thickness of these deposits and the Postglacial sediments, a time interval of at least 50,000 to 60,000 years is suggested. Allowing for compaction of sediment, this value would be over 75,000 years.

Stratigraphy of the Littoral and Alluvial Beds

Alluvial deposits cannot be correlated with either the 10-meter or the 5- to 6-meter stages, even though beach gravels may occur within the littoral facies. In the case of the 8.5-meter beach deposits, a contradiction seems to exist between Mersa Sifein, where estuarine gravels were shot into lagoonal environments by the wadis, and Km 5.5, where no local wadi activity other than silt transport is recorded. At least part of the evaporite beds at Km 5.5 appear to date from an earlier interglacial, however, so that they are probably not contemporary with the estuarine gravels near Mersa Sifein.

Where a single major wadi terrace is developed, as at Km 2.5, Km 5.5, Wadi Ambaut, and Sharm Sheikh, it is in each case younger than the lowest fossil littoral deposits and is separated from them by a period of erosion or of major dissection. In Wadi Alam, the 6.5-meter Low Terrace and the 13-meter Middle Terrace are both later than the 5- to 6-meter sea-level stage. In Wadi Sifein, the 8- to 10-meter Low Terrace and the 12- to 15-meter Middle Terrace are again younger than the 8.5-, 5- to 6-, and 10-meter stages of the Red Sea, and the Low Terrace at Mersa Sifein is at least in part younger than the estuarine conglomerate at the highwater mark. In both Wadis Alam and Sifein, the Low and Middle Terraces were separated by a period of weathering (calcification) and dissection (to below modern wadi floor). Finally, major dissection, in response to a glacial-eustatic regression of some duration, followed deposition of these major wadi fills.

Sporadic occurrences of minor wadi fills, such as the 1- and 4-meter terrace spurs in Wadis Sifein and Ambaut, must remain ignored until more substantial stratigraphy can be found. As it is, there is no certain evidence to indicate whether these terrace fragments postdate or antedate the period of major dissection.

Table 8–3 indicates a tentative scheme of late Pleistocene stratigraphy suggested by analysis of the field evidence. No attempt is made to take into account the sporadic evidence of middle Pleistocene features. As in the Mediterranean Basin (Butzer, 1963a, b), major alluviation postdates the high interglacial sea levels and predates the climax of the glacial-age regressions. Alluviation on the coastal plain at the beginning of a regression can obviously not be attributed to eustatic factors. In this connec-

Table 8–3. Correlation of middle and late Pleistocene wadi and littoral phenomena

Wadi phenomena	Littoral phenomena
Partial wadi aggradation, with drowning of lowermost wadis	Holocene transgression to modern sea level
Major dissection to below sea level	Major regression of sea level
Low Terrace alluviation Dissection and calcification	Regression underway
Middle Terrace alluviation	Estuarine beds at high-tide mark(?)
Dissection	Regressive oscillation
Limited activity	3.5 m stage (moderate duration) Regressive oscillation
Local alluviation	8.5 m stage (moderate duration)
Limited activity	5–6 m stage (long duration) Regressive oscillation 10 m stage (long duration)

tion, the eustatic origin of the Red Sea wadi terraces—as proposed by Sandford and Arkell (1939: 67–68) and Büdel (1952)—clearly does not correspond with the field evidence, since the wadi terraces are quite unrelated to, and distinctly younger than, the interglacial beach deposits. Instead, a climatic impulse to aggradation offers the only possible explanation.

Paleoclimatic Interpretation

CLIMATOLOGICAL DATA

The closest and most representative climatic station is Quseir, located on the coast about 110 kilometers north of Mersa Alam. The long-term (1927–59) mean annual precipitation at Quseir is 4 millimeters, falling primarily between October and December (S. P. Jackson, 1961). The Mersa Alam area lies within this Red Sea winter-rainfall belt. Flohn (1965a) believes the moisture is of local origin, i.e., evaporation over the Red Sea, with precipitation triggered by low pressures associated with upper air troughs in the westerlies. Rain falls primarily during the night hours—at which time there is commonly an offshore land breeze complemented by a downvalley mountain breeze. Dominant wind directions during the autumn months are northerly to northwesterly (Hamed, 1950). The mean January temperature at Quseir is 65° F (18.3° C); the mean daily minimum, 57° F (13.9° C); and the absolute minimum, 39° F (4° C) (1927–45, Hamed, 1950). The August mean is 87° F (30.5° C), the mean daily maximum, 93° F (34.1° C). The mean daily range is highest in January (16°F or 8.8° C) and lowest in September (13° F or 6.9° C). Without climatic stations in the high country, precipitation and temperature values in the Red Sea Hills are a matter of speculation only.

THE CORAL REEFS

Reef corals thrive best at water temperatures of 77° to 86° F (25° to 30° C) but tolerate temperatures as low as 64.5° to 68° F (18° to 20° C). At Quseir (depth 1.5 m), the coldest mean water temperature of 68.9° F (20.5° C) occurs in February (Hamed, 1950: 148). In other words, coral would not have thrived if mean water temperature in February were one or two degrees lower. Consequently, temperatures were not colder and were possibly somewhat warmer than today during the 10-meter, 5- to 6-meter, and 3.5-meter stages. This in itself confirms the interglacial age of the deposits. Conversely, coral did not thrive and may have been absent along the coast of Mersa Alam during the cool glacial-age regressions corresponding to the deep-sea faunal zones "y" and "w" (Herman, 1965).

Only indirect inferences concerning atmospheric precipitation can be made from the presence of coral reefs. Since reef corals hardly tolerate fresh or brackish water (or, for that matter, abundant silt), reefs are commonly interrupted near the embouchures of larger wadis today—despite the highly intermittent character of water flow. At Mersa Sifein and in Wadi Alam, the 3.5-meter-stage and 5- to 6-meter-stage reefs either closed off the wadis entirely or constricted them to a greater degree than in the case of the modern coral platform. It would seem that wadi activity was no greater, and possibly less, than today during formation of these reefs. The 10-meter reef is less well preserved, so that no conclusions can be drawn concerning its former continuity.

The coral genera and molluscan fauna represented in the middle to late Pleistocene reefs are all abundant on the coast today, so that no deductions about basic ecological changes appear possible on the basis of such standards.

Finally, the facies contrast of calcarenite versus coral and algal limestone deserves mention. At least some of the so-called coral limestones examined differ from the calcarenites only in terms of cementation and recrystallization of the carbonates. Finer-grained calcarenites will often be cemented rather earlier than the coarse-grained varieties. In addition, the greater age and the greater degree of recrystallization of the 10-meter reef are obvious. At present, the differences of lithofacies do not provide a substantive argument concerning variations of depositional environment.

BAR-AND-LAGOON DEPOSITS

Shingle ridges and gravel bars are azonal phenomena with few paleoclimatic implications. Although igneous or metamorphic pebbles must ultimately be derived from the Red Sea Hills by fluvial transport, exposures of ancient wadi gravels are available along much of the coast for wave erosion and subsequent redeposition in a variety of littoral facies.

Estuarine gravels or evaporites deposited in lagoonal environments are another matter. Lagoonal evaporites are a standard facies for both arid and semiarid environments, and similar deposits have a long tradition in the local Tertiary record. The fact that hypersaline precipitates are not currently forming may be attributed to a lack of suitable lagoons. The fact that a local wadi—such as that debouching at Km 5.5 during the 8.5-meter stage—contributed little or no detrital sediment to such a lagoon speaks for little or no rainfall on the coastal plain.

THE ABSENCE OF EOLIAN DEPOSITS

Regressional eolianites, composed primarily of lime sands, are fairly typical of carbonate coasts in the arid and semiarid subtropics. Their absence can often be attributed to an abundance of detrital sediment. Eolianites are completely and notably absent from the Pleistocene record of the Red Sea littoral at Mersa Alam. Today, lime sands are present in unlimited quantity in the beach zone, and detrital sediments are rare, being derived intermittently from the wadis or in limited quantity from erosion of older gravels. A regression of sea level should expose ample and suitable sands for deflation. Yet the glacial-eustatic fluctuations of the Pleistocene produced no eolianites. Two explanations are possible. The sediment-binding activity of algae such as *Rivularia* or *Lithothamnion* may have restricted the quantity of loose sand. On the other hand, activation of the local wadis during the regressional stages may have provided large quantities of coarse detritus. Redistribution of such gravel and very coarse sands by beach drifting and longshore currents may have favored development of shingle beaches and gravel bars. This second explanation seems the more reasonable and is supported by the deep-sea cores taken from the southern half of the Red Sea (Herman, 1965). The rate of sedimentation was approximately doubled during the Würm cold interval, with the detrital component increasing from 5% to 25%, five times greater than at present.

Although terrestrial eolian deposits are wholly absent in the Pleistocene record and are represented solely by veneers of eolian sand today, Pleistocene alluvial and littoral beds show a remarkable frequency of frosted quartz grains in the 0.06- to 0.6-millimeter grade. The percentage fluctuates between 70% and 100%, with polished grains accounting for the remainder (see Appendix C). When 90% of the quartz sand (between 0.2 and 0.6 mm) forming the matrix to a marine conglomerate is frosted, it is impossible to accept the standard interpretation of sandblast micropitting. Instead, the conclusions of Kuenen and Perdok (1962) seem applicable, namely, that chemical action by corrosive solution, or by alternate solution and deposition, is primarily responsible for quartz frosting.

WADI ALLUVIATION

Interpretation of alluvial deposits in arid regions can best be made on a comparative basis. If ancient aggradation products imply greater stream transport, competence, or capacity than is recorded by the contemporary bed load, increased wadi activity is indicated. Since vegetation, except for isolated acacias and low thorn shrubs, is totally absent in the Red Sea Hills today, differences of vegetative mat, infiltration factor, or interception cannot explain increased runoff. Finally, if all other factors remain constant, a regression of the Red Sea should induce gradual dissection of the lower wadi courses.

As the evidence stands, alluviation was at least partly contemporary with a gradual drop of base level. In addition, great masses of gravel and cobbles were transported from the hill country across the coastal plain. Such gravel is generally rounded, moderately sorted, and notably flattened. Sieve analyses and field approximations have in each case shown that the Wentworth textural class of the Low and/or Middle Terrace on the coastal plain is a sandy gravel or gravel, compared with modern bed loads of sand or gravelly sands. And the gravel grade of the Pleistocene deposits is inevitably coarser. Cobbles are not transported in the contemporary bed load, and even the dispersed medium gravel present is generally derived from local terraces. Considering the coarse nature of the bed load, flat pebbles moved dominantly by sliding motions (see Butzer, 1964b: 161–63) can only be explained by stream conditions of rather uniform velocity rather than by sporadic, torrential spates.

In other words, the late Pleistocene gravel terraces of the coastal wadis can only be explained by greater runoff and rainfall, i.e., by a pluvial climate. It is interesting that minor wadis—which have catchment areas limited to the coastal plain—transported finer, less homogeneous, and more angular materials than did the large wadis. As today, most of the rains would have fallen over the elevated, rough terrain of the Red Sea Hills. The fact that no fine-grained beds seem to have terminated wadi alluviation suggests that the luxuriance of vegetation even in the Red Sea Hill wadis was never sufficient to impede pluvial erosion and transport. A necessary conclusion is that, even at the climax of the late Pleistocene pluvial phases, the vegetative mat of level surfaces in the hill country was incomplete. Climate presumably was subarid or semiarid, rather than hyperarid as it is today. Given the strong concentration of water in drainage channels with notable gradients, a precipitation value of 100 millimeters or so in the high country could adequately explain the available evidence.

The question may also be raised whether cold-climate weathering and aggradation were effective in the Red Sea Hills during the colder climate of the Würm Glacial. Extrapolating from the Nile Valley and the Red Sea coast, and assuming a lapse rate of 1° F (0.6° C) per 100 meters, the mean January minimum temperature at 1000 meters elevation today may lie between 35° and 42° F (2° and 6° C). A Würm temperature depression of 10° F (6° C) would bring the high country within the zone of moderately effective frost-weathering. But such weathering would hardly be more than an auxiliary agent providing ready detrital materials for stream transport, and pluvial-type overland flow and wadi activity must be considered the dominant geomorphic agency responsible for all alluvial deposits. There is no general or local evidence in the Red Sea Hill country of southern Egypt that requires or suggests a periglacial environment at any time during the Pleistocene. The highest peak near the study area is Gebel Nuqrus (42 km southwest of Mersa Alam) with an elevation of only 1504 meters, so that at this latitude it is unreasonable to expect Pleistocene periglacial phenomena.[2]

A final geomorphic problem is the matter of oversteepened terrace gradients in the kilometer or two adjacent to the modern coast. Three explanations can be provided. First, as has been implicit in the foregoing discussion, aggradation of the coastal plain may have been synchronous with a falling sea level, leading to a rapid change of gradient in the immediate vicinity of the coast. Second, the oversteepened wadi embouchures might be explained as fans accumulated downstream of the constricted wadi courses common to the seaward margins of the coastal plain (see Fig. 8–1). Third, the oversteepened terraces might also be

---

2. The only record of modern cold-climate phenomena comes from Mt. Sinai (2637 m), where Büdel (1954b) and Klaer (1962: 100–10) describe patterned ground, including stone rings (with diameters up to 20 cm) and stone stripes, at elevations above 2600 meters. Vegetation structures reflecting soil frost are found down to 2400 meters. Whereas both of these authors convincingly negate the possibility of Pleistocene glaciation on Gebel Katharina, Klaer (1962: 107 f.) invokes Pleistocene periglacial processes to explain the gross topography of the general area. He believes that the convex, dome-shaped forms developed in the Basement Complex are insufficiently distinctive for a subtropical desert. This curious and uncompelling "argument" is followed up by a hypothesis of intensified Pleistocene frost-weathering, leading to "blockstream-like" talus fans or cones, extending down into the valleys. Such blockstream-like features are neither described nor illustrated, and the text does not make clear whether they are hypothetical or were actually observed. Nonetheless, Klaer gratuitously offers a self-admitted guess that the "lower periglacial limit" on the Sinai Massif was at an elevation of 1800 meters during the Pleistocene glacials.

attributed to intermittent uplift of the coast. The second hypothesis does not fit the field observations. The low terraces do not widen in coastal proximity, and their configuration is not fanlike. In some of the wadis, the oversteepening takes place within the constricted wadi channel, in others it begins upstream. In fact, there is no systematic relationship between the oversteepened terraces and wadi constriction. The tectonic hypothesis depends on whether or not the glacial-eustatic interpretation of late Pleistocene events be accepted. The available isotopic dates indicate that the widespread undeformed high coral reefs of much of the Egypto-Sudanese littoral are a result of glacial-eustatic regressions. Similarly, the geomorphology of the coastal plain is incomprehensible without several major regressions of sea level, allowing the wadis to incise deep, buried channels near the coast. Because there is no evidence for deformation in the late Tertiary and late Pleistocene sediments of the littoral zone at Mersa Alam, the tectonic hypothesis cannot be applied.

The geomorphic evidence strongly favors the first explanation. If the base level falls during a period of aggradation, the balance of scour and fill will be different at midstream and near the lower terminus of the stream. Both short- and long-term scour will be accelerated at the mouth of the stream, in response to steeper gradients. Consequently, net aggradation will be less there than further upstream, a factor that will necessarily affect the longitudinal profile of the stream. As sea level continues to drop, the wadi floodplain will gradually be extended seaward, so that the oversteepened gradient would not travel further upstream, but would remain confined to the emergent shelf. Finally, as wadi activity waned, alluviation would cease, and dissection of fill would be initiated—first near the stream mouth, with the new impetus gradually transmitted upstream.

WADI DISSECTION

The Pleistocene record of the Mersa Alam area indicates that one or more periods of major wadi downcutting in coastal proximity were originally responsible for bedrock dissection to well below sea level. Later on in Pleistocene times, the High, Middle, and Low Terraces were repeatedly dissected to below modern wadi floor or even to below sea level. During the later stages of the Pleistocene, such wadi cutting affected unconsolidated alluvium and, less commonly, consolidated to semicemented reefs. Given an impetus such as a glacial-eustatic fall of base level, the intermittent wadi flows that occur today would probably suffice for dissection of fill. Appreciably greater rainfall would probably lead to aggradation, so that dissection in the geomorphic record implies a climate drier than that of the pluvial phases and probably not much different from that of the present.

WEATHERING

Compared with other parts of Egypt with a similarly complex sedimentary record, the evidence for paleosols on the Red Sea littoral is surprisingly scarce. Most late Pleistocene sediments show a little superficial patination, and one or more phases of calcification can be postulated. The significance of quartz grain corrosion (by dew moisture?) has already been referred to. As can be expected, physical hydration of salt has been of some significance. But pedogenic zones of organic accumulation or oxidation are almost entirely absent. An exceptional example of oxidation horizons is provided by the lagoonal deposits at Km 5.5, but these carry no climatic implications. Nowhere was rubefaction observed *in situ*, and the red soil sediments on the $+20$ to $+22$ meter conglomerate at Km 2.5 are almost unique. The only conclusion possible is that soil development of anything but Yermas (see Kubiena, 1953: 178–87) was minimal, at least during the late Pleistocene and Holocene. Such soil products as did form have been destroyed by denudation or deflation. Away from the immediate littoral, the geomorphic agents at work today and during the later Pleistocene appear to have been almost exclusively mechanical in nature.

Conclusions

The Red Sea littoral at Mersa Alam preserves a complex and interesting record of late Pleistocene events. A sequence of high interglacial shoreline stages at $+10$, $+5$ to $+6$, $+8.5$, and $+3.5$ meters can be identified. These clearly predate one or two major phases of gravel alluviation by the local wadis and, ultimately, by the watercourses draining the Red Sea Hills. This aggradation was contemporary with a sea level lower than that of today, with the result that abrupt breaks of longitudinal wadi gradients developed near the coast. As a result, the wadi terraces within a kilometer or so of the coast were oversteepened and must formerly have continued onto the emerged shelf, at below modern sea level. Such terraces are consequently ascribed to climatic rather than to eustatic factors, i.e., to greater stream transport, competence, and capacity. These pluvial gravels predate the climax of the last, major glacial-eustatic regression and probably pertain to the early and middle parts of the Würm Glacial. By contrast, the Holocene has left no systematic geomorphic record, either of high sea level stages or of wadi phenomena. No climatic changes of any import seem to be recorded.

Older features, recorded by High Terrace spurs in Wadis Alam and Sifein, by "buried" beach gravels, and by coral reefs higher than $+10$ meters, are too fragmentarily preserved for systematic study. They appear to be of middle Pleistocene age.

Broadly speaking, the Mersa Alam record parallels those of the Nile Valley and the Kurkur Oasis. Minor climatic changes do not seem to have left any geomorphic record on the coastal plain, and evidence for red paleosols is almost nonexistent. Thus, as at Kurkur, the last significant pluvial deposits appear to be contemporary with the middle part of the Korosko Formation.

An important inference that can be drawn from the Mersa Alam area is that, contrary to the claims of earlier authors, the alluvial terraces of the coastal plain are *not* interdigited with high coral reefs and were not deposited in response to periods of high glacial-eustatic sea levels. On the contrary, the very existence of coral reefs in the present wadi embouchures precludes much detrital sedimentation. Instead, those deposits which are intercalated suggest that at least the smaller wadis were completely defunct at the time the high coral reefs were formed. The wadi terraces either cut through the reefs, from which they are stratigraphically separated by an interval of dissection, or they rest disconformably on them. Not only are these conclusions significant for an understanding of the Egyptian Pleistocene, but they have a far greater geomorphic implication: namely, that smaller streams in arid regions are far more sensitive to climatic changes than to fluctuations of base level.

9

Towards a History of the Saharan Nile

Cenozoic Evolution of the Saharan Nile

At the beginning of the Tertiary, the waters of the ancestral Mediterranean Sea covered the greater part of Egypt, probably penetrating as far south as the Sudanese border in what are now the deserts west of the Nile (see Figs. 3–15 and 1–2). Marine sediments were deposited over most of this surface during Paleocene and Lower Eocene times (for general references on the Tertiary geology see Papp and Thenius, 1959; Said, 1962). The southern half of the Red Sea Hill country—including the Red Sea—and the Sudan were dry land, for the most part exposing sandstones of Nubian facies. Extensive denudational plains at elevations of 600 to 800 meters were probably cut in the northern and central Sudan at about this time (see Sandford, 1933, 1949). By the terminal stages of the Lower Eocene (Ypresian), the sea had withdrawn northwards into central Egypt (to about latitude 25° 30′ N). This regression continued, with a number of oscillations, until the coastline extended parallel to the modern Mediterranean shores at 29° to 30° N latitude by mid-Oligocene times.

Some limited conclusions concerning the origins of the Proto-Nile can be drawn from the middle Tertiary deposits of Egypt. Vertebrate and plant remains are found in shallow-water marine strata north of the Fayum, dating from the late Upper Eocene (Qasr el-Sagha Formation—Said, 1962: 102–3). These suggest proximity to a river mouth. Estuarine or fluviomarine deposits are subsequently recorded from the Lower Oligocene Qatrani Formation of the Fayum, related to the mouth of an ancestral Nile River (Said, 1962: 103–4, 219 ff.). These beds include a fauna of fishes, giant tortoise, turtle, crocodile, giant ostrich, and a host of mammals, as well as silicified wood swept into the sea by the Proto-Nile. In the Cairo area, the early Oligocene quartz sands include minerals derived from the Nubian Sandstone as well as from the Basement Complex, indicating that some of the igneous and metamorphic basement was already exposed at the time. Extensive spreads of cobble gravel were deposited by the Proto-Nile in the Libyan Desert south of the Fayum, over a hundred kilometers west of the modern Nile Valley. These flint, quartz, and quartzite gravels finally terminate a little northwest of Luxor (Knetsch et al., 1963), suggesting that the Oligocene Proto-Nile followed the contemporary valley in southern Egypt. Possibly, the 450-meter surface that was cut on the Libyan Plateau west of Kurkur (see Chap. 7) dates from this general time.

Further south in the Sudan, there are a number of thin but widespread sediments generally believed to be of Oligocene age. The Hudi Chert is one such formation, probably deposited in a great inland lake that extended 500 kilometers from east to west, from Berber into Wadi Melik (see Fig. 3-15) (Andrew and Karkanis, 1945). These chert beds cover a vertical elevation range from about 350 meters to over 400 meters and are younger than erosional surfaces at about 400 meters elevation, widespread in the northern Sudan (see Sandford, 1949). These surfaces are absent in the central Sudan (Kordofan) and appear to be linked to the regional base level provided by major rivers draining northward toward the Proto-Nile or Wadi Odib. Conceivably, the Hudi inland lake formed during the initial stages of rifting in the Red Sea area, when the eastern periphery of the Nubian Shield was first upwarped or upfaulted in late Oligocene or early Miocene times (see Sestini, 1965). Ultimately, this lake was drained, and chert breccias, cemented by a ferruginous cement, were laid down around the peripheries, with sandstones in the center of the former lake (Andrew and Karkanis, 1945). The Hudi Chert and associated breccias mark the only deposits known from the former Howar Basin (see Fig. 1-1). Reemergence of the area (in Oligo/Miocene times?) may reflect a drier climate or it may be related to renewed cutting of an antecedent drainage route to the Red Sea or, on the other hand, to development of a new exit via the Proto-Nile.

In the Kordofan, extensive lateritic deposits are also thought to be of Oligocene age. This Laterite Series has been described in detail by Kleinsorge and Zscheked (1959). The basal units consist of fine-grained colluvial materials, rich in aluminum and ferric oxides. The upper part of the basal sequence includes vulcanic tuffs deposited in standing water, a feature of regional stratigraphic interest. The upper units of the Laterite Series include (*a*) extensive lateritic crusts, thought to have developed *in situ*, occasionally interrupted by lateritic breccias; (*b*) bands of oolitic iron deposited subaqueously in small, shallow lakes; and (*c*) irregular ferruginous deposits due to impregnation of cracked colluvial or lacustrine beds by iron-rich waters. The various elements of the Laterite Series cap older erosional surfaces over much of the central and northwestern Sudan (see Sandford, 1935). They appear to be younger than the 600-to-800-meter and 400-meter pediplains and partly contemporary with the Hudi Chert.

The Oligo/Miocene tectogeny led to accelerated rifting in the Red Sea area, accompanied by extensive gentle deformations throughout southern Egypt (as described in Chaps. 2 and 5) and the Sudan (Sestini, 1965), leading to downwarping of the Paleo-Sudd Basin (Kleinsorge and Zscheked, 1959). Localized vulcanic activity was initiated in many areas (see Fig. 3–15) during the middle Oligocene, producing a series of basaltic lavas, cinder cones, ash fields, and small craters, scattered from Cairo to the central Sudan. In northern Egypt, all of these vulcanic phenomena predate the Lower Miocene transgression (Said, 1962: 104, 177–78). Lavas are not uncommon on top of both the 600-to-800-meter and the 400-meter denudation surfaces (Sandford, 1935, 1949), and similar basaltic lavas rest on the Hudi Chert and the breccias (Andrew and Karkanis, 1945). The lavas in the Bajuda Steppe of the Abu Hamed bend appear to be related to NNW fractures, paralleling the Red Sea Graben (Putzer, 1958). Thus, the mid-Tertiary vulcanism helps tie together many events in Egypt and the northern half of the Sudan: the first significant deformation of the Red Sea Graben, the draining of the Hudi inland lake, the early phases of the Laterite Series, and the completion of sculpture of the early to middle Tertiary erosional surfaces.

The Kurkur, Ballana, and Aswan Pediplains of Lower Nubia and the Kom Ombo region (Chaps. 2, 5, and 7) mark successive stages in the subsequent development of the Proto-Nile, presumably in Miocene times. Certainly the Aswan denudation surface and probably the Ballana surface as well exhibit no intensive deformation, suggesting that they postdate the major tectogenic spasms of the Oligo/Miocene. Yet the nature of these surfaces, extending tens or hundreds of kilometers from one side of the Proto-Nile Valley to the other, can only be reconciled with long periods of denudation, measured in terms of millions of years. It seems that the influx of waters from the Shait and Kharit basins rivalled or exceeded that of the Nubian Nile at this time, as it still did at the beginning of the Pleistocene. Since deposits of the Miocene Proto-Nile are not preserved (except in the estuarine zone near Cairo), the relationships of the Proto-Nile and Howar basins (Fig. 1–1) cannot be determined.

Prior to the late Pleistocene, the only fluvial sediments known from the

Howar Basin predate some of the basaltic lavas. Near Berber, Andrew and Karkanis (1945) have described the following sequence, from top to bottom: (a) over 9 meters of basalt with a basal, ferruginized quartz conglomerate; (b) 15 meters of quartz grit with a clayey matrix; and (c) 2 meters of Hudi Chert breccia—all resting on siltstones and sandstones of Nubian facies. The conglomerate fossilized by the basalt seems to be fairly common and may be related to an early through-river of the Howar Basin. It would seem that, if the Howar drainage had been forced across a fairly high-level divide towards the Red Sea, fine-grained fluvial or lacustrine beds, postdating the basalts, should be well developed within the basin. Consequently, although Hudi Chert is absent from the autochthonous Nile gravels of Lower Nubia, it is difficult to assume effective isolation of the Proto-Nile and Wadi Howar basins until the late Pleistocene.

During the course of the Pliocene, the Egyptian Nile rapidly entrenched its bed into a small segment of the Aswan Pediplain, cutting down to well below modern base level. The presence of a fairly complete Miocene sequence on the Red Sea littoral precludes the possibility of deep incision of the Nile prior to the terminal stages of the Miocene. On the other hand, the general absence of Lower Pliocene beds along the coastal periphery indicates a marine regression, presumably favorable to Nile incision. The phenomenal vertical dimensions of the buried, entrenched valley of the Nile (to -172 m at Aswan according to Chumakov, 1965) cannot be readily explained by any single factor, such as a major regression of the late Miocene and early Pliocene seas. Whatever the reasons, the Plaisancian sea subsequently invaded this fiordlike valley, extending far up into Lower Nubia, and, during the regressive phases of the late Pliocene, a complex of lagoonal, brackish, and lacustrine beds was deposited at Kom Ombo and Aswan. Presumably, sedimentation of the Umm Ruwaba Series within the Paleo-Sudd Basin (Fig. 1–1) had begun by this time, but here again the impulse, other than epeirogenic deformation, is obscure.

The early and middle Pleistocene brought revitalization to the Proto-Nile Basin. The Pliocene beds were largely eroded, bedrock incision was still common, and the Shait-Kharit basins provided a high proportion of the Nile water and sediment. Modern drainage lines were only established (or reestablished) during the course of the early Pleistocene. The geomorphic record of the first two million years or so of the Pleistocene is one of dissection and gravel aggradation, a series of complex cycles superimposed on a general trend of valley incision. In response to pluvial paleoclimates of modest but significant amplitude, discharge from the local wadis provided the bulk or the entirety of the Nile flow. Wadi and Nile discharge were, as a result, simultaneous and seasonally in phase, presumably reaching flood stage during the winter half-year. Whether or not summer rains played a role, at least in the southern half of Lower Nubia, cannot be determined. All of the gravels deposited by the Nile and its tributaries were either derived from the Basement Complex of the Red Sea Hills and the northeastern Sudan or from the Nubian Sandstone of the Nile Valley peripheries. As a result, the early and middle Pleistocene deposits preserved today are limited to the Proto-Nile Basin. It remains to be proven, however, that there were no summer floods of Ethiopian origin, since suspended-load sediments do not seem to be recorded and conditions probably were inimical to their preservation.

The late Pleistocene ushered in revolutionary changes in the hydrographic regime of the Nile. The Korosko Formation records an intermediate pattern, with winter-season wadi deposits interdigited with coarse sands, marls, and silts deposited by summer floods and derived, at least in part, from the Blue Nile and Atbara basins (see Appendices B and D). Thus, some 50,000 years ago, the Ethiopian link is verified, although characteristic deposits from the Howar Basin remain absent. The succeeding Masmas Formation illustrates that, at the height of Würm Glacial, the Egyptian wadis were as defunct as they are today, while for the first time the Nile floodplain was accreted by annual increments of nilotic muds from the Blue (and the White?) Nile and Atbara basins. Yet Hudi Chert was still absent in southern Egypt. Finally, about 17,000 years ago, the Gebel Silsila stage introduced chert gravels along the length of the Egyptian Nile, while the White Nile was now definitely linked to the Saharan Nile at Khartum (see Chap. 6). The floodplain did not necessarily remain stable in more recent millennia, but the changes that have occurred resemble oscillations of a regime allied to that of the present.

There can be no question that the geomorphic evolution of the Nile Basin has been complex, and, all in all, the modern hydrographic patterns were exceedingly slow in evolving, when compared with those of other major river basins. But it should be emphasized that negative evidence may be circumstantial. The absence of Hudi Chert from the early Pleistocene gravels does not necessarily prove that the Proto-Nile Basin was still isolated from the Upper Nubian Nile. And again, the absence of nilotic silts does not preclude a summer-flood regime during the early and middle Pleistocene—subsidiary to winter wadi activity during pluvial intervals and dominant during nonpluvial intervals. Consequently, the record of southern Egypt can tell no more than one part of the story. The remaining problems must be resolved further upstream, in the Sudan and in Ethiopia.

Cretaceous and Cenozoic Changes of Climate in Egypt and the Northern Sudan

The available evidence suggests that the eastern Sahara has been comparatively dry throughout geological times, a fact which has already been pointed out by Schwarzbach (1953). A brief survey of pertinent data as recorded by the various formations and geomorphic features may serve to illustrate the paleoclimatic trends of the late Cretaceous and the Cenozoic.

The sandstones of Nubian facies have been recently studied by McKee (1963). In the Aswan area, the massive sandstone strata were primarily laid down in standing waters, separated from the sea. Streams, moving to the north, northeast, or northwest, deposited moderately sorted coarse sands into quiet water bodies. The middle sandstone units, with shales and oolitic iron, were apparently laid down in mixed, mildly agitated waters separated from the open sea by a barrier. The bedding characteristics, the presence of plants and of saltwater as well as freshwater molluscs preclude both an eolian and a marine origin. At Kharga and Dakhla, the Nubian Sandstone suggests a fluvial and floodplain origin, with streams carrying moderately sorted medium sands in a northwesterly or northerly direction. The presence of conglomeratic beds, fossil wood, and numerous leaves, as well as specific bedding properties, all indicate a continental or lagoonal, rather than a marine, origin. In the Khartum area of the Sudan, the sandstones are inhomogeneous in terms of grade-size, and individual beds are poorly sorted, while pebbles are abundant. The bedding characteristics further point to nonmarine, subaqueous deposition, probably on the alluvial plains of delta cones. Bedding is to the northwest.

An overall interpretation of the Nubian Sandstones of the eastern Sahara would need to emphasize the fluvial character of these beds, even though deposited in proximity to the coast. This implies considerable discharge from the exposed Upper Cretaceous landmasses, but the overwhelming preponderance of coarse, detrital sediments indicates a dominance of mechanical rather than of chemical weathering. Climatic conditions may have been generally or periodically a little moister than at present, but they remained arid or subarid. Regional variations of climate are not apparent. The presence of plant remains—few of them identified (except for the study of Barthoux and Fritel, 1925), and none of these ecologically interpreted—is compatible with this explanation, since most of the plants or trees were carried into estuarine or lagoonal deposits from zones of galeria vegetation.

The Lower Tertiary marine deposits of Egypt provide little paleoclimatic information. The dominance of calcareous sediments with phenomenally rich foraminiferal life suggests warm waters, with no implications for local moisture conditions. The shale units, with local evaporites or phosphatic beds, are more difficult to interpret. Many of these beds suggest shallow, littoral or lagoonal environments, separated from the open sea. The absence or scarcity of carbonates may therefore reflect only the siliceous nature of the rocks exposed on the landmasses of the Sudan.

The early Tertiary denudational surfaces of the Sudan are compatible with a climate such as that suggested for the later Cretaceous: essentially arid or subarid, but with considerably more runoff and geomorphic activity than at the present time.

The Oligocene, at least in part, varies from the late Cretaceous–early Tertiary pattern. The extensive cobble gravels of the Proto-Nile in the Libyan Desert can only be explained by a vigorous river of considerable competence and capacity. The galeria forests accompanying this torrential river were periodically uprooted and swept out to sea by flood waters derived from the Egyptian drainage basin of the Proto-Nile. The upwarping of the Nubian Shield provided greater relief-energy at this time, but unquestionably there was considerable runoff. Greater moisture is also suggested by the Hudi inland lake, measuring about 500 by 250 kilometers, that was maintained in the Howar Basin. The dominance of chemical and suspended-load sediments, rather than of clastic detritus, are as significant as the depth of the water, which may have exceeded 50 meters. Equally impressive are the lateritic deposits contemporary with, and later than, the Hudi inland lake. The abundance of aluminum and ferric oxides, locally provided by weathering of sandstone uplands, indicates a seasonally wet, generally subhumid, tropical climate, favoring Rotlehm pedogenesis (see Kubiena, 1957). The presence of widespread colluvial deposits and the genesis of allochthonous lateritic crusts also points to the type of geomorphic activity most commonly associated with the African savanna plains (see Büdel, 1958).

In overview, extended intervals of the (middle to late ?) Oligocene brought a rather moist climate to the eastern Sahara. Conditions in the central, and perhaps also in the northern, Sudan resembled those of the moist savannas of West Africa, while a semiarid climate may have prevailed in Egypt. Pluvial conditions of similar duration and intensity were never matched again.

The Miocene deposits of the Red Sea littoral of Egypt suggest that, apart from any changes of oceanic circulation, a long-term climatic cycle may have been experienced in the Red Sea Hill country. The sediments of the early and late Miocene are rich in conglomerates and other detrital beds (see Said, 1962: 178 ff.; Sestini, 1965; also Chap. 8) swept into the sea by streams coursing down the eastern slopes of the hill country. The Middle Miocene Evaporite Series is practically free of such clastic imports, however, ruling out significant wadi activity or rainfall in the Red Sea Hills. Regardless of the tectonic events of the time, runoff is required to transport gravel in the local wadis. Consequently, it can probably be inferred that climate gradually changed from subarid or semiarid in the early middle Miocene to hyperarid and subsequently to subarid in later Miocene times. Minor oscillations were probably superimposed upon this overall cycle, judging by the interdigitation of coarse clastic and fine-grained marine-littoral deposits, even during the early Miocene.

The Miocene geomorphic record of southern Egypt (Chaps. 2, 5, and 7), with its widespread denudational plains, suggests a moister climate than today but much drier conditions than those experienced in Oligocene times. Extensive pedimentation requires abundant, periodic overland flow—an agency next to absent at the present time—but rainfall averages need not have exceeded 100 or 200 millimeters. Thus, the climate associated with pediplantation in southern Egypt was either subarid or semiarid. No climatic fluctuations can be identified, and stratigraphic dating for these erosional surfaces is not precise.

The Lower Pliocene, with rapid vertical cutting of the Nile Valley, offers no paleoclimatic clues, since deposits have not been recognized and the erosional evidence is liable to different kinds of interpretation. The Plaisancian estuarine beds of the Pliocene Gulf are completely lacking in coarse detrital beds, but are rich in montmorillonitic clays and organic matter. Iron is exclusively present in the ferrous form, indicating persistant reducing conditions. This could be explained by a moist climate, an effective vegetation cover, and little or no torrential runoff, so that stream discharge into the estuarine zone would include little or no coarse alluvium. The alternative explanation of complete aridity, with no runoff at all, is unsatisfactory, since the fresh waters that strongly diluted the marine gulf required a significant amount of local precipitation. Consequently, the mid-Pliocene climate of Egypt seems to suggest considerable moisture, distributed fairly evenly over the year, as well as an effective cover of vegetation. This conclusion is compatible with the foraminiferal evidence from the Cairo area (Said, 1955). The transitional Astian unit includes grit and gravel and is characterized by rapid changes of facies. This suggests more torrential discharge into the gulf, with changing environmental conditions reflecting either periodic fluctuations of moisture or gradual but continuing uplift of the Red Sea Hills. The significance of ferric hydrates and the absence of ferrous sulfate are the converse of the Plaisancian beds and point to periodic emergence and oxidation of the gulf deposits. This can only be interpreted climatically and indicates either increased aridity or greater seasonality of rainfall, or both. Thus, late Pliocene climate was transitional between that of the mid-Pliocene and that of the Pleistocene. The localized evaporites of Kom Ombo would be compatible with such a regional interpretation, while further north in Upper and Middle Egypt the massive tufas and conglomerates of the final Pliocene or earliest Pleistocene leave no doubt that the Egyptian climate had become subarid or semiarid, with seasonally concentrated runoff.

During the course of the early and middle Pleistocene, three primary paleoclimatic situations appear recurrently in the geomorphic record of Egypt: (a) aggradation of coarse gravels by wadis and Nile, presumably as a result of subarid or semiarid climate and torrential, winter-season runoff; (b) development of red paleosols in response to moderate or fairly intensive chemical weathering, at times with little surface denudation, a good mat of vegetation, and, possibly, gentle rains; and (c) periods of limited geomorphic activity as a result of hyperarid climate, similar to that of the present day. In our interpretation (see Chap. 2), dissection of wadi fill characterized the periods of waning rainfall that followed the aggradation phases, while dissection of the Nile floodplain may have persisted during the hyperarid phases of limited morphogenesis.

The aggradation phases of the Egyptian Pleistocene include the most active periods of pluvial erosion of upland surfaces and lowland plains; transport of detritus by sheetfloods, wadi spates, or Nile inundations; and, at times, effective lateral or vertical bedrock cutting. During the late Pleistocene, such events were most common and most effective during parts of the early and middle stages of the Würm Glacial, in analogy with geomorphic processes in the Mediterranean Basin (see Butzer, 1963a, b). Presumably, this analogy should also be applicable to the pluvials of the Egyptian early and middle Pleistocene, with the aggradation-planation phases *sensu lato* corresponding to major glacial episodes. Thus, the Wadi Korosko stage might be related to the European Warthe, the

Dakka stage to the Saale, etc. But, in the absence of suitable older alluvia on the coastal plain of Mersa Alam, such correlations could not be demonstrated in the field and must remain highly speculative.

There seem to have been many intervals of red-soil development, of varying intensity, during each stage of the Pleistocene—at least if we take the last 50,000 years as a guide. Yet only a few of the major periods of biochemical weathering have left a lasting imprint on the geomorphic record of the earlier parts of the Pleistocene. In the case of very modest red paleosols, kaolinitic clays seem to have been formed under desert conditions, with surficial rubefaction by extremely limited quantities of hematite. In a few cases, too, the intensity of weathering approaches that of Rotlehm soils, suggesting a long, warm, moist season.

There can be no doubt that the gravel-planation periods and the red pedogenesis indicate two distinct kinds of pluvial period, one promoting geomorphic activity with erosion and deposition, the other leading to soil development and excluding significant soil erosion. These two different types of equilibrium are mutually incompatible. But, whereas the morphodynamic pluvial type can be readily explained, the red paleosols require environmental conditions that are extremely difficult to reconstruct or find analogies for. The only reasonable answer seems to lie in a different rainfall regime, with more even distribution of moisture during the year and with frequent, gentle rains and next to no torrential discharge. The type of paleometeorological patterns which may have produced this geomorphic situation is not known. So, for example, there is no rational explanation from the available northern hemisphere evidence of why a red paleosol should have formed during the sixth millennium B.C. Our lack of explanation for the red paleosols is just as unfortunate as the fact that the temporal or sequential relationships of the morphodynamic pluvials and the periods of red-soil development cannot be determined. In the Mediterranean world, *terra rossas* and Rotlehms developed during several interglacial periods, under warm, seasonally moist conditions (Butzer, 1964*b*: 88 ff., 1964*c*; Fränzle, 1965). In Egypt, however, it is impossible to tell from the Pleistocene record whether the red paleosols developed before, between, or after phases of pluvial aggradation and planation, or whether they represent interruptions of the hyperarid climate that dominated certain interglacial periods.

Seen in retrospect, the net geomorphic impact of the Pleistocene pluvials on the Egyptian landscape has been moderate. Tertiary erosional surfaces remain intact over great expanses of the country, while most of the deeply incised relief forms also appear to antedate the Pleistocene, at least north of Aswan. Significantly, too, all of the late Cenozoic sculpture has left but a single geomorphic imprint on the country, namely, the development of characteristic arid-zone landforms (see Butzer, 1965; Hansen, 1966). Thus, angular, jagged profiles are the rule rather than the exception in any dissected upland. Yet, seen in detail, slope segments everywhere fall into similar patterns of steep, rectilinear midslopes, abruptly offset from upland or lowland surfaces by angular breaks of gradient. Lithological variation plays a conspicuous role in modifying individual details while not altering the large-scale patterns of slope profile or gradient. Drainage density is rather low, and wide expanses of pediment plains remain undissected despite periodic reactivation of fluvial processes.

Although the total result of two million years or more of hyperarid, subarid, or semiarid paleoclimates has been a homogeneous arid-zone landscape, the trend of geomorphic forces has not been unidirectional. The major exceptions have been the enigmatic periods of red-soil development. Whatever the reasons for or stratigraphic position of the red paleosols, the biochemical weathering associated with their formation had considerable geomorphic significance. Soil development produced a mantle of fine-grained soil products at the surface, followed at depth by a partly-weathered and broken down zone of regolith. Bedrock surfaces were thus prepared for subsequent denudation by wind or water or both. Consequently, periodic chemical weathering contributed to the effectiveness of morphodynamic pluvials in landscape sculpture. Similarly, the analogous or distinctive phases of deep patination or impregnation by ferromanganese solutions had the converse geomorphic impact of indurating surface materials and impeding the denudational forces. An idealized cyclical scheme of geomorphic evolution has been proposed by Hansen (1966: 142 ff.) to explain the development of pediment landscapes at Arminna. These notions are in many ways relevant to all the sandstone plains of southern Egypt and can be summed up briefly.

a) During certain periods of moister and more equitable climate, a continuous or subcontinuous vegetative cover was supported, with resulting chemical weathering. A mantle of red soil developed, and surface waters may have become saturated with ferromanganese solutions, causing accelerated and widespread ferruginization. The vegetation mat and soil mantle served to impede areal denudation and promote linear incision, i.e., wadi cutting.

b) With deterioration of the precipitation regime, vegetation was diminished, thus exposing the soil cover to raindrop erosion and surface denudation. Charged with soil sediment and regolith, the streams concentrated their erosional attack on the wadi walls rather than on the valley floors. Rill wash on cliff faces removed slope debris while lateral wadi erosion led to cliff undermining, thus ultimately widening the valley floors. Eventually, the wadis became sufficiently broad that wall-to-wall flooding or the incidence of channel impingement on wadi walls became increasingly less frequent. Rillwash and sheetwash or sheetflooding transported slope detritus across these incipient pediments to a major channel, for subsequent removal. As scarp retreat and pediment enlargement continued, the wadis lost their original identity and became broad, shallow drainage lines flanked by denudational surfaces.

c) Complete desiccation brought pedimentation to a halt, as wind became the dominant geomorphic agent. Truncated relicts of former soil horizons remained on upland surfaces, while soil sediments frequently survived among the wash on the lowland pediment plains, where not completely deflated. Lacking sufficient moisture for chemical weathering or for erosion, backwearing of cliff faces ceased, yielding a relict landscape such as that encountered today.

Whether or not this geomorphic cycle occurred in strict sequence is irrelevant. What matters is that it does provide an effective explanation of arid-zone morphogenesis that is theoretically compatible with the landscapes of Kom Ombo and Lower Nubia. The first situation (a) parallels that of the morphostatic pluvials, while the second (b) corresponds to the morphodynamic pluvials, with gravel aggradation and lateral planation in the major valleys and with pediment development along their peripheries. The third situation (c) describes current morphogenesis in the western deserts. Repeated alternations of these patterns are verified in the Pleistocene record, and they provide a reasonable explanation for the landscapes of southern Egypt.

The fluvial activity recorded during the early and middle Pleistocene of Egypt has been ascribed to winter rainfall, related to westerly disturbances. Proof is impossible to provide at the moment, since there is no seasonal-stratigraphic guide such as that provided by nilotic deposits during the late Pleistocene. But at least there are no indications as yet that summer monsoonal rains of geomorphic significance must be assumed. Presumably, if we are dealing with morphodynamic pluvials that correlate with the stages of glacier growth at higher latitudes, westerly disturbances should have penetrated far into the Sahara, possibly accompanied by an equatorward shift of the summer-rainfall zone. Here too, however, we cross the borderline into the realm of speculation.

The late Pleistocene and Holocene climatic record of southern Egypt is comparatively well understood (Chaps. 3 and 6 to 8). Several pluvial episodes of some significance are recorded during the early and middle

Würm Glacial, in the form of the Wadi Floor Conglomerate and the Korosko Formation in the Nile Valley, the younger Wadi Tufas (III and IV) at Kurkur, and the Middle and Low Terraces of the Red Sea coastal plain. Several further moist intervals are recorded from the terminal stages of the Würm Glacial and from the early to middle parts of the Holocene—in particular, the Ineiba and Shaturma Formations of the Nile Valley. Geomorphologically speaking, these last moist interludes were not very effective. They were nonetheless tangible, however, and they indubitably record winter rains, out of phase with the Nile flood season. Two aspects of these late Pleistocene pluvials deserve emphasis. One is their comparatively short duration, seldom exceeding two or three millennia and interrupted or succeeded by a millennium or more of hyperarid climate. This is remarkably similar to the rhythm of glacial advance, standstill, and retreat recorded by the many morainal substages of the continental glaciers and points towards a hemispheric impetus for the climatic changes involved. Equally significant is the close correspondence between increased Nile discharge and several of the Egyptian pluvial intervals. The Masmas Formation provides a single, striking exception to this pattern—at the time of the Würm Glacial climax, some 20,000 years ago—but, in the case of the Korosko Formation and of each of the substages of the Gebel Silsila Formation, the correspondence of Ethiopian and Egyptian pluvial trends is surprisingly well defined.

This survey of Cenozoic climatic changes once again raises many questions concerning the regions upriver. Why do no Pleistocene gravels appear to be recorded in the Howar Basin? Why does the facies of late Pleistocene nilotic deposits in Lower Nubia and Egypt change repeatedly? Why are some pluvial oscillations synchronous in Egypt and Ethiopia while others are not? And, finally, what geomorphic processes on the Ethiopian Plateau promoted the soil erosion which provided the millions of cubic kilometers of silts that inundated the Gezira Plain and the length of the Saharan Nile Valley during the late Pleistocene? All of these answers, in as far as they can be provided, lie in the Sudan or Ethiopia. Consequently, a discussion of Ethiopia would be useful at this point, if only to underscore what the problems are. This discussion will involve both the contemporary origin of nilotic silts and the Pleistocene record of the Ethiopian Plateau.

Origin and Nature of the "Ethiopian" Flood Silts

The Ethiopian Plateau is generally considered to be a great, complex faultblock, related to the East African Rift system. A basement of metamorphic rocks, intruded with granites, syenites, etc., is overlain by a series of broadly horizontal sedimentary and vulcanic rocks. The following stratigraphic sequence can be recognized (after Mohr, 1964; Grabham and Black, 1925: 34–42; Dainelli, 1943; Büdel, 1954a; and McKee, 1963), resting on the crystalline base:

a) Adigrat Sandstone. Maximum thickness about 1000 meters. White to very-pale-brown, sterile, medium- to coarse-grained sandstones, mainly quartz but with a certain proportion of feldspar. There are coarse-grade, basal conglomerates, while bands of gray to red shale are common in the upper parts of the sequence. A transgressive facies, deposited in a littoral environment, the assumed age decreases from late Triassic to Middle Jurassic toward the northwest.

b) Antalo Limestone. Maximum thickness 800 meters. Light-gray to brown limestone, often shelly, oolitic, or coralline, with occasional bands of shale, marl, sandy limestone, or gypsum near the top. Middle to Upper Jurassic or lower Cretaceous, according to the area.

c) Upper Sandstone. Maximum thickness about 500 meters. A regressive littoral facies, lithologically similar to the Adigrat Sandstone, the age varies from Upper Jurassic in the northwest to Upper Cretaceous in southeastern Ethiopia. Subsequent marine transgressions were limited to the eastern and southeastern part of the country. Elsewhere, the Tertiary was marked by accumulation of the massive Plateau Basalts or Trap Series, traditionally subdivided on lithological grounds.

d) Lower Plateau Basalts (Ashangi Group). 1200 meters. Massive flood basalts, with highly amygdaloid structure, including nodules of agate, stilbite, and zeolite, as well as some interbedded eolian tuffs or sedimentaries with traces of fossilized wood. Eocene (Mohr, 1965).

e) Upper Plateau Basalts (Magdala Group). 2600 meters. Massive, compact basalts associated with conformable masses, distinct flows, or intrusive bosses of resistant felsite or syenite, with obsidian or cryptocrystalline rhyolitic lavas in the uppermost column. Eolian tuffs or sedimentary strata of lacustrine origin are interbedded with the lavas. The dominant basalts vary from rocks consisting almost entirely of soda-lime feldspar, augite, and iron ore to varieties with augite, olivine, and feldspars. Oligocene to early Pliocene.

f) Aden Vulcanic Series. Localized basaltic or rhyolitic lavas, with some tuffs. An early unit, of Plio-Pleistocene age, was followed by further middle to late Pleistocene eruptions, that dammed back Lake Tana and various lakes of the Ethiopian Rift Valley. Finally, there are some volcanoes of Holocene age on the eastern coastal lowlands.

The Mesozoic sedimentaries of Ethiopia are only exposed in the deeper river gorges and along the peripheries of the plateau and should consequently be expected to contribute little to the sediment yield of the Atbara, the Blue Nile, and the Sobat. Similarly, the younger vulcanic series is quite localized and is, in large part, located outside of the Nile watershed. It is the Plateau Basalt—almost exclusively the Magdala Group—which forms the bedrock of the Ethiopian uplands, and it is from this material that much of the suspended sediment of the Nile floods appears to be derived.

Ethiopia has traditionally been subdivided into three physical environments: (*a*) the tropical Kolla region of the strongly dissected river gorges and peripheral foothills, at elevations below 1800 to 2000 meters; (*b*) the subtropical Voina Dega, between 1800–2000 and 2500–2700 meters elevation, a flat, basaltic plateau studded with rhyolitic and syenitic ridges or inselbergs and slightly dissected by shallow, V-shaped valleys; and (*c*) the temperate Dega, at between 2500–2700 and 3700–3800 meters elevation, a region of broad, gently sloping interfluves, cut by deep, steep-sided valleys (see Werdecker, 1955; Semmel, 1963). The primary source areas of Nile sediment can be localized in the Voina Dega, an upland landscape crossed by seasonal and perennial tributaries that drain into Lake Tana and the Blue Nile and the Atbara rivers.

The soils and geomorphic processes of the Voina Dega of central Ethiopia have been described by Semmel (1963, 1964). Flat, lower-level surfaces are found in the form of intermontane plains commonly measuring 8 to 15 kilometers in length, 3 to 7 kilometers in width. Such plains have a spontaneous vegetation of coarse grasses and acacias over a mixed residual-colluvial mantle averaging well over 2.5 meters in depth. The soils of these seasonally flooded savanna plains are of vertisol or margalitic type, with a dark-grayish-brown (10 YR–2.5 Y 4/2) horizon varying between 70 and 180 centimeters in depth, resting on a dense, whitish zone of decomposing basalt. In the dry state, crack networks penetrate to a depth of 80 centimeters, while the soil is swollen and impermeable when wet. The resulting self-mulching process works fresh organic matter and minerals into the soil. Texture is a silty clay with over 60% in the clay size-fraction. Montmorillonite is dominant, while kaolinite and hermatite account for the remainder of the clay minerals. Reaction is moderately acidic (pH 5.0–5.3), and humus content is intermediate (3.6% in the topsoil).

On better drained parts of the Voina Dega plains, particularly along the stream levees, there is a different soil type. Profiles are complicated by alluvial deposition, but, in general, these soils are dark to strong brown (7.5 YR 3–5/2–6), with a silty clay texture and pH values of 4.7 to 5.5. Kaolinite is the dominant clay mineral, with traces of muscovite-illite and hematite. The spontaneous vegetation on these latosolic soils consists of nutritious grasses or of galeria woodland. Finally, the interfluves and gently inclined hillslopes may show a distinctively tropical soil profile with a shallow, brown A-horizon (7.5 YR 4/4 moist) of 15 centimeters in depth, over a (B)-horizon between 1 and 3 meters deep. The (B)-horizon is a silty clay with about 60% clay fraction, a pH of 3.8 to 4.2, and reddish-brown to yellowish-red (5 YR 3/4 moist; 5 YR 4/8 dry) color. Metahalloysite is the major clay mineral, in association with hydrargillite and hematite. A 1-meter (B)C-horizon of white-mottled, partly decomposed basalt underlies the (B)-horizon. These soils are friable Braunlehms in the definition of Kubiena (1953: 266 ff.) or oxisols in that of G. D. Smith et al. (1960: 238 ff.). Although almost completely eradicated today, the original vegetation may have been a moist montane woodland. As a corollary, the soils have been severely eroded through agricultural use. In comparison with the vertisols, the two varieties of latosolic soils are slightly more organic (up to 4% humus) and considerably richer in free iron oxides (7% to 8.5% compared with 2% to 4%). Carbonate contents are invariably very low (less than 1%).

Whereas the Voina Dega has a mean annual temperature over 64° F (16° C) (Werdecker, 1955) and a rainfall of 1250 to 1500 millimeters (see S. P. Jackson, 1961), the higher-lying Dega is much cooler, with a mean annual precipitation of 1400 to 1800 millimeters. Annual temperatures at the upper elevation limit of the Dega lie around 46° F (8° C). Whereas frost is rare on the Voina Dega (mean monthly minimum temperatures in January between 40° and 48° F, or 5° to 10° C, see S. P. Jackson, 1961), it is fairly common on the Dega. As a result, the soils of this higher montane zone differ markedly from those of the Voina Dega (Semmel, 1964). Highly organic (28% humus), acidic (pH 4.0) soils of alpine type are dominant, a fact which matches well with the original, ericaceous vegetation (*Erica arborea, Lobelia*, etc.). Heavy minerals are abundant in these soils, primarily hornblende with a little augite. The dominant clay mineral is metahalloysite, with some montmorillonite, muscovite-illite, and magnetite.

The rainy season of the Ethiopian Plateau is a complicated matter, and a proper understanding is in no way helped by the paucity of good, long-term climatic observations (see British Meteorological Office, 1958; S. P. Jackson, 1961; Hövermann, 1961; Weickmann, 1964; Flohn, 1965b). The catchment areas of the Sobat, the Blue Nile, and the Atbara all experience a rainfall maximum in July and August, although the rainy season decreases in length from six months in the south to four months in the north. In addition, there is a secondary maximum of rainfall in April or May, affecting most of the highlands but not the Sudanese plains or the northerly foothill zone. Consequently, although the levels of the Blue Nile and the Sobat begin to rise in May, ushering in a "pre-flood" season of some importance, the Atbara and Khor Gash drainage basin is not affected by these early rains, and the waters of the Atbara first make their way to the Nile confluence during the month of June. There also appear to be significant regional differences of rainfall type, with thundershowers dominant over the Atbara Basin and the northern part of the Blue Nile Basin and with prolonged rains of lesser intensity characteristic over the Sobat drainage and the southern part of the Blue Nile catchment (see Hövermann, 1961). Brief, intensive showers bring almost all of the summer rainfall to the plains of the central Sudan (July through September, see Oliver, 1965). Although this point has not been explored, it is possible that soil stripping and gullying are more significant along the northern periphery of the plateau, where the rainfall season is short but intensive, and that slow and rapid earth flows and other forms of mass movement will be more important on the central and southern plateau, where soil and regolith are thoroughly soaked during a long rainy season with fairly persistant overcast skies.

It is generally agreed that Ethiopian rivers are highly turbid during flood stage, exhibiting a brick-red color as a result of the considerable amount of suspended material. Sand and gravel are rather limited because of both the paucity of quartz in the exposed bedrock and the ready decomposition of augite, feldspar, hornblende, and olivine sands. Gravel lenses do occur in stream beds but are generally limited to areas just downstream of rock outcrops on the stream floor. Within the savanna plains of the Voina Dega, Semmel (1963) shows that colluvial redeposition on the seasonally inundated lowlands is the major geomorphic activity, with net active erosion confined to the stream channel and to the levee zone. Major dissection is confined to the larger rivers. As a consequence, the strongly dominant clay mineral in alluvial sediments of the Voina Dega is kaolinite, with 85% of the heavy minerals consisting of augite in one sample analyzed by Semmel (1963). In addition to clay minerals, the waters contain calcium and colloidal silica, released through continuing decomposition of freshly eroded lime-soda feldspars. On the margins between the Kolla and the Voina Dega, gradients increase abruptly, and mass movements, including slumping and debris slides, are known to occur (see Kuls and Semmel, 1965).

Although these generalizations are based on limited observations, they do appear to give a reasonable impression of soils, erosive activity, and sediment yield on the Ethiopian Plateau. The comparative observations available to us from the central Sudan are equally sparse. Most informative are the preliminary microscopic examinations of Nile silts by W. F. Hume (in Lucas, 1908: 55–57), as well as the preliminary heavy mineral studies by Shukri (1950, see Appendix B).

The Blue Nile flood silts (deposited July through September) contain abundant quartz grains of sand size (30 to 500 microns), with 12.7% iron oxide and with whitish-brown clayey silt that is gritty rather than plastic to the touch because of the abundance of very fine quartz. The quartz sands are angular in shape, precluding distant transport, and their occurrence near the Ethiopian frontier suggests derivation from the Adigrat sandstones of the Blue Nile gorge and the Basement Complex of the Kolla foothill zone. In addition to some heavy minerals, there is still much quartz in the deposits of the postflood season (October through January), while fine quartz and brownish clayey silts constitute the preflood deposits (February through June). In other words, the sediment yield of the Blue Nile includes a considerable proportion of materials that do not owe their origin to the Ethiopian Plateau. The heavy minerals, which are primarily derived from the uplands, only work their way downstream through saltation and traction, arriving in the Sudan during the postflood season. This emphasizes that the angular quartz grains—in the same grade-sizes—must be of more local origin, since they appear in the sediment load several months earlier. This raises the question of whether a part of the suspended sediments is not obtained in the foothill zone or eroded from older silts of the Gezira Plain (see Fig. 3–15). This is a basic problem, since the silts deposited in Egypt today are primarily characterized by montmorillonite, not by kaolinite or metahalloysite. On the other hand, montmorillonitic vertisols are found all over the Gezira Plain (Finck, 1961). It would seem that several areas have contributed to the modern nilotic sediments of Egypt, which include medium-to-coarse eolian sands along the Saharan Nile Valley; fossil silts of the Gezira Plain; fine-to-medium quartz sands and possibly micas from the Basement Complex and sandstones of the Kolla; and, last of all, clayey soil sediments from the Voina Dega.

The sediments of the White Nile, largely derived from the western periphery of Ethiopia via the Sobat, seem to confirm this impression (Hume in Lucas, 1908: 55–56). Fine quartz sands are significant at all seasons, with a surge of silts particularly rich in ferric oxides (27.6%) during the flood season. In contrast, the Atbara transports little quartz but carries instead a suspended load three times as great as that of the Blue Nile and fifteen times that of the White Nile. The majority of this may have been eroded from the old silt plains that accompany the modern floodplain from the Kolla to the Nile. The level of free oxides is intermediate between that of the Blue Nile and that of the White Nile (17.4%).

In short, the popular concept that the Nile mud is supplied by the Ethiopian uplands is an oversimplification. It is true that the modern flood silts are very largely derived from Ethiopia, but not directly so. Instead, before it reaches the Saharan Nile, the suspended matter is deposited and reeroded several times en route, and much of what is reeroded today was originally removed from the Voina Dega in Pleistocene times. The Ethiopian flood silts are not genetically homogeneous. Each affluent of the Saharan Nile carries a mixed load of different density, material, and derivation. The end-product, as deposited on the Egyptian floodplain, is a composite of eroded materials of different ages, of which perhaps only the smaller part has been derived directly from the Ethiopian Plateau.

Pleistocene Environmental Changes in Ethiopia

An understanding of Pleistocene environmental changes in Ethiopia may help to provide the answers to two important questions concerning the history of the Saharan Nile: (a) whether the late Pleistocene record provides any paleoclimatic or tectonic evidence that may explain the nilotic silt formations of Egypt, and (b) whether the older Pleistocene record contains any clues to the evolution of the Nile and Atbara drainage basins.

Although not glaciated today, the highest peaks (between 4130 and 4580 meters) of the Ethiopian Plateau were glaciated during the (late ?) Pleistocene. Early work by Nilsson (1940) demonstrated the existence of two former glacial stages with climatic snowlines at 3600 to 4100 meters and at 4200 meters elevation. Werdecker (1955) has confirmed Nilsson's results, setting the snowline for the period of major glaciation at 3600 to 3700 meters. A retreat phase, correlated with the terminal Pleistocene, corresponded to a snowline at 4400 meters. Since the modern snowline presumably is situated at 4700 to 4800 meters, just above the highest peaks, the maximum Pleistocene snowline depression is estimated at 1100 meters. All of these rather localized glacial phenomena lie at or above the present upper Dega limit and could hardly have exercised much direct influence on the behavior of the rivers debouching from the uplands.

Periglacial activity may have been significant at lower levels, however. The present lower limit of marked solifluction phenomena and soil-frost activity is situated at 4200 to 4300 meters (Büdel, 1954a; Werdecker, 1955), while restricted solifluction under grassy vegetation has been noted down to 3600 meters. Fossil solifluction deposits down to elevations of 2700 meters—attributed to frost activity and greater humidity—are reported by Büdel (1954a), who implies that the modern Dega formerly lay within the late Pleistocene periglacial zone. It is apparent that at least some of Büdel's solifluction features represent colluvial screes, tropical "stone lines," or other phenomena resulting from mass movement of lubricated clays. Kuls and Semmel (1965) found no evidence for Pleistocene cold climates in central Ethiopia. Shallow, smoothly dimpled, concave valley heads—which Büdel (1954a) attributes to cold-climate mass movements—lack detrital beds but are instead underlain to great depths by intensively weathered, partly decomposed basalt. These features are explained by differential weathering. Kuls and Semmel (1965) believe that, if there had been significant frost activity, there should be detrital accumulations in the river valleys. There are none, however, and stream terraces are also absent. In fact, Semmel (1963, 1964) points out that there is no evidence for bisequential soil development in the Voina Dega of central Ethiopia. It would seem, therefore, that the temperature depressions of the Pleistocene glacials did not exert any significant influence on the geomorphic development of the Dega and Voina Dega regions, at least in central Ethiopia.

Local evidence of greater moisture in late Pleistocene times is inconclusive. Nilsson (1940) has described five high shorelines of Lake Tana up to +125 meters, with a less distinct level at +148 meters. Mohr (1964: 200) indicates, however, that there is only one lake terrace, developed in lacustrine clays and sands, at about +20 meters. An additional 40-meter terrace is confined to stream valleys debouching onto the lake basin and is composed of alluvial deposits. Higher terraces are structural. The existence of other high shorelines, identified by Nilsson (1940) around the smaller lakes of the Ethiopian Rift Valley, is disputed by Büdel (1954a), but many of them have been confirmed by Mohr (1964: 203), who describes a variety of lacustrine beds. Stream terraces and lake sediments have been reported from the Awash Valley, in the northeastern part of the Ethiopian Rift, by Büdel (1954a) and Mohr (1964: 202). These formations are now being studied in detail by Jean Chavaillon and others. Finally, the presence of alluvial terraces in the upper reaches of the Atbara and Khor Gash has been indicated by Mohr (1964: 205). All in all, the available local evidence of pluvial climates on the Ethiopian Plateau is scanty and unsatisfactory, and it will remain difficult to isolate tectonic and climatic factors. Pollen sampling of Pleistocene and younger sediments, recently completed by E. M. Van Zinderen Bakker, offers the best prospects for an understanding of possible environmental changes.

Older Pleistocene deposits, other than vulcanic beds, appear to be very scarce in Ethiopia, and published data is limited to the controversial Yaya beds. These sandstones, clays, and lignites were studied by Nilsson (1940) at five different exposures, and were interpreted as lacustrine sediments. They rest on the Magdala Series and are capped by a younger basalt flow. According to Nilsson (1940), this basalt is of early or middle Pleistocene age, but it forms part of the mid-Tertiary Magdala unit according to Mohr (1964:143, 182–83). Found at between 2400 and 2800 meters elevation today, these beds suggest to Nilsson a former lake with an area of as much as 25,000 square kilometers—provided that all these sediments are contemporary and pertain to a contiguous body of water. The original level of "Lake Yaya" is unknown, since there may have been subsequent uplift. If the lake did exist, and if it was of Pleistocene age—which seems unlikely—it may have been dammed up behind lava barriers, although the original direction of overflow must remain conjectural. On the basis of elevation of the different exposures, Nilsson (1940) emphasizes that the Yaya beds are tilted northwestwards. This dip, however, is only about 400 meters over a distance of 75 kilometers, or roughly 0.5%. This would seem to lie within the margins of error of measurement, available topographic mapping, and sediment preservation.[1]

1. According to Mohr (1964: 143), the lake beds themselves are steeply inclined to the southeast, together with the associated stratoid lavas. On such questionable grounds, Nilsson (1940, 1963) suggests a late Pleistocene tilting of the Ethiopian

The Pleistocene and late Tertiary history of Ethiopia is still very incompletely understood. A simplified sequence of geomorphic evolution seems to emerge from the available evidence, however. At the end of the Mesozoic era, Ethiopia formed a low-lying landmass of horizontal sedimentaries. During the Cretaceous and early Tertiary, these emerged lands were subject to erosion and weathering, with formation of extensive planation surfaces and lateritic crusts (Mohr, 1964: 107-8). Uplift of the Ethiopian swell began at the end of the Eocene, climaxing a period of compressional movements with extrusion of magmas to form shield volcanoes and great fissure flows at the surface (Mohr, 1964: 151 ff.; 1965). These essentially epeirogenic deformations climaxed in several spasms of block faulting during Oligo/Miocene times, with intermittent movements persisting during the Pliocene and early to middle Pleistocene (Mohr, 1964: 159 ff.). As a result, the Ethiopian swell was broken up into two great fault blocks: the Ethiopian and Somalian plateaus. The contact between the Basement Complex and the overlying sedimentary and vulcanic rocks is at about 1,500 to 2,000 meters in northern and western Ethiopia. This indicates that Tertiary uparching and upfaulting of the Ethiopian Plateau along its western edge (see Mohr, 1964: 175), relative to the plains of the central Sudan, was on the order of 1,000 to 1,500 meters. The total flexure of the Basement Complex in Ethiopia is estimated to be over 5,000 meters (Mohr, 1964: 155).

Concomitant with the uplift of the late Tertiary, the runoff of the uplands was concentrated in new or existing drainage lines, and vigorous dissection of the plateau rim was begun. During the subsequent period of geological time, a radial drainage pattern developed which was biased toward the northwest and the southeast, possibly an indication that the fault blocks reached their highest elevations in Eritrea and near the center of Ethiopian Rift, dipping away laterally. The courses followed by the major rivers were dictated in detail by combinations of several factors: preexisting valleys, centers of updoming or downwarping superimposed on the fault blocks, fault systems, extrusive vulcanism, and

---

block to the northwest, resulting in "rejuvenation" of the upper Blue Nile Basin. This would have promoted extensive erosion of Pleistocene and older sediments on the uplands, supposedly providing the surge of flood silts transported to the Gezira Plain and the Saharan Nile. Nilsson further attempts to explain the details of the Nile Valley deposits by climatic oscillations during the last Ethiopian pluvial phase. These climatic oscillations, however, are based on his sequence of Ethiopian lake levels which, apart from the fact that their existence is questioned by Büdel (1954a), are not even stratigraphically dated.

differential erosion. The valleys cut by the Sobat, the Blue Nile, and the Atbara are impressive, that of the Blue Nile attaining a depth of 1,500 to 2,000 meters, cutting vertically through over 1,500 meters of sedimentary rocks and locally through as much as 2,000 meters of the igneous and metamorphic basement (Mohr, 1964: 72–73, 155). These vertical dimensions rival those of the Grand Canyon, although the Blue Nile Gorge is almost twice as long.[2] On these grounds, it is highly probable that the basic outlines of the Sobat, Blue Nile, and Atbara drainage basins had been established by Mio/Pliocene times. And, in the case of the Blue Nile, incision through 2,000 meters of the Basement Complex, across the dip-slope of the surface contours and of the Basement Complex isohypsal lines, strongly suggests an antecedent river (see Mohr, 1964: 155, 183, for a similar opinion). The question that remains is where the discharge of these rivers terminated—in the Sudan or beyond?

During the course of the Pleistocene, vulcanism and tectonic activity recurred sporadically, particularly within the Ethiopian and Red Sea rifts, while fluvial erosion continued to sculpture the uplands, even during the succession of cool and warm climates that followed in the wake of repeated continental glaciation in higher latitudes. Although late Pleistocene glaciation of the highest summits is verified, we still know next to nothing about the geomorphic balance on the uplands during such periods of cooler climate.

Problems and Conclusions

In concluding, several problems or partial answers can be stated concerning the Cenozoic evolution of the Saharan Nile and its southerly tributary systems.

a) The existence of an Egyptian Proto-Nile since late Eocene times is verified, and this drainage system had adopted a general configuration resembling that of the present by Miocene times. Erosional surfaces related to this late Tertiary Nile can be traced through Lower Nubia to beyond the Sudanese border.

b) During Oligocene times, a period of comparatively moist climate in the Sudan and Egypt, a large inland lake was maintained in the center of

---

2. By way of comparison, the Grand Canyon of the Colorado was excavated to its present dimensions during a considerable span of time. The earliest widely recognized erosional level, the Valencia Surface, predates a late Miocene basalt, according to recent results of M. E. Cooley (unpublished 1964 symposium on Colorado River history). Approximately 1,300 meters of vertical incision have been accomplished since formation of the earliest Kaibito surface in late Pliocene times.

the Howar Basin. The Hudi Chert and associated breccias, derived from this lake, first make their appearance in southern Egypt with the Gebel Silsila Formation. The question arises of where the waters of this basin went after the Hudi lake was drained in late Oligocene times. They may have continued to flow towards what is now the Red Sea, via the Wadi Odib gorge (Fig. 1–1), or they may have flowed in a southeasterly direction toward the Tokar gap (Fig. 1–1).[3] In view of the existing topography, a northerly exit to the Proto-Nile seems more reasonable. The lithological argument of the Hudi Chert against such a connection, prior to 15,000 B.C., is not necessarily conclusive, since Ethiopian components (see Appendices B and D) to the Proto-Nile drainage are witnessed as early as the Korosko Formation (50,000 years ago).

c) The Paleo-Sudd Basin formed a center of alluvial or lacustrine deposition in Plio/Pleistocene times, presumably receiving the drainage of the Sobat and other west Ethiopian rivers. Whether or not this basin had external drainage is debatable. If it did, were the waters discharged northward toward the Howar Basin or eastward toward the Tokar gap? Ball (1939: 74–84) suggests a closed basin first drained in late Pleistocene times and attempts to explain the surge of nilotic silts in Egypt through a sudden rejuvenation of the Paleo-Sudd, with large-scale erosion of older sediments. None of the field evidence supports such a dramatic case of river-capture, but the fact remains that there is no record of integrated drainage before the appearance of heavy minerals characteristic of the Bahr el-Ghazal in the sediments of the Korosko Formation (Appendix B). In view of the significance of the Sobat discharge today, it seems rather unlikely that the Paleo-Sudd Basin would remain endorheic for very long. Since the divide between the Blue Nile and the White Nile basins on the Gezira Plain is poorly defined, it would seem that the vicissitudes of both rivers were closely linked, at least for the comparatively brief duration of the Pleistocene.

d) The Blue Nile Basin, on geomorphic grounds, appears to have

3. The Wadi Odib–Wadi Amur Gap, although well defined on the topographic maps, has a modern saddlepoint elevation of about 1,000 meters, some 500 meters above the level of the Hudi Chert beds today. Consequently, a considerable age would be indicated unless rather recent upfaulting of the Red Sea Hills were assumed.

The Tokar gap would be readily negotiable today by the waters of the Atbara or the Khor Gash, following Wadi Langeb through the mountains at ± 400 meters elevation. The Red Sea–Khor Gash divide lies across a flat silt plain, and exceptionally heavy floods from the Khor Gash may spill over to the Atbara as well as to the Red Sea.

drained towards the plains of the Sudan since at least the early Tertiary. As we have emphasized before, the absence of flood silts in the early to middle Pleistocene record of Egypt is no proof that Blue Nile waters did not reach Lower Nubia earlier than the deposition of the Korosko Formation. With the apparent absence of older Blue Nile sediments—presumably buried at depth under the Gezira Plain—there can be no conclusions. As an alternative, the Blue Nile may have flowed eastward from the Gezira Plain, discharging into the Red Sea via the Tokar Gap.

e) The Atbara River, although not an antecedent valley, poses a problem similar to that of the Blue Nile. Of all the components of the Nile Basin, however, the Atbara is the river that could most readily have used another outlet during at least a part of the Pleistocene. But even here argumentation is speculative unless geological examination of the Tokar Gap is undertaken. One reasonably certain assumption is that, at times of greater discharge, the Khor Gash would have formed a tributary to the Atbara. Today its waters lose themselves on the Pleistocene silt plain once deposited by the Atbara.

f) The available evidence from the central Sudan indicates that the White Nile, the Blue Nile, and the Atbara were all linked to the Saharan Nile during the last siltation stage of the late Pleistocene. Presumably, the same would apply to the mid-late Pleistocene, but even here the local evidence is inconclusive, and we must turn to Nubia for corroboration (Appendices B and D). Thus far, sedimentological criteria for differentiating Atbaran and Blue Nile components to the Saharan Nile have not been perfected, and the recognition of White Nile elements in mixed deposits cannot yet be based on an extensive series of quantitative results. Here again is a fruitful field for future research, both in the field and in the laboratory.

g) The mixed nilotic components of the Korosko Formation indicate one or more major sources of Ethiopian discharge. The significance of kaolinite over montmorillonite distinguishes the clay minerals from those of modern nilotic beds, however (Appendix D). This may be due to greater erosion of latosolic soils on the Voina Dega, or it might also be explained by the direct derivation of these beds from Ethiopia, without repeated redeposition on the Gezira Plain. A more reasonable explanation would be a greater influx of Bahr el-Ghazal waters and sediments, a notion supported by the relative significance of titanite and iron ores and by the reduced proportions of pyroxenes and epidote in the heavy mineral spectra of the Korosko Formation (Appendix B). Since the Korosko Formation was deposited by summer rather than winter floods, the turbulence and velocity recorded by these deposits in southern Egypt indicates a much greater subsaharan discharge.

h) The mixed nilotic beds of the Masmas Formation also show an unusually high kaolinite component. Several explanations may again be offered, but the writers suggest that the Bahr el-Ghazal and el-Arab runoff was greater than it has been subsequently. Despite an overall similarity to the modern flood regime, both the exceptionally broad Nile floodplain in southern Egypt and the evidence for frequent crevasses point towards somewhat greater Nile discharge.

i) The Gebel Silsila Formation has about the same clay and heavy mineral spectrum as the present Nile sediments of Egypt have. This suggests that the type of soil erosion and the relative proportions of the subsaharan discharge were similar to those of today. The great preponderance of montmorillonitic clays may indicate that most of these silts were first deposited on the Gezira Plain, where soil-forming processes developed authigenic montmorillonite. These altered sediments were later eroded and were ultimately redeposited along the Saharan Nile. Another possibility is that a certain part of the montmorillonite developed *in situ* on the Egyptian floodplain after final deposition. The braided, rapidly shifting channel of the Gebel Silsilan Nile in Egypt can only mean more vigorous floods, capable of transporting a gravelly bed load a thousand kilometers and more from its source, without any influx of tributary waters. Greater discharge is once more indicated, but, except for heavier summer rains in Ethiopia and the Sudan, no major geomorphic changes, other than an intensified rate of denudation, need be supposed to account for the greater sediment yield. Without a change of the natural vegetation on the Ethiopian Plateau, heavier rainfall—either greater in intensity or in duration—would promote or accelerate mass movements and would thus provide a ready source of sediment to the major river valleys. The colluvial screes and debris slides reported by Büdel (1954*b*) and by Kuls and Semmel (1965) would support this impression.

j) All in all, the late Pleistocene sediments of Egypt do not necessarily imply any major geomorphic changes in Ethiopia and the White Nile Basin, although they would seem to have required appreciable changes in the precipitation amount or regime. Presumably, some of these climatic changes affected some parts of the subsaharan drainage basin of the Nile more than others. Whether or not the pluvial climates of Ethiopia did promote accelerated mass movements and general denudation can only be decided through field studies in the Kolla.

k) Finally, the details of the climatic fluctuations implicit from the Egyptian, the Sudanese, and the Ethiopian records seem susceptible to rational meteorological interpretation. But to attempt to do so before the dynamic climatology of the present summer circulations over Ethiopia and the Sudan has been properly studied would be premature.

Appendices A–J

APPENDIX A

Mechanical and Chemical Sample Analyses

Karl W. Butzer
and Carl L. Hansen

Carbonate Determination

Calcium carbonate content was normally determined by the Chittick gasometric apparatus, which measures carbon dioxide displacement after submergence of a 1.7 gram sample in 20% hydrosulfuric acid. Samples were allowed to stand between fifteen minutes and six hours, depending on the amount of reaction. A mean of two samples was normally taken. In the case of homogeneous samples with high carbonate content (marls, travertines, etc.), sample size was commonly reduced by a half, to accelerate breakdown of all carbonates present. The Chittick apparatus was found to give readings substantially too low when carbonate content exceeded about 40%. In such cases, total loss of solubles (by weight) after application of cold hydrochloric acid was also determined, and an approximation for the true calcium carbonate content was obtained by taking the mean value derived from the two methods.

pH Determination

pH values were determined electrometrically for a paste prepared from 10 to 25 grams of sediment immersed in distilled water.

Sample Preparation for Hydrometer or Wet-Sieve Analysis

Samples were originally selected in the field so that they would be representative of the sediments to be studied. As a result, rather large samples were usually collected and were subsequently reduced in size at the field laboratory by a riffle with 1/2 inch chutes. Each sample was further reduced in size prior to processing, at which time 100 grams were normally selected. After oven drying, the sample was weighed and then subjected to hydrochloric acid treatment until carbonates were removed.

After washing and decanting repeatedly to assure removal of the acid, the sample was again dried and weighed. In the case of samples from the Kurkur Oasis, the noncarbonate residue was seldom more than 10% to 20% of the original sample weight. It would therefore have been more meaningful to determine the grade-size of both the quartz and the lime sands, but these sediments were too well indurated to allow proper separation without carbonate removal.

The final sample was stirred up in a solution with 50 milliliters of 5% sodium pyrophosphate and was allowed to stand for from twelve to twenty-four hours.

Hydrometer Analysis

The standard hydrometer method, used to determine the content (by weight) of particles smaller than 60 to 80 microns, is described by the American Society for Testing Materials (A.S.T.M., 1958: 1119 ff., also 1960 supplement to book of A.S.T.M. Standards, part 4, p. 1151–52). This technique was employed in our hydrometer work, making use of the appropriate correction factors. Readings were taken at intervals to determine the quantity of material in size classes of 20-to-63, 6-to-20, 2-to-6, and under-2 microns. Coarser particles were subsequently graded by wet-sieving.

Wet-Sieve (Sand) Size Analysis

Notably coarse samples with few fine particles under 63 microns were often wet-sieved without hydrometer analyses. These, and all hydrometer samples, were put through a set of U.S. Standard Sieves with the following openings: 63, 210, and 595 microns and 2 and 6.35 millimeters. These size-fractions approximate the modified Atterberg as well as the Wentworth units quite closely. The fractions in each sieve were dried and weighed, and textural classes were subsequently determined (see Chap. 1).

Textural Data for Samples Cited in Text Tables

The grade-size fractions for all samples referred to in the text are listed in Table A-1. The grade units are as follows:

1)	granules	2 to 6.35 millimeters
2)	coarse sand	595 to 2000 microns
3)	medium-coarse sand	210 to 595 microns
4)	medium sand	63 to 210 microns
5)	fine sand	20 to 63 microns
6)	coarse silt	6 to 20 microns
7)	fine silt	2 to 6 microns
8)	clay	under 2 microns

Percentages, in other words, include the granule sand fraction, although granules were normally omitted in determining textural classes in Table 1–3. Components 5, 6, and 7 are collectively referred to as the silt fraction. Since the degree of accuracy attained in hydrometer measurements is not as high as that of the wet-sieve fractions, percentages in the hydrometer size-range have been rounded off to the nearest half digit.

Table A-1. Textural data (in per cent) for noncarbonate residues of samples cited in text tables

Sample number	1	2	3	4	5	6	7	8
				Table 2-1				
19	33.4	17.7	12.1	14.8	4.0	8.0	2.5	7.5
133	4.7	3.5	15.1	23.9	24.5	6.0	4.5	18.0
126	32.4	23.1	8.0	8.5	12.0	6.5	4.0	5.5
129	13.6	2.5	1.6	3.5	31.5	19.5	6.5	21.5
121	42.9	18.1	11.1	10.3	3.9	4.3	1.9	7.5
80	56.1	17.5	7.5	4.7	3.6	7.7	1.4	1.8
124	—	—	0.1	32.2	36.0	12.0	9.5	10.0
125	—	—	0.1	5.4	16.5	28.0	13.5	26.5
89	22.4	10.1	17.4	23.1	10.0	6.0	3.5	7.5
88	2.9	12.8	43.5	26.8	5.0	3.0	2.0	4.0
87	14.9	4.7	23.0	43.4	4.5	3.0	2.0	4.5
86–II	—	1.3	9.4	49.3	10.0	7.0	7.0	16.0
86–I	0.1	2.3	39.0	46.6	6.5	2.5	1.0	2.0
90	—	0.2	1.2	2.6	5.0	20.0	8.0	63.0
				Table 3-2				
39	—	—	0.2	3.6	42.5	19.5	10.5	23.5
40	—	0.1	0.4	0.9	48.0	23.5	8.5	18.5
38	—	—	0.5	16.4	55.5	13.0	5.0	9.5
35	—	—	0.6	86.2	13.2[a]	—	—	—
34	—	—	0.4	6.1	13.5	15.0	11.0	54.0
37	—	9.4	72.6	8.0	10.0[a]	—	—	—
36	—	6.1	46.2	28.8	18.9[a]	—	—	—
43	—	2.3	28.3	31.0	8.5	9.0	5.0	16.0
42	—	0.6	14.9	13.0	22.0	23.0	8.5	18.0
41	—	1.5	16.4	22.5	11.0	11.5	7.0	30.0
				Table 3-3				
6	26.8	44.3	11.2	9.2	2.5	0.5	1.5	4.0
79	0.2	1.1	4.9	24.3	26.0	15.5	11.0	17.0
16	—	—	12.7	52.8	8.5	7.5	5.0	13.5
15	—	0.3	0.6	7.1	23.0	21.0	13.0	35.0
14	0.5	2.3	3.4	15.8	36.5	17.5	8.5	15.5
13	—	0.3	1.0	15.2	47.5	13.0	5.0	18.0
301	6.0	12.3	34.3	15.4	8.0	5.5	3.5	15.0

[a] Total of fines under 63 microns in wet-sieve samples.

(Table continued on following page)

(*Table A–1, continued*)

Sample number	1	2	3	4	5	6	7	8
			(*Table 3–3, continued*)					
300	1.6	0.8	11.8	56.8	5.0	6.5	4.0	13.5
299	—	—	3.8	13.7	27.5	20.5	12.5	22.0
298	—	0.5	5.8	27.2	26.0	17.0	7.5	16.0
322	5.6	19.6	13.4	25.4	15.0	6.0	3.5	11.5
71	—	0.8	0.9	10.3	24.0	17.0	11.0	36.0
72	1.2	0.9	1.0	20.9	33.5	16.0	9.0	17.5
70	—	0.6	1.1	42.3	28.0	12.0	5.5	10.5
69	2.2	9.0	3.1	23.2	29.0	9.0	4.5	20.0
74	5.0	14.0	5.1	25.4	23.5	10.0	5.5	11.5
76	10.1	34.4	12.2	25.3	6.0	4.0	2.0	6.0
			Table 3–4					
94	—	—	0.1	5.9	29.5	17.0	10.5	37.0
95	—	—	0.1	9.4	31.5	33.0	6.5	19.5
101	—	—	0.3	2.2	30.5	19.5	11.5	36.0
229	—	—	1.0	5.5	32.0	22.5	11.5	27.5
98	—	—	0.2	25.9	73.8[a]	—	—	—
97	—	—	0.1	0.9	51.5	26.0	6.5	15.0
96	—	—	0.1	0.4	35.0	28.0	11.0	25.5
			Table 3–5					
99	2.3	0.6	1.7	5.9	36.0	18.0	8.5	27.0
231	—	—	1.9	4.6	40.5	18.0	9.0	26.0
388	—	—	1.7	19.8	35.0	18.5	8.0	17.0
387	0.2	3.3	18.1	12.9	22.0	15.5	9.0	19.0
226	—	9.7	78.6	6.2	5.5[a]	—	—	—
227	0.1	0.9	52.4	19.1	8.0	3.5	2.5	13.5
230	—	3.8	28.1	33.6	10.0	6.0	3.5	15.0
233	5.6	6.7	23.9	19.8	12.0	10.0	5.5	16.5
396	0.2	11.0	37.9	10.4	13.0	8.0	4.5	15.0
228	0.1	4.1	33.4	16.0	12.0	11.0	5.0	18.5
234	—	11.1	85.6	1.9	1.4[a]	—	—	—
232	—	2.2	55.6	5.3	36.9[a]	—	—	—
			Table 3–6					
392	—	—	1.7	1.4	45.0	24.0	7.5	20.5
391	—	0.1	5.5	46.7	47.7[a]	—	—	—
390	—	0.6	7.6	16.3	35.0	15.0	5.5	20.0
			Table 3–8					
128	14.0	10.6	17.3	31.6	7.5	6.0	5.0	8.0
22	0.3	0.8	5.0	57.3	36.6[a]	—	—	—
29–II	—	0.4	47.1	46.3	6.2[a]	—	—	—
58	—	1.2	54.8	42.1	1.9[a]	—	—	—
398	3.5	6.1	6.3	24.6	22.0	13.5	8.5	15.5
85	—	0.9	2.1	13.5	24.0	24.0	13.5	22.0
33	—	—	0.1	—	—	—	—	—

[a] Total of fines under 63 microns in wet-sieve samples.

(Table continued on following page)

(Table A-1, continued)

Sample number	1	2	3	4	5	6	7	8
(Table 3-8, continued)								
25	—	0.1	0.2	87.0	12.7[a]	—	—	—
27	—	0.9	5.2	33.1	60.8[a]	—	—	—
28	7.0	14.8	11.6	26.1	40.5[a]	—	—	—
30	—	0.1	2.0	26.9	71.0[a]	—	—	—
26	0.9	4.6	7.9	10.3	76.3[a]	—	—	—
31	—	0.1	0.7	8.5	90.7[a]	—	—	—
32	—	—	0.1	5.9	23.0	31.0	12.0	28.0
63–I	—	—	0.1	11.4	40.0	15.0	8.5	25.0
141	1.6	13.5	61.2	16.5	7.2[a]	—	—	—
142	—	0.1	3.9	57.5	37.5[a]	—	—	—
144	0.1	0.1	0.5	26.8	37.0	16.5	5.5	13.5
325	1.2	0.7	5.3	18.8	16.5	16.5	8.5	32.5
23	—	0.1	13.3	13.3	26.0	20.0	8.0	19.5
83	—	0.2	2.3	74.0	10.0	1.0	2.0	10.5
326	—	—	0.2	2.8	83.0	6.5	3.0	4.5
328	—	0.5	0.9	11.6	17.0	14.0	11.0	45.0
329	—	1.2	22.7	27.1	18.0	10.5	5.0	15.5
332	—	1.0	2.6	10.9	24.0	14.5	9.0	38.0
77	7.5	22.1	10.1	19.8	20.5	6.5	4.0	9.5
81–I	—	6.3	10.5	19.7	30.0	7.0	5.5	21.0
81–II	—	0.1	0.5	47.4	36.5	1.5	3.0	11.0
84	8.6	23.8	11.3	26.4	16.0	3.5	1.0	9.5
47	1.8	15.5	13.8	34.4	20.5	3.0	3.5	7.5
Table 5-2								
158	13.0	27.4	33.5	12.6	2.0	2.0	1.5	8.0
154	2.7	3.2	9.4	18.7	14.0	8.5	5.0	38.5
172	3.3	10.9	18.2	27.6	8.5	11.5	5.0	15.0
161	1.5	0.9	13.4	27.9	56.3[a]	—	—	—
162	—	1.1	19.7	52.0	27.0[a]	—	—	—
163	—	0.6	3.4	43.6	52.3[a]	—	—	—
Table 6-2								
182	—	—	1.0	49.5	30.5	5.0	2.0	11.0
181	17.0	57.0	18.6	4.8	2.6[a]	—	—	—
177	0.6	2.1	4.7	20.6	25.5	11.0	7.0	28.5
186	3.7	18.4	17.1	24.4	36.3[a]	—	—	—
185	3.3	71.4	15.8	4.6	4.9[a]	—	—	—
184	0.1	12.3	13.4	17.2	19.0	12.5	6.0	19.5
180–II	—	1.1	4.9	25.0	14.5	14.0	8.0	32.5
179	7.3	20.7	35.6	27.3	9.0[a]	—	—	—
170	0.3	2.0	32.8	26.4	12.5	8.0	4.5	13.5
166	0.2	0.5	12.4	69.9	2.0	1.0	3.0	11.0
159	7.3	15.4	33.0	27.5	16.8[a]	—	—	—
157	7.6	12.9	17.5	11.5	12.5	12.5	7.0	18.5

[a] Total of fines under 63 microns in wet-sieve samples.

(Table continued on following page)

(*Table A–1*, continued)

Sample number	1	2	3	4	5	6	7	8
				Table 6–3				
209	11.1	34.3	37.4	14.8	2.4[a]	—	—	—
199	2.8	12.7	43.8	32.2	8.5[a]	—	—	—
200	6.7	34.5	39.9	7.2	11.7[a]	—	—	—
207	0.4	18.6	65.3	10.8	4.9[a]	—	—	—
192	22.3	30.3	32.7	12.1	2.6[a]	—	—	—
198	0.8	15.1	39.6	30.9	13.6[a]	—	—	—
195	6.2	17.2	34.3	20.7	21.5[a]	—	—	—
194	19.6	22.3	25.5	15.0	17.6[a]	—	—	—
189	0.2	7.4	35.6	36.9	19.9[a]	—	—	—
191	12.3	27.7	30.5	19.2	10.3[a]	—	—	—
208	0.2	10.4	34.0	18.4	8.5	10.0	5.5	13.0
196	1.3	11.1	23.0	19.6	14.0	8.0	6.0	17.0
193	—	14.7	64.6	20.1	0.6[a]	—	—	—
201	1.7	11.9	55.1	30.8	0.5[a]	—	—	—
197	4.8	16.9	32.4	25.6	20.2[a]	—	—	—
188	—	0.2	2.6	38.7	23.0	5.0	8.0	22.5
203	—	2.0	13.5	60.7	23.7[a]	—	—	—
206–I	0.2	9.0	63.7	17.7	9.4[a]	—	—	—
206–II	—	5.0	65.1	27.9	2.0[a]	—	—	—
204	5.7	25.0	51.7	12.7	4.9[a]	—	—	—
202	0.3	2.7	8.1	12.8	76.1[a]	—	—	—
				Table 6–4				
234	—	0.7	12.5	11.7	45.0	13.0	5.0	12.0
253	—	0.1	0.7	25.6	34.5	20.0	6.0	13.0
252	0.1	0.1	0.1	10.2	52.0	17.5	5.0	15.0
164	25.4	17.9	33.1	7.6	15.9[a]	—	—	—
153	—	0.5	43.5	46.2	9.8[a]	—	—	—
156	—	1.1	46.8	30.8	21.3[a]	—	—	—
152	3.3	4.9	9.7	29.0	53.1[a]	—	—	—
259	0.1	5.2	29.2	45.5	3.5	4.0	3.5	9.0
260	—	12.9	29.9	38.6	2.5	3.0	2.5	10.5
250	1.9	8.2	39.5	38.8	11.5[a]	—	—	—
249	39.6	10.9	38.6	9.8	1.1[a]	—	—	—
248	0.3	0.1	0.3	7.3	50.5	20.5	6.0	15.0
148	0.1	2.5	3.7	28.2	24.0	21.5	6.0	14.0
222	—	0.3	4.0	59.2	16.0	2.5	2.0	16.0
				Table 7–2				
359	—	0.4	57.2	28.4	2.6	3.1	1.7	6.6
350	—	3.4	44.6	39.9	4.9	0.8	—	6.4
365–II	—	0.1	0.3	3.1	—	—	11.0	85.5
365–I	—	0.1	9.3	9.6	—	—	12.0	68.0
364	—	0.1	9.6	10.8	13.5	17.0	15.0	34.0
358	—	0.7	5.4	9.4	—	—	8.0	78.0
361–I	—	0.5	19.9	8.2	9.5	24.4	14.2	23.3

[a] Total of fines under 63 microns in wet-sieve samples.

APPENDIX B

Heavy Minerals of the Late Pleistocene Nilotic Deposits: A Preliminary Report

Karl W. Butzer
and Carl L. Hansen

Some fifty samples were analyzed for heavy minerals, most of these selected from the three major late Pleistocene nilotic formations. The materials were extracted from the medium-sand fraction (63–210 microns), after removal of the light minerals by bromoform (density 2.9). Mounted in glass slides, the heavy minerals were identified and counted microscopically, a hundred grains for each sample. Identification of minerals was generally confined to the major families represented in the spectrum of the nilotic silts: the opaques, i.e., the iron minerals, dominated by ilmenite and magnetite; the amphiboles, including three varieties of hornblende, together with rare grains of actinolite and tremolite; the pyroxenes, including monoclinic and rhombic varieties, but consisting almost exclusively of augite; epidote; titanite or sphene; biotite; altered minerals; and "others," which include, *inter alia*, garnet, kyanite, monazite, rutile, staurolite, zircon, and members of the zeolite group, as well as unidentified minerals. Since the specific gravity of biotite is approximately that of bromoform, the biotite is partly removed in suspension, and percentages always tend to be too low. The altered minerals are, in all probability, amphiboles and pyroxenes, the chemically most unstable grains present in the nilotic spectrum (see Cailleux and Tricart, 1963: Table 4).

The parent material from which the heavy minerals of the modern Nile are derived is highly varied, although some usable generalizations can be made. Ilmenite is present in basalt, dolerite, rhyolite, and metamorphic rocks, as well as in sedimentaries. Magnetite may also come from these parent materials, as well as from granites and syenites. Consequently, the iron ores represented in the Saharan Nile may be derived from almost any part of the Nile Basin, a fact confirmed by isolated samples studied by Shukri (1950). Relative percentages vary from 12% in the Atbara and 15% in the Blue Nile to 27% in the White Nile, where the Sobat and the Bahr el-Ghazal are the major sources (Table B–1). Hornblende, of the greenish variety dominant in our samples, may be derived from basalt, dolerite, granite, syenite, and metamorphics. It too is rather universal, although Shukri's samples indicate a very low percentage in the Atbara (6%) compared with high values in the Blue and the White Niles (52% and 35% respectively). Augite is largely confined to basalt, dolerite, and rhyolite and is today concentrated in the Atbara (73%) and the Blue Nile (15%), with less than 1% in the Sobat and the other components of the White Nile drainage system (Table B–1). The presence of abundant dolerite, felsite, and serpentine in the Red Sea Hills precludes a strict localization of augite in the case of Pleistocene deposits, however. Epidote is derived from basalt, dolerite, metamorphics, and certain basic dykes. Today it seems to be rare in the Atbara (less than 1%) and plentiful in the Blue and the White Niles (12% and 20% respectively, see Table B–1). Titanite is found in granites, syenites, rhyolites, and metamorphic rocks and is very rare in the Atbara and the Blue Nile while consistently present in low percentages in White Nile sediments (Table B–1). Biotite is found in rocks similar to those in which the greenish varieties of hornblende are found. Finally, some of the other minerals may be of interest. Garnet, kyanite, monazite, and staurolite are all derived from acidic igneous or metamorphic rocks.

Obviously, with a great variety of acidic and basic igneous rocks intruded or extruded in the Basement Complex of the Nubian Shield, heavy mineral statistics from the autochthonous early to middle Pleistocene Nile gravels are meaningless except in the local context of individual wadis—and even here the selective decomposition of the ferromagnesian minerals renders statistical analysis dubious. The heavy minerals within a sandy gravel are closely related to gravel lithology, and heavy-mineral statistics are seldom better than lithological counts on pebbles. In sandy deposits, the heavy-mineral content decreases with age, and, for example, the chemically unstable ferromagnesian minerals are entirely absent from Pliocene sands. This provides no basis for assumptions concerning changes of the hydrographic basin of the Nile or its Egyptian tributaries, such as those attempted by Shukri (1950) or by Shukri and Azer (1952). Heavy-mineral suites may possibly have stratigraphic value in the Tertiary and early Pleistocene record, as indicated by Chumakov (1965), but they have few paleogeographic implications.

Table B–1. Relative percentages of selected heavy-mineral groups from modern sediments of the major Nile drainage basins
(Data approximated from Shukri, 1950: Fig. 2.)

Basin	Iron	Amphiboles	Pyroxenes	Epidote	Other
Atbara	12	6	73	0	9
Blue Nile	15	52	15	12	6
White Nile (above Khartum)	27	35	1	20	17
Sobat	27	42	1	12	18
Bahr el-Gebel (above Sobat confluence)	23	17	1	26	33
Bahr el-Ghazal	43	14	1	2	40

For the late Pleistocene and Holocene, the picture is more optimistic. Significant influx of Red Sea Hill materials into the Nile is less of a problem than for earlier stages of the Pleistocene, although even here caution is required. So, for example, the unusually high pyroxene values for Samples 69, 70, and 74 from the Korosko Formation (Table B–2), at the type site in Wadi Shait, can be safely attributed to local augite derived from breakdown of adjacent dolerite gravels. Barring such local derivatives, there still are difficult statistical problems related to selective sorting of heavy minerals in different depositional environments. So, for example, the Masmas beds within the embouchure of Wadi Or contain almost three times the average value of iron minerals (Table B–3). This is clearly a result of sorting factors, since the underlying wadi sands only contain 4% opaques. With these reservations in mind, it seems preferable to examine fewer grains (100 instead of 300) from a larger number of samples, chosen from a wide range of localities (as opposed to the small numbers of samples studied by Shukri, 1950, or Khadr, 1961). Analysis of 300 grains per sample seems justified only when statistical counts of all mineral subvarieties are made by a competent mineralogist.

The results presented by Tables B–2, B–3, and B–4 are to be considered preliminary only. Identification on a specific level, including the "rare" minerals, remains to be done, and a larger suite of samples should be analyzed for some stratigraphic units. Nonetheless, the preliminary data provide a number of insights into the late Pleistocene evolution of the Nile Basin. Seen in a general context, both the statistical mean of each mineral in each formation and the range of variation lie within the range of variability of the three stratigraphic units combined. In other words, the mineral spectra of all three units are comparable. Furthermore, the average value of each heavy-mineral group generally lies within the typical variability of modern Nile sediments in Egypt as studied by Khadr (1961)—bearing in mind that Khadr's statistics omit iron ores and unknowns. A significant exception is the case of the amphiboles, which are abnormally low in our statistics. Shukri

Table B–2. Preliminary heavy-mineral analysis of sediments from the Korosko Formation (in per cent)

Sample number[a]	Opaque	Amphiboles	Pyroxenes	Epidote	Titanite	Biotite	Altered	Other
70	19	—	36	16	1	5	3	20
69	16	—	27	15	2	—	5	35
74	15	—	25	13	2	8	7	30
179	23	2	9	10	3	—	38	15
166	29	1	8	12	6	1	16	27
197	29	—	19	10	3	3	16	20
188	33	4	12	3	3	1	10	34
203	36	5	11	9	2	—	10	27
202	18	4	10	4	2	3	20	39
152	8	18	15	11	1	1	26	20
222	18	3	10	4	4	1	45	15
Mean	22.2	3.4	16.5	9.7	2.6	2.1	17.8	25.6

[a] Numerical sequence follows Table A–1.

(1950) and Shukri and Azer (1952) also record much higher values of hornblende for late Pleistocene sediments, but they list no altereds or unknowns. In all probability, a very appreciable part of our hornblende falls into these two categories, so that our statistics for the amphiboles have little value. It is readily apparent that the combined total of amphiboles and pyroxenes within Tables B–3 and B–4 varies inversely with the altereds.

The overall significance of Tables B–2, B–3, and B–4 is that the characteristic components of both the Blue and the White Niles are already present in the Korosko Formation in appreciable quantities. This is also the case for the lower member of the Korosko Formation in Wadi Or (No. 202). In other words, the heavy-mineral evidence shows that the Nile Basin was already integrated along modern lines by the Early Würm, *ca.* 50,000 years ago.

Although the basic spectrum of all three stratigraphic units is similar, there are well-defined frequency shifts when these units are averaged out. The percentage of pyroxenes increases markedly from 16.5% in the Korosko Formation to 21.2% in the Masmas Formation and 26.9% in the Gebel Silsila Formation. A slight, but consistent, parallel increase is observable for epidote. Inversely, titanite decreases steadily. There also is a noticeable decline between the iron ores present in the Korosko Formation and in the younger formations. Seen as a general phenomenon, this trend is paralleled in the X-ray diffractograms (Appendix D), where kaolinite decreases steadily in importance through time. The most reasonable, although necessarily tentative, interpretation would be a shift in the relative importance of the major Nile tributary basins. A declining Bahr el-Ghazal influx (see Table B–1),

Table B–3. Preliminary heavy-mineral analysis of sediments from the Masmas Formation (in per cent)

Sample number[a]	Opaque	Amphiboles	Pyroxenes	Epidote	Titanite	Biotite	Altered	Other
299	8	1	27	15	—	1	23	25
298	14	—	35	16	—	3	22	10
94	11	7	25	10	8	3	5	31
95	7	5	30	8	4	6	20	20
101	18	14	21	10	—	3	14	20
229	9	3	36	13	—	—	18	21
98	4	1	18	10	—	7	28	32
97	13	1	3	5	—	1	71	6
96	16	2	5	6	2	—	56	13
325	17	—	20	12	4	3	11	33
328	17	—	29	7	1	—	18	28
329	10	2	36	15	1	3	10	23
332	12	1	27	13	—	—	12	35
184	19	4	16	4	3	7	27	20
170	19	4	10	11	6	—	25	25
208	43	6	12	4	3	—	14	18
196	41	1	6	6	3	1	12	30
248	—	—	25	4	—	11	44	16
Mean	*15.4*	*2.9*	*21.2*	*9.8*	*1.9*	*2.7*	*23.9*	*22.6*

[a] Numerical sequence follows Table A–1.

matched by increased Blue Nile discharge, would adequately explain the decline of iron ores and titanite as well as the increase in pyroxenes and epidotes. An early or middle Würm pluvial period in the Bahr el-Ghazal and Bahr el-Arab basins could also be expected to provide more abundant kaolinite from eroded red and yellow latosols or plastosols on the adjacent interfluves (see the soil catenas described by Morison, Hoyle, and Hope-Simpson, 1948). Although the field evidence for such a pluvial is so far limited to the Sodiri deposits (discussed in Chap. 6) and a variety of alluvial deposits found along the foothills of Jebel Marra (Lebon and Robertson, 1961), the Bahr el-Ghazal basin promises to be of considerable interest for future Pleistocene work.

In conclusion, the heavy minerals of the nilotic sediments verify the existence of a modern Nile hydrography some 50,000 years ago, during the earliest stages of the Korosko Formation. At the same time, the changing ratios of these minerals indicate that the relative importance of the Blue Nile and the Atbara increased during the course of the late Pleistocene, presumably at the expense of Bahr el-Ghazal waters. The frequencies of pyroxenes and certain other minerals in the older, autochthonous Nile gravels of Egypt cannot be interpreted, however, except in the context of wadi influx derived from the igneous and metamorphic rocks of the Red Sea Hills and Nubia. Thus, this exceptionally useful technique is of limited application prior to the late Pleistocene, when "southern" components in the preserved sediment record first assume a prominent role.

Table B-4. Preliminary heavy-mineral analysis of sediments from the Gebel Silsila Formation (in per cent)

Sample number[a]	Opaque	Amphiboles	Pyroxenes	Epidote	Titanite	Biotite	Altered	Other
99	9	17	27	18	1	—	11	17
231	6	1	37	14	—	—	17	25
388	9	—	37	13	2	7	1	31
387	61	—	5	1	2	—	1	30
226	27	8	20	8	1	—	14	22
227	1	3	46	11	—	—	6	33
230	8	3	34	16	1	—	9	29
233	8	3	42	12	—	5	6	24
228	25	3	16	12	3	—	19	22
234	19	—	18	8	3	—	1	51
232	43	1	10	4	1	—	5	36
391	10	—	43	17	—	1	—	29
390	4	—	52	19	—	1	9	15
182	9	3	18	9	—	2	29	30
177	18	3	14	10	2	—	19	34
153	32	1	18	14	2	1	10	22
156	22	—	22	12	—	3	18	24
250	8	1	26	14	—	1	28	22
Mean	17.7	2.6	26.9	11.8	1.0	1.2	11.3	27.6

[a] Numerical sequence follows Table A-1. No. 177 is from the Arminna Member, No. 182 from the Kibdi. All other samples are from the Darau Member.

APPENDIX C

Quartz-Grain Micromorphology

Karl W. Butzer
and Bruce G. Gladfelter

One of the most interesting problems in sediment interpretation of the nilotic formations concerns the origin of the quartz sands. Whereas fine sands, forming part of the "silt" fraction, are prominent in almost every grain-size spectrum, medium or coarse sands are present in rather variable amounts. These coarser-grained particles are seldom carried in suspension and usually form part of the bed load, deposited in areas of rapid water movement. They may be derived (*a*) from local wadi sands, washed into the Nile Valley by periodic spates; (*b*) from eolian sands, blown directly into the Nile or derived from eroded dunes; and (*c*) from older alluvia or bedrock, undermined or otherwise eroded by the Nile. At the same time, such medium to coarse sands may be transported on a local scale only or may be carried many hundreds of kilometers along the length of the Nile Valley. The ultimate source for the bulk of the quartz sands of the nilotic formations is the Nubian Sandstone, exposed over most of the drainage basin of the Saharan Nile. Such sands would necessarily have been subject to repeated resedimentation. In addition, primary sands would have been provided by the Basement rocks exposed in the Nile Valley at Aswan, at Kalabsha, and in Upper Nubia.

In order to gain some insight into the depositional history of these sands, systematic microscopic examination was carried out of quartz grains from the nilotic formations, as well as from the wadi units, from the Kurkur Oasis, and from the Red Sea littoral. By using several of the available techniques, it was hoped some conclusions might be reached concerning the primary sources of the quartz sands.

Cailleux and Tricart (1963: 53–102) describe at length a method of quartz-grain microscopy used by many European workers. In brief, they distinguish:

a) *Subrounded, polished* grains (*émoussés-luisants*), attributed to mechanical abrasion in moving waters and attaining maximum development on grains 1 millimeter in diameter. It is thought that such polish is a common characteristic of sand grains in major rivers, tending to be infrequent in periodic streams.
b) *Rounded, frosted* grains (*ronds-mats*), characterized by minute impact-marks, lending a stippled aspect to the surface and often rendering the grain opaque. These peck-marks, with a density of 50 to 100 per 0.1 square millimeter, are

attributed to mutual mechanical abrasion by wind-transported sand grains. Such "eolian" frosting is best developed in grains with diameters between 0.3 and 1.5 millimeters and attains a maximum at 0.7 millimeter. A subclass of *subrounded, pecked* grains (*émoussés-picotés*) is also recognized, with a dull but often transparent surface formed by minute stipples of smaller size and greater relative density than those of the "rounded, frosted" type. These are attributed to chemical alteration, although no details are given, except by Tricart (1958), and this category is basically ignored in the statistical presentation of results by Cailleux and Tricart (1963). Another subclass of *discolored, rounded, frosted* grains (*ronds-mats sales*) is recognized, with the individual grains showing traces of opaque, reddish cement, either ferruginous or siliceous. This variety, absent from the Egyptian samples, is attributed to derivation of older eolian deposits.

c) *Unworn* grains (*non-usés*), with irregular shapes and angular contours, lacking well-defined polish or frosting and primarily derived directly from igneous or metamorphic rocks.

Recent empirical studies of quartz-grain abrasion and surface morphology by Kuenen (1959, 1960; Kuenen and Perdok, 1962) show that these categories are oversimplifications at best and that such facile interpretations are of dubious value. Our own results strongly corroborate Kuenen's views, as will be further discussed below. In order to examine our quartz grains as objectively as possible, percentages of frosted grains were counted without regard for rounding attributes. Simultaneously, qualitative observations were made of frosting type, i.e., whether the entire grain was frosted, the protuberances only, or recessed surfaces only. It was found that the *mats* and the *picotés* types of Cailleux and Tricart (1963) could not be objectively distinguished, a fact earlier proven by electron microphotographs taken by Kuenen and Perdok (1962).

A number of methods to determine roundness of sand grains have been outlined and discussed in previous studies. One such classification includes four classes of roundness, the limits of which are arbitrary. Classification is accomplished by visual comparison of the grains with a set of illustrations (Russell and Taylor, 1937; Payne, 1942). These classes may be described as follows:

Angular. Irregular grain surface, dominated by edges and corners, all of which are sharp.
Subangular. Irregular surface; edges and corners still prominent but less sharp and possibly slightly curved or rounded.
Subrounded. Grain outline still somewhat irregular, but edges and corners mostly curved.
Rounded. Surface outline smooth, tending towards an oval or circular shape; edges and corners absent.

The roundness scale of Powers (1953), another widely used classification illustrated by plates in Shepard and Young (1961), introduces a "well-rounded" category, which shifts the scale in an obvious fashion. For the sake of counting accuracy, however, the four basic categories of Payne (1942) were found to be preferable for individual grain classification. This was subsequently converted by assigning arbitrary values to each grain of a particular class as follows: angular, 0; subangular, 5; subrounded, 10; and rounded, 15. An index of rounding can then be devised for a 100-grain sample, giving a range of values from 0 to 1,500. To allow for "well-rounded" grains and to obtain correspondence with our gravel-shape classes (see Chap. 1), the following classes of roundness are defined (for samples of 100 grains):

Angular	0–250
Subangular	250–500
Subrounded	500–750
Rounded	750–1,250
Well-rounded	1,250–1,500

Two grain-size fractions were examined: the 210–595 micron and the 63–210 micron. Coarser fractions are poorly represented in most of the nilotic sediments, precluding systematic study. The medium-sand fraction (63–210), although of marginal interest to Cailleux and Tricart (1963), is actually of considerable comparative value in assessing rounding characteristics and types of frosting. But, whereas a minimum of 100 grains was counted for the coarse sands of the 210–595 micron unit, only 50 grains were employed for the medium grade, the rounding index simply being doubled. Altogether, over 14,000 grains were classified. The detailed results are presented in Tables C–2 to C–5, which list the percentage of frosted and of angular, unworn grains, as well as the index of rounding (as defined above). Table C–1 summarizes the data for each formation or region.

In a general way, the data provided by Table C–1 indicates that 65% to 93% of the coarse-sand fraction is frosted and that this grain-size class tends to be "rounded." The medium-sand fraction is more variable, with frosting ranging from 57% to 93% for different groups of samples and averaging almost 13% less than in the case of the coarse fraction. At the same time, the rounding index is systemati-

Table C–1. Quartz-grain micromorphology:
regional or stratigraphic means of frosting and rounding indices

Formation or area	Number of samples	210–595 Microns			63–210 Microns		
		Frosted (%)	Unworn (%)	Rounding index	Frosted (%)	Unworn (%)	Rounding index
Korosko (Nubia)	9	89.2	0.4	911.6	57.2	4.9	757.8
Korosko (Kom Ombo)	5	65.6	0.0	928.0	65.8	0.0	742.1
Masmas (Nubia)	7	88.3	0.7	1025.0	59.4	4.0	748.6
Masmas (Kom Ombo)	11	78.1	0.0	836.8	65.7	1.1	725.0
Gebel Silsila (Nubia)	10	85.8	6.2	895.5	64.6	12.6	650.5
Gebel Silsila (Kom Ombo)	18	82.5	0.2	906.1	72.9	9.2	605.6
Wadi beds (Nubia)	16	93.2	0.2	770.6	93.2	0.4	718.1
Wadi beds (Kom Ombo)	6	68.2	0.6	903.0	64.8	0.0	736.7
Kurkur Oasis	7	84.7	1.6	842.8	76.9	3.7	676.4
Red Sea Coast	5	93.0	6.6	660.0	80.6	1.2	640.0
Mean		*82.9*	*1.7*	*867.9*	*70.1*	*3.7*	*700.1*

cally lower, giving an overall class description of "subrounded." The percentage of angular, unworn grains is also double that of the coarse fraction. By the interpretation of Cailleux and Tricart (1963), these results would "prove" that all of the sands in question were predominantly of eolian origin. Logically enough, eolian modification would be less significant in the case of the medium sands, where fluvial characteristics are more prominent. A closer inspection of the individual samples and microscopic analysis of the frosted grains, however, leaves no doubt that such an interpretation would be quite erroneous.

Statistically, the occurrence of frosted grains among the groups of samples may be systematic but may also be wholly unpredictable. At a particular locality, frosting commonly tends to occur in similar percentages regardless of the stratigraphy. This is so in the case of Wadi Or (Table 6–3) or of Gebel Silsila 2B (Table 3–4, 3–5, 3–6). The result is that regional variability between Kom Ombo and Nubia is greater than any variability among individual formations. Thus, whereas frosting of coarse sand grains increases steadily from the Korosko to the Gebel Silsila units in Nubia, the reverse trend is evident at Kom Ombo. It is equally curious that the percentages of frosting evident for the medium sand are inversely proportional to those for the coarse fraction. Closer inspection of the individual statistics shows further that the medium sands exhibit a bimodal pattern: either frosting is common and comparable to percentages among the coarse sands, or the percentage of frosting is very low. In the latter case, highly polished samples may occur at random in a sedimentary sequence, with no relationship to the variability of frosting among the coarse sands. A final and rather significant paradox is the notable inverse correlation between percentage of frosting and index of rounding evident for both the coarse and the medium sands in Table C–1. The experiments of Kuenen (1960) show that eolian transport leads to rapid rounding of sand grains, at rates on the order of 100 to 1,000 times faster than in a fluvial medium. Thus, the frosted samples, if they are indeed eolian in nature, should be conspicuously more rounded. Instead, the opposite holds true.

Examination of the individual sand grains shows two general characteristics that are incompatible with an eolian interpretation of the frosting present: (a) Practically every frosted grain counted (approximately 11,000) shows distinct evidence of frosting on the totality of minute indentations and deeply recessed surfaces. Kuenen and Perdok (1962) have shown, however, that mechanical eolian abrasion only produces frosting on the corners and edges of a sand grain. (b) Many sand grains exhibit a combination of frosted and polished surfaces. The overwhelming majority of these intermediate types (which were counted as "frosted") have their protruding parts polished, their recessed surfaces frosted. Such selective polish cannot be produced by any chemical agency; yet, both eolian and fluvial abrasion of mechanical type will frost (rather than polish) surface protuberances (Kuenen and Perdok, 1962). The only experimental way these authors could devise to produce polish on protuberances was to roll dry sand in a bottle. This analogy is not particularly helpful in interpreting our intermediate class, which may account for as many as two-thirds of the frosted grains in some samples.

Examination of frosted dune sands by Kuenen and Perdok (1962) indicates that the typical "desert frost" evident may be due to a combination of mechanical pitting and of chemical solution and redeposition through alternate wetting and drying of desert sands by dew, whereby evaporation develops highly concentrated solutions. At any rate, experimental eolian frosting produces coarser surfaces than those observed on true desert sands. Application of sodium hydroxide or hydrofluoric acid solutions at normal temperatures soon produces a stippled surface texture quite comparable to "desert frost." Although the experimental evidence rules out mechanical abrasion as the exclusive or dominant agency for "desert frosting," the fact that frosting may be conspicuously rare on medium sand grains in the case of desert sands raises new problems for a strictly chemical interpretation. In the case of fluvial sands (Kuenen, 1959; Kuenen and Perdok, 1962), chemical attack either produces total frosting or total polish, depending on the chemical properties and equilibrium of the particular water body. Thus, frosting in the case of the Nile Valley samples may also reflect surficial alteration in a subaqueous environment.

In conclusion, some three-quarters of the 14,000 sand grains studied were frosted on all faces by chemical attack before, during, or after final deposition. The fact that an appreciable proportion of these grains was subsequently polished mechanically on exposed edges and corners indicates that much or all of the frosting was accomplished prior to final sedimentation. Presumably, therefore, the frosting is an inherited characteristic, possibly quite unrelated to the final mode of transport and deposition. Consequently, the frosting properties are inconclusive. The occurrence of 85% to 95% frosted grains among beach sands, in well-stratified wadi deposits, or in laminated flood silts is by itself sufficient evidence for questioning the immediate association between frosting and eolian abrasion. And the statistical features, together with the microscopic details, serve to disprove such a relationship in the case of all of the Egyptian samples examined. This underscores the theoretical and experimental conclusions of Kuenen and Perdok (1962). At the same time, it raises serious questions about the conclusions reached by Cailleux and Tricart (1963: Table 12, 83 ff.) concerning the significance of eolian factors in the different geological formations of the Sahara.

The rounding indices are not only of interest for demonstrating a lack of positive correlation between optimal rounding and frosting, but, together with the percentage of angular, unworn grains, they also provide some useful data on possible origins of the sand. Table C–1 shows that the Nubian wadi deposits, all collected from small catchments, are morphologically distinct from all other sample groups in the Nile Valley. Although unworn grains are rare, rounded grains are equally few, so that rounding indices are low in both the coarse and the medium categories. Sands from the major wadis, exemplified by those from the Kom Ombo Plain, have higher indices, particularly in the coarse size-fraction. This is probably due to greater fluvial transport distances, since eolian components should not be affected by size of catchment basin. On these grounds, the sands of the Korosko and the Masmas Formations can be distinguished from the local wadi sands of Nubia. Three possible reasons can be given for this fact: (*a*) the nilotic sands are primarily derived from better-rounded eolian sands; (*b*) the bulk of the nilotic sands was provided by a few major, rather than many minor, tributaries; or (*c*) the sands, whatever their origin, were further rounded during transport through the Nile Valley. Since the highest indices are attained by the coarser Masmas sands of Nubia and since Egyptian wadi activity was minimal at that time, it appears that secondary eolian components are important in at least some of the nilotic formations.

A second interesting roundness characteristic is the greater incidence of unworn sand grains in Nubia than at Kom Ombo. This can be readily explained by the derivation of fresh quartz aggregates from outcrops of Basement Complex or from the erosion of older beds of granitoid or mesocrystalline quartz gravels. Older gravels of such types can be excluded as a source of the unworn grains in the Khor Adindan, Ballana, and Wadi Or sediments belonging to the Korosko and Masmas Formations. The unworn quartz of at least this area can be safely attributed to the long reach of Basement Complex rocks traversed by the Nile in Upper Nubia, south of the Second Cataract. Basement rocks that outcrop further north at Kalabsha and Aswan are far less extensive than those of the Batn el-Hagar. This, and the gradual rounding to be expected with increasing distance from their source, explains the decrease in unworn grains downstream.

A veritable surge of unworn quartz grains makes its appearance with the Gebel Silsila Formation (Table C–1), analogous with the sudden influx of Hudi Chert at the same time. In the medium-sand fraction, these unworn grains are supplemented by very high frequencies of subangular grains, reducing the rounding index conspicuously. Since there can hardly have been a significant hydrographic change at Aswan or Kalabsha at that time, a period of intensive erosion in the Batn el-Hagar (or elsewhere in Upper Nubia) must be postulated, beginning ca. 15,000 B.C. This further indicates that an appreciable proportion of the late Pleistocene nilotic sands of southern Egypt were derived from upstream of the Second Cataract. More difficult to explain is the concentration of the unworn grains in flood silts laid down during the terminal stages of deposition in many areas (Tables C–2, C–3). This may be the effect of sorting.

The rounding indices obtained from the Kurkur samples (Tables C–1, C–4) tend to be fairly uniform and are well below averages from the Nile Valley. This, and the presence of unworn grains, seems to suggest that the quartz sands of the oasis are derived from the local Chalk and Nummulitic and have not been transported far by eolian agencies. These quartz grains certainly have not been subject to much eolian abrasion. In other words, the primary or secondary eolian component of the freshwater limestones of Kurkur is definitely of limited importance.

Finally, the rounding indices for the Red Sea littoral (Table C–5) reflect the abundance of unworn quartz grains supplied to the coast by the wadis emerging from the igneous and metamorphic Red Sea Hill country. Rounding is greatest in the beach deposits, least in the Pleistocene wadi alluvia.

In overview, it seems that morphology (rounding of sand grains) seems to be of greater potential value for environmental deductions than does surface texture (frosting or polish). In all cases, however, the possible significance of inherited traits—reflecting a complex past history—should not be ignored. Because quartz sand is relatively indestructible, multicycle quartz grains from the Nubian Sandstone pose a particular problem in the Nile Valley. Consequently, the Nile Valley samples are much more difficult to interpret than those from Kurkur and the Red Sea coast. Fortunately, the introduction of unworn quartz grains from the Batn el-Hagar provides a useful tool in the case of the nilotic sediments.

Table C-2. Quartz-grain micromorphology (Kom Ombo Plain)

Sample number	210–595 Microns			63–210 Microns		
	Frosted (%)	Unworn (%)	Rounding index	Frosted (%)	Unworn (%)	Rounding index
			Table 3-3			
79	79	—	695	46	—	700
16	59	2	880	89	—	850
15	—	—	—	72	—	580
14	75	—	1010	57	—	805
13	82	—	920	70	7	650
301	70	—	1040	38	—	935
300	70	—	1065	76	—	620
299	82	—	1080	84	—	815
298	72	—	1015	18	—	835
71	63	1	835	68	—	735
72	79	—	790	69	—	795
70	74	—	790	69	—	780
69	59	—	985	68	—	760
74	39	—	975	22	—	750
76	74	—	970	100	—	770
			Table 3-4			
94	37	—	715	82	—	590
95	81	—	755	62	12	585
101	91	—	675	18	—	745
229	81	—	805	79	—	710
98	96	—	745	76	—	650
97	98	—	615	88	—	630
96	77	—	1000	90	—	815
			Table 3-5			
99	82	—	930	88	24	380
231	88	1	885	85	24	490
388	83	—	835	79	23	435
387	88	—	895	64	28	510
226	86	—	1015	82	9	665
227	72	—	980	20	—	720
230	77	—	940	81	—	615
233	83	—	955	83	16	600
396	78	—	1170	53	11	455
228	74	—	1040	64	—	755
234	84	—	920	78	—	655
232	77	—	830	77	—	740
			Table 3-6			
392	98	—	985	74	15	585
391	75	—	780	83	15	480
390	83	—	830	86	—	740
			Table 3-8			
141	90	—	835	44	—	695
142	84	2	755	90	—	665
144	84	—	730	81	—	715

Table C-3. Quartz-grain micromorphology (Nubia)

Sample number	210–595 Microns			63–210 Microns		
	Frosted (%)	Unworn (%)	Rounding index	Frosted (%)	Unworn (%)	Rounding index
Table 5-2						
154	57	9	825	35	43	320
161	84	2	745	78	4	705
162	78	4	760	82	18	625
163	77	—	920	71	5	825
Table 6-2						
182	90	—	1070	77	17	590
181	90	—	925	92	—	1190
177	93	—	985	12	—	815
185	92	—	1040	96	—	875
184	89	—	1020	71	—	935
180-II	85	—	995	88	14	655
179	94	—	1005	98	13	675
170	94	—	1130	96	—	760
166	84	—	870	64	16	930
159	94	—	995	90	—	925
157	94	—	895	89	—	750
Table 6-3						
209	85	2	775	93	—	630
199	88	—	710	91	—	620
200	99	—	685	98	—	670
207	99	—	665	91	—	610
192	100	—	915	94	—	665
198	97	—	720	96	—	760
Table 6-3						
195	91	—	670	87	—	615
194	94	—	770	84	2	630
189	91	1	740	83	—	745
191	93	—	845	92	—	700
208	79	—	1055	15	—	720
196	89	4	1055	8	—	775
201	98	—	655	92	—	645
197	96	2	935	56	—	795
188	77	—	885	30	—	870
203	93	—	1210	5	—	785
206-I	90	—	690	79	2	665
206-II	99	—	750	88	2	710
204	91	—	690	85	—	655
202	87	—	1145	24	—	745
Table 6-4						
252	93	42	805	65	50	290
164	96	2	1255	60	21	615
153	85	2	890	60	4	685
156	100	—	815	85	4	765
152	91	—	815	70	13	640
250	57	6	665	55	13	530
249	69	15	700	70	11	680
248	—	—	—	49	14	645
148	81	—	775	72	6	610
222	83	2	685	76	2	735

Table C-4. Quartz-grain micromorphology (Kurkur Oasis) (See Table 7-3)

Sample number	210–595 Microns			63–210 Microns		
	Frosted (%)	Unworn (%)	Rounding index	Frosted (%)	Unworn (%)	Rounding index
359	91	—	755	87	—	700
350	72	—	900	74	2	695
365–II	81	7	795	54	15	535
365–I	89	—	825	85	—	695
364	84	—	890	70	—	740
358	94	—	915	84	—	665
361–I	82	4	820	84	9	705

Table C-5. Quartz-grain micromorphology (Red Sea Coast)

Sample location and stratigraphy	210–595 Microns			63–210 Microns		
	Frosted (%)	Unworn (%)	Rounding index	Frosted (%)	Unworn (%)	Rounding index
Modern wadi bed, Wadi Ambaut	100	1	685	72	2	650
Low terrace matrix, Wadi Alam	83	32	430	78	4	615
Offshore gravel bar (+ 8.6 m), Km 5.5 South	90	—	690	81	—	720
Derived beach sand in + 8.6 m bar, Km 5.5 South	100	—	760	96	—	715
Marine gypsum (+ 8.5 m), Km 5.5 South	92	—	735	76	—	500

APPENDIX D

Clay Minerals

Karl W. Butzer
and Carl L. Hansen

Procedure

The clay minerals of the nilotic deposits and paleosols are of considerable importance in a study of this kind. As a result, detailed X-ray studies were carried out, basically following the procedure outlined by M. L. Jackson (1956, 1964). A 50-gram sample was first cleaned of carbonates by acetic acid, buffered with sodium acetate to a pH of 4.5 to 5.0. Organic matter was removed by hydrogen peroxide, and, after washing with methanol and with water, the sample was saturated with Na^+ by using NaCl. After separation of the clays in suspension, clay layers were built up on porous slides of ceramic tile and were then saturated with Mg^{++} by a 2N solution of $Mg\ Cl_2$. Finally, after removal of excess $Mg\ Cl_2$, the clay was solvated with 10% glycerol and was X-rayed. For samples saturated with K^+, the procedure is similar, except that KCl is substituted for $Mg\ Cl_2$ and the glycerol treatment is omitted. After the first X-ray run at 25° C, the K^+-saturated samples were heated at 300° C for two hours and at 550° C for two hours and were X-rayed again after each heating.

Altogether, forty-three samples were analyzed. The X-ray diffractograms (Mg-saturated at 25° C) are presented as stratigraphic groups in Figs. D–1 to D–6. The diffractograms for K-saturation and at higher temperatures are not shown. Although X-ray analyses do not yield quantitative results and intensity of peaks is partly related to degree of crystallinity, the diffractograms nevertheless give a fairly accurate picture of the relative significance of the different clay components. The spectrum of minerals present in the clay size-fraction is the same for practically all of the Egyptian samples analyzed and includes kaolinite, montmorillonite, illite/mica, quartz, and feldspars. There is considerable variability among the peaks displayed by the different diffractograms, however, primarily reflecting the different relative importance of these clay minerals. At the same time, recognizable patterns emerge clearly for the three late Pleistocene nilotic formations, for the red paleosols, and for the late Pleistocene to middle Holocene wadi alluvia. Hematite, which seldom appears on X-ray diffractograms unless present in quantities exceeding 5%, does not show up distinctly on any of the graphs. Similarly, halloysite and hydrargillite, both members of the kaolin family, could not be recognized on any of the nilotic diffractograms. Their characteristic peaks lie very close to those of kaolinite, however, so that the presence of some halloysite or hydrargillite is not definitely precluded in at least some of these samples.

The Korosko Formation

Seven samples were analyzed, all of them from Nubia (Nos. 152, 166, 179, 188, 202, and 222). Kaolinite and montmorillonite are the major clay minerals present (Fig. D–1). Sample 202, from the lower member, corresponds closely with the remaining samples from the classical or upper member.

Montmorillonite is identified by a peak at 18 A when the sample is saturated with Mg and solvated with glycerol. Saturation with K results in a collapse of the montmorillonite to about 12.6 A. Heating the K- or Mg-saturated samples to 550° C causes the montmorillonite to collapse to a broad 10-A peak. Lack of precise definition of the peaks at both low and high temperatures suggests a certain amount of interlayering of the montmorillonite with sesquioxides or with hydroxyl Al.

Illite/mica is indicated by a 10 A peak with Mg-saturation and glycerol solvation. These peaks are somewhat more intense than those of the Masmas Formation, except in Samples 152 and 179, where they are sufficiently modest to allow recognition of a low-intensity peak in the 8.9–9.1 A range—suggesting a second-order montmorillonite peak.

Kaolinite peaks are evident at 7.15 A and 3.57 A (first- and second-order respectively), and their intensity in relation to the peaks of the Masmas beds is on the order of 2:1. Samples treated with KCl show kaolinite peaks of even better definition and stronger intensity.

Quartz is indicated by a well-defined 3.34-A peak of low to pronounced intensity and is confirmed by a 4.26-A peak. The intensity of the latter is enhanced with K-saturation. The presence of some feldspars in all samples except 179 is also indicated by distinct peaks of low intensity in the 3.18–3.24 A range.

The Masmas Formation

Six samples were analyzed, both from the Kom Ombo Plain and from Nubia (Nos. 95, 96, 328, 170, 184, and 248). The clay minerals are dominated by montmorillonite (Fig. D–2), although the proportions of kaolinite are still significant when compared with those in younger nilotic sediments.

The presence of montmorillonite is indicated by an 18-A peak of high intensity. Asymmetry and broadening of this peak—even after heating results in collapse to 10 A—indicate interlayering of the montmorillonite, probably with sesquioxides.

In Samples 95 and 170, the 10-A illite/mica peak displays a low intensity, whereas the remaining samples show this same peak with moderate intensity and clearer definition. The sharpness of the illite/mica peak is partly related to the proportion of mica, which appears to be higher in Samples 96, 328, 184, and 248.

The first- and second-order kaolinite peaks at 7.2 A and 3.6 A are found in all samples, at slightly greater intensities than in the case of the Gebel Silsila Formation.

Next to the montmorillonite group, the 3.34-A quartz peak is the most prominent feature of the Masmas diffractograms—with the exception of Sample 328, which is not as clear-cut and is of rather low intensity. For all samples, the 4.26-A quartz peak is of low intensity but is rather sharply defined. Some feldspars are present in all samples, as indicated by peaks of very low intensity and poor definition in the 3.18–3.24 A range.

The Gebel Silsila Formation

Four samples from the classical Darau Member at Kom Ombo (Nos. 227, 231, 233, and 392), one from the same member in Nubia (No. 250), two from the Arminna Member (Nos. 177 and 252), and a last from the Kibdi Member (No. 182) were analyzed. All of these diffractograms are very similar and are characterized by very sharply defined and intensive montmorillonite peaks (Fig. D–3). These peaks are found in the 18-A range with Mg saturation and glycerol solvation. They lose their clear-cut definition in the presence of K and after heating to 300° C, which suggests the presence of some interlayered material.

10-A peaks of modest intensity and broad definition indicate the presence of a moderate illite/mica fraction. The frequent extension of this peak, or even its culmination in the 8.5–10 A range, suggests its possible displacement through interlayering of montmorillonite and illite.

Clear but very-low-intensity peaks near 7.2 A and 3.55 A show the presence of kaolinite, while distinct 3.32-A peaks of variable intensity—as well as very-low-order 4.26-A peaks—indicate quartz. The 3.32-A quartz peaks are a little more intensive in the Arminna and Kibdi samples. The presence of some feldspars in all samples is indicated by very-low-intensity peaks of rather poor definition in the 3.18–3.24 A range.

Implications of the Nilotic Diffractograms

Several generalizations can be made about the X-ray diffractograms from the late Pleistocene nilotic sediments (Figs. D–1 to D–3).

a) Kaolinite decreases strongly in relative proportions between the Korosko and the Masmas units, with further decline in the Gebel Silsila Formation.
b) Montmorillonite increases steadily in relative importance during the course of the late Pleistocene.
c) Illite/mica is of moderate significance in most of the diffractograms from all three stratigraphic units, although the peaks are subject to considerably greater variability than those of either kaolinite or montmorillonite.
d) Quartz in the clay size-fraction is prominent in the Korosko, and particularly in the Masmas, beds and is fairly insignificant in the Gebel Silsila units.
e) Some feldspars are present in almost all of the samples, presumably in very small quantities.
f) A remarkable uniformity of diffractograms is evident for each formation, regardless of their respective subdivisions. Consequently, the diffractograms have a well-defined stratigraphic potential.

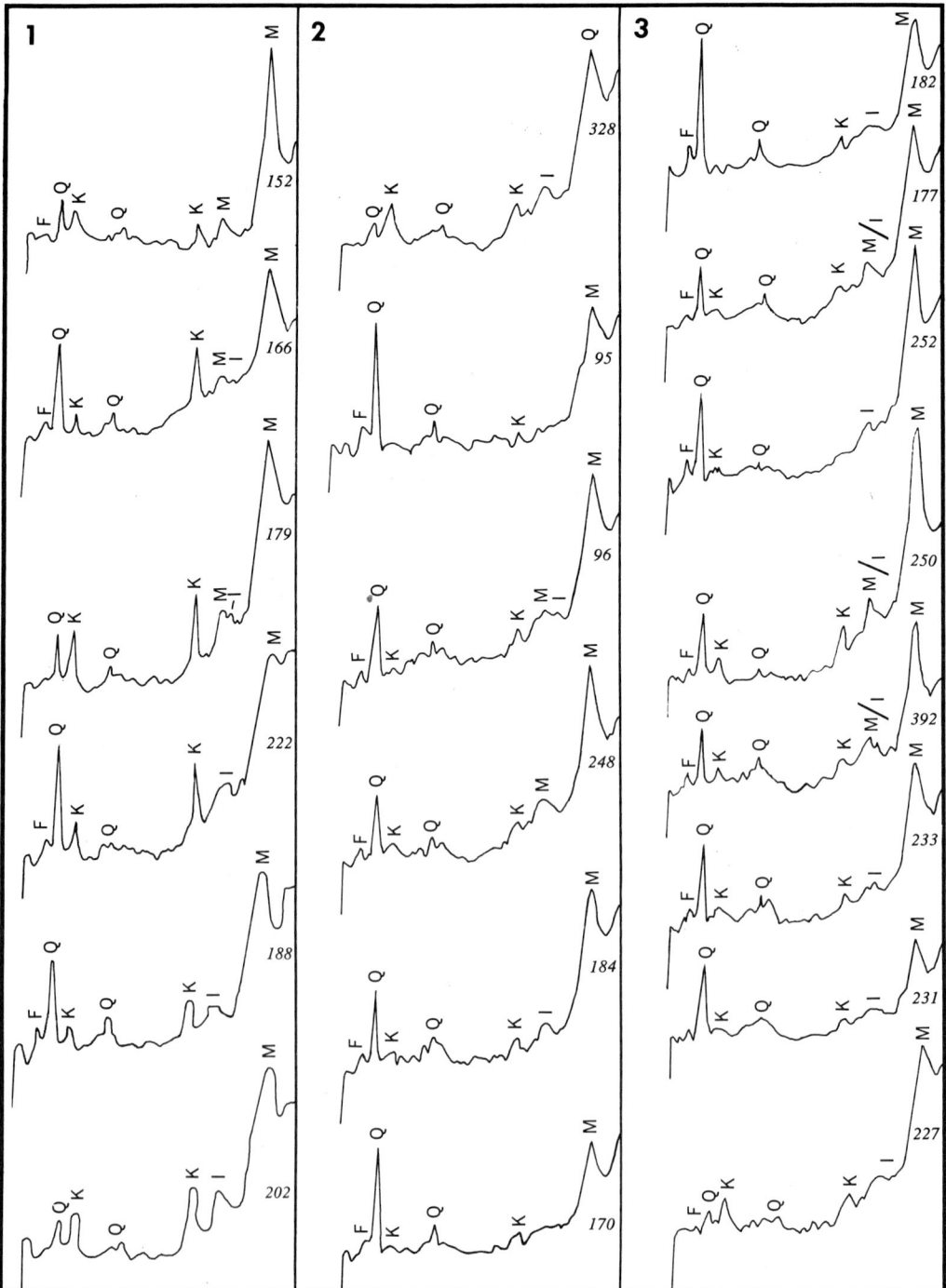

Fig. D–1. X-ray diffractograms from the Korosko Formation. Magnesium-saturated and solvated in glycerol, at 25° C. Simplified. The characteristic peaks of the major clay minerals present are identified by letters: M (montmorillonite), I (illite/mica), K (kaolinite), Q (quartz), and F (feldspars). Sample numbers are the same as those used elsewhere in the text. Graph: UW Cartographic Lab.

Fig. D–2. X-ray diffractograms from the Masmas Formation (Mg-saturated, 25° C). Graph: UW Cartographic Lab.

Fig. D–3. X-ray diffractograms from the Gebel Silsila Formation (Mg-saturated, 25° C). Graph: UW Cartographic Lab.

Interpreting these broad generalizations, several suggestions can be made or conclusions drawn.

1) The clay-mineral spectrum of the suspended Nile sediments remained fairly constant over many millennia, only to change rather markedly—during a time span of 1,000 or 2,000 years—a little after 23,000 B.C. and again shortly before 15,000 B.C. This can be more readily explained by changes in the relative importance of the major subsaharan tributaries than by gradual environmental changes on the Ethiopian Plateau.

2) Kaolinite seems to be one of the most important soil constituents in the White Nile drainage today, particularly in the Sudd area, the Bahr el-Ghazal, and the Bahr el-Arab (see Morison, Hoyle, and Hope-Simpson, 1948). Montmorillonite appears to be dominant in the Blue Nile, and particularly the Atbara, waters (see Finck, 1961) and, to a lesser extent, in the Sobat also. Thus, the clay-mineral shift may be a result of a decline in the relative importance of White Nile influx, particularly that from the central and western parts of the Paleo-Sudd Basin. Before this hypothesis can be considered as established, however, systematic clay-mineral analyses must be carried out on recent sediments from the major subsaharan tributaries.

3) The presence of very fine quartz particles in all but the Gebel Silsila Formation is significant and must be attributed to the subsaharan drainage basin, especially since there is no lack of quartz sands in the Gebel Silsila beds. Quartz in the clay fraction is today supplied by all of the tributaries and is apparently derived from the Basement Complex or the Adrigat Sandstone rather than from the extrusive vulcanics of the Ethiopian Plateau (see Chap. 9). Possibly the decline in very fine quartz after 15,000 B.C. resulted from a decline of precipitation on the plains of the central and southern Sudan, at least in relation to the Ethiopian Plateau. In connection with the quartz patterns in the clay diffractograms, it should be emphasized that Sample 328, from New Ballana I, does not fit the Masmas spectrum as well as it does the Gebel Silsila Formation. This supports our argument that these deposits form a part of the latter unit.

4) Almost all of the clay minerals in the nilotic sediments were undoubtedly derived from the subsaharan drainage of the Nile, and sedimentary concentration of montmorillonite in alluvial deposits by slowly moving waters is an established fact. The high exchange capacity and the water-holding properties of montmorillonitic soils makes them to some extent self-perpetuating, however (M. L. Jackson, 1959). Consequently, some of the montmorillonite, particularly in the case of the vertisols, may indeed be authigenic.

The Ineiba and Shaturma Formations in Wadis Kharit and Shait

The late Pleistocene to middle Holocene wadi deposits of the Kharit-Shait area were studied as a group, basically similar in terms of origin and catchment area. Sample 128 was analyzed from the Shaturma Formation; Samples 16, 58, 78, and 301 from the Sinqari Member; Samples 62, 75, 85, and 127 from the Malki Member; and Samples 57 and 68 from the basal gravel unit of the Malki. In addition, two samples (22 and 398) were run from the red paleosol (Omda Soil) developed in beds of the Sinqari Member. The spectra are dominated by kaolinite, montmorillonite, and quartz, generally in that order of importance (Fig. D–4).

In the case of the Sinqari sediments, montmorillonite exhibits a strong, but seldom sharply defined, peak of moderate intensity at 17.5–17.9 A, and an ill-defined, second-order peak at about 9 A may also be present. Lack of precise definition of the montmorillonite peaks at low and high temperatures suggests interlayering. Presence of some illite and micas is supported by a disordered peak at about 9.8–10 A and by another peak of low intensity at 4.98 A. With K-saturation, the 9.8–10 A peak loses much of its identity, becoming an irregularity on the flank of an irregular 13-A peak. The 4.98-A peak is slightly intensified with K-saturation. Clearly defined kaolinite peaks appear at 7.15 A and 3.57 A, and, together with their rather strong intensity, this suggests good crystallization. Better definition and stronger intensity are achieved with K-saturation. Peak intensities are similar for all samples except No. 301, where intensities are unusually low. Quartz is indicated by a slight peak at 4.26 A and by a more intense and clearly defined one at 3.34 A. Finally, peaks of low intensity in the 3.15–3.24 A range suggest the presence of small amounts of feldspars.

The clay-mineral constituency of the Malki samples is very much like that of the Sinqari Member, with the diffractograms exhibiting peaks in the same relative positions. The major difference between the two groups is in the intensity of the first- and second-order kaolinite peaks. These are rather less intense than those of the Sinqari sediments, usually on the order of 2:3.

No interpretation can be offered for the greater relative significance of kaolinite in the Sinqari Member. The fact that the finer facies of the Malki Member frequently exhibits cracking lends support to the evidence of the diffractograms, namely, that montmorillonite is of greater relative importance. And, since the clay content of the Malki deposits may be several times greater than in the Sinqari Member, the absolute quantities of kaolinite present in the Malki may be at least as great as those in the Sinqari. Presumably, almost all of these clay minerals are derived from the Red Sea Hills and from weathered, local sediments of Pleistocene age. There is, however, a possibility that some authigenic montmorillonite has developed *in situ* in the clayey floodplain deposits of Wadis Kharit and Shait.

The Omda Soil samples taken from rubefied Sinqari sediments in Wadi Kharit (Nos. 22 and 398) show kaolinite peaks equal to or greater than those of the parent material, so that considerable authigenic formation of kaolinite must be assumed. At the same time, the montmorillonite peaks at 18 A are far less clearly defined than those of the Malki or the Sinqari beds, indicating greater interlayering of the montmorillonite, presumably with sesquioxides. Finally, the feldspar peaks are rather more subdued, possibly suggesting that the feldspars have been

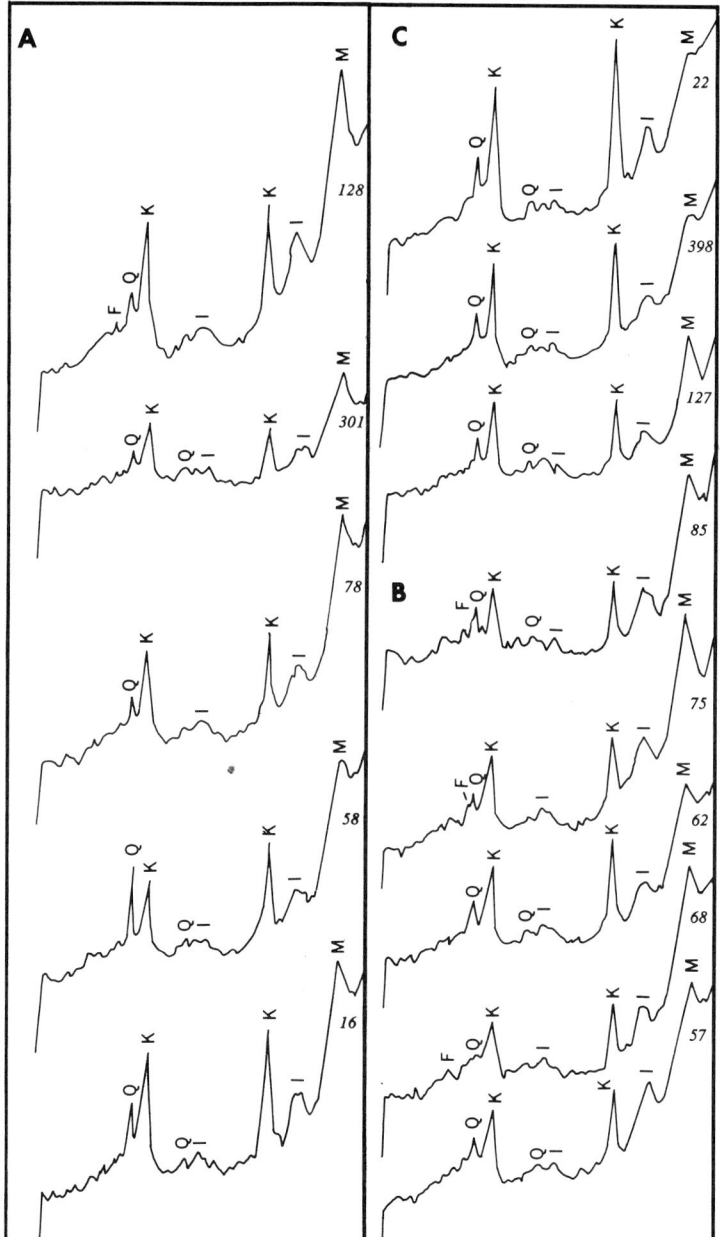

Fig. D–4. X-ray diffractograms from the Ineiba Formation: (A) the Sinqari Member; (B) the Malki Member; and (C) the Omda Soil. Sample 128 represents the Shaturma Formation (Mg-saturated, 25° C). Simplified. The characteristic peaks of the major clay minerals present are identified by letters: M (montmorillonite), I (illite/mica), K (kaolinite), Q (quartz), and F (feldspars). Sample numbers are the same as those used elsewhere in the text. Graph: UW Cartographic Lab.

altered to kaolinite. In comparing Nos. 22 and 398, it should be noted that No. 22 alone pertains to a typical (B)-horizon and that these properties of kaolinite enrichment and sesquioxide formation are more clearly defined for this sample.

The Red Paleosols

In addition to the two samples of the Omda Soil in Wadi Kharit, a range of other red paleosols or reddish soil sediments was examined. These include Samples 200 and 207, Omda paleosols developed on the Upper Ineiba unit in Wadi Or; No. 121, a red paleosol developed on Low Terrace I at New Shaturma; No. 154, a deep, red soil on the Dihmit stage gravels near Arminna Temple; No. 162, a (B) C-horizon on Nubian Sandstone above Abu Simbel; Nos. 259 and 260, colluvial deposits of late Pleistocene age incorporating red soil wash in bedrock depressions northwest of Arminna Station; and No. 172, a colluvial soil collected in swales on the Allaqi stage gravels near Adindan East. In comparison with the nilotic sediments, the diffractograms of this group (Fig. D–5) are fairly heterogeneous, reflecting differences in age, original weathering intensity, parent material, subsequent mechanical eluviation of clays, and, possibly, derivation from older beds or soils.

Montmorillonite peaks of irregular and asymmetric aspect are generally found in the 17.5–18.4 A range. Collapse of this peak to about 10 A after heating to 550° C confirms the presence of montmorillonite. The lack of precise definition of this new peak serves to emphasize that this clay fraction is interlayered with other minerals, probably sesquioxides. Whereas most of the samples have fairly-well-developed peaks of moderate to high intensity, Samples 162, 172, and 259 show peaks of low intensity and marked disruption in the 18-A range. This may indicate stronger interlayering, lack of crystallinity, or weathering. By exception, Sample 154 lacks a montmorillonite peak.

All of the samples show a recognizable illite/mica peak in the 9.8–10.4 A range. A lack of sharp definition to these peaks suggests that the mixture is dominated by illite or that the micas have been altered.

Kaolinite is present in all samples, as indicated by peaks at 7.15 A and 3.57 A, which collapse on heating to 550° C. Whereas Samples 154, 162, 200, and 207 show clearly defined and very intense peaks of both orders, the intensity of the remaining samples is lower. In the case of Nos. 259 and 260, it is on the order of two-thirds of the former group: in the case of Nos. 121 and 172, intensities are from one-fifth to one-third those of the first group.

With the exception of Samples 121 and 200, a fairly high percentage of quartz is indicated by peaks in the 3.33–3.35 A range and near 4.26 A. The presence of a little feldspar in all samples except No. 121 is indicated by noticeable but low-intensity peaks in the 3.18–3.24 A range.

Synthesizing this data, the Omda Soil, as recorded in Wadi Or (Fig. D–5), is basically comparable with the contemporary paleosol of Wadi Kharit (Fig. D–4), including intensive kaolinite and moderate montmorillonite peaks. Except for a somewhat lower montmorillonite peak, the red soil developed in bedrock above Abu Simbel is also similar. The deep, Rotlehm-like soil at Arminna Temple stands out because montmorillonite is absent. The red paleosol on the Low Terrace near Kom Ombo is difficult to explain: feldspars and quartz are absent or scarce, while both

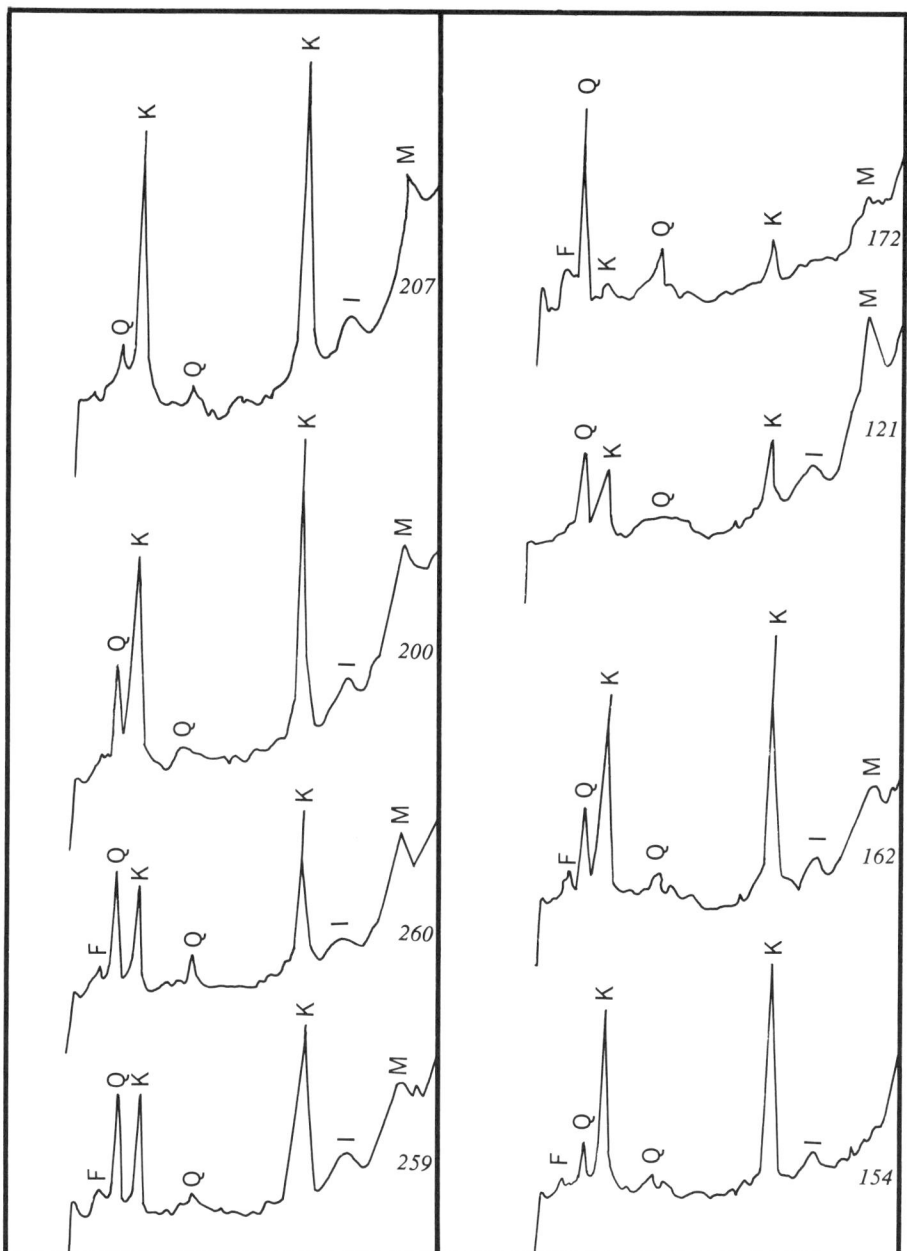

Fig. D–5. X-ray diffractograms from various red paleosols and colluvial soils (Mg-saturated, 25° C). Simplified. The characteristic peaks of the major clay minerals present are identified by letters: M (montmorillonite), I (illite/mica), K (kaolinite), Q (quartz), and F (feldspars). Sample numbers are the same as those used elsewhere in the text. Graph: UW Cartographic Lab.

montmorillonite and kaolinite are abundant. Finally, the colluvial soils are, understandably, a mixture of all available clay components and are correspondingly ill defined.

In general, it would appear that kaolinite is the most characteristic authigenic clay mineral in the Egyptian paleosols. The remarkably poor definition of the montmorillonite peak also indicates that significant interlayering of the montmorillonite with other minerals, such as sesquioxides, is the rule rather than the exception. At least some of the evidence indicates that the kaolinite is developed, in part, from such quantities of feldspars as may be present. All of the red paleosols under consideration are well drained and highly permeable, favoring rapid leaching of metallic cations and other substances produced by weathering, and consequently leaving the soil somewhat depleted in cations and silica. In middle latitudes, where micas are present in abundance, such conditions may favor the formation of beidellite, a variety of montmorillonite (M. L. Jackson, 1959). Depending on the relative abundance of micas, feldspars, and carbonates, clay-mineral formation in Pleistocene Egypt may possibly have tended towards either montmorillonite or kaolinite in the ever-permeable terrestrial soils. Unfortunately, studies of this kind are seriously impeded by the probability of polygenetic weathering in most relict paleosols and by the mechanical eluviation of clays common to soils developed on the Pleistocene gravels.

Kurkur Oasis

Two samples were analyzed from Kurkur: No. 361 from the marls of Tufa IIIa in Northwest Wadi and No. 365–I from Tufa IIIc near North Well (Fig. D–6). Both samples are dominated by very intense, rather-well-defined montmorillonite peaks at approximately 18 A. A slight asymmetry in these peaks suggests a certain amount of interlayering of the montmorillonite. A double peak of moderate intensity in the 9.3–10.7 A range probably indicates an admixture of illite/mica and interlayered montmorillonite. Low-intensity, first- and second-order peaks at about 7.15 A and 3.57 A show the presence of some kaolinite. Quartz is recorded by peaks of moderate intensity near 3.36 A and 4.28 A, and some feldspars are indicated by low-intensity peaks in the 3.18–3.24 A range.

The clay minerals present in these Kurkur samples represent a combination of derived clay impurities from the local limestone bedrock, derived clay minerals from older surficial materials, and some authigenic clays. This would seem to explain the abundance of montmorillonite, as well as some kaolinite. The presence of even minute quantities of feldspars is problematic, however.

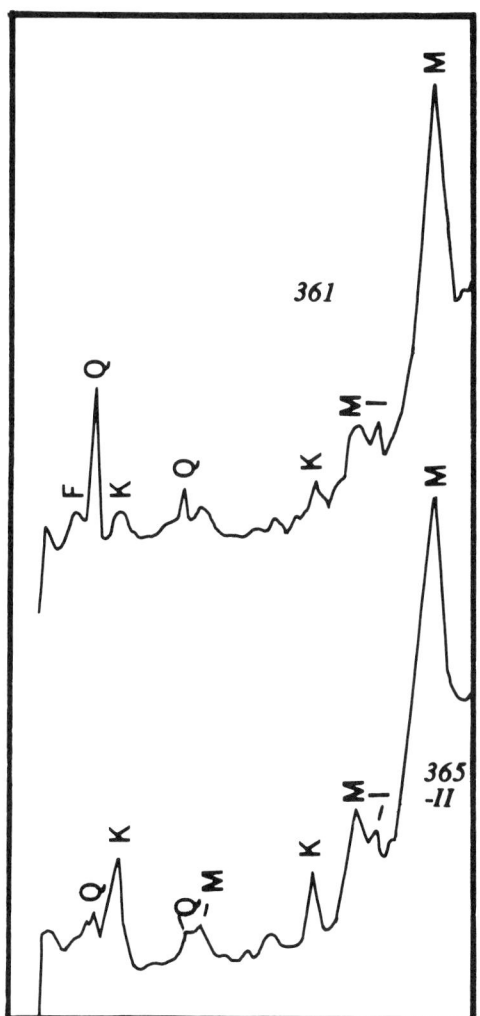

Fig. D–6. X-ray diffractograms from Wadi Tufa III, Kurkur Oasis (Mg-saturated, 25° C). Simplified. The characteristic peaks of the major clay minerals present are identified by letters: M (montmorillonite), I (illite/mica), K (kaolinite), Q (quartz), and F (feldspars). Sample numbers are the same as those used elsewhere in the text. Graph: UW Cartographic Lab.

APPENDIX E

Isotopic Dates

Karl W. Butzer
and Carl L. Hansen

Nine radiocarbon determinations by Isotopes, Inc., Westwood, New Jersey, are relevant to the late Pleistocene chronology of the Nile Valley and the Kurkur Oasis. The samples were collected in 1962–63, submitted in 1966, and processed under the supervision of C. S. Tucek. All of the samples were pretreated with dilute acid to leach superficial carbonates. Bark or charcoal were further treated with sodium hydroxide to remove organic acids and other organic contaminants. The list of dates is as follows:

Isotopes, Inc., Sample Number	$-\delta\ C^{14}$	*Age in years* B.P.	B.C./A.D. *date*
I–2061	966 ± 4	27,200 + 1000 / − 900	25,250 B.C.
I–2060	897 ± 4	18,300 ± 300	16,350 B.C.
I–2178	881 ± 6	17,100 ± 400	15,150 B.C.
I–2179	885 ± 4	17,400 ± 300	15,450 B.C.
I–2567	440 ± 7	4,660 ± 100	2,710 B.C.
I–2561	101 ± 13	860 ± 115	1,090 A.D.
I–2063	> 993	> 39,900	
I–2064	> 993	> 39,900	
I–2062	981 ± 4	31,800 + 1900 / − 1500	29,850 B.C.

Sample I–2061 was a marl from bed *a* at Km 1.53 in Wadi Or, North Branch, corresponding to Sample 188 (Table 6–3). It comes from the lower half of the upper or "classical" member of the Korosko Formation. This is a nilotic deposit, and the carbonates are derived from the subsaharan drainage basin of the Nile. Except in Ethiopia, limestones are absent from the Nile Basin upstream of Nubia. Carbonates are also not present in the Wadi Or drainage.

Sample I–2060 was a dense, clayey marl from bed *c*, Pit 2, at New Korosko (Kom Ombo Plain) (see Fig. 3–3). It comes from a marly lateral subfacies of the uppermost Masmas Formation as recorded in Wadi Shait. Limestones are absent in Wadi Shait and in the Nile Basin upstream of Kom Ombo, so that the carbonates of this nilotic deposit must be derived from subsaharan drainage.

Sample I–2178 was a porous, marly sandy silt from bed *b*, Pit BH, at New Shaturma (Fig. 3–4; No. 14, Table 3–3), intercalated between typical beds of the Korosko and Ineiba Formations. This sediment corresponds very closely to the marly facies of the Masmas Formation in Pit 21 at New Korosko (Chap. 3; No. 72, Table 3–3). The sediment was rich in *Planorbis, Bulinus,* and *Valvata* shells, but these were far too few to provide an adequate radiocarbon sample. Consequently, the shells were run together with the sandy-marly matrix. The resulting value is lower than that for I–2179—which was determined from the overlying Ineiba Formation at another, nearby locality. A glance at the sedimentology of the BH profile (Table 3–3) indicates that the overlying bed is rich in carbonates, making it more than likely that a considerable quantity of younger material has penetrated the rather permeable Masmas sediment. Consequently, the date is more representative of the Ineiba than of the Masmas Formation. Allowing that the age is about 10% too young, this sample would correspond with I–2060.

Sample I–2179 was a compact, silty marl from bed *E*, Pit 36 at New Korosko (see Figs. 3–2 and 3–3). It comes from the lowermost, typical, fine-grained alluvium of the Malki Member but is younger than the basal gravel underlying the Ineiban unit in nearby exposures. These beds are flood silts deposited by Wadi Shait (see Chap. 3), and limestones are absent in the drainage basin upstream of New Korosko.

Samples I–2060, I–2061, I–2178, and I–2179 were all deposited subaqueously in waters containing no "dead" carbonates derived from nearby limestones or other carbonate-rich rocks. In each case, there should have been ample opportunity for equilibrium between free atmospheric carbon and the carbon of the water bodies in question. Any contamination of these three samples would be a result of subsequent carbonate illuviation or re-solution in the soil zone. In fact, all show evidence of recrystallization. Consequently, most of these radiocarbon ages will be too young—for example, I–2060, I–2061, and particularly I–2178.

Sample I–2063 was a typical homogeneous marl from bed *a* of the standard section at Km 0.24 of Northwest Wadi, Kurkur Oasis (Fig. 7–4). It corresponds to Sample 361–I (Table 7–3) and comes from Tufa IIIa. The facies in question is a fluvial, rather than a spring or a lacustrine, deposit.

Sample I–2064 was a typical marl from bed *b*, exposed west of former North Well at the Kurkur Oasis (Figs. 7–5, 7–22, and 7–23). It corresponds to Sample 365–II (Table 7–2) and comes from the type section of Tufa IIIc. The facies suggests a subaqueous deposit.

Sample I–2062 was a partly cemented calcite crust 15 centimeters below the surface at Km 1.61 in North Wadi, Kurkur Oasis (Figs. 7–5, 7–25, and 7–26), and belongs to Tufa IV. The carbonates appear to have impregnated a sandy fluvial deposit *per descensum* during the terminal stages of sedimentation.

Samples I–2062, I–2063, and I–2064 all pertain to the deposits of carbonate-rich streams or ponded waters, largely derived from seepage water or springs emerging from a groundwater reservoir in the local Chalk bedrock. Obviously, a fair proportion of the carbonates were derived from the "dead" limestone, and there was little opportunity to achieve a complete equilibrium with free atmospheric carbon over such short distances and during the brief time that elapsed between emergence from the Chalk and deposition. Consequently, the ages would be too high. Two of the samples are completely "dead," however, implying that there are no fresh carbonates younger than 40,000 years. In other words, the determinations given for I–2063 and I–2064 are valid. In the case of I–2062, there are younger carbonates present. Contamination with "dead" carbonates only rarely leads to dates up to 20% too old. Assuming even such a degree of contamination, however, Sample I–2062 would still give a possible age of about 23,500 B.C. In other words, Tufa IIIc is probably a correlative of the upper Korosko Formation.

Samples I–2567 and I–2561 comprise the noncarbonate materials analyzed by Isotopes, Inc. Both were pretreated with hydrochloric acid as well as with sodium hydroxide.

Sample I–2567 consisted of pieces of *Acacia* bark, embedded in superficial, semiconsolidated sands next to a sunken hearth, some 11 kilometers east of New Arminna, in the bed of Wadi Kharit. The bark probably was contemporary with the hearth (see Chap. 4) and was selected for radiocarbon purposes because the hearth charcoal was powdery and therefore unsuitable for Na OH pretreatment. The age obtained (2850 B.C. by the new half-life value) is very probably relevant both for the period of late prehistoric settlement amply attested to in lower Wadi Kharit and for Member I of the Shaturma Formation.

Sample I–2561 comes from the bed of Wadi Qena, 4 kilometers northeast of Qena town, in upper Egypt. Charcoal was removed from a hearth zone about 60 centimeters below the top of a massive, light-yellowish-brown marl with a noncarbonate residue of clayey silt. This homogeneous, clayey marl, a source for the clay used in the Qena *qulleh* ware, underlies two beds of younger deposits, each about a meter thick. The younger beds are less homogeneous and contain derived potsherds of incised ware and straw-tempered, ridged ware—of late Coptic or early Islamic age—as well as more recent, very coarse ware (kindly identified by George T. Scanlon). The clay source bed exceeds 1.2 meters in thickness and probably accumulated over a period of at least a century or two. It appears to be contemporary with the post-Byzantine phase of alluviation evident in Middle Egypt (Butzer, 1959b: 67 ff.) and may therefore provide an approximate date for Member II of the Shaturma Formation.

Isotopic determinations made for the Yale Expedition by the Yale Natural Radiocarbon Laboratory (Minze Stuiver), the Niedersächsisches Landesamt für Bodenforschung (M. A. Geyh), and the Lamont Geological Observatory (David Thurber) will be reported on by the respective laboratories.

APPENDIX F

Modern and Fossil Marine Mollusca of the Red Sea Littoral at Mersa Alam

Egbert G. Leigh, Jr.

Introduction

The following is an account of some of the associations of marine molluscs living today in the neighborhood of Mersa Alam and of the similar fossil associations to be found along the coast.[1] We seek to infer the environmental settings of the fossil associations from the distribution patterns of the modern ones. The account of the modern associations is based on four days of intensive collecting in late March, 1963, followed by another brief visit in April. The fossil materials were collected by Butzer in December, 1962, and March, 1963.

In many ways, the modern mollusca and reefs at Mersa Alam (25° N) recall those of the Florida Keys (24° 40′–25° 10′ N), and it seems useful to point out the similarities and contrasts between the two areas. The features of molluscan distribution common to the two regions will hardly be the result of accident and will thus provide a far sounder basis for paleoecologic inference than the Red Sea data considered alone. When more is known of the functional significance of shell forms and of the ecological characteristics of the various molluscan taxa, we will be able to discover which features of a molluscan distribution relate to its substratum and environment, and in what way; now, however, we are restricted to clumsier and less direct inferences from comparisons.

The Setting of the Modern Coral Reef

Along most of the coast near Mersa Alam, a coral bench extends out from shore some tens or hundreds of meters to the open reef front, from whence the coral

---

1. The writer is grateful to the departments of molluscs of the National Museum, Washington, and of the Academy of Natural Sciences, Philadelphia, for use of their collections and for help with the identifications. He is also deeply grateful to Charles A. Reed for providing an opportunity to revisit the Mersa Alam area.

surface drops off fairly sharply. The outer margin of the bench is overgrown with the gelatinous, concentrically ridged semicircles of a blue-green alga, probably *Rivularia*. Much of the bench is often exposed at low tide, or nearly so. Tidal range is slight, usually on the order of 60 centimeters. Behind the shore, remains of Pleistocene reefs rise several meters as a series of raised reef platforms, and this juxtaposition of modern and fossil reef is the source of the peculiar interest of this locality. The situation is similar to that in the upper Florida Keys, themselves the remains of a Pleistocene reef which now forms the Key Largo Limestone. In Florida, however, the modern reef stands several miles offshore, and it is almost entirely subtidal and mostly dead: although the Key Largo limestone forms a fossil coral platform on the seaward side of Key Largo, most of the flats near shore are not coral bench at all, but turtlegrass flats or expanses of nubbly coral growths like *Porites*. The tidal range in the Keys is roughly twice as great as that at Mersa Alam.

The bench or platform surface is commonly pitted with potholes and fissures, sometimes opening into catacombs extending laterally some distance beneath the bench surface. These openings shelter octopi and the spectacular fishes characteristic of tropical reefs everywhere, and the irregularities of the rock surfaces harbor a variety of sea urchins. Stormy seas also cast up occasional rocks onto the seaward portions of the bench surface, and between storms these rocks and boulders have the opportunity to build up quite a fauna. They harbor sea urchins, a variety of encrusting organisms, and an occasional eel. One often finds turtlegrass patches where sand has accumulated in the bench hollows; other sandy depressions will hide a small ray. The watermark itself will sometimes be formed by a beach, sometimes by a nip or small cliff cut into fossil coral rock, and sometimes by broken coral rock.

In the immediate neighborhood of Mersa Alam, indentations in the shore have resulted in the formation of two lagoons between reef and watermark, and these lagoons provide a variety of habitats not found on the open bench. The first of these lagoons is situated at the settlement of Mersa Alam itself. Its northern shore, where the land bends back away from the reef front, is quite steep and gravelly. After the shoreline turns southward and parallel to the reef front, the shore slopes less and less steeply, and one finds more and more mud mixed with the gravel. As one passes farther south along the lagoon shore, one finds occasional patches of turtlegrass or expanses of coral bench, and areas where the watermark is formed by fluvial conglomerate and fossil reef, providing an unusual number of small tide pools. Soon the coral bench becomes continuous and extends further and further seaward until it merges with the outer reef. Along this part of the shore, the bench is perhaps a kilometer wide: the watermark is a sandy beach which slopes to a gravelly, grassy bottom, soon merging into a great smooth expanse of white bench extending to the reef front. The outer reef front is overgrown with the usual *"Rivularia"* and other, fuzzier algal growths, but, along the lagoonal margins of the bench, the *"Rivularia"* is lacking, although here the bench is greenish and fuzzy with other algal growths for some distance back.

About two kilometers south of the Mersa Alam embayment, one encounters a second lagoon, separated from the reef by a sandbar which bears a turtlegrass meadow on its crest and a rich population of large sea anemones, pink and white, on its steep landward slope: the sandbar itself seems to be merely an extension of the beach which forms the watermark north of this lagoon. Behind the promontory, the lagoon extends back to a flat of soft, brownish mud. Along the shoreward (western) side of the lagoon, a rock-lined coast confronts a muddy bottom. Here one finds quite a number of oysters and a few mussels. Further south is an area where the sea has carved a low cliff in the fossil reef, and this somewhat sheltered rocky coast bears a peculiar fauna of its own. Finally, the coast and reef front rejoin, and one encounters the usual conformation of a shore confronted by perhaps 50 meters of coral bench. I collected most of the lagoonal habitats I could wade into at low tide and some of the surrounding bench. In the next section, I will describe the results of these collections.

The Modern Fauna

We shall now consider the associations to be found in the various habitats, beginning from the north. The open reef front north of the first lagoon is characterized by a great growth of "*Rivularia*" and by a number of coral fragments scattered over the bench, most of which are also grown over. Here one finds *Vasum turbinellum* and the large *Trochus dentatus* out on the bench, along with a quantity of the large *Conus sumatrensis* and *Conus flavidus*, a very few of the small *Conus coronalis*, and more of the similar *Conus fulgetrum*. One also finds *Turbo radiatus* here—larger ones out in the open and small brown ones among the growths in crevices. One finds a number of the very small *Conus papillosus* among the "*Rivularia*," and a few of the larger, heavier *Conus taeniatus*. Underneath the more overgrown rocks, one finds the collumbellid *Engina mendicaria*, the thaid *Nassa francolina*, and *Nerita albicella*; more occasionally, one finds *Clanculus pharaonis* or the mussel *Modiolus moduloides* and, most rarely of all, the orange *Cypraea carneola*.

On the northernmost large segment of coral bench in the first lagoon, there was a curious row of rocks which, when turned over, yielded a wealth of forms, including *Trochus erythraeus*, small *Turbo radiatus*, an unidentified *Astraea*, *Engina mendicaria*, a number of *Stomatella*, and a single *Cypraea nebrites*, along with the usual array of *Nerita albicella* and assorted *Drupa* characteristic of such places.

South of this segment of coral bench, the bottom becomes a mixture of darkbrown mud and gravel, and the watermark is formed by a fluvial conglomerate which sometimes merges further down into fossil coral bench. One finds *Nerita albicella, Nerita polita*, a few *Nerita undata*, some *Planaxis sulcata*, and the *Monodonta* characteristic of the region. At low tide, pools form in the depressions of the fossil bench. Here one finds *Nerita albicella; Cerithium dorsuosum; Cerithium caeruleum;* occasional limpets (*Cellana rota*) and chitons; *Conus taeniatus; Nassarius* sp.; more occasionally, a *Morula fiscella* or *Polinices melanostoma* (the latter were usually found fully extended and quite limp, for some reason); and, very rarely indeed, a small *Natica marochiensis*. This association is somewhat reminiscent of associations one can find in rocky intertidal places in the Florida Keys (but not further north), where three species of nerite and various periwinkles live higher up, and chitons, a limpet, Cerithiums, and occasionally a *Natica* quite similar to *marochiensis* live lower down. For some reason, a single limpet seems to be characteristic of such places: very wave-beaten coasts, such as those of California, harbor many more species because the limpet form takes the waves so well, but, even in the

Table F–1. Modern marine mollusca of the Mersa Alam Littoral

Haliotidae
  Haliotis varia L.

Trochidae
  Trochus dentatus Forskal
  T. erythraeus Brocchi
  Clanculus pharaonis L.
  Monodonta sp.

Patellidae
  Cellana rota Gmelin

Turbinidae
  Turbo radiatus Lamarck
  Astraea sp.

Littorinidae
  Nodilittorina sp.
  Melaraphe sp.

Neritidae
  Nerita polita L.
  N. albicella L.
  N. undata L.

Cerithiidae
  Cerithium dorsuosum Menke
  C. nodulosum Bruguiere
  C. caeruleum Sowerby

Hipponicidae
  Sabia conica Schumacher

Cypraeidae
  Cypraea carneola L.
  C. isabella L.
  C. grayana Schilder
  C. nebrites Melvill

Lambidae
  Lambis truncata sebae Kiener

Strombidae
  Strombus tricornis Lamarck
  S. mutabilis Swainson
  S. erythrinus Dillwyn
  S. fasciatus Gmelin
  S. gibberulus alba Morch

Naticidae
  Natica marochiensis Gmelin
  Polinices mamilla L.
  P. melanostoma Gmelin

Bursidae
  Bursa granifera Roding

Muricidae
  Drupa morum Roding
  D. ricina L.
  Morula triangulata Pease
  M. uva Roding
  M. granulata Duclos
  M. fiscella Gmelin
  Nassa francolinus Bruguiere
  Thais aculatea savignyi

Columbellidae
  Engina mendicaria L.

Melongenidae
  Melongena pyrum Gmelin

Nassariidae
  Nassarius pullus L.
  Nassarius sp.

Vasidae
  Vasum turbinellum L.

Mitridae
  Strigatella litterata Lamarck

Conidae
  Conus textile L.
  C. sumatrensis Hwass
  C. virgo L.
  C. arenatus Bruguiere
  C. fulgetrum Sowerby
  C. papillosus Kiener
  C. coronalis Roding
  C. rattus Hwass
  C. sanguinolenta Quoy and Gaimard
  C. flavidus Lamarck
  C. nussatella L.
  C. monachus L.
  C. tesselatus Bornemann
  C. taeniatus Hwass

Turridae
  Turris sp.

Planaxidae
  Planaxis sulcatus Bornemann

Pyramidellidae
  Otopleura sp.

Arcidae
  Barbatia sp.
  Arca sp.

Mytilidae
  Modiolus moduloides Roding

Pinnidae
  Pinna muricata L.

(*Table F-1*, continued)

Ostreidae	Tellinidae
Ostrea sp.	*Arcopagia rugosa* Bornemann
Lucinidae	Stomatellidae
Codakia tigerina L.	*Stomatella* sp.
Chamidae	Tridacnidae
Chama limbula Lamarck	*Tridacna maxima* Roding
Veneridae	
Gafrarium lentiginosum Gray	Mactridae
Periglypta puerpera L.	*Atactoides glabrata* Gmelin

associations we are presently concerned with, the ordinary coiled form does not seem to be the best for all gastropod niches.

I collected the mud and gravel bottom below the watermark once at dusk as the tide was going out. On the exposed but wet surface, I found a vast quantity of *Strombus fasciatus*, *Strombus gibberulus*, and *Melongena pyrum*, all quite active; two active species of *Nassarius*; *Cerithium dorsuosum*; and *Cerithium caeruleum*. These *Strombus* have a very well-developed foot musculature, so that they are able to use their somewhat long and narrow operculum as a claw to fight with. The operculum of *Strombus gibberulus* has a serrate edge, which suggests that selection has developed it as a fighting instrument, perhaps as an aid in securing prey. I rarely found either *Strombus* or *Melongena* quite so active again and could not determine the cycle of their activity. Similarly, one could find *Conus textile* here occasionally in the early morning, but not at other times. Once I also found a *Strombus tricornis* here.

There are occasional sandy, grassy patches on this bottom, which exhibit higher concentrations of *Nassarius* and *Strombus fasciatus;* a quantity of the small *Strombus mutabilis;* an occasional *Cerithium nodulosum;* and, buried so that only its posterior margin is exposed, an occasional *Pinna muricata*. A *Pinna* lives in similar settings in the Florida Keys.

That part of the bench fronting the lagoon provided the richest fauna. On the higher parts of the bench margin (and for many yards back from the actual edge), one would find *Vasum turbinellum*, occurring fairly commonly in large groups; occasional smaller *Strombus;* and, sometimes in the early morning, a *Conus textile* (one of which was caught eating a *Strombus gibberulus*). Certain of the lower parts of the bench margin bore a very well-developed, small-cone fauna: a great many *Conus fulgetrum* (tens to the square foot); rarer *coronalis* and *papillosus;* and, towards the very edge, a number of *Chama limbula* and an occasional *Tridacna* or juvenile *Lambis*. In other areas, nearer the outer reef and often with a heavier growth of larger algae, one finds a number of larger cones, mostly *Conus flavidus*, but with some *sumatrensis* and a rare *virgo*—and with these *Turbo radiatus*. On the edges of potholes and fissures, one finds occasional *Morula fiscella* and *Cerithium nodulosum* and, more frequently, *Trochus dentatus* and *Turbo radiatus*. In the sand floors of such potholes, one sometimes finds *Conus arenatus;* more rarely, *Conus textile;* more rarely still, *Conus nussatellus;* and, at night, small but active *Nassarius*.

If the sand floor has been grown over by turtlegrass, one finds *Strombus mutabilis* and perhaps *fasciatus*, *Nassarius pullus*, sometimes a *Pinna*, and more rarely a *Bursa granifera*. As one passes out to the actual reef front, the large cone–large *Turbo* association persists, *Conus textile* becomes more common (and can be found at more times of day), and *Trochus dentatus* appears more frequently on the edges of potholes. Along the reef front, there are occasional, scattered coral fragments up to half a meter long and well overgrown. When turned over, these yield *Nassa francolina*; *Nerita albicella*; *Engina mendicaria*; *Conus papillosus*; often *Clanculus pharaonis*; occasionally a *Barbatia*; and, much more rarely, a *Cypraea carneola* or *nebrites*.

Passing back from the reef front south of the first lagoon by the most direct route to dry land, one encounters first a rather monotonous flat expanse of white bench which bears *Strombus gibberulus* in some abundance; an occasional *Melongena pyrum*; a quantity of the *Thais aculatea savignyi*; *Morula uva*; *Morula granifera*; occasional *Drupa ricina*; and *Nerita albicella*. A *Thais* was once found here drilling a *Nerita albicella*.

Nearer shore, this clean bench merges into a seemingly dirtier substratum supporting some turtlegrass. Here *Strombus fasciatus* is more common than *Strombus gibberulus*, and one finds *Cerithium caeruleum* in abundance, as well as some *Melongena pyrum* and *Nassarius*. The turtlegrass flat leads in turn into a sandy beach where one still finds *Melongena pyrum*; at least occasional specimens of the clam *Atactoides glabrata* buried in the sand (the shells of this clam were quite common on the beach); and, very rarely, a *Polinices mamilla* half buried in the sand.

This network of bench associations would perhaps be the most unfamiliar to the Floridian. Individual elements of the fauna would make him feel at home. He is, after all, familiar with rays and strangely colored fishes, octopi, and warm-water sea urchins; he knows a *Vasum*, which occurs in the sandier portions of the bench off Key Largo, and sometimes finds *Melongena* at the foot of sandy beaches; he will recognize *Pinna* and *Barbatia* as occurring where they did at home and also some of the Thaids. But the larger grazers, *Trochus* and *Turbo*, and the array of large carnivorous cones will be new to him. Nor will he find analogues of the larger Floridian predators, *Fasciolaria* and *Pleuroploca*, which depend on the agility of their muscular feet rather than on a poisonous bite for capturing their food. Yet the two faunas have some curious points in common. Both Florida and the Red Sea support large and active carnivorous snails in the intertidal zone, whereas in the cooler waters of New Jersey or California one finds only small Muricid drills, or an occasional *Polinices* which as often as not functions as a drill itself. In New Jersey, a large whelk occasionally lumbers into the intertidal zone, but such animals are not notoriously agile. Moreover, the masses of mussels and barnacles of the temperate intertidal are not to be found in either the Florida Keys or the Red Sea: grazing seems to be favored over filter-feeding among the mollusca, and even *Tridacna* sees fit to grow symbiotic algae.

The turtlegrass bar bounding the second lagoon harbored a number of *Strombus tricornis*, which somewhat resembles the *Strombus raninus* characteristic of similar habitats in Florida. With *tricornis*, one finds *Strombus fasciatus*, *Strombus mutabilis*, and, rarely, a *Strombus erythrinus*. Here one also finds *Cerithium nodulosum* and assorted mitridae never identified.

Beyond this point, the collections were sketchier, and we mention only the high points. That part of the lagoonal shore where the modern bench converged with a low cliff cut into fossil coral rock bore *Nodilittorina pyramidalis* and *Melaraphe* spp. One of the two species of *Melaraphe* resembled *Littorina ziczac* of the Florida Keys quite closely. Here one also finds, a bit lower down, *Nerita albicella*, *Nerita undata*, and a *Monodonta*. Similar associations occur in similar habitats in Florida, only one finds no *Monodonta* there.

The collections on the open bench south of the second lagoon were made under windy conditions, and most things were under rocks. Under large rocks, one would find *Barbatia*; *Nassa Francolina*; *Nerita albicella*; *Engina mendicaria*; sometimes *Conus papillosus*; more rarely *Bursa granifera*, *Drupa ricina*, or *Drupa morum*; more rarely still a cowry—*Cypraea grayana* or *Cypraea carneola*. Sometimes one also found an *Arca*, reminiscent in both form and ecology of the *Arca umbonata* of the Florida Keys. There was also a great profusion of hermit crabs under these rocks, and the shells they bring in must distort fossil associations.

About 5.5 kilometers south of Mersa Alam, there is a gap in the reef at the mouth of a wadi, cradling a beach of lime sand. There was a huge quantity of large *Lambis* on the neighboring bench. This is somewhat unusual, because, like the similar-sized *Strombus gigas* of the Caribbean, they are taken for food, and they are only common in out-of-the-way places where they have not yet been fished. For this reason, present distribution patterns are no key to the past. Potholes in the bench often cradled a *Tridacna maxima* which, when rooted out, was often found to bear a small abalone, *Haliotis varia*, on its dorsal surface. Under rocks, one found *Barbatia*, the above-mentioned *Arca*, and, occasionally, a small *Modiolus*, along with the usual assemblage characteristic of such places. On the bench, one could find *Turbo radiatus*, *Trochus dentatus*, *Conus sumatrensis*, and an occasional *Conus textile*. The sand beach yielded occasional *Codakia tigerina* and a *Periglypta puerpera*.

There were associations of importance I never found alive. In some places, a great many *Gafrarium lentiginosum* had accumulated on the beach, often with *Arcopagia rugosum*. I even found occasional live ones there. Likewise, there were places where the beach was thick with *Modiolus* shells, sometimes with an occasional *Tonna* or *Dolium* mixed in. These presumably lived in off-shore sands or other such places that a wader could not reach.

The Fossil Mollusca

Recent work on the fossil mollusca of the Red Sea coast has been confined to the 1962 University of Khartoum expedition to the Sudanese Red Sea littoral, and Berry, Whiteman, and Bell (1966) have provisionally identified thirteen species of gastropods and eleven pelecypods from coral reefs dated from the last interglacial transgression(s) of the Pleistocene. Newton (1900) reported on several earlier and rather small collections of Pleistocene mollusca from the Quseir area, and W. J. Arkell (1928) made more extensive observations there. Arkell noticed that species which could be found in the Quseir fossil reefs but which were not living there now were always alive somewhere else in the Indopacific, and he concluded that differences in species composition of successive fossil strata were quite as likely to be caused by migrations and changes of distribution patterns as by evolutionary modifications. It has only been much more recently that the study of distribution patterns and their changes has become a major tool in the elucidation of conditions in fossil environments.

No information has yet been forthcoming from the southerly parts of the Egyptian Red Sea coast.

The individual collections made near Mersa Alam in 1962–63 are listed in stratigraphic and geographic order below. Species or genera not encountered in the modern molluscan associations are marked (x).

A) Coral Reefs and Gravel Beaches at + 5 to + 6 m.

1) *Mersa Sifein, 500 m south of wadi.* Fossil coral reef to + 5.1 m.
   Turbo radiatus Lamarck
   Cypraea erosa L. (or *nebrites* Melvill)
   Lambis sp.[2]
   Strombus gibberulus alba Morch
   S. fasciatus Gmelin
   Polinices melanostoma Gmelin
   Nassarius pullus L.
   Nassarius sp.
   (x) *Conus striatus*
   C. virgo (?) L.
   (x) *C. achatinus* Chemnitz
   C. papillosus Kiener
   Trachycardium sp.
   Barbatia sp.
   Chama sp. (large)
   Arcopagia rugosa Bornemann
   Tridacna maxima Roding [= *T. elongata* Lamarck]

2) *Km 2.5 south of Mersa Alam.* Fossil coral reef to + 4.6 m.
   (x) *Cypraea* sp. (*arabica* complex)
   Conus sumatrensis Hwass
   C. textile L.
   (x) *C. achatinus* Chemnitz

3) *Km 5.5 south of Mersa Alam.* Fossil coral reef to + 5.3 m.
   (x) *Cypraea* cf. *arabica* L.
   (x) *C. erosa* L.
   (x) *Colubraria maculosa* Gmelin
   Conus monachus L.
   Conus sp.
   (x) *Trachycardium leucostomum* Bornemann
   Periglypta puerpera L.
   Tridacna maxima Röding

---

2. All of the *Lambis* sp. listed pertain either to the common spider conch *L. lambis* L. or to the giant spider conch *L. truncata*.

4) *Km 5.5 south of Mersa Alam.* Beach gravels to + 4.5 m.
   - (x) *Cypraea lynx* L.
   - *Strombus tricornis* Lamarck
   - (x) *Trachycardium leucostomum* Bornemann
   - *Trachycardium* sp.
   - *Ostraea* sp.
   - (x) *Gafrarium divaricatum* Gmelin
   - *Dosinia* sp.
   - (x) *Anadara scapha* L.
   - *Cyrenoida* sp.

B) Gravel Beaches at + 6 to + 8.5 m.

1) *Mersa Sifein, 500 m south of wadi.* Base of estuarine gravels to + 8 m.
   - *Nerita albicella* L.
   - *Cypraea erosa* L. (or *nebrites* Melvill)
   - *Strombus fasciatus* Gmelin
   - *Trachycardium leucostomum* Bornemann
   - (x) *Anadara scapha* L.
   - *Atactoides glabrata* Gmelin
   - *Tridacna maxima* Röding
   - *Dosinia* sp.

2) *Mersa Sifein, 200 m south of wadi.* Upper estuarine gravels, to + 6 m.
   - *Cypraea* sp. (of *arabica* complex)
   - *C. erosa* L. (or *nebrites* Melvill)
   - *Lambis* sp.
   - *Vasum turbinellus* L.
   - *Strombus fasciatus* Gmelin
   - *S. gibberulus alba* Morch
   - *Conus flavidus* Lamarck (or *virgo* L.)
   - *C. textile* L.
   - *Trachycardium* sp.
   - *Anadara scapha* L.
   - *Dosinia* sp.

C) Coral Reefs at + 3.5 m.

1) *Mersa Sifein, 50 m south of embouchure.* Fossil coral reef to + 3.2 m.
   - *Cellana rota* Gmelin
   - *Cypraea* cf. *annulus* L.
   - *Lambis* sp.
   - *Strombus fasciatus* Gmelin
   - *S. gibberulus alba* Morch
   - *Conus arenatus* Bruguiere
   - (x) *Asaphis deflorata* L.
   - (x) *Anadara scapha* L.
   - *Barbata virescens* Reeve
   - *Codakia tigerina* L.
   - (x) *Gafrarium divaricatum* Gmelin
   - *Arcopagia rugosa* Bornemann
   - *Atactoides glabrata* Gmelin

D) Estuarine Beds at High-Tide Mark.

1) *Mersa Sifein, 50 m south of embouchure.* Estuarine conglomerate to + 0.3 m.
   Trochus dentatus Forskal
   Gafrarium lentiginosum Gray

E) Deposits of Unknown Age (Late Pleistocene or Holocene)

1) *Km 2.5 south of Mersa Alam.* Beach conglomerate at modern sea level.
   Lambis sp.
   Strombus fasciatus Gmelin
   S. gibberulus alba Morch
   Cypraea erosa L. (or nebrites Melvill)
   Polinices mamilla L. (?)
   Trachycardium sp.
   Ostraea sp.
   Tridacna maxima Röding

Interpretation of the Fossil Collections

Altogether, forty-two species or genera were identified from the fossil collections, compared with seventy-four identified in the modern associations of Table F–1. Of these, some thirteen species or genera, i.e., 31% of the Pleistocene mollusca, are apparently absent or at least are rather uncommon in the Mersa Alam area today. All of these species are present elsewhere along the Red Sea coasts, however. Nonetheless, it is interesting that some differences are apparent, even though these cannot yet be explained in ecological terms. It is probably significant that the more recent beach fauna of collection E.1 is most similar to the modern associations.

Turning to the individual collections, the fauna of the 5-meter coral reef near Mersa Sifein (A.1) is a rather typical platform assemblage. To a lesser extent, the same is probably true of the same reef at Km 5.5 south of Mersa Alam (A.3). The fauna of the 4.5-meter reef at Km 2.5, such as it is (A.2), suggests a reef crest or an outer platform assemblage. More difficult to interpret is the fauna of the 3.2-meter reef at Mersa Sifein (C.1), which includes both beach clams and bench gastropods.

By comparison, the fauna of the estuarine beds at Mersa Sifein (B.1 and B.2) suggests a more protected environment and more tranquil waters than the nearby 5-meter reef. The whole association of the 4.5-meter beach gravels at Km 5.5 (A.4) suggests a typical beach accumulation washed together from further out and embedded within the shore gravel.

The beach conglomerate at modern sea level found near Km 2.5, and believed to be of more recent age, poses a problem. The fauna was collected from beds deposited at the base of a well-developed nip, at a locale where the fringing reef now converges with the shoreline. Yet the fauna is typical of sheltered waters, possibly the shore side of a lagoon or an inshore section of the coral platform where the bench is very wide.

APPENDIX G

Fossil Mollusca from the Kom Ombo Plain

Egbert G. Leigh, Jr.
and Karl W. Butzer

Mollusca were collected from late Pleistocene nilotic and wadi deposits of the Kom Ombo Plain by various members of the Yale Expedition during 1962–63. No systematic study was carried out, however, and the lists and comments presented here are incomplete and make no claim to authoritative identification. A fairly extensive collection of shells made by Leigh was kindly identified by J. P. E. Morrison. This list was used to verify or supplement lists of tentative field identifications made by either Leigh or Butzer on the basis of plates published by Gardner (1932a, see also Sandford and Arkell, 1933: Plate 43).[1]

The following inventory of late Pleistocene mollusca can be presented for the Kom Ombo Plain (Table G–1), in general employing the nomenclature of Gardner (1932a). In the taxonomy of the Corbiculae, we have followed Llabador (1962: 265–66, 268–69) and other recent authors who believe that all the Corbiculae of Africa and southwestern Asia are identical on the specific level. Thus, *Corbicula vara*, *Corbicula artini*, and *Corbicula consobrina* would each only be subspecies of *Corbicula fluminalis* Müller. Distinguishing the Lymnaeae on the specific level proved very difficult, since only broken, incomplete shells are preserved.

With the exception of *Cleopatra bulimoides*, all of the freshwater gastropods of Table G–1 are today primarily associated with muddy habitats or with stagnant or slowly moving waters such as canals, pools, or marshy tracts (Gardner, 1932a; Harris, 1965). By contrast, *Cleopatra* as well as the suite of freshwater clams are largely confined to the channels of larger streams with clear or more rapidly moving waters. These ecological preferences are amply borne out in the provenance of the fossil shells, with the one group almost exclusively extracted from flood silts or semilacustrine deposits and the other from channel sands or levee beds. All of the freshwater gastropods, other than the Lymnaeae, are abundant in Egypt, as is the

---

1. We are deeply indebted to J. P. E. Morrison of the Smithsonian Institution, Washington, for identifying a large part of the mollusca and for valuable discussion. We appreciate equally the collaboration of Charles A. Reed.

clam *Caelatura nilotica* and, to a lesser extent, the Nile-oyster *Etheria* (see Gardner, 1932a). *Corbicula* is also abundant, although the subspecies *vara* is extinct (Gardner, 1932a: 107). The Unionidae, however, are absent from contemporary Egypt. *Lymnaea* is now restricted to Europe and southwestern Asia (Gardner, 1932a: 10), and its presence in late Pleistocene Egypt may reflect a cooler climate.

Neither of the terrestrial snails, *Zootecus* and *Pupoides*, appear to be living in Egypt today. As in the case of the Kurkur Oasis (see Appendix H), they would suggest wetter conditions than at present, despite the fact that they are desert forms.

The distribution of the various mollusca obtained *in situ* from the sequence of late Pleistocene deposits is given in the following lists. A qualitative scale of "abundant" (a), "common" (c), and "rare" (r) has been applied to the individual sites in the case of the Gebel Silsila Formation, or to the different species or genera in the case of the other stratigraphic units. Because of the difficulties of identification, *Unio*, *Corbicula*, and *Lymnaea* are not distinguished on the specific level.

A) The Korosko Formation

Corbicula ssp. (r). New Korosko, Pit 21; New Shaturma, Pit 46.

Lymnaea sp. (r). New Korosko, Pit 21; New Shaturma, Pit 52; New Sebua, Pits 1, 2.

Planorbis ehrenbergi (a). New Korosko, Pits 16, 18, 21, H; New Shaturma, Pit 52; New Sebua, Pits 1, 2.

Bulinus truncatus (c). New Korosko, Pits 16, 21, H; New Shaturma, Pits 46, 52; New Sebua, Pits 1, 2.

Table G–1. Late Pleistocene mollusca from the Kom Ombo Plain

Freshwater Pelecypods

    Unio (Potamida) willcocksi Bullen Newton (= *U. vignardi* Pallary)
    Unio (Potamida) sp.
    Caelatura (Reneus) nilotica Cailliaud
    Etheria elliptica Lamarck
    Corbicula fluminalis vara Gardner
    C. fluminalis artini Pallary
    C. fuminalis consobrina Cailliaud

Freshwater Gastropods

    Lymnaea (Radix) stagnalis L.
    Lymnaea (Radix) pesegra Muller (?)
    Lymnaea (Stagnicola) palustris Muller (?)
    Planorbis (Gyraulus) ehrenbergi Beck
    Bulinus truncatus Audouin
    Valvata nilotica Jickeli
    Cleopatra bulimoides Olivier

Terrestrial Gastropods

    Zootecus insularis Ehrenberg
    Pupoides coenopictus Hutton (?)
    P. sennaariensis Pfeiffer

Valvata nilotica (a). New Korosko, Pits 16, 18, 21; New Shaturma, Pits 46, 52, 56; New Sebua, Pits 1, 2.

B) The Masmas Formation

Unio sp. (r). 2 km north of New Ibrim (stratigraphy uncertain).

Corbicula ssp. (r). 2 km north of New Ibrim (stratigraphy uncertain); New Korosko, Pit 21.

Lymnaea sp. (r). New Korosko, Pits 10, 12(?), 21; New Wadi el-Arab.

Planorbis ehrenbergi (a). Darau South; New Masmas; el-Nuqu; El-Nasser; New Korosko, Pits 2, 7, 10, 21, 22, 24, E; New Shaturma, Pit BH; New Wadi el-Arab.

Bulinus truncatus (c). Darau South; New Masmas; El-Nasser; New Korosko, Pit 21; New Shaturma, Pit BH; New Wadi el-Arab.

Valvata nilotica (a). Darau South; New Masmas; el-Nuqu; El-Nasser; New Korosko, Pits 2, 7, 10, 21, 22, 24, E; New Shaturma, Pit BH; New Wadi el-Arab.

Cleopatra bulimoides (r). 2 km north of New Ibrim.

Zootecus insularis (r). New Masmas.

Pupoides sennaariensis (r). New Masmas.

C) The Gebel Silsila Formation

Unio sp. Sebil (a); Fatira (c); Gebel Silsila 2A (c); Gebel Silsila 2B (a); Silsila Gap (a); Khor el-Sil (c).

Etheria elliptica. Darau South quarries (r); Dar es-Salaam (r).

Corbicula ssp. New Ballana I (a); Sebil (c); Fatira (c); Gebel Silsila 2B (c); Khor el-Sil (r); 1.5 km northwest of New Abu Simbel (c).

Cleopatra bulimoides. Sidi Hamuda (c); Darau South quarries (c); Dar es-Salaam (c); Sebil (r); Gebel Silsila 2A (a); Gebel Silsila 2B (r); Silsila Gap (c); Khor el-Sil (c).

Unidentified pulmonate gastropod, Khor el-Sil (c).

D) Sinqari Member, Ineiba Formation

Planorbis ehrenbergi (derived). New Masmas (r); New Shaturma, Pit 56 (r).

Bulinus truncatus (derived). New Masmas (r).

Valvata nilotica (derived). New Masmas (r).

Zootecus insularis (a). 3 km northeast of New Qustul; New Masmas; Wadi Ellawi; New Korosko, Pits 12, 36; New Shaturma, Pits 53, 56; New Sebua, Pits K, P; 9 km east of New Sebua, Wadi Shait terrace; New Wadi el-Arab.

Pupoides coenopictus (?) (r). 3 km northeast of New Qustul.

Pupoides sennaariensis (r). 3 km northeast of New Qustul; New Masmas; Wadi Ellawi; New Korosko, Pits 12, 36; New Shaturma, Pit 56; 9 km east of New Sebua, Wadi Shait terrace.

APPENDIX H

Fossil Mollusca from the Kurkur Oasis

Egbert G. Leigh, Jr.

Fossil mollusca from the Kurkur Oasis were first mentioned by Ball (1902), who identified *Pupa* sp. from calcareous tufas near the oasis. During the field work in 1963, I collected a number of terrestrial and freshwater snails from the various Pleistocene deposits (Chap. 7).[1] These fossils can be listed in stratigraphic sequence as follows:

A) Tufa IIIa (exposures in Northwest Wadi and nearby Central Wadi)
Melanoides tuberculata Müller

B) Tufa IIIb (exposures in Northwest Wadi and nearby Central Wadi)
Pupoides coenopictus Hutton
Melanoides tuberculata Müller

C) Tufa IIIc (lacustrine marl in Central Wadi)

Freshwater molluscs
Biomphalaria (Planorbis) boissyi Potiez and Michaud
Gyraulus (Planorbis) ehrenbergi Beck
Segmentina angusta Jickeli
Bulinus truncatus Audouin
Radix (Lymnaea) natalensis caillaudi Bourguignat

Terrestrial molluscs
Lamellaxis gracilis Hutton (?)

D) Tufa IV (slope wash and pan silts, partly of uncertain age, in North and Central Wadis)
Zootecus (Pupa) insularis Ehrenberg
Pupoides coenopictus Hutton

---

1. I am indebted to J. P. E. Morrison of the Smithsonian Institution, Washington, for the identifications and a valuable discussion. Permission to reprint this appendix, with minor changes, from *The Canadian Geographer*, Vol. 8 (1964), pp. 138–39 is also gratefully acknowledged.

The fauna of Tufa IV seems to indicate a rather dry environment, but one definitely wetter than that of the present day in Kurkur. According to Gardner (1935), *Zootecus insularis* is a typically desert snail, ranging from North Africa to Arabia and India, although it is at present very uncommon or absent in Egypt (see also Reed in Wendt, 1966). The presence of *Pupoides* is consistent with this interpretation, and the absence of other forms of land snail in this deposit supports it.

The freshwater fauna of the Tufa IIIc (lacustrine marl) is appropriate to still water containing decaying vegetation. According to Gardner (1935), *Segmentina angusta* is found on decaying reed stems, and the *Gyraulus* also is characteristic of decaying vegetation. None of the other species present are inconsistent with this interpretation. The general aspect of this fauna bears a striking resemblance to a fauna in a small, nearly stagnant cove of Lake Carnegie, Princeton, New Jersey, which the writer has studied. This fauna contained a large planorbid, a small *Gyraulus* which was found on the stems of decaying vegetation, and the *Physa* which seems to play the ecological role in North America that *Bulinus* does in Egypt; the cove was characterized by still water and a quantity of decaying vegetation.

The *Melanoides* of the Tufa III a/b complex do not permit any ecological conclusions of importance, although the fact that this species is parthenogenetic is of interest.

APPENDIX I

Fossil Pollen from Late Tertiary and Middle Pleistocene Deposits of the Kurkur Oasis

Madeleine Van Campo, Philippe Guinet, and Jacqueline Cohen

Palynological study of several samples from late Tertiary and middle Pleistocene deposits of Kurkur, collected by Karl W. Butzer and Carl L. Hansen in 1963, has provided phytogeographical information of considerable interest. Of a total of six samples, two from the Plateau Tufa in False Wadi (Nos. U–14 and U–15) were too poor in pollen for systematic study, while one sample of Wadi Tufa I sediments in False Wadi (No. U–17) remains to be studied. The provenance of the three samples reported on here is as follows:

No. U–8. *Plateau Tufa* (upper member) from Central Wadi, east of North Well (see Fig. 7–22), at 341 meters elevation. Collected from travertine facies at the base of a 7- to 15-meter-thick column of Plateau Tufa.

No. U–4. *Plateau Tufa* (upper member) from False Wadi, south of the main channel of Wadi Abu Gorma (see Fig. 7–14). Collected from organic tufa facies in the middle of a 10-meter-thick column of the upper member, at 371 meters elevation. Although facies correlations from False Wadi to Central Wadi are not secure, Sample U–4 appears to be somewhat younger in the Plateau Tufa aggradation phase than Sample U–8.

No. U–16. *Wadi Tufa I* from False Wadi (see Fig. 7–14), travertine facies. 328 meters elevation.

Madeleine Van Campo is Director of Research, Centre National de la Recherche Scientifique; Director, Laboratory of Palynology, École Pratique des Hautes Études; and Editor and founder of *Pollen et Spores:* Muséum National d'Histoire Naturelle, Paris. Philippe Guinet is Assistant Director, Laboratory of Palynology, École Pratique des Hautes Études, and Jacqueline Cohen is Assistant, Laboratory of Palynology, École Pratique des Hautes Études.

The results of these pollen investigations are given in Tables I–1 and I–2. For a number of reasons, we believe that the sediments and their pollen are of local origin. So, for example, available studies of nilotic alluvium, be it at the mouth of the Nile or even in secondary deposition along the coast of Israel (see Rossignol, 1961, 1962), always show a large number of fern spores of equatorial origin. By contrast, the percentage of spores in the Kurkur pollen is only about 0.5%. This conclusion is confirmed by the geomorphological evidence. All of the deposits are of local origin, derived from a fairly restricted watershed near and west of the modern oasis (see Chap. 7). Furthermore, there are no high mountains in the general area, from which montane pollen could be derived. Instead, the problem of how accurately the fossil pollen reflects local vegetation—contemporary with these particular sediments—revolves around (a) possible derivation from older sediments at Kurkur and (b) possible long-distance wind transport from the southern margins of the Sahara or from the Mediterranean littoral.

The question of derivation from older sediments can be readily answered, at least in the case of the Plateau Tufa samples. The youngest strata of the Kurkur region which antedate the Plateau Tufa are the marine Nummulitic Limestones of early Eocene age. The spectra recorded are all wholly incompatible with such an origin, so that we may be confident that the Plateau Tufa pollen were not derived from older deposits. In the case of Wadi Tufa I, there is, of course, a possibility that some of the pollen grains were derived from the Plateau Tufa. Nonetheless, the spectrum of U–16 is distinctive, compared with Samples U–8 and U–4, and, at the same time, the internal composition of the pollen appears quite logical.

Long-distance transport by wind can hardly have played a role in derivation of the "Tertiary boreal" and "montane mediterranean" elements present, i.e., of those elements of greatest interest in the spectra. The Mediterranean littoral is over 850 kilometers away, with Cyrenaica, Crete, Greece, and Anatolia providing the nearest highlands within the quadrant of prevailing northerly to northwesterly winds. Yet these highlands are 1,400 to 1,500 kilometers distant at their closest points. Instead, it would seem that the prevailing winds from the heart of the Libyan Desert should have accentuated the xerophytic elements of the local pollen rain.

For these reasons, the pollen spectra of Tables I–1 and I–2 may be considered to reflect the contemporary local floras of Kurkur in a significant way. The percentages calculated for Table I–1 cannot be rigorously applied, since only rarely were pollen grains determined down to the specific level. As a result, many types were only identified at the generic or even the family level. Yet, even though the distribution of percentages between the mediterranean and xerophytic classes may be criticized in its details, such criticism is unwarranted in view of the great numbers of pollen counted.

It would have been more significant to subdivide the group of tropical and subtropical elements (item 12 of Tables I–1, I–2) into two ecological groups: the xerophytes and the mesohygrophytes. This was impeded, however, by the difficulty of identifying the pollen grains, in certain cases, even down to the generic level. Had this been possible, it would have permitted a more precise assessment of the Nile Valley as a corridor for exchange of tropical and mediterranean floras. A large part of the mesohygrophytic tropical flora has persisted in the Nile Valley to this day, demonstrating the permanence of such a corridor of floral exchange. On the slopes of the Hoggar Mountains in the western Sahara, on the other hand, the mesohygrophytic tropical floras have not maintained the role of interaction they once played before the hydrography became intermittent.

If we compare the spectra of U–8, U–4 (both Pliocene), and U–16 (mid-Pleistocene), it is immediately apparent that the vegetation contemporary with deposition of the Plateau Tufa at Kurkur was subject to considerable variation. By inference, climate must have fluctuated considerably during the long period of time represented by the Plateau Tufa (Chap. 7). Sample U–8 offers certain analogies to the percentages of the major pollen groups known from early Pleistocene deposits of the Hoggar (see Van Campo et al., 1964, 1965, 1966; Rognon, 1967) and those more recently discovered in terminal Tertiary sediments of Tibesti (Van Campo, unpublished). It has been suggested that a relict Tertiary flora of boreal character—in part still found on the mountain ranges of the eastern Mediterranean Basin today—was present in parts of Africa at the beginning of the Pleistocene: so, for example, in the Hoggar Mountains (22° N) at 2300 meters elevation (Van Campo et al., 1964). Sample U–8, from comparatively low elevations at Kurkur (24° N), with 16% of its pollen from this phytogeographical group and 22.5% from among the lowland Mediterranean elements, lends strong support to this hypothesis.

Sample U–4 would suggest a rather severe decimation of the boreal and mediterranean arboreal elements at Kurkur, with a corresponding increase of xerophytic elements from 18% to 42.5%. This information serves as a reminder that late Tertiary environments of the Sahara were not uniformly moist but, rather, were also subject to significant change.

The mid-Pleistocene sample, U–16, from Wadi Tufa I, coming from an identical travertine facies as Sample U–8, suggests that the local vegetation included rather more xerophytic elements during the height of the middle Pleistocene pluvial phases than during the moister phases of Plateau Tufa aggradation. Xerophytic elements account for 53%, boreal or montane mediterranean elements for 4%, lowland mediterranean elements for only 9%. This indicates a permanent decimation of the Tertiary relict flora during the course of the Pleistocene. It further suggests that the Pleistocene pluvials were more modest than those of the late Tertiary, in corroboration of the geomorphological evidence (see Chap. 7).

Table I–1. Fossil pollen spectra from the Kurkur Oasis

Pollen classes	Plateau Tufa (No. U–8)		Plateau Tufa (No. U–4)		Wadi Tufa I (No. U–16)	
	Pollen	%	Pollen	%	Pollen	%
Tertiary boreal element and montane elements of eastern Mediterranean region	37	16	10	1	7	4
Subtropical and tropical xerophytic or mesohygrophytic elements	19	7.5	232	29	9	6
Mediterranean elements	51	22.5	61	8	14	9
Xerophytic and halophytic elements	44	18	342	43	79	53
Gramineae	83	34	133	16.5	25	17
Hydrophytic elements	5	2	20	2.5	17	11
Total	*239*	*100*	*798*	*100*	*151*	*100*

Table I-2. List of fossil pollen from the Kurkur Oasis

	U-8	U-4	U-16
Tertiary boreal elements and montane elements of the eastern Mediterranean region			
cf. *Carya*	2	1	—
Alnus	9	3	1
Corylus	8	—	—
Salix	—	2	1
Ostrya	3	—	—
cf. *Betula*	3	—	—
Tilia	2	—	—
Aesculus	1	—	—
Carpinus	3	—	—
cf. *Quercus*	1	4	2
Platanus	1	—	—
cf. Ulmaceae	4	—	3
Total	*37*	*10*	±*7*
Subtropical and tropical xerophytic and mesohygrophytic elements			
Podocarpus cf. *montana*	1	—	—
Ericaceae (nonmediterranean)	2	1	2
cf. *Aphania* (?) (Sapindaceae)	1	—	—
Loranthoideae	1	—	—
Phyllantus	2	—	—
Combretaceae	—	1	—
Melastomaceae	—	1	—
Ilex	—	1	—
Celastrales	—	3	—
Sapotaceae	—	2	—
Hymenocardia	—	3	1
Phoenix (including *Chamaerops*)	—	80	3
cf. *Phoenix* spp.	9	—	—
cf. *Hyphaene*	1	45[a]	—
Acacia (*etbaica–raddiana* group)	—	87	2
Capparidaceae	1	8	1
Ficus	1	—	—
Total	*19*	*232*	*9*
Mediterranean elements			
Pinus (including *P. halepensis*)	20	12	5
Cupressaceae	—	9	1
Caryophyllaceae	8	6	2
Urticaceae	6	4	—
Olea (?)	1	3	2
Oleaceae	—	11	—
Rhamnaceae	3	1	1
Cistaceae	1	1	1
Umbelliferae	2	9	—
Pistacia	3	—	—
Celtis	5	1	—
Callitris	2	—	—

(Table I–2, continued)

	U–8	U–4	U–16
Ranunculaceae	—	3	1
Euphorbia cf. *amygdaloides*	—	1	—
Labiatae	—	—	1
Total	*51*	*61*	*14*
Xerophytic and halophytic elements			
Compositae (including *Artemisia*)	30	76	29
Chenopodiaceae	9	124	27
Amaranthaceae	—	1	—
Cruciferae	2	6	4
Leguminosae	2	9	13
Resedaceae	1	11	—
Tamaricaceae	—	7	—
Trichodesma	—	1	—
Echium	—	—	1
Zygophyllum	—	89	2
Fagonia	—	1	—
Zygophyllaceae	—	14	2
Ephedra	—	3	—
Total	*44*	*342*	*79*
Gramineae			
Total	*83*	*133*	*25*
Hydrophytic elements			
Total	*5*	*20*	*17*
Varia (excluded from percentages in Table I–1)			
Campanulaceae	—	1	—
Plantago	—	1	—
Rumex	2	1	—
Rubiaceae	—	1	—
Euphorbiaceae	—	—	3
Rosaceae	1	1	—
Unidentified	(?)	95[b]	23[c]

[a] 2 species.
[b] Including 17 species.
[c] Including 6 species.

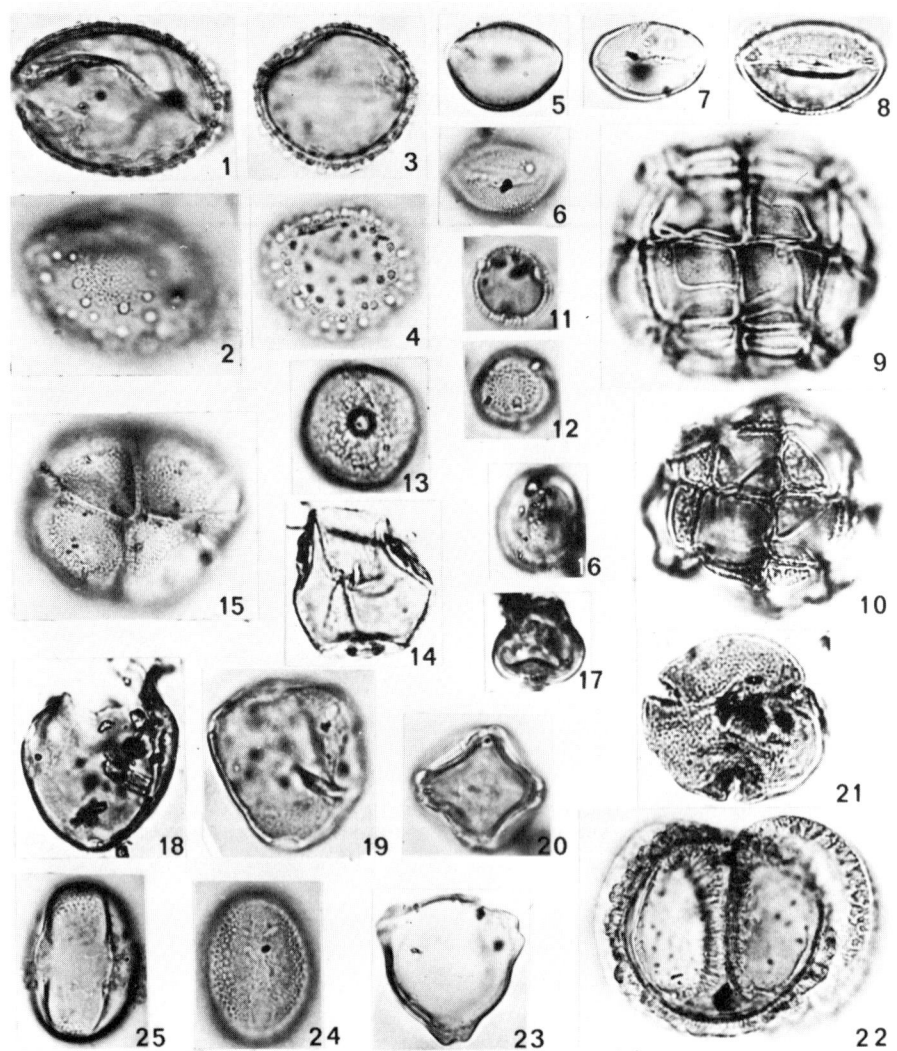

Fig. I–1. Some fossil pollen grains from the Plateau Tufa, Kurkur Oasis. *1, 2:* cf. *Hyphaene* (Palmae), type 1; *3, 4:* cf. *Hyphaene*, type 2; *5, 6:* cf. *Phoenix* (Palmae), type 1; *7:* cf. *Phoenix*, type 2; *8:* cf. *Phoenix*, type 3; *9: Acacia* (Mimosaceae), *etbaica* group; *10: Acacia, raddiana* group; *11: Phyllantus* sp. (Euphorbiaceae); *12: Phyllantus* sp.; *13: Hymenocardia* sp. (Euphorbiaceae); *14: Hymenocardia* sp. (pollen poorly preserved); *15: Typha* sp. (Typhaceae); *16, 17: Aesculus* sp. (Hippocastanaceae) (pore barely visible on photograph); *18: Carya* sp. (Juglandaceae); *19: Carya* sp.; *20: Alnus* sp. (Betulaceae); *21: Tilia* sp. (Tiliaceae); *22: Pinus* cf. *halepensis; 23: Carpinus* cf. *orientalis* (Betulaceae); *24, 25: Fagonia* sp. (Zygophyllaceae).

APPENDIX J

Pliocene Ingression into the Nile Valley According to New Data

I. S. Chumakov
translated by Richard G. Klein

Until recently, marine Pliocene deposits in the Nile Valley had only been established in Lower Egypt. Many investigators have described marine-littoral deposits (limestones, sandy marls, sandstones, and conglomerates with abundant macrofauna and microfauna) of Calabrian age from exposures in Wadi Natrun, near Giza, in the Helwan area, and near Beni-Suef. Higher up in the geological column, these beds pass over into continental, gritty gravels (Villafranchian), found to elevations of 160 to 180 meters. The base of the Upper Pliocene deposits has been established by drill core at Cairo to be near 300 meters below sea level. According to data of Said (1962), sandy marls of the Middle Pliocene rest beneath these deposits. Thus, the total thickness of Upper Pliocene beds in Lower Egypt approaches 500 meters.

In the course of geological investigation in the High Dam area at Aswan (Upper Egypt), Pliocene deposits of marine and of continental facies were uncovered by drill core and by mining operations and were studied in detail. Three distinct series can be recognized in the Pliocene profile here.

The Lower Series (Marine-Estuarine) fills the deepest parts of the buried valley of the Proto-Nile, occurring from absolute elevations of -172 meters to -35 meters. The beds consist of gray, montmorillonitic clays with thin lenses of fine-grained, polymictic sands and sandy loams, rich in vegetative detritus. The persistantly uniform suite of authigenic minerals includes a considerable amount of glauconite, zeolite (of the stilbite group), pyrite, and siderite. The sands are partially cemented by secondary calcite. Rare ostracod remains belong to the genera *Cypridea, Cyprinotus, Limnocythere, Eucypris,* and *Candoniella,* the majority of

---

This Appendix is a translation of an article from Byulletin' Moskovskogo obshchestva ispetalel'ej prirody, Geological Section, Vol. 4 (1965), pp. 111–12. The detailed sections are described by Chumakov (1967).

which are widespread in brackish sediments of the Ponto-Caspian Miocene and Pliocene.

The Middle Series (Transitional Facies) has only been established near the mouths of east-bank wadis (Khor Umm Buweirat, Khor Kundi), where Pliocene deposits were preserved from early Pleistocene erosion and occur at absolute elevations of 7 to 116 meters. The deposits consist of polymictic sands and sandy loams mixed with grit and gravel and interbedded with clay lenticles. The sands and clays of the lower part of this unit are gray to yellowish-gray, and, in contrast to the Lower Series, zeolite is completely absent; glauconite is rare, and vegetative detritus and pyrite are only abundant in certain lenses. The upper part of the Middle Series is characteristically yellow or brown because of omnipresent ferric hydrates, while glauconite, pyrite, and organic matter are completely absent.

The Upper Series (Alluvial) is represented by red gritty gravels found as isolated patches or broad spreads on both banks of the Nile on the highest erosional surface at 160 to 200 meters elevation, the valley bottom of the Paleo-Nile. Quartz is strongly predominant, with comparatively rare, intensively weathered pebbles or cobbles of extrusive igneous rocks. Up to 10 or 15 meters thick, these beds are mainly derived from the crystalline massif of the Central Sudan and to a lesser extent from the mountains of the Eastern Desert.

Thus, the tripartite subdivision of the Pliocene is established in Upper Egypt as it is elsewhere through the whole Mediterranean Basin (Italy, France, Spain, the Maghreb). The Lower Series at Aswan corresponds to the maximum marine transgression (Plaisancian); the Middle Series to the regressive phase, with accumulation in a more or less freshwater basin (Astian); while the Upper Series attests to alluviation in the Nile Valley (Villafranchian). In Lower Egypt, these coarse alluvial beds appear to correspond to two different facies, the marine-littoral Calabrian and the continental Villafranchian. The maximum elevations of the Pliocene marine sediments in Upper Egypt as well as the gradient of the bedrock floor of the buried valley of the Nile above Aswan indicate that the Pliocene sea formed a deep estuary in the Nile Valley, penetrating much further inland than previously supposed and reaching upstream of Wadi Halfa in the Sudan. It is possible to conclude that the narrow Upper Egyptian valley of the Proto-Nile (0.5 to 2 or 3 kilometers wide) was formed in the Upper Miocene, a period of general uplift in Egypt and regression of the sea. At the same time, this valley was cut into the bottom of an older, broad valley of the Paleo-Nile (Oligo/Miocene?) up to 70 or 100 kilometers wide, which had developed in Nubia within the rocks of the Basement Complex, the sandstones of Nubian facies (Carboniferous to Cretaceous?), and the limestones of the Libyan Desert plateau (Upper Cretaceous to Eocene).

Appendix K

Geological Map of
Egyptian Nubia (1:166,000)

Simultaneously with this volume, the publishers are issuing a set of maps in a format larger than is possible within the book itself. These include the ten maps of Egyptian Nubia at 1:41,500 and the following maps from elsewhere in the book:

Fig. 2–5	Geomorphology of the Aswan–Kom Ombo area	1:250,000
Fig. 2–12	Late Tertiary to Middle Pleistocene geology of the Kom Ombo region	1:100,000
Fig. 3–1	Late Pleistocene and Holocene deposits of the Kom Ombo Plain	1:100,000
Fig. 5–4	Geomorphology of Egyptian Nubia	1:500,000
Fig. 7–11	Surficial deposits of the Kurkur Oasis	1:25,000

Offered for sale as a set, but not separately, these maps may be ordered from the publishers while they remain in print.

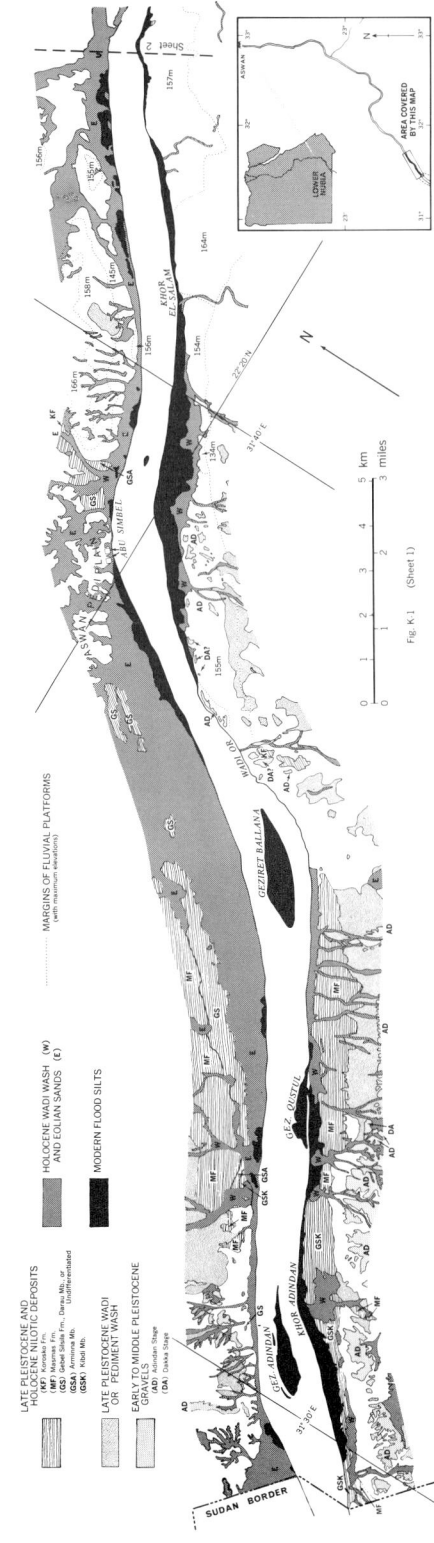

Fig. K-1. Surficial geology of the Nile Valley from Adindan to Abu Simbel. Map: UW Cartographic Lab.

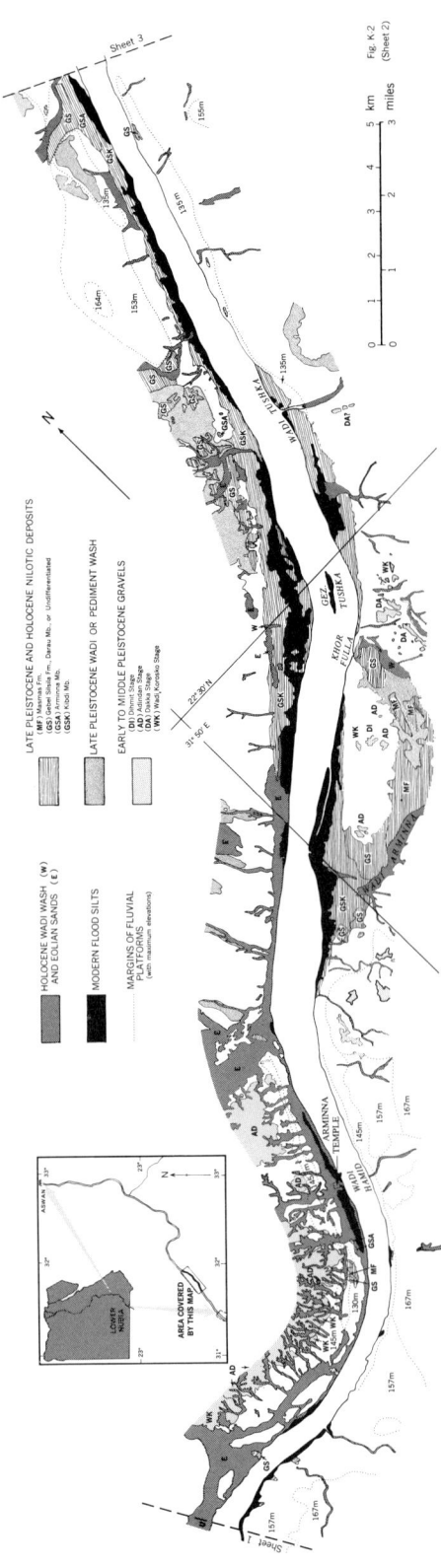

Fig. K-2. Surficial geology of the Nile Valley from Arminna to Tushka. Map: UW Cartographic Lab.

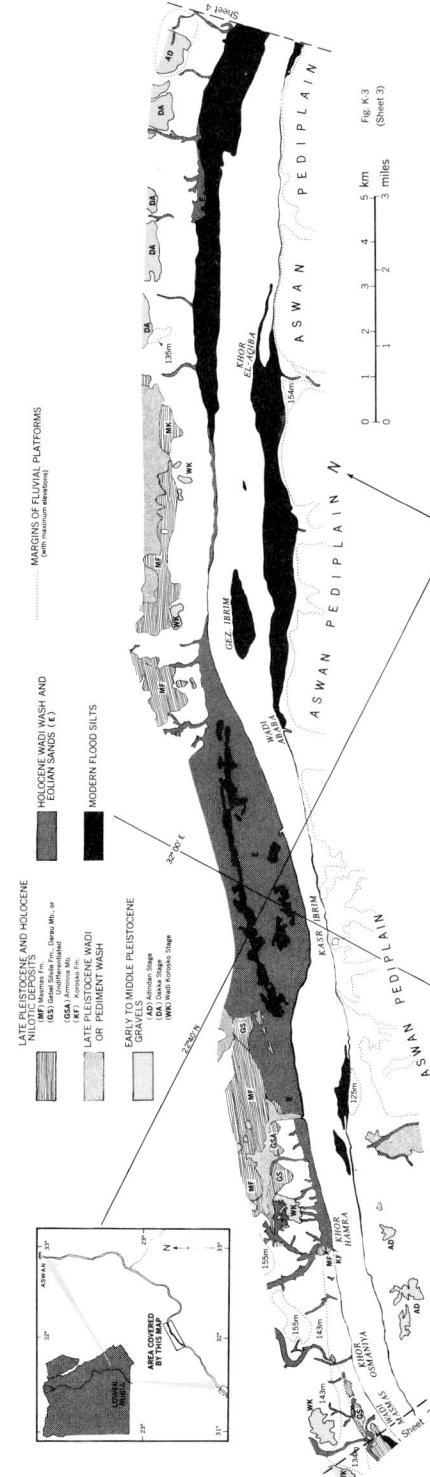

Fig. K-3. Surficial geology of the Nile Valley from Masmas to Qatta. Map: UW Cartographic Lab.

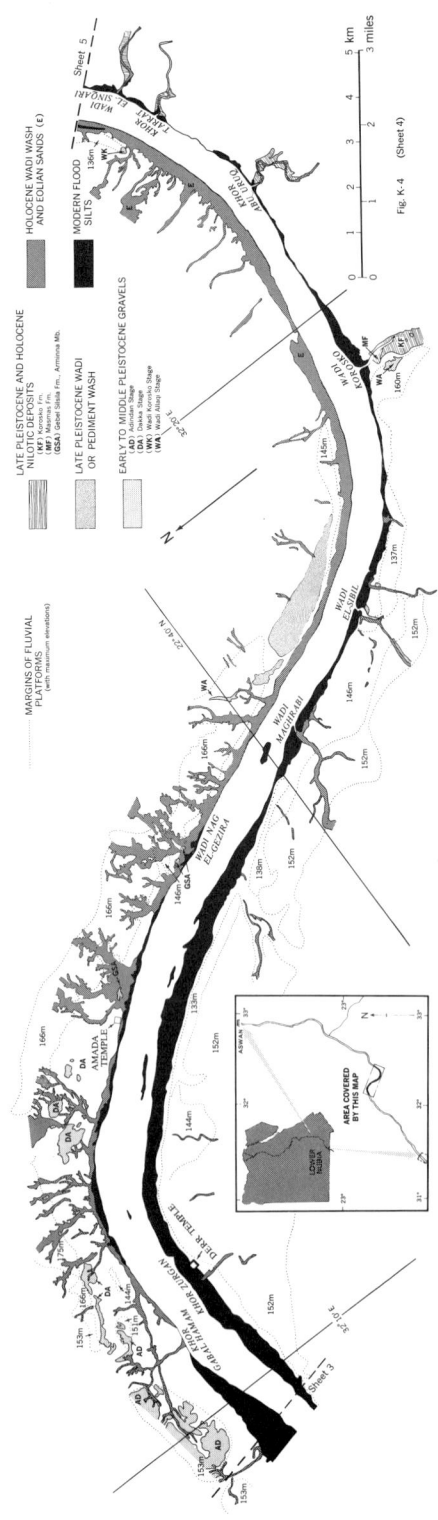

Fig. K-4. Surficial geology of the Nile Valley from Tumas to Sinqari. Map: UW Cartographic Lab.

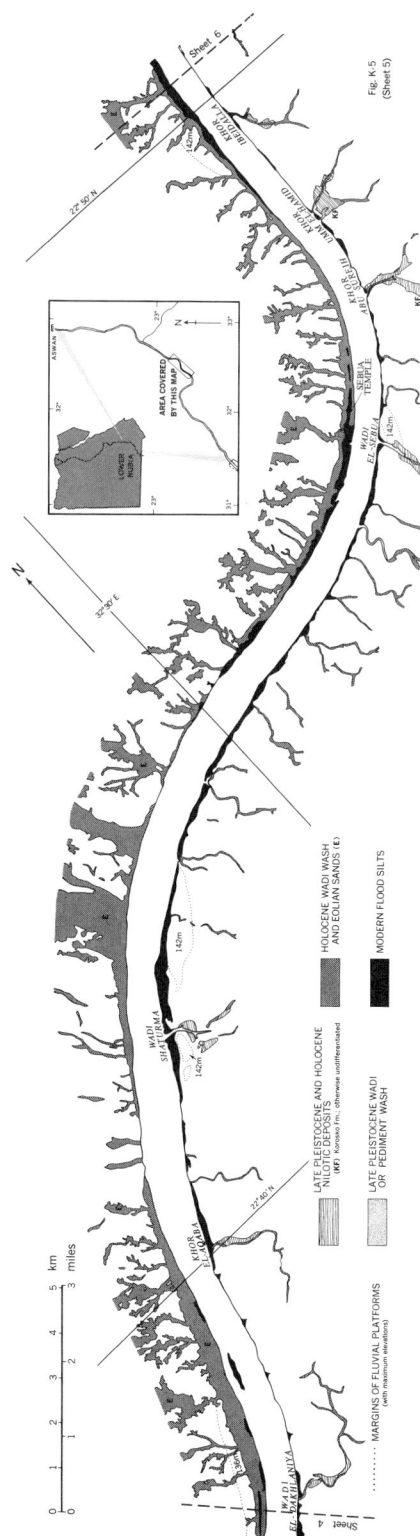

Fig. K-5. Surficial geology of the Nile Valley from Malki to Sebua. Map: UW Cartographic Lab.

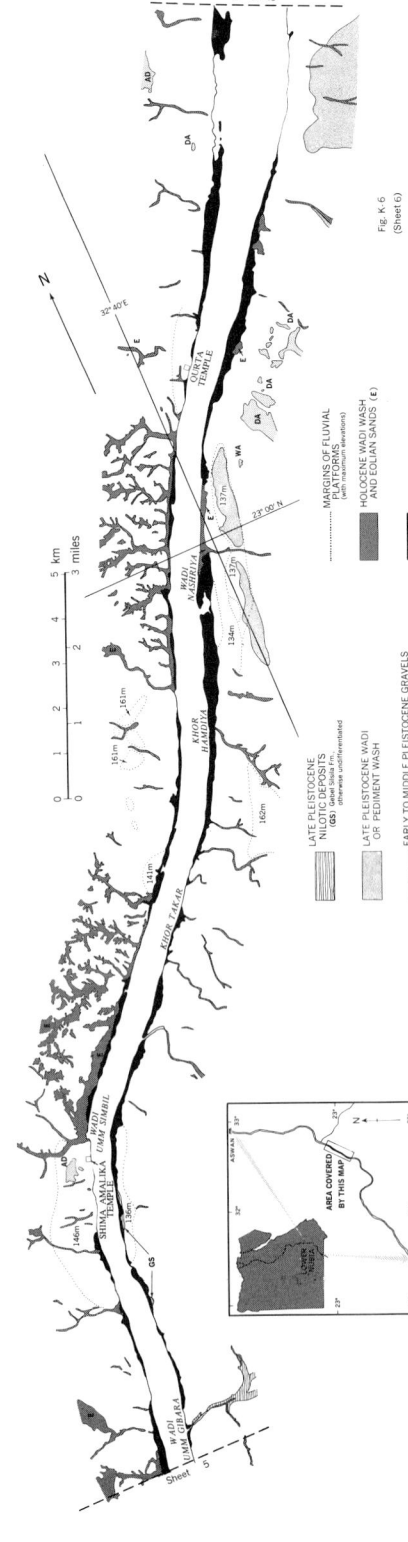

Fig. K-6. Surficial geology of the Nile Valley from Madiq to Qurta. Map: UW Cartographic Lab.

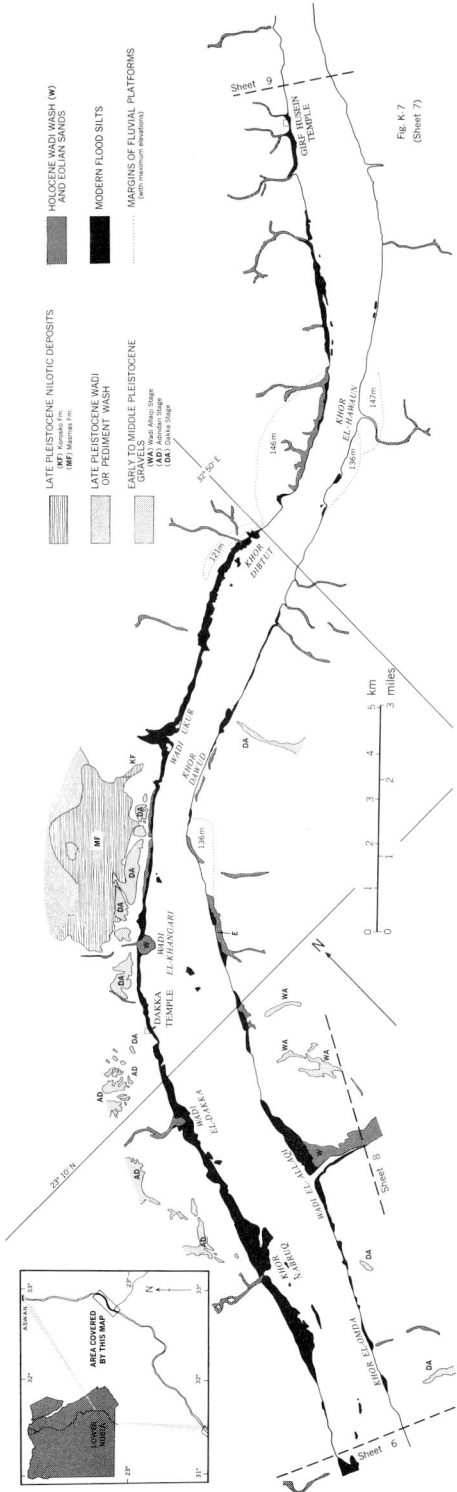

Fig. K-7. Surficial geology of the Nile Valley from Allaqi to Girf Husein. Map: UW Cartographic Lab.

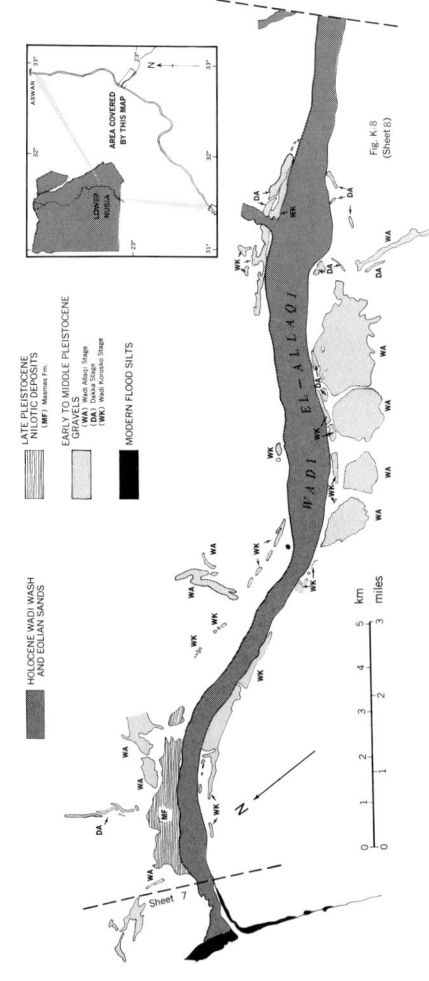

Fig. K–8. Surficial geology of lower Wadi Allaqi. Map: UW Cartographic Lab.

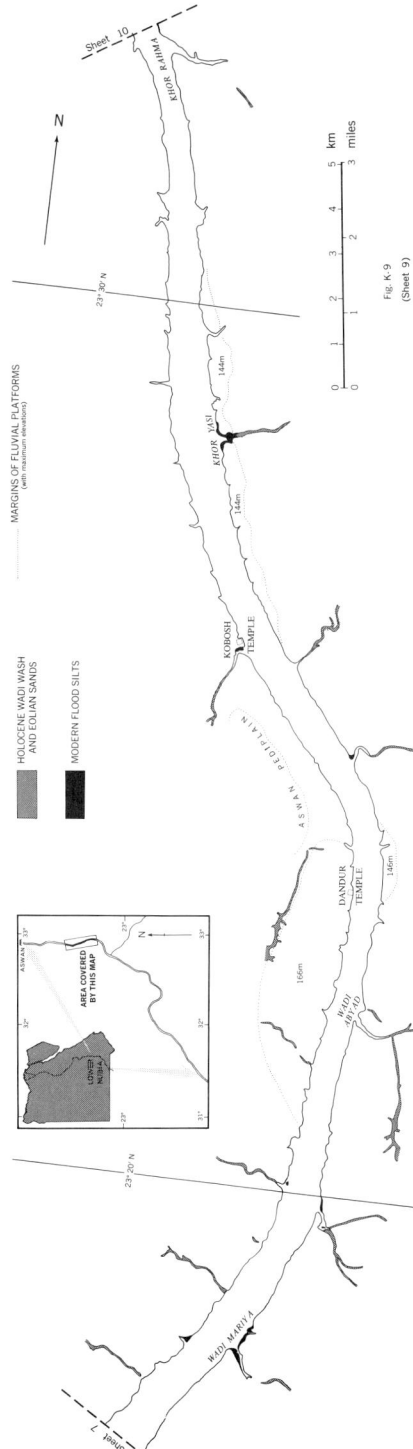

Fig. K–9. Surficial geology of the Nile Valley from Mariya to Abu Hor. Map: UW Cartographic Lab.

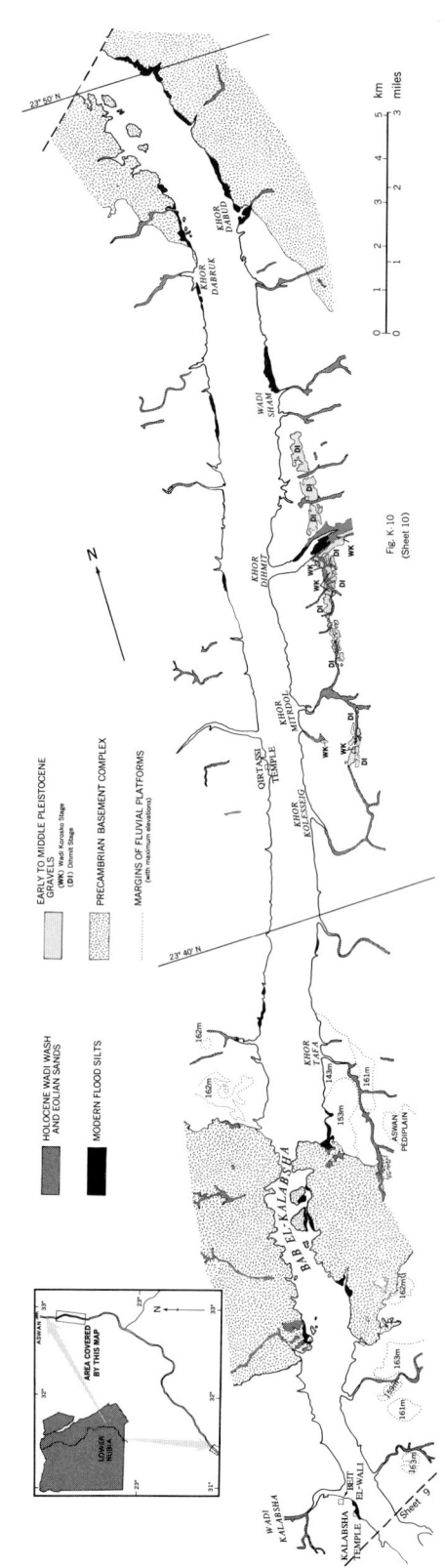

Fig. K-10. Surficial geology of the Nile Valley from Kalabsha to Dabud. Map: UW Cartographic Lab.

Bibliography and Index

Bibliography

Adams, A. L., with S. P. Woodward and H. Falconer. 1864. Notes on the geology of a portion of the Nile Valley north of the Second Cataract in Nubia: *Quarterly J., Geol. Soc. of London* 20: 6–19.

Aguirre, E. de, ed. 1964. *Actas, II. Reunión del Comité del Neógeno Mediterráneo* (Madrid-Sabadell, Sept. 1961). Curs. Confs. Inst. Lucas Mallada Invest. Geol. 9: 318 pp.

Alimen, H., J. Chavaillon, and S. Duplaix. 1964. *Mineraux lourds des sédiments quaternaires du Sahara nord-occidental.* Cent. Nat. Réch. Sci., Paris. 73 pp.

American Society for Testing Materials. 1958. *Procedures*, Part 4, pp. 1119–29.

Amin, M. S. 1955. *Geology and mineral deposits of Umm Rus sheet.* Geol. Surv. Egypt, Government Press, Cairo. 51 pp. and 1:100,000 map of Pre-Cambrian geology.

Andrew, G. 1954. Geology of the Sudan. In J. D. Tothill, ed. *Agriculture in the Sudan.* 2nd ed. Oxford Univ. Press, London. Pp. 84–128.

———, and G. Y. Karkanis. 1945. Stratigraphical notes, Anglo-Egyptian Sudan. *Sudan Notes Rec.* 26: 157–66.

Arambourg, C. 1948. *Mission scientifique de l'Omo 1932–1933.* Mus. Nat. Hist. Natur., Paris. Vol. 1, Fasc. 3, 406 pp. and 40 pl.

Arkell, A. J. 1949a. *The Old Stone Age in the Anglo-Egyptian Sudan.* Sudan Antiq. Serv., Occ. Paper No. 1, 51 pp.

———. 1949b. *Early Khartoum.* Oxford Univ. Press, London. 116 pp.

———. 1953. *Shaheinab.* Oxford Univ. Press, London. 114 pp.

———, and P. J. Ucko. 1965. Review of Predynastic development in the Nile Valley. *Curr. Anthropol.* 6: 145–66.

Arkell, W. J. 1928. Aspects of the ecology of certain fossil coral reefs. *J. Ecol.* 16: 134–49.

Armelagos, G. J. 1964. A fossilized mandible from near Wadi Halfa, Sudan. *Man* 64: 12–13.

———, G. H. Ewing, D. L. Greene, and M. L. Papworth. 1965. Physical anthropology of the prehistoric populations of the Nile Valley. Paper presented at Seventh INQUA Congr., Boulder, Colo. Sept. 1, 1965.

Atlas of Egypt: a series of maps with descriptive text illustrating the orography, geology, meteorology and economic conditions. 1928. Geol. Surv. Egypt, Government Press, Cairo. 31 pl. and 11 pp.

Attia, M. I. 1950. *Geology of the iron ore deposits of Egypt.* Geol. Surv. Egypt, Government Press, Cairo. 34 pp.

———. 1954. *Deposits in the Nile Valley and the Delta.* Geol. Surv. Egypt, Government Press, Cairo. 356 pp.

Attia, M. I. 1955. *Topography, geology and iron-ore deposits of the district east of Aswan.* Geol. Surv. Egypt, Government Press, Cairo. 247 pp.

Ball, J. 1902. *On the topographical and geological results of a reconnaisance-survey of Jebel Garra and the Oasis of Kurkur.* Geol. Surv. Egypt, Government Press, Cairo. 40 pp.

———. 1903. The Semna Cataract or rapid of the Nile: a study in river erosion. *Quarterly J., Geol. Soc. of London* 59: 64–79.

———. 1907. *A description of the First or Aswan Cataract of the Nile.* Geol. Surv. Egypt, Government Press, Cairo. 113 pp.

———. 1939. *Contributions to the geography of Egypt.* Government Press, Cairo. 308 pp.

Barthoux, J., and P. H. Fritel. 1925. Flore crétacée du Grés de Nubie. *Mém., Inst. Égypte* 7: 65–119.

Bate, D. M. A. 1951. The mammals from Singa and Abu Hugar. *Fossil Mammals of Africa*, No. 2 (Brit. Mus. Natur. Hist.), pp. 1–28.

Bauer, E. E. 1958. Recent developments in the hydrometer methods as applied to soils. *A.S.T.M. Spec. Tech. Publ.* 234: 89–97.

Baumgartel, E. J. 1965. Predynastic Egypt. 39 pp. Fasc. 38 in *Cambridge Ancient History*. 2nd ed. Cambridge Univ. Press.

Baumhoff, M. A. 1963. Ecological determinants of aboriginal California populations. *Univ. Calif. Publ. in Amer. Archeol. and Ethnol.* 49: 155–236.

———. 1965. Excavations on the Kom Ombo Plain. Paper presented at Seventh INQUA Congr., Boulder, Colo., Sept. 1, 1965.

Beadnell, H. J. L. 1905. The relations of the Eocene and Cretaceous systems in the Esna-Aswan reach of the Nile Valley. *Quarterly J., Geol. Soc. of London* 61: 667–78.

———. 1924, Report on the geology of the Red Sea coast between Qoseir and Wadi Ranga. *Petrol. Res. Bull.* (Cairo) 13: 37 pp.

Berggren, W. A. 1964. Biostratigraphy of the Paleocene–Lower Eocene of Luxor and nearby Western Desert. In F. A. Reilly, ed. *Guidebook to the Geology and Archaeology of Egypt.* Petrol. Explor. Soc. Libya, Sixth Annu. Field Conf., Tripoli. Pp. 149–76.

———. 1965. Paleocene—a micropaleontologist's point of view. *Bull., Amer. Ass. Petrol. Geol.* 49: 1473–81.

Berry, L. 1961a. Large-scale alluvial islands in the White Nile. *Rev. Géomorphol. Dynam.* 12: 105–9.

———. 1961b. The physical history and development of the White Nile. *Hydrobiol. Res. Unit, Univ. Khartum, Eighth Annu. Rep.* Pp. 14–19.

———, A. J. Whiteman, and S. V. Bell. 1966. Some radiocarbon dates and their geomorphological significance: emerged reef complex of the Sudan. *Z. Geomorphol.* 10: 119–43.

Bietak, M., and R. Engelmayer. 1963. Eine frühdynastische Abrisiedlung mit Felsbildern aus Sayala-Nubien. *Denkschr., Öst. Akad. Wiss.* (Vienna), Phil.-Hist. Kl. 82: 50 pp.

Blanckenhorn, M. 1901. Das Pliocän- und Quartärzeitalter in Aegypten ausschliesslich des Rothen Meergebietes. *Z., Deut. Geol. Ges.* 53: 307–502.

———. 1921. *Aegypten. Handbuch der regionalen Geologie*, Vol. 7, Sect. 9. C. Winters, Heidelberg. 244 pp.

Boinet, A. 1899. *Dictionnaire géographique de l'Égypte*. Government Press, Cairo. 649 pp.

Boulos, L. 1966. A natural history study of Kurkur Oasis. Part IV: The vegetation. *Postilla* (Peabody Mus. Natur. Hist.), No. 100, 22 pp.

British Meteorological Office. 1958. *Tables of temperature, relative humidity and precipitation for the world. Part IV: Africa*. Her Majesty's Stationery Office, London. 220 pp.

Büdel, J. 1952. Bericht über klima-morphologische und Eiszeitforschungen in Niederafrika. *Erdkunde* 6: 104–32.

———. 1954a. Klimamorphologische Arbeiten in Äthiopien im Frühjahr 1953. *Erdkunde* 8: 139–56.

———. 1954b. Sinai, die Wüste der Gesetzesbildung. *Raumforsch. Landesplan.* (*Mortensen–Festschrift*), Abh. 28, pp. 63–85.

———. 1958. Die Flächenbildung in den feuchten Tropen und die Rolle fossiler solcher Flächen in anderen Klimazonen. *Abh., Deut. Geogr.*, Würzburg, 1957. F. Steiner, Wiesbaden. Pp. 89–121.

Buol, S. W. 1965. Present soil-forming factors and processes in arid and semiarid regions. *Soil Sci.* 99: 45–49.

Butzer, K. W. 1959a. Contributions to the Pleistocene geology of the Nile Valley. *Erdkunde* 13: 46–67.

———. 1959b. Die Naturlandschaft Ägyptens während der Vorgeschichte und dem dynastischen Zeitalter. *Abh., Akad. Wiss. Lit.* (Mainz), Math.-Naturw. Kl., 1959, No. 2, 81 pp.

———. 1960a. Archeology and geology in ancient Egypt. *Science* 132: 1617–24.

———. 1960b. Dynamic climatology of large-scale European circulation patterns in the Mediterranean area. *Meteorol. Rdsch.* 13: 97–105.

———. 1960c. On the Pleistocene shorelines of Arabs' Gulf, Egypt. *J. Geol.* 68: 626–37.

———. 1961. Archäologische Fundstellen Mittel- und Oberägyptens in ihrer geologischen Landschaft. *Mitt., Deut. Archäol. Inst., Abt. Kairo* 17: 54–68.

———. 1963a. Climatic-geomorphologic interpretation of Pleistocene sediments in the Eurafrican subtropics. *Viking Fund Publ. in Anthropol.* 36: 1–27.

———. 1963b. The last "pluvial" phase in the Eurafrican subtropics. *Arid Zone Res.* (UNESCO) 20: 211–21.

———. 1964a. Pleistocene palaeoclimates of the Kurkur Oasis. *Can. Geogr.* 8: 125–41. With an appendix on the fossil mollusca by E. G. Leigh.

———. 1964b. *Environment and archeology: an introduction to Pleistocene geography*. Aldine, Chicago, and Methuen, London. 524 pp.

———. 1964c. *Pleistocene geomorphology and stratigraphy of the Costa Brava region, Catalonia. Abh., Akad. Wiss. Lit.* (Mainz), Math.-Naturw. Kl., 1964, No. 1, 51 pp.

———. 1965. Desert landforms at the Kurkur Oasis, Egypt. *Ann., Ass. Amer. Geogr.* 55: 578–91.

———. 1966a. Geologie und Paläogeographie archäologischer Fundstellen bei Sayala (Unternubien). *Denkschr., Öst. Akad. Wiss.*, Phil.-Hist. Kl. 92: 89–98.

———. 1966b. Climatic changes in the arid zones of Africa 8000–0 B.C. In J. S. Sawyer, ed. *World climate 8000–0 B.C.* Roy. Meteorol. Soc., London. Pp. 73–84.

Butzer, K. W. 1967. Late Pleistocene deposits of the Kom Ombo Plain, Upper Egypt. *Alfred Rust–Festschrift, Fundamenta* (Cologne) B/2: 213–27.

———, and J. Cuerda. 1962. Coastal stratigraphy of southern Mallorca and its implications for the Pleistocene chronology of the Mediterranean Sea. *J. Geol.* 70: 398–416.

———, and C. L. Hansen. 1965. On Pleistocene evolution of the Nile Valley in southern Egypt. *Can. Geogr.* 9: 74–83.

———, and ———. 1967. Upper Pleistocene stratigraphy in southern Egypt. In W. W. Bishop and J. D. Clark, eds. *Background to African evolution.* Univ. Chicago Press. Pp. 329–56.

Cailleux, A., and J. Tricart. 1963. *Initiation à l'étude des sables et des galets.* Cent. Doc. Univ., Paris. 2nd ed. Vol. 1. 368 pp.

Caton-Thompson, G. 1946. The Levalloisian industries of Egypt. *Proc., Prehist. Soc.* 12: 57–120.

———. 1952. *The Kharga Oasis in prehistory.* Athlone Press, London. 213 pp. and 128 pl. Geology by E. W. Gardner. Pp. 1–14.

———, and E. W. Gardner. 1932. The prehistoric geography of the Kharga Oasis, *Geogr. J.* 80: 369–409.

———, and ———. 1934. *The Desert Fayum.* Roy. Anthropol. Inst., London. 2 vols. 167 pp. and 114 pl.

———, and ———. 1939. Climate, irrigation and early man in the Hadhramaut. *Geogr. J.* 93: 18–38.

Chavaillon, J. 1964. *Étude stratigraphique des formations quaternaires du Sahara nord-occidental.* Cent. Nat. Réch. Sci., Paris. 393 pp.

Chmielewski, W. 1965. Archaeological research on Pleistocene and Lower Holocene sites in northern Sudan: preliminary results. In F. Wendorf, ed. *Contributions to the prehistory of Nubia.* S. Meth. Univ. Press, Dallas. Pp. 147–64.

Christophe, L. A. 1963. Remarques sur l'économie de la Basse Nubie égyptienne. *Bull., Soc. Géogr. Égypte* 35: 77–128.

Chumakov, I. S. 1965. Pliozenovaya ingressiya v dolinu Nila (po novim dannim). *Byull. mosk. Obschch. Ispȳt. Prir.*, Geol. Sect. 4: 111–12. (English translation included as Appendix J above.)

———. 1967. *Pliozenovaye i Pleistozenovaye otlojeniya dolinu Nila v Nubiya i Verkhnem Egipte. Trudy Akademiya Nauk SSSR* 170: 115 pp.

Collet, L. W. 1926. L'oasis de Kharga dans le Désert Libique. *Ann. Géogr.* 35: 528–34.

Cooke, R. U. 1965. Desert pavement. *Mineral Inf. Serv.* 18: 197–200.

Coque, R. 1962. *La Tunésie présaharienne: étude géomorphologique.* A. Colin, Paris. 476 pp.

Crané, H. R., and J. B. Griffin. 1965. University of Michigan radiocarbon dates X. *Radiocarbon* 7: 123–52.

———, and ———. 1966. University of Michigan radiocarbon dates XI. *Radiocarbon* 8: 256–85.

Cuvillier, J. 1935a. Contribution a la géologie du Gébel Garra et de l'Oasis de Kourkour (Desert Libyque). *Bull. Soc. Roy. Géogr. Égypte.* Vol. 19, No. 1, pp. 127–53. See also geographical record in *Ibid.*, Vol. 18, Nos. 3–4, pp. 34–38.

———. 1935b. Les Kurkurstufe. *Bull., Inst. Égypte* 17: 117–22.

Dainelli, G. 1943. *Geologia dell'Africa Orientale*. Centro Studi per l'Africa Orientale Italiana, Rome. Vol. I. 464 pp. and 1: 2,000,000 geological map.
Degens, E. T. 1963. Geochemische Untersuchungen von Wässern aus der ägyptischen Sahara. *Geol. Rdsch.* 52: 625–39.
De Heinzelin, J. 1963. A tentative paleogeographic map of Neogene Africa. *Viking Fund Publ. in Anthropol.* 36: 648–53.
——. 1964. Le sous-sol du temple d'Aksha. *Kush* 12: 102–10.
——. 1967. Pleistocene sediments and events in Sudanese Nubia. In W. W. Bishop and J. D. Clark, eds. *Background to African Evolution*. Univ. Chicago Press. Pp. 313–28.
——, and R. Paepe. 1965. The geological history of the Nile Valley in Sudanese Nubia: preliminary results. In F. Wendorf, ed. *Contributions to the prehistory of Nubia*. S. Meth. Univ. Press, Dallas. Pp. 29–56.
Donn, W. L., W. R. Farrand, and M. E. Ewing. 1962. Pleistocene ice volumes and sea-level lowering. *J. Geol.* 70: 206–14.
Drake, C. L., and R. W. Girdler. 1964. A geophysical study of the Red Sea. *Geophys. J.* 8: 473–95.
Dumanowski, B. 1960. Notes on the evolution of slopes in an arid climate. *Z. Geomorphol.*, Suppl. Vol. 1: 178–89.
Dunbar, J. H. 1941. *The rock-pictures of Lower Nubia*. Serv. Antiq. Égypte, Government Press, Cairo. 100 pp.
Durand, J. H. 1959. *Les sols rouges et les croûtes en Algérie*. Alger-Birmandreis, Serv. étud. sci. 188 pp.
Edmonds, J. M. 1942. The distribution of the Kordofan Sand (Anglo-Egyptian Sudan). *Geol. Mag.* 79: 18–30.
Edwards, I. E. S. 1964. The Early Dynastic period in Egypt. 74 pp. Fasc. 25 in *Cambridge Ancient History*. 2nd ed. Cambridge Univ. Press.
El-Fandy, M. G. 1946. Barometric lows of Cyprus. *Quarterly J., Roy. Meteorol. Soc.* 72: 291–306.
——. 1948. The effect of the Sudan monsoon low on the development of thundery conditions in Egypt, Palestine and Syria. *Quarterly J., Roy. Meteorol. Soc.* 74: 31–38.
——. 1950. Troughs in the upper westerlies and cyclonic disturbances in the Nile Valley. *Quarterly J., Roy. Meteorol. Soc.* 76: 166–72.
Emery, W. B. 1965. *Egypt in Nubia*. Hutchinson, London. 264 pp.
Engel, C. G., and R. P. Sharp. 1958. Chemical data on desert varnish. *Bull., Geol. Soc. Amer.* 69: 487–518.
Erhart, M. 1956. *La genèse des sols en tant que phénomène géologique: ésquisse d'une théorie géologique et géochimique*. Masson, Paris. 90 pp.
——. 1965. Le témoignage paléoclimatique de quelques formations paléopediques dans leur rapport avec la sédimentologie. *Geol. Rdsch.* 54: 15–23.
Evans-Pritchard, E. E. 1935. A trip to Dunqul Oasis. *Bull., Fac. Arts Egypt. Univ.*, Vol. 3, No. 1, pp. 24–56.
Fairbridge, R. W. 1963. Nile sedimentation above Wadi Halfa during the last 20,000 years. *Kush* 11: 96–107.
Felix, J. 1904. Studien über tertiäre and quartäre Korallen und Riffkalke aus Aegypten und der Sinaihalbinsel. *Z., Deut. Geol. Ges.* 56: 168–206.

Fernea, R. A., and J. G. Kennedy. 1966. Initial adaptation to resettlement: a new life for Egyptian Nubians. *Curr. Anthropol.* 7: 349–54.

Finck, A. 1961. Classification of Gezira clay soils. *Soil Sci.* 92: 263–67.

Flohn, H. 1965a. Klimaprobleme am Roten Meer. *Erdkunde* 19: 179–91.

———. 1965b. Studies on the meteorology of tropical Africa. *Bonner Meteorol. Abh.* 5: 57 pp.

Franz, H. 1958. Beitrag zur Kenntnis der Stratigraphie und Klimatologie des Quartärs in Tschadbecken. *Mitt., Paläontol. Ges. Wien* 51: 19–65.

———. 1967. On the stratigraphy and evolution of Quaternary climate in the Chad Basin. In W. W. Bishop and J. D. Clark, eds. *Background to African evolution.* Univ. Chicago Press. Pp. 273–83.

Fränzle, O. 1965. *Die pleistozäne Klima- und Landschaftsentwicklung der nordlichen Po-Ebene im Lichte bodengeographischer Untersuchungen. Abh., Akad. Wiss. Lit.* (Mainz), Math.-Naturw. Kl., 1965, No. 8, 144 pp.

Gaillard, C. 1934. *Contribution à l'étude de la faune préhistorique de l'Égypte. Arch., Mus. Hist. Natur. Lyon* 14: 126 pp.

Gardner, E. W. 1932a. *Some lacustrine mollusca from the Faiyum Depression: a study in variation. Mém., Inst. Égypte* 18: 123 pp.

———. 1932b. Some problems of the Pleistocene hydrography of the Kharga Oasis, Egypt. *Geol. Mag.* 69: 386–421.

———. 1935. The Pleistocene fauna and flora of Kharga Oasis, Egypt. *Quarterly J., Geol. Soc. of London* 91: 479–518.

Gindy, A. R. 1954. The plutonic history of the Aswan area, Egypt. *Geol. Mag.* 91: 484–97.

Grabham, G. W., and R. P. Black. 1925. *Report of the mission to Lake Tana 1920–21.* Min. Pub. Works, Government Press, Cairo. 207 pp.

Guichard, J., and G. Guichard. 1965. The Early and Middle Palaeolithic of Nubia: a preliminary report. In F. Wendorf, ed. *Contributions to the prehistory of Nubia.* S. Meth. Univ. Press, Dallas. Pp. 57–116.

Guilcher, A. 1958. *Coastal and submarine geomorphology.* Methuen, London. 274 pp.

Gumprecht, F. 1850. Die Reise des Pater Krump nach Nubien in den Jahren 1700–1702 und dessen Mittheilungen über Abyssinien. *Mber. Verh. Ges. Erdk. Berlin,* N.S. 7: 39–88.

Hamed, H. 1950. *Climatological normals for Egypt.* Meteorol. Dep., C. Tsoumas, Cairo. 157 pp.

Hammond, E. H. 1964. Analysis of properties in land form geography: an application to broad-scale land form mapping. *Ann., Ass. Amer. Geogr.* 54: 11–19.

Hansen, C. L. 1966. *Pediment landscapes at Arminna West, Egyptian Nubia, and at el-Faliq, Red Sea Foothills: a comparative analysis.* Univ. Wis., Ph.D. Diss. 193 pp.

Harris, S. A. 1957. Mechanical constitution of certain present-day Egyptian dune sands. *J. Sedim. Petrol.* 27: 421–34.

———. 1958. Differentiation of various Egyptian aeolian microenvironments by mechanical composition. *J. Sedim. Petrol.* 28: 164–74.

———. 1965. Ecology of the freshwater mollusca of Iraq. *Can. J. Zool.* 43: 509–26.

Heinrich, E. W. 1965. *Microscopic identification of minerals.* McGraw-Hill, New York. 414 pp.

Helbaek, H. 1955. Ancient Egyptian wheats. *Proc., Prehist. Soc.* 21: 91–93.

Herman, Y. Rosenberg 1965. Études des sédiments quaternaires de la Mer Rouge. Thèse, Fac. Sci., Paris, Ser. A–1123: 339–415.

Hester, J. J., and P. M. Hobler. 1965. Settlement patterns in the Nubian Desert, Egypt. Paper presented at Seventh INQUA Congr., Boulder, Colo., Sept. 1, 1965.

Hewes, G. W. 1964. Gezira Dabarosa: report of the University of Colorado Nubian Expedition, 1962–63 season. *Kush* 12: 174–87.

———, H. Irwin, M. Papworth, and A. Saxe. 1964. A new fossil human population from the Wadi Halfa area, Sudan. *Nature* 203: 341–43.

Horton, A. W. 1964. The Egyptian Nubians. *Amer. Univ. Field Staff Rep. Serv.*, N.E. Afr. Ser., Vol. 11, No. 2, pp. 283–302.

Hövermann, J. 1961. Über Witterung und Klima in Abessinien. *Abh. Braunschw. Wiss. Ges.* 13: 100–27.

Huckriede, R., and H. Venzlaff. 1962. Über eine pluvialzeitlich Molluskenfauna aus Kordofan (Sudan). *Paläontol. Z. (H. Schmidt Festschrift)*, pp. 93–109.

Hull, E. 1896. Observations on the geology of the Nile Valley and the evidence of the greater volume of the river at a former period. *Quarterly J., Geol. Soc. of London* 52: 308–19.

Hume, W. F. 1908. The southwestern desert of Egypt. *Cairo Sci. J.* 2: 314–25.

———. 1925. *Geology of Egypt.* Vol I: *The surface features of Egypt.* Surv. Egypt, Cairo. 408 pp.

———, and O. H. Little. 1928. Raised beaches and terraces of Egypt. *Int. Geogr. U.*, First Rep., Comm. on Pliocene and Pleistocene Terraces, Paris. Pp. 9–15.

Hurst, H. E. 1944. *A short account of the Nile Basin.* Phys. Dep., Government Press, Cairo. 77 pp.

Huzayyin, S. A. 1941. *The place of Egypt in prehistory. Mém., Inst. Égypte* 43: 474 pp.

Jackson, M. L. 1956. Soil chemical analysis—advanced course. Mimeographed course notes, Dep. Soil Sci., Univ. Wis., Madison.

———. 1959. Frequency distribution of clay minerals in major great soil groups as related to the factors of soil formation. *Proc., Sixth Nat. Conf. on Clays and Clay Minerals* (1957). Pergamon, New York. Pp. 133–43.

———. 1964. Soil clay mineralogical analysis. In C. I. Rich and G. W. Kunze, eds. *Soil clay mineralogy.* Univ. N. C. Press, Chapel Hill. Pp. 245–94.

Jackson, S. P., ed. 1961. *Climatological Atlas of Africa.* Government Printer, Pretoria. 55 pl.

Jacotin, M. 1826. *Description de l'Égypte. Carte topographique de l'Egypte* (*1:100,000*). 2nd ed. Panckoucke, Paris. Sheets 2 (Kom Ombo) and 3 (Edfu).

Kaiser, W. 1957. Zur inneren Chronologie der Naqadakultur. *Archaeol. Geogr.* 6: 69–77.

Kassas, M., and M. Imam. 1954. Habitat and plant communities in the Egyptian Desert: the wadi bed ecosystem. *J. Ecol.* 42: 424–41.

Kerr, P. F. 1959. *Optical mineralogy.* 3rd ed. McGraw-Hill, New York. 442 pp.

Khadr, M. 1961. Heavy residues of some Egyptian soils. *Geol. Mijnb.* 40: 11–25.

King, L. C. 1962. *The morphology of the earth.* Hafner, New York. 699 pp.

Klaer, W. 1962. *Untersuchungen zur klimagenetischen geomorphologie in den Hochgebirgen Vorderasiens. Heidelberger Geogr. Arb.* 11: 135 pp.

Kleinsorge, H., and K. Kreysing. 1960. Über ein Vorkommen von oolithischen

Eisenerzen in der Nubischen Serie der Provinz Kordofan, Republik Sudan. *Z., Deut. Geol. Ges.* 112: 267–77.

———. and J. G. Zscheked. 1959. Eisenerze im westlichen Kordofan, Republik Sudan. *Geol. Jb.* 77: 121–42.

Klinge, H. 1958. Eine Stellungnahme zur Altersfrage von Terra-Rossa-Vorkommen. *Z. Pfl. Ernähr., Düng., Bodenk.* 81: 56–63.

Knetsch, G. 1954. Allgemein-geologische Beobachtungen aus Ägypten (1950–53). *Abh., Neues Jb. Geol. Paläontol.* 99: 287–97.

———. 1957. Eine Struktur-Skizze Ägyptens und einiger seiner Nachbargebiete. *Geol. Jb.* 74: 75–86.

———. 1960. Über aride Verwitterung unter besonderer Berücksichtigung natürlicher und künstlicher Wände in Ägypten. *Z. Geomorphol.*, Suppl. Vol 1: 190–205.

———, and E. Refai. 1955. Über Wüstenverwitterung, Wüstenfeinrelief und Denkmalzerfall in Ägypten. *Abh., Neues Jb. Geol. Paläontol.* 101: 227–56.

———, with A. Shata and M. M. Shazly. 1963. Untersuchungen an Grundwassern der Ost-Sahara. *Geol. Rdsch.* 52: 587–610, 640–50.

Kolbe, H. 1957. Zur Geologie der Eisenerzvorkommen Ägyptens. *Geol. Jb.* 74: 611–28.

Kubiena, W. L. 1953. *The soils of Europe.* Murby, London. 317 pp.

———. 1955. Über die Braunlehmrelikte des Atakor (Zentral-Sahara). *Erdkunde* 9: 115–32.

———. 1957. Neue Beiträge zur Kenntnis des planetarischen und hypsometrischen Formenwandels der Böden Afrikas. *Lautensach Festschrift, Stuttg. Geogr. Stud.* 69: 50–64.

Kuenen, P. H. 1959. Experimental abrasion III. Fluviatile action on sand. *Amer. J. Sci.* 257: 172–90.

———. 1960. Experimental abrasion IV. Aeolian action. *J. Geol.* 68: 427–49.

———, and W. G. Perdok. 1962. Experimental abrasion V. Frosting and defrosting of quartz grains. *J. Geol.* 70: 648–58.

Kuls, W., and A. Semmel. 1965. Zur Frage pluvialzeitlicher Solifluktionsvorgänge im Hochland von Godjam (Äthiopien). *Erdkunde* 19: 292–97.

Lacaille, A. D. 1951. The stone industries of Singa–Abu Hugar. *Fossil Mammals of Africa* (Brit. Mus. Natur. Hist.), No. 2, pp. 43–50.

Langbein, W. B., and S. A. Schumm. 1958. Yield of sediment in relation to mean annual precipitation. *Trans., Amer. Geophys. U.* 39: 1076–84.

Lebon, J. H. G., and V. C. Robertson. 1961. The Jebel Marra, Darfur, and its region. *Geogr. J.* 127: 30–49.

Leopold, L. B., M. G. Wolman, and J. P. Miller. 1964. *Fluvial processes in geomorphology.* Freeman, San Francisco. 522 pp.

Leuchs, K. 1913a. Eine Reise in der südlichen Libyschen Wüste: Gebel Garra, Oase Kurkur, Gebel Borga. *Petermann's Mitt.* 59: 190–91.

———. 1913b. Geologisches aus der südlichen Libyschen Wüste: Gebel Garra, Oase Kurkur, Gebel Borga. *Neues Jb. Mineral. Geol.* 2: 33–48.

Llabador, F. 1962. Résultats malacologiques de la mission scientifique du Ténéré. In H. J. Hugot, ed. *Missions Berliet: Ténéré-Tchad: documents scientifiques.* Arts Métiers Graph., Paris. Pp. 234–70.

Lucas, A. 1908. *The chemistry of the River Nile.* Surv. Dep., Government Press, Cairo. 78 pp.

Lucas, A., and J. R. Harris. 1962. *Ancient Egyptian materials and industries*. E. Arnold, London. 523 pp.

Lyons, H. G. 1894. On the stratigraphy and physiography of the Libyan Desert of Egypt. *Quarterly J., Geol. Soc. of London* 50: 531–47.

Lyubin, V. P. 1964. Nizhnij Paleolit v rajone Dakki-Koshtamny. In B. Piotrovsky, ed. *Drevnyaya Nubiya*. Izdatel'stvo "Nauk," Moscow-Leningrad. Pp. 32–68.

Mackin, J. H. 1937. Erosional history of the Big Horn Basin, Wyoming. Bull., *Geol. Soc. Amer.* 48: 813–94.

McKee, E. D. 1963. Origin of the Nubian and similar sandstones. *Geol. Rdsch.* 52: 551–87.

Meckelein, W. 1959. *Forschungen in der zentralen Sahara. Klimageomorphologie*. Westermann, Braunschweig. 181 pp.

Meredith, D. 1958. *Tabula Imperii Romani: Coptos* (NG 36). Society of Antiquaries of London, Oxford. 18 pp. and 1:1,000,000 map.

Mohr, P. A. 1964. *The Geology of Ethiopia*. Univ. Coll. Addis Ababa Press, Asmara. 286 pp.

———. 1965. Re-classification of the Ethiopian Cainozoic volcanic succession. *Nature* 208: 177–78.

Morgan, J. de. 1897. *Récherches sur les origines de l'Egypte*. Leroux, Paris. Vol. 2, 393 pp.

Morison, C. G. T., A. C. Hoyle, and J. F. Hope-Simpson. 1948. Tropical soil-vegetation catenas and mosaics: a study in the southwestern part of the Anglo-Egyptian Sudan. *J. Ecol.* 36: 1–84.

Münnich, K. O., and J. C. Vogel. 1963. Untersuchungen an pluvialen Wässern der Ost-Sahara. *Geol. Rdsch.* 52: 611–24.

Munsell Color Co. 1954. *Munsell Soil Color Charts*. Baltimore.

Murray, G. W. 1925. The Roman roads and stations in the Eastern Desert of Egypt. *J. Egypt. Archeol.* 11: 138–50.

———. 1939. The road to Chephren's quarries. *Geogr. J.* 44: 97–114.

———. 1967. Trogodytica: the Red Sea littoral in Ptolemaic times. *Geogr. J.* 133: 23–33.

———, and D. E. Derry. 1923. A Predynastic burial on the Red Sea coast of Egypt. *Man* 23: 129–31.

Myers, O. H. 1958. Abka re-excavated. *Kush* 6: 131–41.

———. 1960. Abka again. *Kush* 8: 174–81.

Newton, R. B. 1900. Pleistocene shells from raised beach deposits of the Red Sea. *Geol. Mag.*, 4 decade, 7: 500–14, 544–60.

Nilsson, E. 1940. Ancient changes of climate in British East Africa and Abyssinia. *Geogr. Annlr.* 22: 1–79.

———. 1963. Pluvial lakes and glaciers in East Africa. *Stockh. Contr. Geol.* 2: 21–57.

Oliver, J. 1965. The climate of Khartoum Province. *Sudan Notes Rec.* 46: 1–40.

Papp, A., and E. Thenius. 1959. *Tertiär (Handbuch der stratigraphischen Geologie)*. Enke, Stuttgart. 2 Vols., 411 pp. and 328 pp.

Passarge, S. 1955. *Morphologische Studien in der Wüste von Assuan*. Abh. Geb. Auslandsk. (Hamburg), Vol. 60, Series C, No. 17, 61 pp.

Payne, T. G. 1942. Stratigraphical analysis and environmental reconstruction. *Bull. Amer. Ass. Petrol. Geol.* 26: 1697–1770.

Pettijohn, F. J. 1957. *Sedimentary rocks*. 2nd ed. Harper & Row, New York. 718 pp.

Pfannenstiel, M. 1953. Das Quartär der Levante II. Die Entstehung der ägyptischen Oasendepressionen. *Abh., Akad. Wiss. Lit.* (Mainz), Math.-Naturw. Kl., 1953, No. 7, pp. 335–411.

Porter, B., and R. L. B. Moss. 1936–52. *Topographical bibliography of ancient Egyptian hieroglyphic texts, reliefs and paintings.* Oxford Univ. Press, London. Particularly Vol. 5, 1937, 292 pp.; Vol. 6, 1939, 264 pp.; Vol. 7, 1952, 453 pp.

Posnjak, E. 1940. Deposition of calcium sulfate from sea-water. *Amer. J. Sci.* 238: 539–68.

Powers, M. C. 1953. A new roundness scale for sedimentary particles. *J. Sedim. Petrol.* 23: 117–19.

Putzer, H. 1958. Das Vulkanfeld Bir Sani—Hosh ed Dalam in der Republik Sudan. *Z., Deut. Geol. Ges.* 110: 109–16.

Rathjens, K., and H. von Wissmann. 1933. Morphologische Probleme in Graben des Roten Meeres. *Petermann's Mitt.* 79: 113–17, 183–87.

Reed, C. A. 1959. Animal domestication in the prehistoric Near East. *Science* 130: 1629–39.

———. 1964. A natural history study of Kurkur Oasis. Part I: Introduction. *Postilla* (Peabody Mus. Natur. Hist.), No. 84, 20 pp.

———. 1965a. A human frontal bone from the Late Pleistocene of the Kom Ombo Plain, Upper Egypt. *Man* 65: 101–4.

———. 1965b. Paleo-zoology of Nubia. Paper presented at Seventh INQUA Congr., Boulder, Colo. Sept. 1, 1965.

———. 1966. The Yale University prehistoric expedition to Nubia, 1962–1965. *Discovery* (Peabody Mus. Natur. Hist.), Vol. 1, No. 2, pp. 16–23.

———, M. A. Baumhoff, K. W. Butzer, H. Walter, and D. S. Boloyan. In press. Preliminary report on the archaeological aspects of the research of the Yale University Prehistoric Expedition to Nubia 1962–1963. *Ann. Serv. Antiq.*

Resch, W. F. E. 1963a. Neue Felsbilderfunde in der ägyptischen Ostwüste. *Z. Ethnol.* 88: 86–97.

———. 1963b. Eine vorgeschichtliche Grabstätte auf dem Ras Samadai. *Mitt., Anthropol. Ges. Wien* 93: 119–21.

———. 1964. Kulturhistorische Erwägungen zur Herkunft der Rinderzucht in Nordostafrika. *Paideuma* 10: 1–10.

Robinson, P. 1966. Fossil occurrence of murine rodent in the Sudan. *Science* 154: 264.

———, and G. W. Hewes. In press. 1967. Comments on the late Pleistocene geology of the Wadi Karagan, Murshid District, Northern Province, Sudan. *Kush* 14: 1

Rognon, P. 1967. Climatic influences on the African Hoggar during the Quaternary, based on geomorphologic observations. *Ann., Ass. Amer. Geogr.* 57: 115–27.

Rossignol, M. 1961. Analyse polliniques de sédiments marins quaternaires en Israel. I: sédiments récents. *Pollen Spores* 3: 303–24.

———. 1962. Analyse polliniques de sédiments marins quaternaires en Israel. II: sédiments pleistocènes. *Pollen Spores* 4: 121–48.

Russegger, J. von. 1841–49. *Reisen in Europa, Asien und Afrika.* E. Schweizerbart, Stuttgart. 4 Vols. Particularly Vol. 2, Part 1, 1843, 636 pp.; Part 2, 1844, 778 pp.; and Part 3, 1849, 360 pp.

Russell, R. D., and R. E. Taylor. 1937. Roundness and shape of river sands. *J. Geol.* 45: 225–67.

Rutte, E. 1958. Kalkkrusten in Spanien. *Abh., Neues Jb. Geol. Paläontol.* 56, No. 1, pp. 52–138.

Said, R. 1955. Foraminifera from some Pliocene rocks of Egypt. *J., Wash. Acad. Sci.* 45: 8–13.

———. 1962. *The Geology of Egypt.* Elsevier, Amsterdam and New York. 377 pp.

———, and B. Issawy. 1965. Preliminary results of a geological expedition to Lower Nubia and to Kurkur and Dungul Oasis, Egypt. In F. Wendorf, ed. *Contributions to the Prehistory of Nubia.* S. Meth. Univ. Press, Dallas. Pp. 1–28.

Sandford, K. S. 1929. The Pliocene and Pleistocene deposits of Wadi Qena and the Nile Valley between Luxor and Assiut (Qau). *Quarterly J., Geol. Soc. of London* 85: 493–548.

———. 1933. Geology and geomorphology of the southern Libyan Desert. *Geogr. J.* 82: 213–19.

———. 1934. *Paleolithic Man and the Nile Valley in Upper and Middle Egypt.* Univ. Chicago Orient. Inst. Publ. 18: 131 pp.

———. 1935. Geological observations on the northwest frontiers of the Anglo-Egyptian Sudan. *Quarterly J., Geol. Soc. of London* 91: 323–81.

———. 1949. Notes on the Nile Valley in Berber and Dongola. *Geol. Mag.* 86: 97–109.

———, and W. J. Arkell. 1929. *Paleolithic Man and the Nile-Fayum divide.* Univ. Chicago Orient. Inst. Publ. 10: 77 pp.

———, and ———. 1933. *Paleolithic man and the Nile Valley in Nubia and Upper Egypt.* Univ. Chicago Orient. Inst. Publ. 17: 92 pp.

———, and ———. 1939. *Paleolithic man and the Nile Valley in Lower Egypt, with some notes upon a part of the Red Sea littoral.* Univ. Chicago Orient. Inst. Publ. 46: 105 pp.

Säve-Söderbergh, T. 1941. *Ägypten und Nubien.* Hakan Ohlssons, Lund. 276 pp.

———. 1964. Preliminary report of the Scandinavian Joint Expedition: archeological investigations between Faras and Gamai, November 1962–March 1963. *Kush* 12: 19–39.

Schattner, I. 1961. Weathering phenomena in the crystalline of the Sinai in the light of current notions. *Bull., Res. Coun. Israel* 10–G: 247–66.

Schmidt, W. 1923. Die Scherms an der Rotmeerküste von el-Hedschas. *Petermann's Mitt.* 69: 118–21.

Schokalskaja, S. J. 1953. *Die Böden Afrikas.* Berlin, Akad. Verlag. 408 pp.

Schumm, S. A. 1965. Quaternary paleohydrology. In H. E. Wright and D. G. Frey, eds. *The Quaternary of the United States.* Princeton Univ. Press. Pp. 783–94.

Schwarzbach, M. 1953. Das Alter der Wüste Sahara. *Mh., Neues Jb. Geol. Paläontol.,* 4: 157–74.

Schweinfurth, G. 1901. Am westlichen Rande des Nilthales zwischen Farschut und Kom Ombo. *Petermann's Mitt.* 47: 1–10.

Semmel, A. 1963. Intramontene Ebenen im Hochland von Godjam (Äthiopien). *Erdkunde* 17: 173–89.

———. 1964. Beitrag zur Kenntnis einiger Böden des Hochlandes von Godjam (Äthiopien). *Mh., Neues Jb. Geol. Paläontol.* 8: 474–87.

Sestini, J. 1965. Cenozoic stratigraphy and depositional history, Red Sea coast, Sudan. *Bull., Amer. Ass. Petrol. Geol.* 49: 1453–72.

Shata, A. 1962. Remarks on the geomorphology, pedology and ground water potentialities of the southern entrance to the New Valley. Part I: The Lower Nuba [sic.] Area, Egypt, U.A.R. *Bull., Soc. Géogr. Égypte* 35: 273–99.

Shepard, F. P., and R. Young. 1961. Distinguishing between beach and dune sands. *J. Sedim. Petrol.* 31: 196–214.

Shiner, J. L. 1965. Upper Paleolithic and Mesolithic of Nubia. Paper presented at Seventh INQUA Congr., Boulder, Colo., Sept. 1, 1965.

Shukri, N. M. 1960. The mineralogy of some Nile sediments. *Quarterly J., Geol. Soc. of London* 105: 511–34.

———, and M. K. Ayouty. 1963. The mineralogy of the Nubian Sandstone in Aswan. *Bull., Inst. Désert Égypte*, Vol. 3, No. 2, pp. 65–88.

———, and N. Azer. 1952. The mineralogy of Pliocene and more recent sediments in the Faiyum. *Bull. Inst. Désert Égypte*, Vol. 2, No. 1, pp. 10–53.

Simaika, Y. M. 1940. *The suspended matter in the Nile*. Phys. Dep., Government Press, Cairo. 70 pp.

Smith, G. D., et al. 1960. *Soil classification: a comprehensive system*. U.S. Dep. Agric., Washington. 265 pp.

Smith, P. E. L. 1964a. Expedition to Kom Ombo. *Archaeology* 17: 209–10.

———. 1964b. Radiocarbon dating of a Late Palaeolithic culture from Egypt. *Science* 145: 811.

———. 1966. The Late Paleolithic of northeast Africa in the light of recent research. *Amer. Anthropol.*, Vol. 68, No. 2, Part 2, pp. 326–55.

Smith, W. S. 1962. The Old Kingdom in Egypt. *Cambridge Ancient History*. 2nd ed. Cambridge Univ. Press. Fasc. 5, 72 pp.

Solecki, R. S., ed. 1963. Preliminary statement of the prehistoric investigations of the Columbia University Nubian Expedition in Sudan, 1961–62. *Kush* 11: 70–92.

Springer, M. E. 1958. Desert pavement and vesicular layer of some soils of the desert of the Lahontan Basin, Nevada. *Proc., Soil Sci. Soc. Amer.* 22: 63–66.

Stearns, C. E., and D. L. Thurber. 1965. Th^{230}–U^{234} dates of late Pleistocene marine fossils from the Mediterranean and Moroccan littorals. *Quaternaria* 7: 29–42.

Stuiver, M., and H. E. Suess. 1966. On the relationship between radiocarbon dates and true sample ages. *Radiocarbon* 8: 534–40.

———, et al. In press. Yale natural radiocarbon measurements XI. *Radiocarbon* 10.

Täckholm, Vivi. 1956. *Students' flora of Egypt*. Anglo-American Bookshop, Cairo. 649 pp.

Thenius, E., ed. 1960. *Verhandlungen, I. Tagung des Comité du Néogène Méditerranéen* (Vienna, July 1959). *Mitt., Geol. Ges. Wien* 52.

Tothill, J. D. 1946. The origin of the Sudan Gezira clay plain. *Sudan Notes Rec.* 27: 153–83.

Tricart, J. 1958. Méthode améliorée pour l'étude des sables. *Rev. Géomorphol. Dynam.* 9: 43–54.

———, and A. Cailleux. 1960–61. *Le modelé des régions sèches*. Cent. Doc. Univ., Paris. Vol. 1, 129 pp.; Vol. 2, 179 pp.

Trigger, B. G. 1965. *History and settlement in Lower Nubia*. Yale Univ. Publ. in Anthropol. 69: 224 pp.

———. 1966. The languages of the northern Sudan: an historical perspective. *J. Afr. Hist.* 7: 19–25.

Uhden, R. 1929. Das Formenbild der ägyptischen Wüste. *Z. Geomorphol.* 4: 221–40.
———. 1930. Reise von Debod in Unternubien nach den Oasen von Kurkur und Dungul. *Petermann's Mitt.* 76: 184–88.
Van Campo, M. 1964. Représentation graphique de spectres polliniques des régions sahariennes. *C. R. Acad. Sci.* (Paris) 258: 1873–76.
———, G. Aymonin, P. Guinet, and P. Rognon. 1964. Contribution à l'étude du peuplement végétal quaternaire des montagnes sahariennes. Part I: L'Atakor. *Pollen Spores* 6: 169–94.
———, J. Cohen, P. Guinet, and P. Rognon. 1965. Contribution à l'étude du peuplement végétal quaternaire des montagnes sahariennes. Part II: Flore contemporaire d'un gisement de mammifières tropicaux dans l'Atakor. *Pollen Spores* 7: 361–71.
———, and R. Coque. 1960. Palynologie et géomorphologie dans le sud tunésien. *Pollen Spores* 2: 275–84.
———, P. Guinet, J. Cohen, and P. Dutil. 1966. Nouvelle flore pollinique des alluvions pleistocènes d'un bassin versant Sud du Hoggar. *C. R. Acad. Sci.* (Paris) 263: 487–90.
Vignard, E. 1923. Une nouvelle industrie lithique: le "Sébilien." *Bull., Inst. Fr. Archéol. Orient.* 22: 1–76; and *Bull., Soc. Préhist. Fr.* 25: 200–20.
———. 1935. Le paléolithique en Égypte. *Mélanges Maspero. Mém., Inst. Fr. Archéol. Orient.*, Vol. 66, Fasc. 1, pp. 165–75.
———. 1954. Un gisement du Paléolithique inferieur en couche à Bayarah près Kom Ombo. *Bull., Soc. Préhist. Fr.* 51: 272–80.
———. 1955a. Menchia, une station aurignacienne dans le nord de la plaine de Kom Ombo. *Congr. Préhist. Fr.* (Strasbourg-Metz, 1953). Pp. 634–53.
———. 1955b. Les stations Sébiliennes du Burg el-Makkazin. *Bull., Soc. Préhist. Fr.* 52: 437–52.
———. 1955c. Un kjoekkenmödding sur la rive droit du Wadi Shait dans le nord de la plaine de Kom Ombo. *Bull., Soc. Préhist. Fr.* 52: 703–8.
Vinogradov, A. V. 1964. Sebil'skaya kultura v rajone Dakki. In B. Piotrovsky, ed. *Drevnyaya Nubiya*. Izdatel'stvo "Nauk," Moscow-Leningrad. Pp. 69–82.
Voute, C. 1963. Some geological aspects of the conservation project for the Philae temples in the Aswan area. *Geol. Rdsch.* 52: 665–75.
Waechter, J. 1965. A preliminary report on four Epi-Levallois sites. In F. Wendorf, ed. *Contributions to the prehistory of Nubia*. S. Meth. Univ. Press, Dallas. Pp. 117–45.
Weickmann, L., Jr. 1964. Mittlere Lage und vertikale Struktur grossräumiger Diskontinuitäten im Luftdruck-und Strömungsfeld der Tropenzone zwischen Afrika und Indonesien. *Meteorol. Rdsch.* 17: 105–12.
Wells, L. H. 1951. The fossil human skull from Singa. *Fossil Mammals of Africa* (Brit. Mus. Natur. Hist.), No. 2, pp. 29–42.
Wendorf, F. 1965. Two Mesolithic graveyards in Nubia. Paper presented at Seventh INQUA Congr., Boulder, Colo., Sept. 1, 1965.
———, and R. Said. 1967. Paleolithic remains in Upper Egypt. *Nature* 215: 244–47.
———, J. L. Shiner, and A. E. Marks. 1965. Summary of the 1963–1964 field season. In F. Wendorf, ed. *Contributions to the prehistory of Nubia*. S. Meth. Univ. Press, Dallas. Pp. ix–xxxv.

Wendt, W. E. 1966. Two prehistoric archeological sites in Egyptian Nubia. *Postilla* (Peabody Mus. Natur. Hist.), No. 102, 46 pp.

Wentworth, C. K. 1922. A scale of grade and class terms for clastic sediments. *J. Geol.* 30: 277–92.

Werdecker, J. 1955. Beobachtungen in den Hochländern Äthiopiens auf einer Forschungsreise 1953/54. *Erdkunde* 9: 305–17.

Wheat, J. B., and H. T. Irwin. 1965. Results of the University of Colorado excavations of Paleolithic and Mesolithic sites in Nubia. Abstracts, Seventh INQUA Congr., Boulder, Colo., Sept. 1, 1965. P. 503.

Willcocks, W. 1894a. *Report on perennial irrigation and flood protection for Egypt.* Nat. Print. Off., Cairo. 313 pp. Appendix 7, pp. 3–16.

———. 1894b. *Egyptian irrigation.* 2nd ed., 1899. Spon, London. 485 pp.

Williams, M. A. J. 1966. Age of alluvial clays in the western Gezira, Republic of the Sudan. *Nature* 211: 270–71.

———, and D. N. Hall. 1965. Recent expeditions to Libya from the Royal Military Academy, Sandhurst. *Geogr. J.* 131: 482–501.

Winkler, H. A. 1938–39. *Rock drawings of southern Upper Egypt.* Egypt Exploration Society, London. Vol. 1, 44 pp. and 51 pl. (1938); Vol. 2, 40 pp. and 61 pl. (1939).

Wissmann, H. von. 1951. Über seitliche Erosion. *Colloquium Geogr.* (Bonn) 1: 71 pp.

Wright, H. E. 1951. Geologic setting of Ksar Akil, a paleolithic site in Lebanon—preliminary report. *J. Near Eastern Studies* 10: 115–19.

Yallouze, M., and G. Knetsch. 1953. Linear structures in and around the Nile Basin. *Bull., Soc. Géogr. Égypte* 27: 168–207.

List of Recent Intermediate- to Large-Scale Topographic Maps of Southern Egypt

a) 1:500,000 Series 1404, Edition 1–GSGS, Published by War Office and Air Ministry (London), 1960. Compiled 1958.

    Sheets 544–C (Aswan)
                544–D (El-Kharga)
                545–D (Umm Laj)
                567–A (El-Diwan)
                567–B (Bir Abu Hashim)
                567–D (Wadi Halfa)

    Contours: 100 m, 200 m, 300 m, 400 m, 600 m, 900 m, 1200 m, 1500 m.

b) 1:250,000 Series P 502. Published by Army Map Service (Washington), 1958. Compiled 1952–54.

    Sheets NF 36–2   (Wadi Kalabsha)
                NF 36–5   (Dubayrah)
                NF 36–6   (El-Diwan)
                NG 36–12 (Gebel Igla el-Iswid)
                NG 36–14 (Aswan)
                NG 36–15 (Wadi Shait)
                NG 36–16 (Gebel Hamata)

    Contour Interval: 100 m. Data on some sheets incomplete.

LIST OF RECENT TOPOGRAPHIC MAPS · 551

c) 1:100,000 Nile Valley Series. Published by Survey of Egypt, 1940–44. Surveyed 1937–43.
   Sheets 20/78   (Silwa Bahari)
           16/78   (Kom Ombo)
           12/78   (Aswan)
            8/78   (Kalabsha)
            4/78   (El-Allaqi)
           0/72–78   (Seiyala)
          S 96/72   (Korosko)
           0/66–72   (El-Dirr)
          S 96/66   (Tushka)
          S 92/60–66 (Adindan)
   Contour Interval: 30 m in desert areas; variable, smaller interval on floodplain. Supplemented by hachures. Data on some sheets incomplete.

d) 1:100,000 Western Desert Series. Published by Survey of Egypt, 1940. Surveyed 1938–39.
   Sheets 16/72 (Gebel el-Barqa)
           12/72 (Kurkur)
   Contour Interval: 30 m. Data largely incomplete for Libyan Tableland.

e) 1:100,000 Eastern Desert Series. Published by Survey of Egypt. Republished with some minor corrections 1950.
   Sheets 33 (Umm Rus)   1945
           37 (Wadi Gemal) 1930
   Contour Interval: 100 m or hachures. Data incomplete.

f) 1:100,000 Series P 677. Published by Army Map Service (Washington), 1960–61. Compiled 1959.
   Sheets 5874 (Gebel Silsila)
           5974 (Wadi Shait)
           5873 (Aswan)
           5973 (Khor Qibli)
           5772 (Kurkur)
           5872 (Kalabsha)
           5871 (El-Allaqi)
           5670 (Masmas)
           5770 (El-Diwan)
           5870 (Seiyala)
           5569 (Dibeira Sharq)
           5669 (Arminna)
           5769 (Wadi Hamid)
           5869 (Wadi Kurusku)
           6275 (Gebel Igla el-Iswid)
           6374 (Gezirat Wadi Gemal)
   Contour Interval: 30 m or less. Supplemented by hachures. Data incomplete on some sheets.

g) 1:25,000 Nile Valley Series. Published by German General Staff (Berlin), 1941.
   Sheets 21/795–810 (Gebel el-Silsila)
           21/810–825 (Tiret Cassel Nord)
           20/795–810 (Iqlit)

    20/810–825 (Tiret Cassel)
    19/795–810 (Kom Ombo)
    19/810–825 (Ezbet el-Khor)
    18/795–810 (El-Tiweisa)

Contour Interval: 1 m for floodplain. No desert relief. Published with desert topography by Survey of Egypt 1944–50.

h) 1:10,000 Nubian Series. Published by UNESCO for the Ministry of Labor, United Arab Republic. Produced stereotopographically from air-photo coverage of Egyptian Nubia, 1959. Forty sheets from Adindan to the Aswan Dam, covering the Nile Valley and with an average width of 3 km.

Contour Interval: 10 m.

Index

Abu Hamed, 10, 433
Abu Hugar, 325 f.
Abu Simbel, 155, 159, 190, 197, 199, 201, 203 ff., 220 f., 222 f., 229, 233 f., 238, 250, 253, 258, 267, 300, 308, 314, 490
—Plateau, 211, 220
Acacias, 70, 191 f., 260, 344, 407, 497
Acheulian industries, 13, 22, 59 f., 154 ff., 201, 238, 242, 255, 370, 376, 390
Acheulio-Levalloisian industries, 13, 64, 156, 370
Adams, A. Leith (1820–82), 21, 200, 320
Adda Soil, 304, 319, 328
Addax, 185
Adindan, 225, 228, 231 f., 232 ff., 236, 240, 253 f., 263, 274 ff., 284 ff.
—Stage, 232 ff., 236 ff., 240 f., 241 ff., 244 ff., 252 ff., 263, 265
A-Group culture, 19, 189 f., 192, 278, 289, 321 f.
Aidhab, 10, 20 f., 395
Aiyinat Uplands, 211, 217
Aklit, 17 f., 41, 58 f., 88, 145, 155, 160
—Channel, 88, 143 ff.
Allaqi Plain, 211, 215
Allaqi Station, 190, 198, 201, 212, 233 f., 245 ff., 272, 323
Alluvial deposits, 8 f., 43 ff., 52 ff., 78 ff., 116 ff., 226 ff., 321 ff., 260, 262 ff., 349
Alluviation, 66 f., 69 f., 78 ff., 149 f., 260, 263, 293, 306 f., 318 f., 328 ff., 363, 367, 384 ff., 389, 421 ff., 426 ff., 439 f.
Amada Temple, 189 f., 218, 224 f., 229, 233 f., 241 ff., 267, 276, 307, 309, 314
Ambukol Plain, 211 f., 214
Amratian (Nagada I) culture, 189 ff.
Antelope, 180, 192, 322, 325 f.
Argo, 112
Arkell, William Joscelyn (1904–58), 22, 27, 33, 39, 59 f., 63 f., 145, 146 f., 156 f., 166, 200 ff., 236, 240 ff., 243, 247, 249, 292, 310, 315, 323, 396 f., 399, 420, 423, 505
Arkin, 154 f., 159, 190

—Formation, 188, 276 f., 324
Arkinian industry, 188
Armant, 10, 147
Arminna: type site Arminna Member, Gebel Silsila Formation, 198, 212, 221, 225, 227, 229, 233, 236 ff., 252 f., 260 ff., 267, 271 f., 274 ff., 308 ff., 441, 490
Ashkeit, 155, 190, 278, 289
Aswan, 3, 6, 10, 17, 20, 23 f., 27 f., 42 f., 49, 51 f., 59, 63, 66 ff., 71 f., 74, 109, 112, 124, 134, 155, 164, 166, 185, 190, 193 f., 196 ff., 200 ff., 204 ff., 221 f., 231 f., 233 f., 247 ff., 250 ff., 258 ff., 334, 348, 362, 440, 473, 478, 521 f.
—High Dam, 7, 23, 196, 199 f., 202, 222, 521
—Old Dam, 18, 20 f., 198, 231 f., 320
—Pediplain, 29 f., 32, 41, 49 ff., 67, 211, 213 ff., 221 f., 224 f., 252, 263 ff., 353, 433 f.
—Reservoir, 7, 21, 198 f., 200, 232, 250, 259
Asyut, 10, 334, 361
Atbara–Khor Gash Basin, 3, 258, 327, 435, 447 ff., 453, 454 ff.
Atbara River, 3 f., 8, 71, 112, 254, 445, 467 f., 471
Aterian, 13, 159, 383, 394
Atmur Nuqra Plain, 25, 32, 52
Augite. See Pyroxenes
Austrian Nubian Expedition, 319 ff.
Autochthonous Mountain Dwellers, 193

Badarian culture, 192, 396
Bahan, 189 f.
Bahariya Oasis, 10, 334
Bahr el-Arab: basin and river, 4, 112, 257, 456 f., 487
Bahr el-Ghazal: basin and river, 4, 112, 454 ff., 467 f., 470 f., 487
Bahr el-Jebel, 112
Bajuda Steppe, 10, 110, 433
Ball, John (1872–1941), 200, 323, 335, 337, 343

Ballana, 155, 167 f., 232 ff., 266, 272, 274 ff., 279, 284, 290 ff., 478
—Pediplain, 211, 223 f., 264, 290, 353, 433
Barbary sheep (*Ammotragus lervia*), 185
Baro River, 8, 112
Barramiya, 155, 158, 190, 395
Basal Calcarenite and Detrital Series: Miocene formation, 399 f., 403
Basal Sands and Marls, 87. See also Korosko Formation
Basement Complex, 5, 9, 23 f., 26, 27 f., 44, 51, 66, 71, 112, 203 ff., 205 ff., 210 f., 214, 215, 226 ff., 247 ff., 265, 339, 400 f., 415, 427, 431, 443 f., 453, 468, 473, 478, 487, 522
Batn el-Hagar, 197, 265, 278, 333, 478
Bayara, 22, 36, 41, 59, 88, 145, 152, 155, 157, 160, 181
Beach conglomerates: Pleistocene, 396 ff., 404, 406, 409 ff., 413 ff., 416 ff., 419 f., 421 f., 424 f., 429 f., 505 ff.
Beadnell, Hugh John Llewellyn (1874–1944), 21 f., 396
Beit el-Wali Temple, 197
Berenice, 19 f., 395
Bimban, 18 ff., 41, 57, 189
—Channel, 30, 55, 66 f.
Bir Abu Hashim, 19 f.
Bir Quleib, 20
Black clays: Pleistocene, 256 f.
Blanckenhorn, Max (1861–1947), 200
Blemyes, 20, 197
Blue Nile: Basin, 3 ff., 435, 447 ff., 453, 487
—River, 3 ff., 8, 71, 95, 107, 112, 254 ff., 325 ff., 445, 467 f., 470 f.
Bos, 113, 115, 176, 180, 185, 192 f.
Braunerde, meridional: 84
Bubalis, 113, 148, 176, 185
Bulinus truncatus (freshwater snail), 91, 95 f., 102, 112, 125 ff., 185, 200, 275, 291, 323, 496, 510 ff., 513 f.
Burg el-Makhazin hills, 18, 21 f., 30, 32, 40 f., 44, 49, 58 ff., 88, 115, 165

Cairo, 3, 10, 109, 134, 331, 431, 521
Caracalla (211–17 A.D.), 20
Carboniferous, 5, 24, 522
Catfish Cave: prehistoric site opposite Ibrim, 167, 276, 314
Caton-Thompson, Gertrude, 60, 156, 158, 163, 165 f., 187, 390
Central Wadi. See Wadi Kurkur

C-Group culture, 196, 389 f., 394
Chalk: Paleocene formation, 5, 22, 25 f., 28, 36, 339 ff., 342 ff., 346 ff., 349, 359 f., 362 ff., 369, 371, 373 ff., 390, 393, 478, 496 f.
Chephren Quarries, 203, 233
Chephren Swell, 206, 208 f.
Clay minerals, 53, 61, 71 f., 84, 95, 103, 108, 118, 121, 130, 148, 150, 204 f., 228, 235, 239 f., 252 f., 268, 291, 304 ff., 438, 445 ff., 455 ff., 483 ff.
Cleopatra bulimoides (freshwater snail), 102, 112, 131, 133, 147, 178, 180, 200, 275 f., 287, 308, 325, 509 ff.
Climatic change, 6, 8 ff., 74, 149 ff., 178, 182, 192, 254, 257 f., 263 f., 328 ff., 363 ff., 383 ff., 388 f., 393 f., 423 f., 425, 426 ff., 434 f., 436 ff., 450 ff., 455 ff.
Colocynthis vulgaris, 70, 120, 287
Combined Prehistoric Expedition to Nubia (originally Columbia University–Museum of New Mexico Expedition), 166 f., 272, 274, 292, 311, 390
Congo Basin, 4 f.
Consolidation: types of, 13
Contra Ombos. See Ragaba
Coral Reefs: Pleistocene, 396 ff., 403, 406 f., 409 ff., 413 ff., 416 f., 418 ff., 421 f., 423 f., 429 f., 505 ff.
Corbicula fluminalis ssp. (freshwater bivalve), 21, 91, 95 ff., 99, 102, 112, 115, 119, 125 ff., 130, 136 f., 147, 164, 171 f., 175, 180, 186, 200, 260, 273, 275 f., 287, 293, 308, 310 f., 313, 323, 509 ff.
Cretaceous, 5, 24, 26, 224, 338 ff., 436 f., 444, 452, 522
Crocodile, 113, 148, 176, 192
Cuvillier, Jean (1899—), 336, 338 ff.

Dabud Hills, 210 ff.
Dabud Temple, 190, 197, 233 f., 247 ff., 320, 336
Dakhla Oasis, 10, 334, 348, 436
Dakhla Shale, 26, 338 ff., 347 f., 370
Dakka, 155, 160, 190, 197 f., 200 ff., 205 ff., 215, 233 f., 244 ff., 250, 260, 262 f., 267, 272, 274, 319, 323
—Basin, 206 ff., 215, 221, 223, 224 f., 228 f., 231 f., 244 ff., 247
—Plain, 211, 215
—Stage, 225, 232, 234, 236 ff., 240 f., 241 ff., 244 ff., 252 ff., 263, 265, 440
Dandur Temple, 197, 233 f., 247 ff.

Danian, 26, 339, 341. *See also* Cretaceous
Darau, 17 ff., 30, 32, 41 f., 52, 55, 88, 108, 110, 112, 128 ff., 131 f., 166, 181, 194 f., 267, 334, 348, 511
Dar es-Salam: Nubian resettlement village, 21, 88, 181; prehistoric site, 88, 181, 511
Debba, 116
Desert Institute of Egypt, 202, 336
Dibeira, 155, 159, 168, 188, 190, 201 f., 212, 233, 278
Dibeira-Jer Formation. *See* Khor Musa Formation
Dibeira Plain, 211, 220
Differential erosion, 9, 204 ff.
Dihmit stage, 225, 232 ff., 236 ff., 247 ff., 252 ff., 263, 265
Dihmit Station, 190, 201, 228, 231, 233 f., 247 ff., 250, 267, 320, 334
Diocletian (285–305 A.D.), 20
Dissection: of bedrock, 67, 81 ff., 250 ff., 263 f., 313, 332, 363, 387, 389, 418, 434, 441 f., 445, 453 f.; of fill, 66 f., 81 ff., 87 f., 149 f., 250 ff., 260, 263, 293, 306 f., 318 f., 324, 328 ff., 363, 387, 389, 421 ff., 428 f.
El Diwan (El-Derr), 21, 197, 212, 258
Dungul Oasis, 10, 155, 203, 233, 334 ff., 360 f., 388 f., 394

Earliest Hunters, 192 f.
Early Nile Dwellers, 193
Eastern Desert, 20, 29 ff., 49, 53, 64, 66 f., 68, 71 f., 74, 78 ff., 185, 203, 205, 226, 259 ff., 332. *See also* Red Sea Hills
Edfu, 10, 34 f., 63, 146, 152, 155 f., 190, 193 f., 335, 395
Egypt: Middle, 123, 147 f., 497; Upper, 29, 63 ff., 78 ff., 123, 146 ff., 189, 192, 334, 521 f.
Eighteenth Dynasty (1567–1320 B.C.), 19 f., 189 ff., 197
Elephant, 182, 192 f., 322, 326
Eocene, 5 f., 26, 28, 341, 360, 431, 453, 522
Eolian: deposits, 72 f., 112, 181, 255, 257 f., 261, 291, 308, 321 ff., 324, 347 ff., 356 f., 363, 374, 383, 385, 425, 473 ff.; forms, 30, 73, 133, 136 f., 142, 146, 180 f., 220, 257 f., 261, 292, 339, 383; processes, 30, 72 ff., 181, 252 ff., 257 f., 261, 322 f., 368 f., 425, 476 f.
Epi-Levalloisian, 13, 394
Epi-Paleolithic, 13, 188, 191 ff., 277, 292 f., 383
Equus, 113, 148, 176, 180, 185, 325

Esna, 10, 20, 22, 32, 34 f., 43, 49, 51, 147, 152, 155, 166, 334
—Shale, 26, 341
Etbai Uplands, 29 ff., 77
Etheria elliptica (Nile oyster), 21, 133, 185, 200, 276, 308, 511
Ethiopian Plateau: geomorphic evolution of, 6, 443 ff., 449 ff.; hydrographic regime of, 8 f., 71, 95, 150, 326, 330 ff., 443 ff., 449 ff., 453 ff.
Evaporite Series: Miocene formation, 399 ff., 407, 438

False Wadi (Wadi Abu Gorma), 348, 351, 354 ff., 364 f., 367, 370, 515 ff.
Farafra Oasis, 10, 334
Faras, 190
Faris, 17 f., 26, 33, 41
Farqanda (Nag), 219, 234, 237 f.
Farshut, 10, 20, 334
El-Fashn, 10, 147
Fatira, 17 f., 21, 26, 36 ff., 43, 58 f., 72, 80, 88, 177, 511
—beds, 36, 38 f., 46 f., 57 ff., 60, 75, 79 f., 97, 157, 183, 247
—Channel, 88, 109, 114, 132 ff., 149, 168 ff., 170 ff., 177
Fayum Depression, 10, 33, 146, 148, 187 f., 432
Ficus (fig), 336, 358, 374, 386
First Cataract (Aswan), 155, 187, 189, 194, 200, 228, 249, 251
First Dynasty (*ca.* 3100–2900 B.C.), 194, 395
Fish, 114, 139, 170, 178 f., 185
French Survey of Upper Egypt (1798–99), 20
Frost-weathering, 67, 76, 182, 322, 427

Gallaba Plain (Darb El-Gallaba Plain), 30, 32, 44 f., 49, 53, 56, 67, 213
Gallaba Terrace: stage, 32, 41 f., 43 ff., 50, 57, 63, 65, 67, 222, 226, 247 ff.
Gazelle (*Gazella* sp.), 113, 148, 176, 180, 185, 192 f., 322, 325 f., 391
Gebel el-Barqa, 27 f., 30, 45, 335, 350, 360
Gebel Gharra, 30, 45, 336, 338 ff., 342, 350
Gebel Miyahi, 26
Gebel Silsila (Gebel es-Silsila, ancient *Silsilis*), 17 ff., 21 f., 23, 25, 27, 30 f., 33, 36, 40, 41 f., 45 ff., 52 f., 68, 73, 75 ff., 79, 88, 96, 108, 113, 132 ff., 155, 161, 165, 189 f., 192, 194, 267

—1: complex of prehistoric sites (excavated by P. E. L. Smith), 88, 114, 146, 165 f.
—2A: area of prehistoric sites, 88, 109, 132 ff., 151, 168 ff., 181, 511
—2B: area of prehistoric sites, 88, 103, 109, 127, 132 ff., 139, 142, 168, 170 ff., 177, 511
—Formation: Arminna Member, 188, 237, 239, 242, 274 ff., 279, 281, 282, 285, 287 ff., 292 ff., 305, 307, 308 ff., 312, 324 ff., 327 ff., 471; classical or Darau Member, 37, 46, 48, 87 ff., 92 f., 105, 107 ff., 113, 116, 119, 127 ff., 130 f., 132 ff., 136 ff., 139, 141 ff., 143 ff., 146 f., 148 ff., 151, 166 f., 177 f., 182, 186, 188 f., 195, 235, 237, 239, 267, 271, 274 ff., 279, 281 f., 284 f., 286 ff., 291 ff., 303, 307, 308 ff., 320, 324 ff., 327 ff., 435, 456, 470 f., 476, 478, 485 ff.; Kibdi Member, 188 f., 236 f., 239, 274, 276 ff., 279, 285, 288 ff., 292 f., 310, 324, 328, 332 f., 471
—Station: type site of Darau Member, Gebel Silsila Formation, 88, 100 f., 124, 132, 165
Gemini IV orbital photography (June 1965), 32, 203, 205 ff., 210, 215, 217, 261, 265
Geoistrazivanja (Yugoslav drilling company), 26
Geological Survey of Canada, 23
Gerenuk (*Lithocranius walleri*), 193
Gerzean (Nagada II) culture, 189 ff., 322
Gezira Plain, 257, 325 ff., 443, 448 f., 455 f.
Gineina and Shibbak, 233, 241
Giraffe, 182, 192 f., 322, 326
Girf Husein Temple, 190, 197, 210, 233 f., 247 ff.
Gravel-shape analyses, 16, 79 f., 91, 271, 322, 331, 358, 371, 384, 404

Hadaiyib Uplands, 211, 215
Halfan industry, 167 f.
Hamid Swell, 206, 208 f., 217, 221
Hamid Uplands, 211, 221, 263
El-Hammam, 19, 190, 192
Hatshepsut (1503–1482 B.C.), 19, 194
Heavy minerals, 71 f., 78, 95, 103, 110, 131, 150, 204, 308, 311, 445 ff., 455 ff., 467 ff.
Hierakonpolis, 147, 152, 189 f., 322
High shorelines of Red Sea, 396 ff., 403 f., 406 f., 408 ff., 413 ff., 416 ff., 418 ff., 421 f., 423 f., 429 f., 505 ff.
High terraces: Kom Ombo Plain, 32, 41, 50, 52 ff., 58, 62, 64 f., 66, 156, 161, 235, 265; Red Sea Littoral, 407 f., 428, 430

Hippopotamus, 113, 137, 148, 176, 185, 192 f., 315, 326
Hod Bimban, 19, 194 f.
Hoggar Mountains, 84, 366, 516 f.
Holocene, 11 f., 119 ff., 121 ff., 130, 148 f., 152, 188, 294, 305, 318, 326, 328, 333, 389, 397, 401, 422, 429 f., 442, 445, 483 ff.
Homioceras (Bubalus), 113, 115, 176, 185, 325 f.
Homo sapiens, 167 f., 170, 325
Hosh, 190, 192 f.
Howar Basin, 6, 254, 264, 432 ff., 443, 454
Hudi Chert, 112, 150, 274 f., 324, 330 f., 437, 454 f.
Hull, Edward G. (1829–1917), 21
Hume, William Fraser (1867–1949), 332
Hyaena, 113

Ibex, 185, 192 f.
Ibrim, 233 f., 240 f., 267, 272, 276, 314
Ineiba, 190, 198, 201, 219, 233 f., 240 f., 267, 272, 277, 292, 310 f.
—Formation, 54, 62, 88 f., 91 ff., 105, 108, 116 ff., 117, 120, 124, 125 ff., 128, 149, 151, 165, 182 ff., 244, 267, 271, 281 ff., 285, 288, 290, 295 f., 300, 303 ff., 312 f., 317 ff., 328, 488 ff.

Jebel Brinikol, 154 f.

Kagera River, 3
Kalabsha Gorge (Bab El-Kalabsha), 112, 155, 203 ff., 212 ff., 231, 233 f., 247 ff., 250 ff., 263, 473, 478
Kalabsha Plain, 30, 32, 211, 212 ff., 338, 342
Kalabsha Temple, 197, 199, 201, 233, 320
Karst, 345, 347, 362 ff., 369 f., 384 ff.
Kasr Ibrim, 201, 216, 218, 221, 224, 233
Kerma, 197
Khargan industry, 13, 381, 383, 390, 392 ff.
Kharga Oasis, 10, 155 f., 158, 160, 334, 337, 348, 350, 360 f., 365, 370, 376, 381, 383, 386 ff., 390, 436
Khartum, 3, 112, 156, 188, 254, 256, 258, 325 ff., 435 f.
Khashm el-Girba, 112, 255
Khor Abu Anga: prehistoric site, 156, 254 f.
Khor Abu Sureih, 233, 267, 272, 283, 316 f.
Khor (and Wadi) Abu Uruq, 202, 233, 267, 272, 283, 315
Khor Adindan, 208 f., 232 ff., 267, 272, 273, 274 ff., 277 ff., 283, 284 ff., 285, 333, 478

Khor Dawud, 155, 160
Khor (and Wadi) Dihmit, 206, 212, 214 f., 228, 247 ff., 250, 323
Khor Gash, 4, 112
Khor Hamra, 95, 310 ff.
Khor Ibeidalla, 233, 244, 314
Khor Musa Formation (now called Dibeira-Jer Formation, equivalent to Masmas Formation in the northern Sudan), 160, 167, 274, 324
Khormusan industry, 167, 274
Khor Nabruq, 166
Khor el-Sil: area of prehistoric sites (excavated by P. E. L. Smith), 73, 88, 100 f., 112, 142 ff., 151, 166, 177 ff., 511
—Drain (ancient wadi), 18, 142 ff., 180
Kobosh Temple, 233, 247
Kom Ombo [town] (ancient Ombos), 10, 17 ff, 21 f., 26, 28, 30, 33, 41, 51, 67 f., 88, 155, 189 f.
—Graben, 25, 27 f., 30 ff., 39, 40, 42 f., 50 ff., 55, 59, 60, 117, 209
—Plain: 7, 17 ff., 86 ff., 153 ff., 199, 221 f., 226, 228, 231 f., 247, 252 ff., 260 f., 264 f., 266 ff., 272 f., 274 ff., 282 f., 327, 333, 362, 442, 476, 479, 509 ff.; contemporary climate, 8, 67 f., 73 f., 76; contemporary hydrography, 68 ff.; early geological work, 21 ff.; geology, 23 ff.; geomorphic evolution, 27 ff., 33 ff., 44 ff., 52 ff., 57 ff., 60 ff., 66 f., 78 ff., 86 ff., 148 ff.; historical précis, 17 ff., 194; Nubian resettlement, 7, 21, 108, 199; Pleistocene hydrography, 34 f., 44 ff., 52 ff., 57 ff., 60 ff., 66 f., 78 ff., 86 ff., 97 ff., 107 ff., 116 ff., 121 ff., 132 ff., 141 ff., 143 ff., 148 ff., 182 ff.; Pliocene hydrography, 33 ff.; prehistoric cultures, 9 f., 21, 23, 54, 59 f., 132 ff., 141 ff., 143 ff., 157, 160 ff., 164 ff., 189 ff.; structure, 25, 27 f., 30, 32, 50 f., 67. *See also* High terraces, Low terraces, Middle terraces
Kordofan sands, 211, 257 f.
Korosko Formation: classical or upper member, 50, 54, 86 ff., 102, 117, 124 f., 127 f., 149 ff., 151, 161 f., 239, 244, 251, 265, 266 ff., 271, 279, 283 ff., 291 ff., 294 ff., 300, 308 ff., 312, 315 ff., 318, 321 f., 324 f., 327 ff., 375, 377, 381, 430, 435, 443, 454 f., 469 ff., 475 ff., 486 ff., 495 ff.; lower member in Wadi Or, 271, 295, 298 ff., 306, 329

Korosko Hills, 211, 216 f., 219, 221, 262 f., 275
Korosko Station, 155, 190, 201, 205 f., 209, 212, 216, 222, 228, 231 f., 233 f., 241 ff., 244 ff., 262, 267, 272, 283, 314 ff.
Kosti, 112, 255 ff.
Kubaniya 189 f., 193 f.
—Nile, 34 f., 43, 45, 55, 67
Kurkur Foreland, 30, 339, 342, 358, 370, 373, 375, 378
Kurkur Oasis: 9 f., 20, 112, 155 f., 203, 206, 211, 233, 334 ff., 430, 432, 443, 473, 475, 478, 481, 492 f., 495 ff., 513 f., 515 ff.; contemporary climate, water resources, and vegetation, 335 f., 343 ff., 393; early exploration, 335 ff.; fossil pollen, 515 ff.; geology, 338 ff.; geomorphic evolution, 350 ff., 355 ff., 361, 362 ff., 366 ff., 370 ff., 378 ff., 381 f., 383 ff., 388 f.; history, 334 f.; prehistoric occupation, 371, 377 f., 380, 389 ff.; structure, 339, 341 f., 350, 352
Kurkur Pediments, 339, 342, 350 ff., 361 ff., 368 f., 374, 388. *See also* Kurkur Pediplain
Kurkur Pediplain, 211, 223 f., 252, 264, 343 ff., 353, 363, 433
Kurkurstufe: terminal Cretaceous formation, 338 ff., 342, 345 ff., 348, 351, 364, 373, 375
Kush, 197
Kushtamna, 155, 160, 166, 207, 233 f., 247 ff., 274

Laganum-Clypeaster Series: Pliocene or Pleistocene formation, 396, 399 ff., 417, 420
Lake Tana, 3 f., 112
Lake Victoria, 3
Landslide breccias, 263 ff., 287
Leuchs, Kurt (1881–1949), 336 f., 360
Levalloisian, 13, 60, 97, 114, 148, 154, 160 f., 163, 324 f., 376, 381
Libyan Desert (Western Desert), 9, 10, 22, 49, 226, 230, 259 ff., 334 ff., 388, 394, 437
Libyan Tableland, 30, 211, 213, 339, 343, 350, 360, 363, 432, 522
Limestone Cuesta, 30, 339, 342, 350 ff., 354, 360 ff. *See also* Sinn el-Kaddab
Limicolaria (land snail), 254, 325 f.
Local relief, 31 ff., 210 ff., 221 ff., 342 f., 350 ff., 402

Lower Nubian Plain, 29, 32, 202 f., 207, 211, 213 f., 218 ff., 261
Lower Wadi Alluvium, 87, 116. *See also* Ineiba Formation, Malki Member, Sinqari Member
Low terraces: Kom Ombo Plain, 41, 54, 57 f., 59, 60 ff., 63 ff., 66, 79 f., 87, 90, 94, 160, 265; Red Sea Littoral, 402 ff., 406 ff., 409 ff., 418, 421 ff., 426 ff.
Lucas, Alfred (1867–1945), 335, 343
Luxor, 10, 63 f., 67, 82, 147, 155, 158, 432
Lymnaea sp. (freshwater snail), 91, 95 ff., 112, 510 f., 513
Lyons, Sir Henry (1864–1944), 200

El-Madiq, 233, 267, 272, 314 f., 320
Madiq Hogback, 206 ff.
Malki, 233 f., 244 ff.
—Member (Ineiba Formation), 89, 91 ff., 116, 119 ff., 125 ff., 128, 281 f., 290, 303 ff., 313, 316 ff. 328, 488 ff., 496
El-Manshiya, 88
Manshiya Channel, 88, 141 ff., 177 ff.
Mansuriya, 17 f., 194 f.
Masmas Formation, 37, 48, 50, 57, 62, 87 ff., 92 f., 97 ff., 108 f., 115, 117, 124 ff., 127, 128 ff., 133 f., 138, 141 ff., 148 f., 150 f., 157, 160, 175, 183, 195, 267, 272 ff., 278 f., 285, 286 ff., 291 ff., 296, 300, 302 ff., 308 ff., 312, 317 ff., 323, 324 ff., 333, 435, 443, 456, 469 ff., 475 ff., 484 ff., 496
Masmas Station, 212, 233 f., 240 f., 267, 275, 277, 292, 310
El-Matana 88, 145 ff., 180
Mediterranean Sea, 3, 12, 420
Melanoides tuberculata (freshwater snail), 371 ff., 513 f.
Menchian industry (named after village of Manshiya), 165
Meroe, 197
Meroites, 197
Mersa, 402
Mersa Alam, 9, 10, 51, 63, 396 ff., 402 ff., 413 ff., 417, 419 ff., 423 ff., 440, 473, 475, 478, 481, 499 ff.
Mersa (and Wadi) Samadai, 396, 401, 413, 416, 419
Mersa Sifein, 401, 408 f., 410 ff., 419 f., 424, 507 f.
Mesolithic, 13, 315, 326
Mesozoic, 5, 112, 384, 452
Middle terraces: Kom Ombo Plain, 41, 50, 54, 57 ff., 62, 64 f., 66, 79 f., 94, 161, 265; Red Sea Littoral, 402 ff., 407 ff., 412, 413 ff., 418, 421 ff., 426 ff.
Miocene, 5, 11, 28, 50, 67, 210, 223, 252, 264, 353, 361 ff., 396, 399 ff., 409, 432 ff., 438, 444, 452 f., 522
Morgan, Jacques de (1857–1924), 21
Mt. Ruwenzori, 5
Mousterian, 13, 60, 147, 157 ff.
El-Muglad, 112, 255 ff.
Muharraka Temple, 197, 205, 233 f., 244 ff.
Muneiha, 33, 36, 41, 43 ff., 58 f., 60, 155, 157
Muneih Foreland, 30
Murray, George W. (1885–1966), 335 f.
Murwan, 233 f., 247 ff.

Nag el-Darira (Darau), 19, 88, 132
Nag Ibeidalla, 233, 267, 316
Nag el-Kibdi: type site of Kibdi Member, Gebel Silsila Formation. *See* Adindan
Nag el-Shibeika, 19, 189
Nag el-Shima, 233 f., 244 ff.
Nag Umm Simbil, 207, 233
National Museum of Canada Expedition, 23, 191 f.
Neolithic, 187 ff., 278, 390
New Abu Simbel, 18, 37, 41, 62, 88, 105 ff., 116, 119, 151, 155, 160, 511
New Allaqi, 18, 48
New Arminna, 18, 88, 120, 122, 191
New Ballana, 18, 87 f., 98, 115 f., 129, 151, 487, 511
New Farqanda, 18, 50
New Ineiba: type site of Ineiba Formation. *See* New Masmas, 18, 124 ff.
New Korosko: type site of Korosko Formation, 18, 40 f., 55 f., 61, 69, 79 f., 87 ff., 99, 105, 117, 123 ff., 151, 161, 496, 510 f.
New Malki: type site of Malki Member, Ineiba Formation. *See* New Shaturma, 18
New Masmas: type site of Masmas Formation, 18 f., 50, 87 f., 91, 124 ff., 191, 511
New Sebua, 18, 37, 40 f., 60 f., 69, 79 f., 87 f., 92 ff., 118 ff., 151, 281, 510 f.
New Shaturma: type site of Shaturma Formation, 18, 37, 52, 79 f., 88, 92 ff., 124 f., 127 f., 151, 155, 161 f., 496, 510 f.
New Sinqari: type site of Sinqari Member, Ineiba Formation. *See* New Shaturma, 18
New Tushka, 18 f., 29, 58, 62, 122, 191
Nile Floodplain, 88, 117, 149, 162, 194 f., 234, 236 f., 239, 332

INDEX · 559

Nile River: flood regime, 3, 5, 8 f., 71 f., 86, 182, 264 f., 289, 330 ff., 449; modern sediment load, 8, 71 f., 455 ff.; tertiary evolution, 5 f., 11 f., 431 ff.
Nilotic silts, 8 f., 86 ff., 97 ff., 107 ff., 128 ff., 132 ff., 141 ff., 143 ff., 146 ff., 219, 330 ff., 443 ff., 455 ff., 467 ff.
Nobatae, 20
Norden, Frederik Ludvig (1708–42), 198
North Wadi. See Wadi Kurkur
North Well: Kurkur Oasis, 343, 347, 349, 356 f., 372, 374, 376, 390, 515 ff.
Northwest Wadi. See Wadi Kurkur
Nubia (Egyptian): contemporary climate, 8, 258 f.; contemporary hydrography, 259 ff., 332; early geological work, 200 ff.; geology, 203 ff.; geomorphic evolution, 203 ff., 205 ff., 221 ff., 224 ff., 226 ff., 231 ff., 250 ff., 252 f., 262 ff.; historical précis, 196 ff.; Nubian evacuation, 198 f.; Pleistocene hydrography, 224 ff., 226 ff., 250 ff., 252 f., 262 ff.; prehistoric cultures, 153, 159 f., 166 ff., 189; structure, 205 ff., 210 ff., 223, 224 f., 263 f., 339
Nubia (Sudanese): geomorphic evolution, 254 ff.; prehistoric cultures, 153 ff., 158 ff., 166 ff., 187 ff.
Nubians, 7, 21, 23, 194, 195 ff.
Nubian Sandstone, 9, 22, 24 ff., 46 ff., 54, 57, 62, 76 ff., 98, 103, 129, 154, 158 ff., 178 f., 200, 203 ff., 205 ff., 210 ff., 235 ff., 239, 242, 244, 253 f., 260 ff., 278 f., 295 f., 309, 318, 338 f., 350, 436 f., 522
Nummulitic Limestone, 26, 339, 341 ff., 350, 478, 516
Nuq Muneih, 30, 335
El-Nuqu, 88, 117, 151, 511

Older Floodplain Silts, 87, 97. *See also* Masmas Formation
Old Kingdom (*ca.* 2686–2160 B.C.), 19, 189 f., 196
Oligocene, 28, 50, 67, 210, 223, 264, 350, 353, 363, 431 ff., 437 f., 444, 452 ff., 522
Ombos. See Kom Ombo
Omda Soil, 282, 304 ff., 328, 333, 488 ff.
Omochoerus, 556
Oriental Institute: University of Chicago Survey (1926–33). See K. S. Sandford
Oryx, 185, 325
Ostraea-Pecten Series: Miocene or Pliocene formation, 399 ff., 407, 416 f.
Ostrich, 114, 188, 192 f., 322

Paleocene, 5, 341, 431
Paleolithic: late, 10, 13, 114, 118, 143, 145, 148, 157, 162 ff., 181 ff., 187 f., 190, 192, 202, 275, 279, 287 f., 291 ff. 305, 308, 310, 314 ff., 319 f., 378, 383, 390, 393; lower, 13, 59 f., 153 ff., 157; middle, 13, 52, 54, 60, 155, 157 ff., 160 ff., 163, 236, 245 f., 255, 260 f. 272, 275, 281, 284, 301, 303, 305, 308, 311, 390, 393 f.
Paleosols: calcified (Calcorthid), 110, 139, 297, 304, 328, 333, 408, 429; red, 52 f., 55, 58 f., 61 ff., 66 f., 78, 83 ff., 87 f., 120 f., 149, 161, 233 ff., 236, 238 f., 241, 243, 246 f., 252 ff., 257, 263, 265, 281 f., 283 f., 299 f., 304 ff., 328 f., 333, 359, 362 ff., 369 f., 379, 381 ff., 387, 389, 416, 429, 439 ff.; vertisol, 104 ff., 107, 119, 149, 256, 273, 286, 290, 292, 313, 325, 328, 333
Paleo-Sudd Basin, 4 f., 255 ff., 433 f., 454
Palms, 344 f.
Pan-Grave culture, 320
Pan Silts. See Red Silts
Passarge, Siegfried (1867–1958), 74
Patina: desert varnish, 74 ff., 97, 204 f., 262, 263 f., 369 f., 381, 391, 393, 441
Pediments, 29 ff., 49 ff., 210 ff., 221 ff., 224 ff., 233 ff., 237, 244, 260 f., 263 f., 269, 279, 347, 401, 441 f.
Pediplains, 29 ff., 49 ff., 210 ff., 221 ff., 224 ff., 263 f., 441 f.
Periglacial phenomena, 427, 450 f.
Philae, 20, 190, 197
Planorbis sp. (freshwater snail), 91, 95 f., 102, 112, 125 ff., 185, 200, 275, 496, 510 f., 513
Plateau Tufa, 339 ff., 343, 352, 353 ff., 356 f., 361 ff., 364, 369, 374, 376, 382, 384 f., 391, 515 ff.
Pleistocene: Basal and Lower, 11 f., 57, 66 f., 79 ff., 210, 224 ff., 263, 363, 434 ff., 439 ff.; Middle, 11 f., 35, 63, 66, 79 ff., 86, 226 ff., 263, 313, 315, 321, 331, 363, 396 ff., 419, 422, 434 f., 439 ff., 515 ff.; Upper, 5 f., 9 f., 11 f., 34, 63, 86 ff., 328 f., 389, 296 ff., 401, 422, 426 ff., 435, 439 ff., 483 ff.
Pliocene: 5, 11, 26, 28, 33 ff., 43 f., 45, 49, 50 ff., 59, 63, 67, 256, 263, 341, 353, 361 ff., 396 f., 399 ff., 409, 434, 438 f., 445, 452 f., 468, 521 f.; deposits at Aswan, 63, 67, 250 f., 434, 438 f., 521 f.;

deposits of Kom Ombo region, 33 ff., 49 ff., 54, 58, 67, 439; gulf in Nile Valley, 33 ff., 52, 67, 223, 361 f., 430 f., 521 f.
Pluvials, 8, 74, 80, 82 f., 148 ff., 257, 329, 331, 337, 363, 381, 386, 389, 434, 438 ff.
Precambrian. See Basement Complex
Predynastic cultures, 174, 187, 189 ff., 396
Proto-Nile: river and basin, 4, 6, 9, 22, 43 ff., 67, 222, 254, 264 f., 431 ff., 437 ff., 453 f., 521 f.
Pseudo-eskers, 229 f., 232, 236, 240 f., 244 ff.
Ptolemies (323–30 B.C.), 19 f., 197, 395
Pupoides sp. (land snail), 120, 374, 511, 513 f.
Pyroxenes, 71 f., 110, 467 ff.

Qadan industry, 167 f., 186, 188, 278
Qadrus Formation, 278, 324
Qatta, 233 f., 240 f.
Qau, 10, 147 f.
Qena, 10, 43, 63 ff., 497. See also Wadi Qena
Qirtassi, 197, 233, 247, 320
Quartz-sand micromorphology, 369, 425, 429, 473 ff.
Quffa Plain, 211, 214 f.
Quseir, 10, 68, 395, 396 ff., 417, 420, 423, 505 ff.
Qustul, 232 ff., 237, 253, 272, 274 ff., 292 ff.

Radiocarbon, 11 f., 96 ff., 105, 114 ff., 118, 121, 123, 130, 136, 140, 142, 146, 149, 166 f., 180, 187 f., 192, 261, 271, 274, 276 ff., 282, 302, 305, 314, 326, 372, 377 f., 381, 386, 389 f., 394 ff., 398
Raghama, 41, 58, 88
Raghayim el-Bid, 19, 28, 41, 88, 191
Rambla: soil type, 303
Ramses II (1304–1237 B.C.), 20
Raqaba (Contra Ombos), 18 ff.
Ras Samadai, 401, 416
Red Breccia (*brocatelli*), 359 f., 363 f., 382, 387, 391 f.
Red Sea, 6, 9 f., 51, 164, 395 ff., 454, 499 ff. See also High shorelines of Red Sea
—coast at Mersa Alam: contemporary climate, 423; early geological work, 396 ff.; geology, 399 f.; geomorphic evolution, 400 ff.; structure, 399 f.
—Hills: Eastern Desert, 10, 44 f., 66 f., 68 ff., 71 f., 78 ff., 118, 149, 182, 185, 214 f., 217, 224, 264, 402 ff., 423, 424 f., 426 ff., 431, 468, 488

—Littoral. See High terrace, Low terrace, Middle terrace
—mollusca: fossil and modern, 398, 412 ff., 499 ff.
Red Silts (*limons rouges*), 381 ff., 387, 389, 391 ff.
Refuf Pass: escarpment above Kharga Oasis, 155, 365, 370
Rhinoceros, 182, 193, 326
Rift system: East African, 5 f., 27 f., 67, 433, 444 ff.
El-Riqa, 201 f., 230, 233 f., 241 ff.
Riqa Dome, 206 ff., 216, 223
Riqa Hills, 211, 216
Rock drawings, 74, 190 ff.
Rofa fault, 27, 30, 206 f., 339, 352, 362
Rotlehm, 84, 239, 252, 437, 440
Russegger, Joseph (1802–63), 200

Safaga, 10, 396 ff.
Sahaba Formation: equivalent to Darau Member of Gebel Silsila Formation in the northern Sudan, 167, 275, 324
Sahure (*ca.* 2488–2475 B.C.), 395
Salt-weathering, 53, 75 f., 77, 152, 182, 261 f., 429
Sandford, Kenneth Stuart (1899—), 22 f., 27, 33, 39, 59 f., 63 f., 145, 146 ff., 156 ff., 166, 200 ff., 236, 240 ff., 243, 247, 249, 292, 310, 315, 320, 323, 396 f., 399, 420, 423
Sangoan industry, 159, 255
Saoura Valley, 83
Schweinfurth, Georg (1836–1925), 21
Sebekian industry, 114, 165 f.
Sebil: village, 22, 88, 114, 143 ff., 148, 152, 163 ff., 180 f., 267, 511
—Channel, 88, 143 ff., 149, 163 ff., 180 f.
Sebilian industry, 9 f., 13, 22 f., 114 f., 146 f., 163 ff., 169, 172, 180 f., 185, 187, 278
Sebilian Silts, 22 f. See also Masmas Formation, Gebel Silsila Formation
Sebua Temple, 155, 166, 197 f., 207
Second Cataract (Wadi Halfa), 72, 189, 197, 200, 227, 265, 478
Sediment properties, 13 ff., 461 ff., 467 ff., 473 ff., 483 ff.
Seiyala, 189 f., 199, 201, 207 f., 230, 233 f., 244 ff., 267, 271, 274, 279 f., 284, 319 ff.
—Hogback, 206 ff., 215
Selima Oasis, 10, 334
Semainian (Nagada III) culture, 189 ff.
Semna Cataract, 10, 197, 200 f.

Shaheinab, 326
Shait Pits. See New Korosko, New Sebua, New Shaturma
Shamarkian industry, 188
Sharm Sheikh, 402, 416 ff., 419
Shaturma Formation, 46, 54, 57, 87 f., 89, 92 f., 98, 103, 108, 117, 120, 121 ff., 124 ff., 127 f., 131, 143 f., 149, 151 f., 269, 277, 281 f., 285, 289 f., 296, 298, 300, 304 ff., 317 ff., 328, 331 ff., 488 ff., 497
Sheb (well), 10, 334
Sheikh Timai, 10, 147
Shellal, 63, 189 f., 201 f., 233 f., 247 ff., 267, 320, 323
Shima Amalika Temple, 233 f., 244 ff.
Shurum, 402
Es-Sibaiya: prehistoric site, 155 f.
Silsila Channel, 88, 172
Silsila Fault, 25, 28, 30
Silsila Gap, 21, 48, 73, 101, 115, 133, 151, 511
Silwa Qibli, 87 f.
Singa, 325 f.
Sinn el-Kaddab: escarpment, 27, 30
Sinqari, 233, 272
—Member (Ineiba Formation), 89, 91 ff., 117, 123, 125 ff., 128, 149, 191, 281 f., 290, 295 f., 304 ff., 316 ff., 328, 488 ff., 497
Siwa Oasis, 10, 334
Slope classes: generalized, 29 ff., 77, 210 ff., 221 ff., 342 f., 350 ff., 402
Slope forms, 9, 29 ff., 76 ff., 210 ff., 221 ff., 261 f., 342 f., 350 ff., 402
Sneferu (ca. 2613–2587 B.C.), 194, 196
Sobat River, 8, 71, 112, 258, 445, 447 ff., 453, 467 f., 471, 487
Sodiri, 112, 257
Soil: horizons, 16; properties, 14 ff.
Sorting: classes of, 15
South Wells: Kurkur Oasis, 339 f., 343 ff., 347, 349, 351, 356 f.
Stratification: types of, 13
Stratigraphic frameworks, 11 f., 65 ff., 148 ff., 250 ff., 293, 306 ff., 318 f., 324, 328 f., 363, 388 f., 418 ff., 421 f.
Structure: soil and sediment, 15
Sudd Swamps, 7, 112. See also Paleo-Sudd Basin
Suez, 10, 396 f.
Sus, 148

Tafa, 197, 233 f., 247 ff., 334
Tamarisks, 70, 176, 344 f.

Tectonics. See Kom Ombo Plain, structure; Kurkur Oasis, structure; Nubia (Egyptian), structure
Terra rossa, 359, 362 ff., 369 f., 387 f., 440
Tertiary: 4 ff., 7, 11 f., 24, 26 ff., 33 ff., 66 f., 81, 112, 210, 221 ff., 250 ff., 265, 341, 363, 399 f., 412, 424, 428, 431 ff., 437, 452 ff., 515 ff.; vulcanics, 110, 112, 205, 433, 444 f., 452 f.
Texture, 14 f., 462 ff.
Thebes, 155, 158
Third Cataract (Argo-Kerma), 6, 227, 254
Thorium-Uranium, 398, 414, 420
Thutmosis III (1482–50 B.C.), 19, 194
Tokar Gap, 4, 454 f.
Tortoise, 113, 148, 185
Tufa Wadi. See Wadi Kurkur
Tumas, 197, 201, 204, 206, 209, 218, 221, 223 f., 233 f., 241 ff., 263
—Upland, 211, 218, 220
Tushka, 155, 168, 201 f., 219, 225, 233 f., 240 f., 267, 271 f., 274 ff., 309 ff.
—Plain, 211, 219

Uhden, Richard (1900–39), 336
Umm Naqa Uplands, 211, 214
Umm Ruwaba, 112, 255 f., 434
Unio sp. (freshwater bivalve), 21, 112, 115, 133, 136 f., 147, 164 f., 171 f., 175, 178 ff., 185 f., 200, 275 f., 287, 308, 510 ff.
University of Colorado Expedition, 167, 265, 280
Upper Wadi Alluvium, 87, 121. See also Shaturma Formation

Valvata nilotica, 91, 95 f., 102, 112, 125 ff., 511
Vertigo (freshwater snail), 257
Vignard, Edmond (1885–), 9, 22, 69, 157, 160 f., 163 ff., 172, 180 f., 185

Wadai, 84
Wadi Abbad: alluvial terraces, 34, 63 f.
Wadi Abu Domi, 34, 45, 53
Wadi Abu Gorma, 351, 353 ff., 364, 515. See also False Wadi
Wadi Abu Haggag, 24
Wadi Abusku, 211, 215, 250
Wadi Abu Subeira, 24, 28, 30, 43, 45, 55 f., 67, 77, 164, 185, 189 f., 251, 267
Wadi Abu Uruq. See Khor (and Wadi) Abu Uruq
Wadi Alam-Khariga, 82, 401, 402 ff., 419, 422, 427, 481

Wadi Allaqi, 166, 197 ff., 203, 206, 209, 215, 226, 228, 233 f., 245 ff., 250, 262; stage, 225, 232 ff., 241 ff., 244 ff., 252 ff., 263, 265
Wadi Ambaut, 416 f., 419, 421 f., 481
Wadi el-Arab, 201, 233
Wadi Ayed, 87 f., 267
Wadi Dihmit. See Khor (and Wadi) Dihmit
Wadi Ellawi, 35, 41, 54, 69, 76, 88, 116 f., 122, 151, 511
Wadi Floor Conglomerate, 251, 283 ff., 290, 294 ff., 306, 315 ff., 327 ff., 375, 443
Wadi Gabgaba, 196, 206, 209
Wadi Gemal, 19 f.
Wadi el-Gimmeiza, 34, 53
Wadi Guhr el-Daba, 244, 315 f.
Wadi Halfa, 10, 71, 112, 154 ff., 158 ff., 162, 167 f., 186, 188, 194, 200 ff., 205, 232, 258 f., 265, 272, 280, 522
Wadi Hamid, 211, 223, 238, 250
Wadi Howar, 6
Wadi Ibeidalla, 64
Wadi Kalabsha, 211, 213, 250, 262
Wadi Kharit, 18 f., 21 f., 28, 30, 34 f., 40 ff., 44 f., 50 f., 52 ff., 57, 60 ff., 72, 78, 88, 97, 101, 115 ff., 119 f., 121 f., 145, 149, 162, 182 ff., 190 ff., 267, 433 f., 488, 490
Wadi Kom Ombo Estate (Company), 18, 20 f., 23, 26, 68
Wadi Korosko, 82, 206, 209, 233, 242 ff., 251, 253, 272, 314 ff.
—stage, 232, 234, 236 ff., 240 f., 241 ff., 244 ff., 247 ff., 252 ff., 263, 265, 321, 329, 370, 439
Wadi el-Kubaniya, 30, 34 f., 43, 45, 53, 63, 213, 251. See also Kubaniya Nile
Wadi Kurkur, 30, 213, 250, 334, 339 f., 345 ff.
Wadi Madamud, 64
Wadi Melik, 111, 432
Wadi Natash, 32, 34 f., 44 f., 51, 66 f., 72, 78
Wadi Odib, 4, 6, 432, 454
Wadi Or, 82, 96, 121, 236 ff., 251, 253, 267 ff., 271 f., 281 f., 283 f., 294 ff., 300, 333, 469 f., 476, 478, 490, 495
Wadi Qena: alluvial terraces, 123, 497
Wadi el-Quffa, 211, 215, 250
Wadi Rofa, 30
Wadi Samadai. See Mersa (and Wadi) Samadai
Wadi Shait, 18 f., 21 f., 28, 30, 32, 34 f., 40 ff., 44 f., 51, 52 ff., 57, 60 ff., 78, 88, 94, 97, 101, 115 ff., 119 f., 121 f., 149, 162, 165, 182 ff., 191 f., 267, 433 f., 469, 488, 511
Wadi Shatt Rigal, 19, 190, 192
Wadi Shaturma, 206, 209, 233 f., 244 ff., 314
Wadi Shurafa, 69, 87 f., 97 ff., 102 f., 128 ff., 151
Wadi Sifein, 401, 406 ff., 413, 422
Wadi Tufa I, 356 f., 362, 364 ff., 367 ff., 387, 389, 515 ff.
Wadi Tufa II, 256 f., 266 ff.
Wadi Tufa IIIa, 256 f., 258 f., 270 ff., 287, 289 f., 443, 492, 495 ff., 513 f.
Wadi Tufa IIIb, 356 f., 369, 370 ff., 376, 389 ff., 513 f.
Wadi Tufa IIIc, 356 f., 373, 376 ff., 389, 392, 495 ff., 513 f.
Wadi Tufa IV, 256 ff., 359, 373 ff., 378 ff., 383 f., 387, 389 ff., 395 ff., 413 f., 443
Wadi Tushka, 211, 275, 292, 310 f.
Wadi Umm el-Hamid, 270, 283, 309, 317 ff.
Wadi Umm Mataris, 41, 88
Wadi Umm Rukba, 44, 57, 88
Wadi Umm Seiyala, 340, 350 f., 354
Wadiyein, 64, 147
Warthog, 180, 185, 326
Weathering, 9, 76 ff., 204 f., 262, 400, 402, 441 f. See also Frost-weathering, Paleosols, Salt-weathering
Western Desert. See Libyan Desert
White Nile: Basin, 3 ff., 71, 254 ff., 435, 447 ff., 454 ff., 487
—River, 3 ff., 8, 95, 326 f., 467 f., 471
White Silts, 379 f., 389, 391
Willcocks, Sir William (1852–1932), 21, 200, 232
Willow (*Salix* sp.), 308 ff.
Winkler, Hans Alexander (1900–43), 192 f.
Woodward, Samuel (1821–65), 200
Würm (Last Glacial), 149 f., 328 f., 394, 420, 427, 430, 439, 443

X-Group culture, 292, 294, 311

Yale Prehistoric Nubia Expedition, 10, 23, 33, 73, 113, 166, 171, 191, 202 f., 337, 509
Yerma: soil type, 75 f., 139, 168 f., 178 f., 182, 304, 312, 429
Younger Channel Silts, 87, 107. See also Gebel Silsila Formation

Zootecus insularis (land snail), 120, 154, 260, 282 ff., 291, 302, 304 f., 315, 317, 319 f., 370, 379, 382, 391 f., 511, 513 f.